WORLD HEALTH ORGANIZATION

INTERNATIONAL AGENCY FOR RESEARCH ON CANCER

NATIONAL CANCER INSTITUTE OF THE USA

UNIVERSITY OF CALIFORNIA
LAWRENCE BERKELEY LABORATORY

THE ROLE OF CYCLIC NUCLEIC ACID ADDUCTS IN CARCINOGENESIS AND MUTAGENESIS

Proceedings of a meeting organized by the IARC
and co-sponsored by the US National Cancer Institute,
and the Lawrence Berkeley Laboratory at the University of California,
held in Lyon, 17–19 September 1984

EDITORS

B. SINGER   H. BARTSCH

IARC Scientific Publications No. 70

INTERNATIONAL AGENCY FOR RESEARCH ON CANCER
LYON
1986

The International Agency for Research on Cancer (IARC) was established in 1965 by the World Health Assembly, as an independently financed organization within the framework of the World Health Organization. The headquarters of the agency are at Lyon, France.

The Agency conducts a programme of research concentrating particularly on the epidemiology of cancer and the study of potential carcinogens in the human environment. Its field studies are supplemented by biological and chemical research carried out in the Agency's laboratories in Lyon, and, through collaborative research agreements, in national research institutions in many countries. The Agency also conducts a programme for the education and training of personnel for cancer research.

The publications of the Agency are intended to contribute to the dissemination of authoritative information on different aspects of cancer research.

Distributed for the International Agency for Research on Cancer by
Oxford University Press, Walton Street, Oxford OX2 6DP

London   New York   Toronto
Delhi   Bombay   Calcutta   Madras   Karachi
Kuala Lumpur   Singapore   Hong Kong   Tokyo
Nairobi   Dar es Salaam   Cape Town
Melbourne   Auckland

Oxford is a trade mark of Oxford University Press

Distributed in the USA
by Oxford University Press, New York

ISBN 92 832 1170 7
ISSN 0300-5085

The authors alone are responsible for the views expressed in the signed articles in this publication. All rights reserved. No part of this publication may be reproduced, stored in a retrieval system, or transmitted, in any form or by any means, electronic, mechanical, photocopying, recording, or otherwise, without the prior permission of Oxford University Press.

© International Agency for Research on Cancer 1986
150 cours Albert Thomas, 69372 Lyon Cedex 08, France

PRINTED IN SWITZERLAND

CONTENTS

Foreword vii
Introduction ix
List of Participants xi

I. OCCURRENCE, EPIDEMIOLOGY AND CARCINOGENIC EFFECTS

Keynote note address: The role of cyclic nucleic acid base adducts in carcinogenesis and mutagenesis
H. Bartsch 3
Carcinogenicity of selected vinyl compounds, some aldehydes, haloethyl nitrosoureas and furocoumarins: an overview
H. Vainio & R. Saracci 15

II. CHEMISTRY AND FORMATION OF CYCLIC AND OTHER ADDUCTS

A. Vinyl Halides, Carbamate Esters and Metabolites

Substituted ethenoadenosines and ethenocytidines
N.J. Leonard & K.A. Cruickshank 33
The structure and properties of 7-(2-oxoethyl)guanine: a model for a key DNA alkylation product of vinyl chloride
P. Politzer, R. Bar-Adon & B.A. Zilles 37
Neutral reactions of haloacetaldehydes with polynucleotides: Mechanisms, monomer and polymer products
B. Singer, S.R. Holbrook, H. Fraenkel-Conrat & J.T. Kuśmierek 45
Reaction kinetics and cytosine adducts of chloroethylene oxide and chloroacetaldehyde: direct observation of intermediates by FTNMR and GC-MS
I. O'Neill, A. Barbin, M. Friesen & H. Bartsch 57
Chemical modification of adenine and cytosine residues with chloroacetaldehyde at the nucleoside and the tRNA level: the structural effect of chloroacetaldehyde modification
W.J. Krzyżosiak, M. Wiewiórowski & M. Jaskólski 75
Cyclic adduct formation at structural perturbations in supercoiled DNA molecules
D.M.J. Lilley 83

The role of cyclic base adducts in vinyl-chloride-induced carcinogenesis: studies on nucleic acid alkylation *in vivo*
R.J. Laib.. 101

Modification of DNA and metabolism of ethyl carbamate *in vivo*: formation of 7-(2-oxoethyl)guanine and its sensitive determination by reductive tritiation using ^{3}H-sodium borohydride
E. Scherer, H. Winterwerp & P. Emmelot... 109

B. Haloalkylnitrosoureas

Isolation and characterization of electrophiles from 2-haloethylnitrosoureas forming cytotoxic DNA cross-links and cyclic nucleotide adducts and the analysis of base site-selectivity by *ab initio* calculations
J.W. Lown, R.R. Koganty, U.G. Bhat, S.M.S. Chauhan, A.-M. Sapse & E.B. Allen... 129

Formation of cyclic adducts in nucleic acids by the haloethylnitrosoureas
D.B. Ludlum... 137

Investigation of 6-thiodeoxyguanosine alkylation products and their role in the potentiation of BCNU cytotoxicity
W.J. Bodell.. 147

DNA cross-linking by chloroethylating agents
K.W. Kohn & N.W. Gibson... 155

C. Bifunctional Aldehydes

Reactions of nucleosides with glyoxal and acrolein
R. Shapiro, R.S. Sodum, D.W. Everett & S.K. Kundu.................................... 165

The role of cyclic nucleic acid adducts in the mutational specificity of malondialdehyde and β-substituted acroleins in *Salmonella*
L.J. Marnett, A.K. Basu & S.M. O'Hara... 175

Reaction of 2-nitroimidazole metabolites with guanine and possible biological consequences
G.F. Whitmore, A.J. Varghese & S. Gulyas.. 185

Structure-activity relationships of α,β-unsaturated carbonylic compounds
D. Henschler & E. Eder... 197

Formation of cyclic nucleic acid adducts from some simple α,β-unsaturated carbonyl compounds and cyclic nitrosamines
F.-L. Chung, S.S. Hecht & G. Palladino... 207

Reaction of guanosine with glycidaldehyde
B.T. Golding, P.K. Slaich & W.P. Watson.. 227

D. Aromatic Compounds

Electrophilic attack at carbon-5 of guanine nucleosides: structure and properties of the resulting guanidinoimidazole products
R.C. Moschel, W.R. Hudgins & A. Dipple.. 235

Formation of a cyclic adenine adduct with a 4-nitroquinoline *N*-oxide metabolite model
  N. Tohme, M. Demeunynck, M.F. Lhomme & J. Lhomme......................... 241
Isolation and characterization of psoralen photoadducts to DNA and related model compounds
  J. Cadet, L. Voituriez, F. Gaboriau & P. Vigny............................ 247

III. METABOLISM

Metabolism and covalent binding of *vic*-dihaloalkanes, vinyl halides and acrylonitrile
  F.P. Guengerich, L.L. Hogy, P.B. Inskeep & D.C. Liebler..................... 255
Metabolic activation of vinyl chloride, formation of nucleic acid adducts and relevance to carcinogenesis
  H.M. Bolt.. 261
Covalent binding of reactive intermediates to hemoglobin in the mouse as an approach to studying the metabolic pathways of 1,2-dichloroethane
  K. Svensson & S. Osterman-Golkar.. 269

IV. BIOLOGICAL EFFECTS

Methylglyoxal in beverages and foods: its mutagenicity and carcinogenicity
  M. Nagao, Y. Fujita, T. Sugimura & T. Kosuge............................ 283
Evaluation of the mutagenicity of 1,N^6-ethenoadenine- and 3,N^4-ethenocytosine-nucleosides in *Salmonella typhimurium*
  K. Negishi, K. Oohara, H. Urushidani, Y. Ohara & H. Hayatsu............... 293
Genetic effects of DNA mono- and diadducts photoinduced by furocoumarins in eukaryotic cells
  D. Averbeck & D. Papadopoulo... 299
Induction of sister chromatid exchange by ethyl carbamate and vinyl carbamate
  M.K. Conner... 313
Adaptive response to alkylating agents in mammalian cells
  F. Laval... 321
Aspects concerning the study of the quantitative carcinogenic effects of chemotherapeutic agents in man
  J.M. Kaldor & N.E. Day... 327

V. MECHANISTIC APPROACHES

Reaction of chloroacetaldehyde with poly(dA-dT) and poly(dC-dG) and its effect upon the accuracy of DNA synthesis
  R. Saffhill & J.A. Hall... 339
Mutagenic and promutagenic properties of DNA adducts formed by vinyl chloride metabolites
  A. Barbin & H. Bartsch... 345

Replication and transcription of polynucleotides containing ethenocytosine, ethenoadenine and their hydrated intermediates
  B. Singer & S.J. Spengler 359
Repair of etheno DNA adducts by N-glycosylases
  F. Oesch, S. Adler, R. Rettelbach & G. Doerjer 373
Repair of cyclic nucleic acid adducts and adverse effects of apurinic sites
  J. Laval & S. Boiteux 381
Termination of synthesis resulting from modifying bases in DNA
  B. Strauss, K. Larson, S. Rabkin, D. Sagher & J. Sahm 387
Mutagenesis and repair of O^6-substituted guanines
  J.M. Essigman, E.L. Loechler & C.L. Green 393

VI. SENSITIVE METHODS FOR DETECTION OF NUCLEIC ACID ADDUCTS

Monoclonal antibody-based immunoanalytical methods for detection of carcinogen-modified DNA components
  J. Adamkiewicz, O. Ahrens, G. Eberle, P. Nehls & M.F. Rajewsky 403
Detection and identification of mutagens by the adducts formed upon reaction with guanosine derivatives
  H. Kasai, Z. Yamaizumi & S. Nishimura 413
Three-dimensional fluorometry for the detection of DNA adducts
  G. Doerjer, E. Nies, J. Mertes & F. Oesch 419
Quantitation of etheno adducts by fluorescence detection
  M.A. Bedell, M.C. Dyroff, G. Doerjer & J.A. Swenberg 425
Detection of DNA base damage by ^{32}P-postlabelling: TLC separation of 5'-deoxynucleoside monophosphates
  M. Hollstein, J. Nair, H. Bartsch, B. Bochner & B.N. Ames 437

AUTHOR INDEX 449
SUBJECT INDEX 451

FOREWORD

It is now commonly agreed that modification of DNA by chemical carcinogens is a critical step in the initiation of many neoplasias in experimental animals, and probably also in humans. Chemotherapy of human cancers using antineoplastic agents, many of which exert carcinogenic side effects, is also believed to involve a DNA modification that is related to their cytotoxic action. Because of the fundamental role of DNA interactions, research over the past three decades has been directed to the characterization and study of the consequences of the macromolecular adducts produced by a variety of known carcinogens in human and animal cells. During the past decade, increasing interest has been directed to a number of carcinogenic and mutagenic agents that share the common property of being able to react with nucleic acid bases to form one or more additional ring systems, i.e., exocyclic DNA base adducts. Although the structure and formation of such adducts with industrial carcinogens, chemotherapeutic agents and products formed by metabolic processes, like bifunctional aldehydes, have been reported increasingly, only the multidisciplinary group gathered at this conference has addressed the question of whether the cyclic adducts have any biological relevance. At this meeting, common areas of, as well as gaps in, knowledge in this field were identified, and it became clear that further chemical, biochemical and immunological studies are needed to assess how this group of widely distributed compounds produces cancer.

This volume comprises the proceedings of a meeting held at IARC in Lyon in September 1984. It appears to be the first comprehensive overview on cyclic adducts formed from a variety of compounds, and includes sections on occurrence, epidemiology, carcinogenic effects, metabolism and sensitive methods for detection.

Unlike most conference proceedings, the papers in this volume underwent peer review. The IARC would like to thank the following for reviewing critically the contents of the papers: M.K. Conner, F.P. Guengerich, N.J. Leonard, J.W. Lown, R. Shapiro, B.S. Strauss and J.A. Swenberg.

I should like to thank B. Singer, Laboratory of Chemical Biodynamics, University of California, Berkeley, California, one of the co-organizers, and the National Cancer Institute, NIH, Bethesda, Maryland, for co-sponsoring this meeting.

L. Tomatis, M.D.
Director

INTRODUCTION

There are a substantial number of human and animal carcinogens, widely used in industry, chemotherapy, or formed by metabolic processes or combustion, which have in common the ability to form cyclic nucleic acid derivatives. These include vinyl halides, alkyl carbamates, bifunctional aldehydes, 2-haloalkylnitrosamines and cyclic nitrosamines, acrylonitrile, etc., all of which, or their metabolites, react with exocyclic amino groups, particularly that of guanine.

Progress in understanding how these chemicals exert their biological effects has been impeded by the perception of a lack of a critical mass of scientists and of data. This situation, in reality, no longer exists as shown by this meeting. However, prior to this conference there was no unifying focal point. Chemists studied a single type of chemical reaction for its own sake, while biologists tested another array of compounds for mutagenicity, and metabolic pathways were investigated by other scientists, but the products were not necessarily related to those studied by chemists or biologists. It could be said that tunnel vision impeded elucidation of the general problem.

As a result of this multidisciplinary conference, some common areas of knowledge and of concern were identified. Although cyclic $1,N^2$-guanine derivatives were readily identified *in vitro*, there is little evidence for their occurrence in DNA *in vivo*, but some evidence in RNA. Thus RNA adducts should also be studied *in vivo*. There are, in addition to the one type of guanine cyclic adduct, a multiplicity of minor adducts found *in vitro*. These should also be identified, and ultrasensitive methods for detection of all such derivatives *in vivo* need to be developed. Most of the chemicals discussed also form cross-links as a result of their bifunctional character, and the biological role of such cross-links needs further investigation. It also became apparent that repair has not been demonstrated unequivocally for any cyclic adduct. Finally, the carcinogenicity of most mutagenic aldehydes has not been established, due either to a lack of testing or to poor testing.

It is now clear that future chemical, biochemical and enzymological studies are needed to assess whether and how this group of widely distributed chemicals affect human health.

The Editors

LIST OF PARTICIPANTS

J. Adamkiewicz
Institut für Zellbiologie
Tumorforschung
Universität-Klinikum
Hufelandstrasse 55
4300 Essen 1, FRG
Tel. (0201) 7991 3110

D. Averbeck
Institut Curie
Section de Biologie
26 rue d'Ulm
75231 Paris Cedex 05, France
Tel. 329 12 42 (ext. 3365)

A. Barbin
International Agency for Research on Cancer
Unit of Environmental Carcinogens and Host Factors
150 cours Albert Thomas
69372 Lyon Cedex 08, France
Tel. (7) 875 81 81 (ext. 538)

H. Bartsch
International Agency for Research on Cancer
Unit of Environmental Carcinogens and Host Factors
150 cours Albert Thomas
69372 Lyon Cedex 08, France
Tel. (7) 875 81 81 (ext. 393)

W.J. Bodell
University of California
Department of Neurological Surgery
Brain Tumor Research Center HSW-783
3rd & Parnassus Avenue
San Francisco, CA 94143, USA
Tel. (415) 666-3083

H.M. Bolt
Institut für Arbeitsphysiologie
Universität Dortmund
Abteilung Toxikologie & Arbeitsmedizin
Ardeystrasse 67
D-4600 Dortmund 1, FRG
Tel. 0231/1084348

E. Bresnick
University of Nebraska Medical Center
Eppley Institute
42nd and Dewey Avenue
Omaha, Nebraska 68105, USA
Tel. 402-559-4238

J. Cadet
Centre d'Etudes Nucléaires de Grenoble
DRF-CH, 85 X
F-38041 Grenoble cedex, France
Tel. (76) 97 41 11 (ext. 37 51)

F.-L. Chung
Naylor Dana Institute
Dana Road
Valhalla, NY 10595, USA
Tel. (914) 592 2600 (ext. 361)

PARTICIPANTS

M.K. Conner
Department of Industrial Environmental Health Sciences
Graduate School of Public Health
University of Pittsburgh
130 DeSoto Street
Pittsburgh, PA 15261, USA
Tel. (412) 624-5471

N.E. Day
International Agency for Research on Cancer
Unit of Biostatistics
150 cours Albert Thomas
69372 Lyon Cedex 08, France
Tel. (7) 875 81 81 (ext. 426)

G. Doerjer
Institut für Toxikologie
Johannes Gutenberg-Universität
Obere Zahlbacherstrasse 67
D-6500 Mainz 1, FRG
Tel. 06131/233616

G. Eberle
Institut für Zellbiologie
Tumorforschung
Universität-Klinikum
Hufelandstrasse 55
4300 Essen 1, FRG
Tel. (0201) 7991 3110

E. Eder
Institute of Toxicology
University of Würzburg
Versbacher Strasse 9
D-8700 Würzburg, FRG
Tel. 0931/201 3990

J.M. Essigmann
Bldg E18-561
Massachussetts Institute of Technology
Department of Nutrition and Food Science
50 Ames Street
Cambridge, MA 02139, USA
Tel. (617) 253-6227

H. Fraenkel-Conrat
Department of Molecular Biology
University of California
Berkeley, California 94720, USA
Tel. 415/486-4000

B. Golding
Department of Organic Chemistry
Institute of Chemistry
University of Newcastle
Newcastel upon Tyre NA1 7AU, UK
Tel. (0632) 328511 (ext. 2959)

F.P. Guengerich
Department of Biochemistry
Vanderbilt University
School of Medicine
21st Avenue South & Garland
Nashville, TN 37232, USA
Tel. (615) 322-2261

H. Hayatsu
Faculty of Pharmaceutical Sciences
Okayama University
Tsushima
Okayama 700, Japan
Tel. 0862-52-1111 (ext. 995)

D. Henschler
Institut für Toxikologie
Universität Würzburg
Versbacher Strasse 9
8700 Würzburg, FRG
Tel. (0931) 201 39 80

E. Heseltine
International Agency for Research on Cancer
Editorial and Publications Services
150 cours Albert Thomas
69372 Lyon Cedex 08, France
Tel. (7) 875 81 81 (ext. 347)

PARTICIPANTS

M. Hollstein
International Agency for Research on Cancer
Unit of Mechanisms of Carcinogenesis
150 cours Albert Thomas
69372 Lyon Cedex 08, France
Tel. (7) 875 81 81 (ext. 556)

J. Káldor
International Agency for Research on Cancer
Unit of Biostatistics
150 cours Albert Thomas
69372 Lyon Cedex 08, France
Tel. (7) 875 81 81 (ext. 436)

P. Karran
Mutagenesis Laboratory
Imperial Cancer Research Fund
Mill Hill Laboratories
Burtonhole Lane
London NW7 1AD, UK
Tel. 01 9593236

H. Kasai
Biology Division
National Cancer Center Research Institute
Tsukiji 5-1-1
Chuo-ku
Tokyo 104, Japan
Tel. 03-542-2511 (ext. 664)

K.W. Kohn
National Institute for Public Health
Bldg 37
Room SA17
Bethesda, MD 20205, USA
Tel. (301) 496 2769

W. Krzyżosiak
Institute of Bioorganic Chemistry
Polish Academy of Sciences
Noskowskiego 12/14
61-704 Poznań, Poland
Tel. 585-03

R. Laib
Institut für Arbeitsphysiologie
an der Universität Dortmund
Ardeystrasse 67
D-4600 Dortmund 1, FRG
Tel. (0231) 1084-1-353

J. Laval
Groupe "Réparation des Lésions Radio & Chimio Induites"
Institut Gustave Roussy
94805 Villejuif, France
Tel. 1 726 46 58 (ext. 643)

F. Laval
Groupe "Radiochimie de l'ADN"
Institut Gustave Roussy
94805 Villejuif, France
Tel. 559 44 54

N.J. Leonard
Roger Adams Laboratory
School of Chemical Sciences
University of Illinois
Urbana, IL 61801, USA
Tel. (217) 333-0363

J. Lhomme
Departement de Chimie Organique Biologique
Université de Lille I
59655 Villeneuve d'Ascq, France
Tel. (20) 91 92 22

D.M.J. Lilley
Department of Biochemistry
University of Dundee
Dundee DD1 4HN, Scotland, UK
Tel. 382 23181 (ext. 500)

J.W. Lown
Department of Chemistry
University of Alberta
Edmonton, Alberta, Canada T6G 2G2
Tel. (403) 432-3646

PARTICIPANTS

D.B. Ludlum
Division of Oncology
Department of Medicine
Albany Medical College
Albany, New York 12208, USA
Tel. 518-445-5412

C. Malaveille
International Agency for Research on Cancer
Unit of Environmental Carcinogens and Host Factors
150 cours Albert Thomas
69372 Lyon Cedex 08, France
Tel. (7) 875 81 81 (ext. 519)

L.J. Marnett
435 Departement of Chemistry
Wayne State University
Detroit, MI 48202, USA
Tel. 313 577 2777

R. Montesano
International Agency for Research on Cancer
Unit of Mechanisms of Carcinogenesis
150 cours Albert Thomas
69372 Lyon Cedex 08, France
Tel. (7) 875 81 81 (ext. 461)

R.C. Moschel
National Cancer Institute
Frederick Cancer Research Facility
P.O. Box B
Bldg 538
Frederick, MD 21701, USA

M. Nagao
National Cancer Center Research Institute
1-1, Tsukiji 5 chome
Chuo-ku
Tokyo 104, Japan
Tel. (03)-542-2511 (ext. 656)

E. Nies
Institut für Toxikologie
der Johannes Gutenberg-Universität
Obere Zahlbacher Strasse 67
D-6500 Mainz 1, FRG
Tel. 06131/17-3197

F. Oesch
Institut für Toxikologie
der Johannes Gutenberg-Universität
Obere Zahlbacher Strasse 67
D-6500 Mainz 1, FRG
Tel. 06131/172283

P. Okano
Chemical & Physical Carcinogenesis Branch
Division of Cancer Etiology
National Cancer Institute
Landow Building
Room 9C-18
7910 Woodmont Avenue
Bethesda, MD 20205, USA
Tel. (301) 496 4141

I.K. O'Neill
International Agency for Research on Cancer
Unit of Environmental Carcinogenesis
150 cours Albert Thomas
69372 Lyon Cedex 08, France
Tel. (7) 875 81 81 (ext. 396)

S. Ostermann-Golkar
Department of Radiobiology
Wallenberg Laboratory
University of Stockholm
S-10691 Stockholm, Sweden
Tel. 08/163672

P. Politzer
Department of Chemistry
University of New Orleans
New Orleans, Louisiana 70148, USA
Tel. (504) 286-6850

PARTICIPANTS

N. Sabadie
G.E.R.A.P.
Université de Perpignan
Chemin Passio Vella
66025 Perpignan Cedex, France
Tel. 67 12 13

R. Saffhill
Paterson Laboratories
Christie Hospital and Holt Radium
Institute
Manchester, M20 9BX, UK
Tel. 061-445 8123

R. Saracci
International Agency for Research on
Cancer
Unit of Analytical Epidemiology
150 cours Albert Thomas
69372 Lyon Cedex 08, France
Tel. (7) 875 81 81 (ext. 373)

E. Scherer
Division of Chemical Carcinogenesis
The Netherlands Cancer Institute
Plesman Laan 121
1066 CX Amsterdam, The Netherlands
Tel. 20 512 2480

R. Shapiro
Department of Chemistry
New York University
4 Washington Place
New York, N.Y. 10003, USA
Tel. (516) 466-6396

M. Shariaty
Department of Medical Genetics
Cancer Research Institute
Teheran University
P.O. Box 13-145-617
1365 Teheran, Iran

B. Singer
University of California
Laboratory of Chemical Biodynamics
135 Calvin Hall
Berkeley, CA 94720, USA
Tel. (415) 642 0637

B.S. Strauss
Department of Microbiology
The University of Chicago
920 E. 58th Street
Chicago, Illinois 60637, USA
Tel. (312) 363 4233

K. Svensson
Department of Radiobiology
Wallenberg Laboratory
University of Stockholm
S-10691 Stockholm, Sweden
Tel. 08/163995

J.A. Swenberg
C.I.I.T.
P.O. Box 12137
Research Triangle Park, NC 27709, USA
Tel. 919-541-2070

L. Tomatis
Director
International Agency for Research on
Cancer
150 cours Albert Thomas
69372 Lyon Cedex 08, France
Tel. (7) 875 81 81 (ext. 576)

D. Umbenhauer
110 Mahatongo Drive
Pottsville, PA 17901, USA

H. Vainio
International Agency for Research on
Cancer
Department of Carcinogen
Identification and Evaluation
150 cours Albert Thomas
69372 Lyon Cedex 08, France
Tel. (7) 875 81 81 (ext. 382)

PARTICIPANTS

P. Vigny
Institut Curie
Section de Physique et de Chimie
Laboratoire Curie
11 rue Pierre et Marie Curie
75231 Paris Cedex 05, France
Tel. 329 12 42 (ext. 3107)

W. Watson
Shell Research Ltd
Sittingbourne Research Centre
Sittingbourne, Kent ME9 8AG, UK
Tel. (0) 795 244 44

G. F. Whitmore
Physics Division
Ontario Cancer Institute
500 Sherbourne Street
Toronto, Ontario, Canada M4X 1K9
Tel. (416) 924 0671 (ext. 5163)

M. Wiessler
Institute for Toxicology and
Chemotherapy
German Cancer Research Center
Im Neuenheimer Feld 280
D-6900 Heidelberg, FRG
Tel. 06221/484311

F. Zajdela
Unité de Physiologie Cellulaire de
l'Institut National de la Santé et de la
Recherche Médicale (U-22)
Fondation Curie
Institut du Radium
Bâtiment 110
91405 Orsay, France
Tel. (6) 907 64 67

I. OCCURRENCE, EPIDEMIOLOGY AND CARCINOGENIC EFFECTS

THE ROLE OF CYCLIC NUCLEIC ACID BASE ADDUCTS IN CARCINOGENESIS AND MUTAGENESIS

KEYNOTE ADDRESS

H. BARTSCH

International Agency for Research on Cancer,
Division of Environmental Carcinogenesis, Lyon,
France

Firstly, I want to thank the sponsoring parties who made this conference possible and all the participants who responded to our invitation with such enthusiasm. We are also particularly honoured to have with us several pioneers in the field, who started work on cyclic nucleotide adducts more than two decades ago.

Given the fact that this is the first conference at which a multidisciplinary group has gathered to discuss exocyclic nucleic acid adducts, I would like to make a few remarks on the major themes that should dominate our conference.

1) We are all aware that there is an increasing list of cyclic nucleic acid base adducts being reported with a variety of carcinogens and mutagens of diverse structures.

2) This phenomenon is due to the underlying common denominator of such carcinogens and mutagens, i.e., the ability to form intermediates, *per se* or *via* metabolism, which carry two reactive sites that allow them to act as bifunctional alkylating agents.

3) The major unresolved question we must answer is whether the cyclic adducts thus formed have any biological relevance.

4) If we think they do, then we must decide on the best approaches to assess their genetic, toxic and carcinogenic consequences not only in experimental systems, but also in man, whom we know is exposed to some of these agents from both exogenous and endogenous sources.

Let us now look back over three decades to the time when the first reports appeared on cyclic purine and pyrimidine adducts[1]. The earliest report was of the formation of cyclic nucleoside salts (Fig. 1, I), prepared by Clark and his coworkers in 1951 by

---

[1] This review is not meant to be exhaustive.

Fig. 1. Structures of some known cyclic nucleic acid base adducts

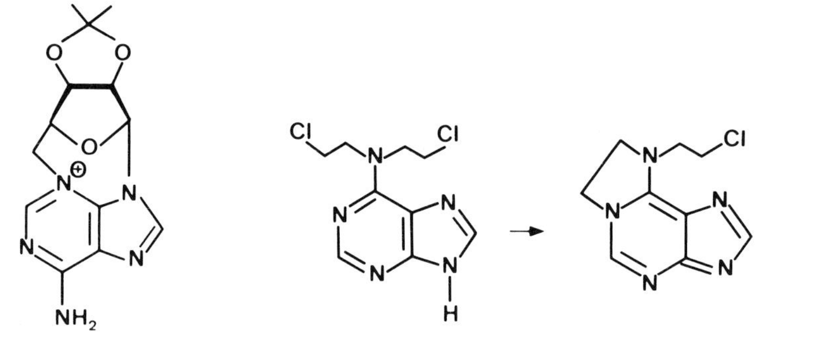

I II III 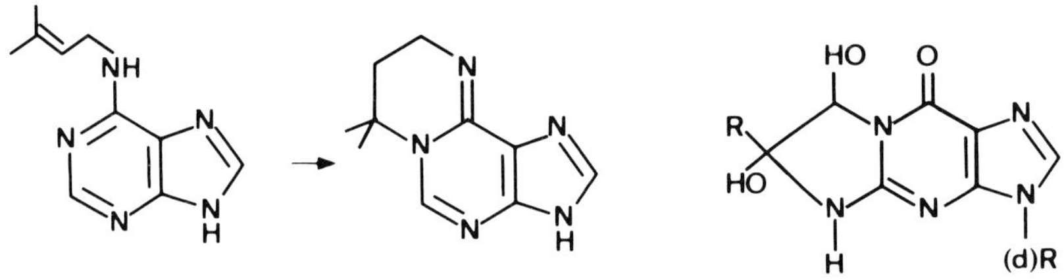 IV V VI

internal cyclization of a tosyl derivative of adenosine after heating. We now know that such derivatives are without relevance to biological systems. In 1962, Johnston *et al.* described a reaction for a nitrogen mustard derivative of adenine (Fig. 1,II,III). In 1966, Leonard *et al.*, on the basis of their earlier work on triacanthine, reported that a similar 1,N^6-cyclic adduct (Fig. 1, V) was formed from N^6-isopentenyl-adenine; these authors were also among the first to examine the effect of this cyclization reaction on biological activity, by showing that the development of cytokinin activity by 6-substituted purine derivatives was inactivated by this cyclization reaction.

Moving to reactive 1,2-dicarbonyl compounds, it was Shapiro and Hachman (1966) who described the reactions of glyoxal and methylglyoxal with guanine to form stable cyclic guanine adducts (Fig. 1, VI; R = H; CH_3). Surprisingly, more than 15 years later, these adducts became of interest again: firstly, due to the work of Varghese and Whitmore (1983), who showed that 2-nitroimidazole can produce this cyclic adduct as well, *via* its metabolic product, glyoxal, and secondly, due to the finding that the mutagenic activity of coffee is partially linked to the presence of methylglyoxal (Nagao *et al.*, these proceedings[2]). Back to chronological order: Van Duuren's group (Goldschmidt *et al.*, 1968) reported the formation of two isomeric cyclic products

---

[2] See p. 283.

Fig. 2. Formation of a bifunctional intermediate from 2-methylfuran, and its subsequent reaction with amino groups (XIII–XV)

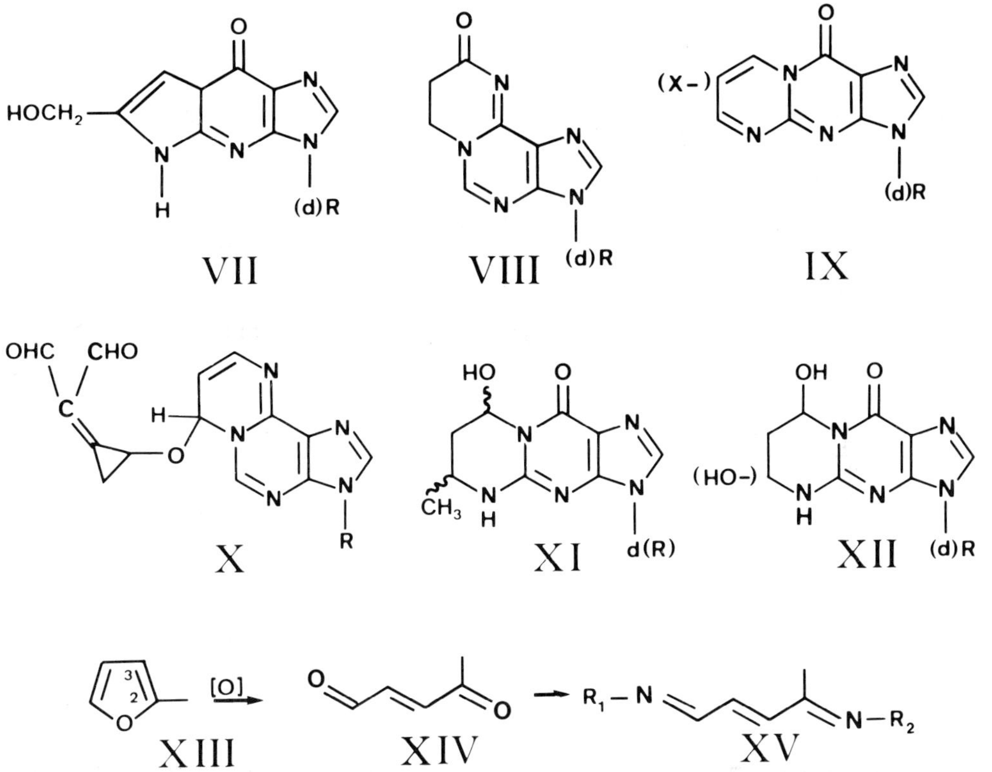

from the reaction of glycidaldehyde with guanine; one of these (Fig. 2, VII) was confirmed by Nair and Turner in 1984. Another alkylating carcinogen, β-propiolactone, was shown as early as 1963 to react with DNA bases, but the formation of a cyclic adenine adduct (Fig. 2, VIII) was proposed only in 1977 (Maté et al., 1977) and confirmed in 1981 (Chen et al., 1981).

To continue with diketo compounds, Moschel and Leonard (1976), and later Seto et al. (1981, 1983), found that malondialdehyde and substituted derivatives – the former being a product of lipid peroxidation and prostaglandin endoperoxide metabolism – can react with guanine to produce a fluorescent 1:1 cyclic adduct (Fig. 2, IX); in addition, a 3:1 adduct with adenine, having an unusual structure (Fig. 2, X), was identified by Nair et al. (1984). Recently, Chung and Hecht (1984) reported that crotonaldehyde and acrolein react in a very similar fashion with guanine to form the tricyclic 1,N^2-derivatives XI and XII, respectively (shown in Fig. 2). Most interesting is the finding that an adduct of the same type as adduct XI is formed from intermediates released after α-hydroxylation of cyclic N-nitrosamines, like N-nitrosopyrrolidine (Chung & Hecht, 1983). The list of bifunctional diketo compounds that react with nucleic acids may not yet be complete. As recently reported (Ravindranath et al., 1984), the toxic 2-methylfuran (Fig. 2, XIII) is oxidized by microsomes to the diketo intermediate, acetylacrolein (Fig. 2, XIV). This compound has already been shown to react with the amino groups in a bifunctional manner (Fig. 2, XV), and may form cyclic DNA adducts as well. In addition, there is a long waiting list of α,β-unsaturated carbonyl compounds, haloacroleins and haloaldehydes which could potentially add another ring to one or the other of the nucleic acid bases, but which have not yet been investigated to any great extent.

Not generally so well-known are bulky cyclic adducts, which have also been reported to be formed from a number of carcinogens having aromatic ring systems. For example, benzo[a]pyrene reacts at the K-region with cytosine in the presence of ultraviolet light to form a cyclobutane derivative (Rice, 1964). A putative metabolite of N,N-dimethyl-4-aminoazobenzene forms a cyclic adduct with cytosine (Roberts & Warwick, 1966) and N-acetoxy-N-4-acetamidostilbene forms one with adenine (Gaugler et al., 1979). An ultimate metabolite of 4-nitroquinoline-N-oxide also reacts with adenine (Tohme, these proceedings[3]). The relevance of most of these bulky cyclic adducts is poorly understood, except in the case of the furocoumarins, represented here by 8-methoxypsoralen (Fig. 3, XVI). Following activation by ultraviolet light on either side of the molecule, this compound can form either one of two cyclic monoadducts with thymine (Fig. 3, XVII), which can give rise upon further irradiation to a diadduct. Data on these adducts and their importance in mutagenesis will be presented at this conference.

Another group of compounds for which human exposure exists are bis-haloalkylnitrosoureas; they are used in the treatment of neoplasia, but are unfortunately associated with a carcinogenic potential. These compounds produce a number of DNA lesions, including exocyclic ring adducts (Ludlum et al., 1975; Tong & Ludlum, 1979). For example, both 1,3-bis-(2-chloroethyl)-1-nitrosourea (BCNU) and the bis-(2-fluoroethyl) analogue (BFNU) form $1,N^6$-ethanoadenine (Fig. 3, XVIII) and $3,N^4$-ethanocytosine (Fig. 3, XIX) adducts (Ludlum et al., 1975; Tong & Ludlum, 1979). Recently, another important cyclic adduct was discovered (Tong et al., 1983), which has provided clues about the reaction mechanism of these nitrosoureas. The O^6-fluoroethyl monoadduct formed initially from the reaction of BFNU with guanine cyclizes to give $1,O^6$-ethanoguanine (Fig. 3, XX), which is unstable and reacts further to yield secondary products such as cross-links.

The real renaissance of cyclic nucleic acid base adducts began in 1972 when Leonard's group (Barrio et al., 1972), on the basis of a report by Kochetkov et al. (1971), found that it was very easy to attach an etheno bridge to adenine, cytosine or guanine simply by chloroacetaldehyde treatment, yielding the highly fluorescent etheno derivatives (Fig. 3, XXI–XXIV) of adenine, cytosine and guanine (Sattsangi et al., 1977). For the latter, there are two possible isomers, N^2,3-ethenoguanine and $1,N^2$-ethenoguanine. Later, Hayatsu's group (Kayasuga-Mikado et al., 1980) introduced bromoacetaldehyde, an even more reactive and specific reagent to react

---

[3] See p. 241.

Fig. 3. Formation of a cyclic monoadduct from 8-methoxypsoralen and thymine (XVI–XVII)

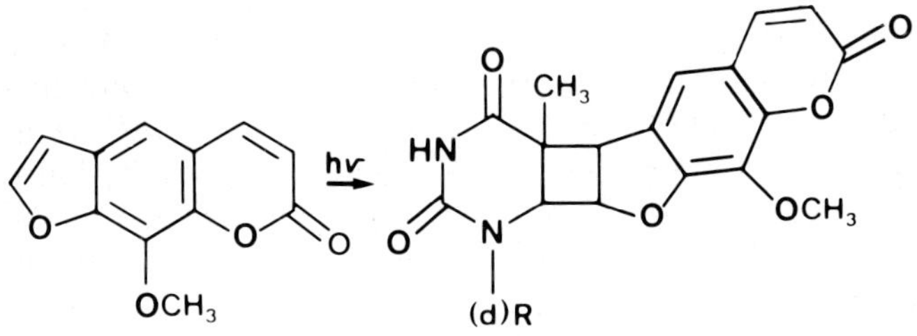

XVI   XVII

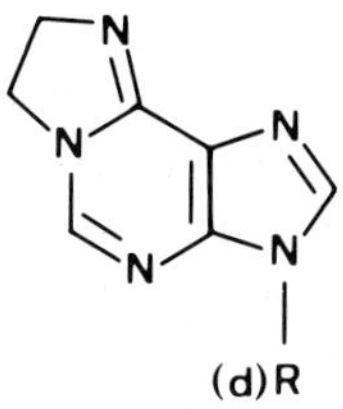

XVIII 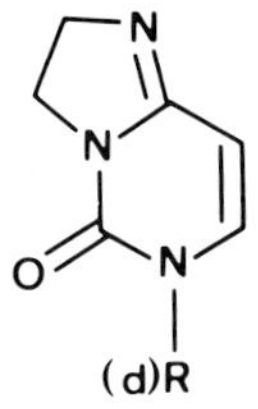   XIX 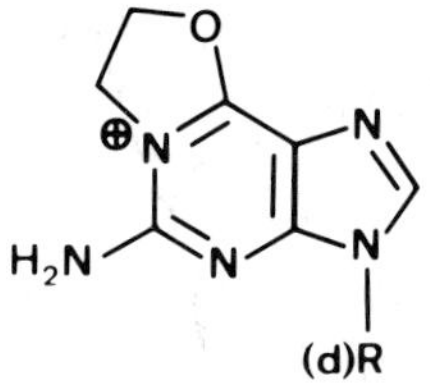   XX

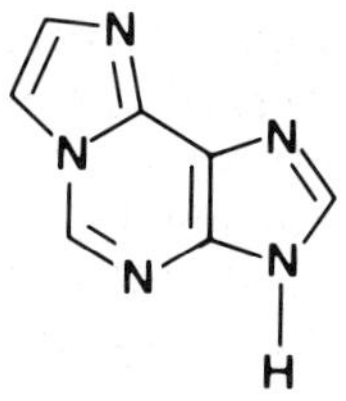

XXI 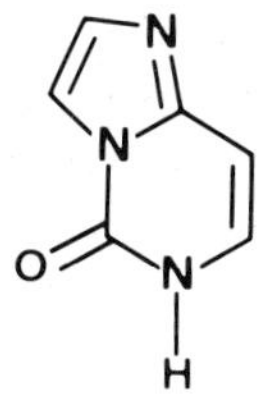   XXII 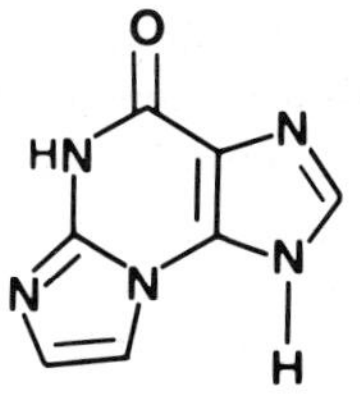   XXIII 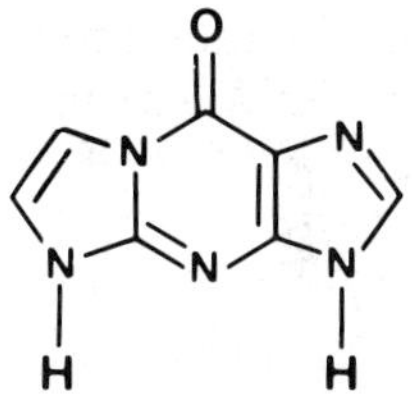   XXIV

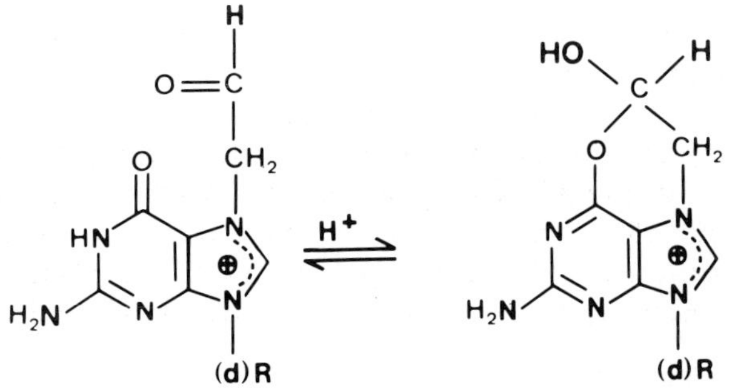

XXV   XXVI

with adenine and cytosine in single-stranded nucleic acids (Kohwi-Shigematsu *et al.*, 1983; Lilley, 1983).

Our group at IARC became interested in cyclic base adducts exactly ten years ago. In September 1974, we attempted to demonstrate that vinyl chloride, which had just been identified as a human carcinogen (Creech & Johnson, 1974), was also mutagenic in *Salmonella typhimurium*. But whatever assay procedure (liquid incubation or plate assay) we used to treat the bacteria, they survived happily and would not mutate. Finally, we chose a way of exposing the bacteria which was not dissimilar to that by which humans are exposed, (e.g., workers cleaning polyvinyl chloride reaction vessels). We put our petri dishes in a dessicator filled with vinyl chloride and air. To our great satisfaction, after only two hours of exposure, we could detect plenty of mutant colonies, and a clear dose-response relationship was evident (Bartsch *et al.*, 1975).

Next, to prove that vinyl chloride (or vinyl bromide) is oxidized to an epoxide, we constructed a syphon, a type of apparatus which contained rodent liver microsomes at the bottom and a trap at the top. When vinyl chloride and air were flushed through, vinyl chloride metabolites were released from this microsomal system and reacted with a classical reagent for alkylating agents 4-(*p*-nitrobenzyl)pyridine (NBP), which was inside the trap (Barbin *et al.*, 1975). The resulting adduct had a spectrum identical to that obtained when synthetic chloroethylene oxide, but not chloroacetaldehyde, was reacted with the same reagent. When adenosine was put into the trap instead, and again a mixture of vinyl chloride and air was passed through the microsomal system, a fluorescent product was formed, which had chromatographic properties and an ultraviolet spectrum that matched those reported for authentic ethenoadenosine (Barbin *et al.*, 1975). Proof that epoxidation of vinyl chloride leads to its ultimate reactive metabolite, chloroethylene oxide, was obtained in our subsequent carcinogenicity experiments (Zajdela *et al.*, 1980): upon intracutaneous injection, chloroethylene oxide produced tumours in mice, while chloroacetaldehyde did not.

Progress in the elucidation of a (possibly) cyclic vinyl chloride adduct was reported by Osterman-Golkar *et al.* (1977) and later by Scherer *et al.* (1981) and Laib *et al.* (1981). The major vinyl chloride-DNA adduct was found to be 7-(2-oxoethyl)guanine (Fig. 3, XXV), which, under acidic conditions, was shown to form a cyclic hemiacetal *in vitro* (Fig. 3, XXVI) with expected miscoding properties. Although Green and Hathway (1978) have reported the presence of ethenocytosine in the liver DNA of vinyl-chloride-treated rats, their results have not yet been confirmed. Laib and Bolt (1977, 1978) found ethenoadenine and ethenocytosine only in the liver RNA, but not in DNA from vinyl-chloride-exposed rats. Other vinyl compounds that have similar metabolic pathways, like acrylonitrile, would also be expected to form etheno base adducts. Indeed, Guengerich *et al.* (1981) showed that ethenoadenosine was formed by the reaction of the synthetic epoxide of acrylonitrile with adenosine. Miller's group (Ribovich *et al.*, 1982) detected ethenoadenosine and ethenocytidine in liver RNA of mice exposed to urethane, and took this observation as strong evidence that urethane and vinyl chloride are metabolized *via* similar pathways. Oesch and Doerjer (1982) reported that N^2,3-ethenoguanine was formed in DNA *in vitro* after treatment with chloroacetaldehyde.

Fig. 4. Concomitant formation of monoadducts and secondary lesions in the reactions of bifunctional agents with nucleic acid bases

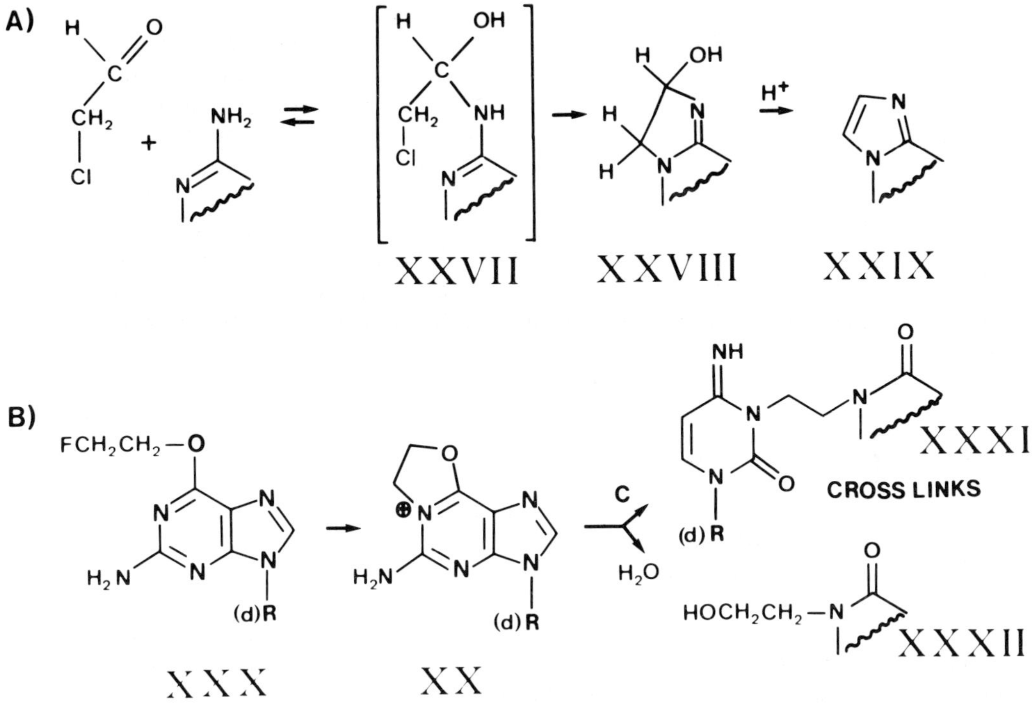

I shall now touch on the most important question of this conference, namely, the possible consequences of nucleic acid bases carrying such exocyclic adducts. Since 1972, a number of studies have been carried out based on the finding that etheno bridges could be introduced into mono-, di- and polynucleotides; these were easily detected because of their fluorescent properties. Most studies were intended to characterize enzyme binding-sites, e.g., in ATPase, or to investigate structural perturbations of transfer RNAs (for a review, see Leonard & Tolman, 1975; Leonard, 1984), and surprisingly little is known about the effect of cyclic adducts in DNA during replication or transcription. To mention a few possibilities, cyclic nucleic acid adducts could mispair directly; they could induce SOS repair; they could form secondary lesions like apurinic/apyrimidinic sites; or they could lead to structural perturbations of DNA. For most of the compounds discussed at this meeting, it is not known whether the cyclic adducts that are formed lead to such consequences.

Because of its metabolic conversion to chloroacetaldehyde, vinyl chloride has received most attention. But there is a dilemma with regard to cyclic vinyl chloride-DNA adducts: ethenoadenine and ethenocytosine have not been detected, except in one report (Green & Hathway, 1978), in double-stranded DNA from vinyl chloride-exposed rodents. Some possible reasons for this have been proposed (Ribovich et al., 1982): the vinyl chloride metabolites react in DNA only with unpaired adenine and cytosine; proteins in DNA shield these base-paired positions; ethenoadenine and ethenocytosine are formed, but are rapidly removed by repair enzymes. However, another alternative should also be kept in mind: cyclic adducts may only be 'indicators' of other concomitantly formed monoadducts or of secondary lesions, like intra- and interstrand DNA cross-links, which could be of equal or greater importance. This possibility is illustrated in Figure 4, taking chloroacetaldehyde and 1,3-bis-(2-fluoroethyl)-1-nitrosourea (BFNU) as prototypes of compounds that carry two reactive sites. Formation of any of the cyclic DNA adducts like ethenoadenine or ethenocytosine (Fig. 4A, XXIX) requires at least two, and probably more, steps. In such a reaction sequence, monoadducts (Fig. 4A, XXVII) must be formed as transient intermediates (Biernat *et al.*, 1978; Kuśmierek & Singer, 1982), which themselves could also lead to genetic changes in a replicating cell. The second example (Fig. 4B), which comes from work on bis-haloalkylnitrosoureas, adds further weight to this idea. The cyclic 1,O^6-ethanoguanine adduct (XX) was shown (Tong *et al.*, 1983) to lead to cross-links (XXXI) or hydrolysis products (XXXII). The monoadduct, XXX, may be responsible for the mutagenicity of this class of compounds, and XXXI for cell-killing. These examples illustrate that those adducts that could be formed either before or after the cyclization reaction cannot be ignored. Our discussions must therefore question the importance of the bifunctional nature of such agents. It is appropriate in this regard to give credit to a scientist who greatly stimulated this field, namely, Loveless (1966), who made a very provocative statement: 'For induction of gene mutations in lower fungi, functionality is irrelevant in determining mutational effectiveness, rather only chemical reactivity seems to be the determinant of quantitative effectiveness'.

Finally, let me give my personal opinion on future approaches. In view of the essential lack of data for assessing the role of cyclic nucleic acid base adducts today, progress in studies of carcinogenesis/mutagenesis may be made in the following areas: (1) improved spectro-analytical, immunological and postlabelling methods for adduct identification and characterization; (2) structure-activity relationships, correlating levels of individual adducts with biological endpoints, in both experimental systems and man; (3) use of bacteria, yeasts and mammalian cell lines with known mutational specificity and repair deficiencies; (4) site-directed mutagenesis, using defined cyclic nucleic acid adducts. Lastly, because humans are exposed to many of these compounds and because several are recognized as being carcinogenic in man, we ought to find out whether cyclic DNA adducts have anything to do with the induction of human cancer. The answer can be obtained only by integrated laboratory/epidemiological investigations, an approach which combines molecular dosimetry, effect-monitoring and risk evaluation in exposed human subjects – for example, those treated with alkylating cytostatic agents (see Kaldor & Day, these proceedings[4]). Our task may not be a simple one as we may have to deal with unstable DNA adducts, as hypothesized earlier. Even if this conference only serves to point out gaps in our knowledge and is unable to establish the biological role of cyclic nucleic acid adducts

---

[4] See p. 327.

in carcinogenesis, I am confident that future directions and approaches will become clear. Let us give credit to the pioneering work of R. Shapiro who reported in 1966 and 1969 the cyclic nucleic acid adducts formed from glyoxal. I quote the concluding statement he made (Shapiro, 1969) at a meeting of the New York Academy of Science at that time: 'As additional reactions of related compounds with nucleic acids are discovered, the biological importance of this class of compounds will undoubtedly grow'. Numerous presentations at this conference in 1984 give proof that his predictions made fifteen years ago were right.

ACKNOWLEDGEMENTS

The author wishes to thank his co-workers, A. Barbin and C. Malaveille, for their research contributions in this area, E. Heseltine for editorial assistance and Y. Granjard for secretarial help.

REFERENCES

Barbin, A., Brésil H., Croisy A., Jacquignon, P., Malaveille, C., Montesano, R. & Bartsch, H. (1975) Liver microsome-mediated formation of alkylating agents from vinyl bromide and vinyl chloride. *Biochem. biophys. Res. Commun.*, **67**, 596–603

Barrio, J.R., Secrist, J.A., III & Leonard, N.J. (1972) Fluorescent adenosine and cytidine derivatives. *Biochem. biophys. Res. Commun.*, **46**, 597–604

Bartsch, H., Malaveille, C & Montesano, R. (1975) Human, rat and mouse liver-mediated mutagenicity of vinyl chloride in *S. typhimurium* strains. *Int. J. Cancer*, **15**, 429–437

Biernat, J., Ciesiołka J., Górnicki P., Adamiak, R.W., Krzyżosiak, W.J. & Wiewiórowski, M. (1978) New observations concerning the chloroacetaldehyde reaction with some tRNA constituents. Stable intermediates, kinetics and selectivity of the reaction. *Nucleic Acids Res.*, **5**, 789–804

Chen, R.F., Mieyal, J.J. & Goldthwait, D.A. (1981) The reaction of β-propiolactone with derivatives of adenine and with DNA. *Carcinogenesis*, **2**, 73–80

Chung, F.-L. & Hecht, S.S. (1983) Formation of cyclic $1,N^2$-adducts by reaction of deoxyguanosine with α-acetoxy-*N*-nitrosopyrrolidine, 4-(carbethoxynitrosamino)butanal, or crotonaldehyde. *Cancer Res.*, **43**, 1230–1235

Chung, F.-L. & Hecht, S.S. (1984) Formation of cyclic $1,N^2$-propanodeoxyguanosine adducts in DNA upon reaction with acrolein or crotonaldehyde. *Cancer Res.*, **44**, 990–995

Clark, V.M., Todd, A.R. & Inssman J. (1951) Nucleotides, Part VIII: Cyclonucleoside salts. A novel rearrangement of some toluene-*p*-sulfonylnucleotides. *J. chem. Soc.*, **4**, 2952–2958

Creech, J.L. & Johnson, M.N. (1974) Angiosarcoma of the liver in the manufacture of polyvinyl chloride. *J. occup. Med.*, **16**, 150–151

Gaugler, B.J.M., Neumann, H.G., Scribner, N.K. & Scribner, J.D. (1979) Identification of some products from the reaction of *trans*-4-aminostilbene metabolites and nucleic acids *in vivo*. *Chem.-biol. Interactions, 27*, 335–342

Goldschmidt, B.M., Blazej, T.P. & Van Duuren, B.L. (1968) The reaction of guanosine and deoxyguanosine with glycidaldehyde. *Tetrahedron Lett., 13*, 1583–1586

Green, T. & Hathway, D.E. (1978) Interactions of vinyl chloride with rat liver DNA *in vivo*. *Chem.-biol. Interactions, 22*, 211–224

Guengerich, F.P., Geiger, L.E., Hogy, L.L. & Wright, P.L. (1981) In-vitro metabolism of acrylonitrile to 2-cyanoethylene oxide, reaction with glutathione, and irreversible binding to proteins and nucleic acids. *Cancer Research, 41*, 4925–4933

Johnston, T.P., Fikes, A.L. & Montgomery, J.A. (1962) The structure of the tricyclic purine derived from the purin-6-yl analogue of nitrogen mustard. *J. org. Chem., 27*, 973–976

Kayasuga-Mikado, K., Hashimoto, T., Negishi, T., Negishi, K. & Hayatsu, H. (1980) Modification of adenine and cytosine derivatives with bromoacetaldehyde. *Chem. pharm. Bull., 28*, 932–938

Kochetkov, N.K., Shibaev, V.N. & Kost, A.A., (1971) New reaction of adenine and cytosine derivatives, potentially useful for nucleic acids modification. *Tetrahedron Lett., 22*, 1993–1996

Kohwi-Shigematsu, T., Gelinas, R. & Weintraub, H. (1983) Detection of an altered DNA conformation at specific sites in chromatin and supercoiled DNA. *Proc. natl Acad. Sci. USA, 80*, 4389–4393

Kuśmierek, J.T. & Singer, B. (1982) Chloroacetaldehyde-treated ribo- and deoxyribopolynucleotides. I. Reaction products. *Biochemistry, 21*, 5717–5722

Laib, R.J. & Bolt, H.M. (1977) Alkylation of RNA by vinyl chloride metabolites *in vitro* and *in vivo*: formation of $1,N^6$-ethenoadenosine. *Toxicology, 8*, 185–195

Laib, R.J. & Bolt, H.M. (1978) Formation of $3,N^4$-ethenocytidine moieties in RNA by vinyl chloride metabolites *in vitro* and *in vivo*. *Arch. Toxicol., 39*, 235–240

Laib, R.J., Gwinner, L.M. & Bolt, H.M. (1981) DNA alkylation by vinyl chloride metabolites: etheno derivatives or 7-alkylation of guanine? *Chem.-biol. Interactions, 37*, 219–231

Leonard, N.J. (1984) Etheno-substituted nucleotides and coenzymes: fluorescence and biological activity. *CRC Crit. Rev. Biochem., 15*, 125–199

Leonard, N.J. & Tolman, G.L. (1975) Fluorescent nucleosides and nucleotides. *Ann. N.Y. Acad. Sci., 255*, 43–58

Leonard, N.J., Achmatowicz, S., Loeppky, R.N., Carraway, K.L., Grimm, W.A.H., Szweykowska, A., Hamzi, H.Q. & Skoog, F. (1966) Development of cytokinin activity by rearrangement of 1-substituted adenines to 6-substituted aminopurines: inactivation by N^6,1-cyclization. *Proc. natl Acad. Sci. USA, 56*, 709–716

Lilley, D.M.J. (1983) Structural perturbation in supercoiled DNA: hypersensitivity to modification by a single-strand-selective chemical reagent conferred by inverted repeat sequences. *Nucleic Acids. Res., 11*, 3097–3112

Loveless, A. (1966) *Genetic and Allied Effects of Alkylating Agents*, London, Butterworth

Ludlum, D.B., Kramer, B.S., Wang, J. & Fenselau, C. (1975) Reaction of 1,3-bis-(2-chloroethyl)-1-nitrosourea with synthetic polynucleotides. *Biochemistry,* **14,** 5480–5485

Maté, U., Solomon, J.J. & Segal, A., (1977) In-vitro binding of beta-propiolactone to calf thymus DNA and mouse liver DNA to form 1-(2-carboxyethyl)adenine. *Chem.-biol. Interactions,* **18,** 327–336

Moschel, R.C. & Leonard, N.J. (1976) Fluorescent modification of guanine: reaction with substituted malondialdehyde. *J. org. Chem.,* **41,** 294–300

Nair, V. & Turner, G.A. (1984) Determination of the structure of the adduct from guanosine and glycidaldehyde. *Tetrahedron Lett.,* **25,** 240–250

Nair, V., Turner, G.A. & Offerman, R.J. (1984) Novel adducts from the modification of nucleic acid bases by malondialdehyde. *J. Am. chem. Soc.,* **106,** 3370–3371

Oesch, F. & Doerjer, G. (1982) Detection of N^2,3-ethenoguanine in DNA after treatment with chloroacetaldehyde *in vitro*. *Carcinogenesis,* **3,** 663–665

Osterman-Golkar, S., Hultmark, D., Segerbäck, D., Calleman, C.J., Göthe, R., Ehrenberg, L. & Wachtmeister, C.A. (1977) Alkylation of DNA and proteins in mice exposed to vinyl chloride. *Biochem. biophys. Res. Commun.,* **76,** 259–266

Ravindranath, V., Burka, L.T. & Boyd, M.R. (1984) Reactive metabolites from the bioactivation of toxic methylfurans. *Science,* **224,** 884–886

Ribovich, M.L., Miller, J.A., Miller, E.C. & Timmins, L.G. (1982) Labeled 1,N^6-ethenoadenosine and 3,N^4-ethenocytidine in hepatic RNA of mice given [ethyl-1,2-^{3}H]- or [ethyl-1-^{14}C]-ethyl carbamate (urethane). *Carcinogenesis,* **3,** 539–546

Rice, J.M. (1984) Photochemical addition of benzo[a]pyrene to pyrimidine derivatives. *J. Am. chem. Soc.,* **86,** 1444–1446

Roberts, J.J. & Warwick, G.P. (1966) Azo-dye carcinogenesis. The reactions of 4-hydroxymethylaminoazobenzene with cytosine derivatives. *Int. J. Cancer,* **1,** 107–117

Sattsangi, P.D., Leonard, N.J. & Frihart, C.R. (1977) 1,N^2-ethenoguanine and N^2,3-ethenoguanine. Synthesis and comparison of the electronic spectral properties of these linear and angular triheterocycles related to the Y bases. *J. org. Chem.,* **42,** 3292–3296

Scherer, E., Van Der Laken, C.J., Gwinner, L.M., Laib, R.J. & Emmelot, P. (1981) Modification of deoxyguanosine by chloroethylene oxide. *Carcinogenesis,* **2,** 671–677

Seto, H., Akiyama, K., Okuda, T., Hashimoto, T., Takesue, T. & Ikemura, T. (1981) Structure of a new modified nucleoside formed by guanosine-malonaldehyde reaction. *Chem. Lett.,* 707–708

Seto, H., Okuda, T., Takesue, T. & Ikemura, T. (1983) Reaction of malonaldehyde with nucleic acid. I. Formation of fluorescent pyrimido-(1,2-a)purin-10(3H)-one nucleosides. *Bull. chem. Soc. Jpn.,* **56,** 1799–1802

Shapiro, R. (1969) Reactions with purines and pyrimidines. *Ann. N.Y. Acad. Sci.,* **163,** 624–630

Shapiro, R. & Hachmann, J. (1966) The reactions of guanine derivatives with 1,2-dicarbonyl compounds. *Biochemistry,* **5,** 2799–2807

Tong, W.P. & Ludlum, D.B. (1979) Mechanism of action of the nitrosoureas. III. Reaction of bis-chloroethylnitrosourea and bis-fluoroethylnitrosourea with adenosine. *Biochem. Pharmacol., 28,* 1175–1179

Tong, W.P., Krik, M.C. & Ludlum, D.B. (1983) Mechanism of action of the nitrosoureas. V. Formation of O^6-(2-fluoroethyl)guanine and its probable role in the cross-linking of deoxyribonucleic acid. *Biochem. Pharmacol., 32,* 2011–2015

Varghese, A.J. & Whitmore, G.F. (1983) Modification of guanine derivatives by reduced 2-nitroimidazoles. *Cancer Res., 43,* 78–82

Zajdela, F., Croisy, A., Barbin, A., Malaveille, C., Tomatis, L. & Bartsch, H. (1980) Carcinogenicity of chloroethylene oxide, an ultimate reactive metabolite of vinyl chloride, and bis(chloromethyl)ether after subcutaneous administration and in initiation-promotion experiments in mice. *Cancer Res., 40,* 352–356

CARCINOGENICITY OF SELECTED VINYL COMPOUNDS, SOME ALDEHYDES, HALOETHYL NITROSOUREAS AND FUROCOUMARINS: AN OVERVIEW

H. VAINIO & R. SARACCI

International Agency for Research on Cancer, Lyon, France

VINYL COMPOUNDS

Vinyl chloride

In 1930, shortly after the commercial introduction of vinyl chloride, the acute toxic effects of this compound were reported in experimental animals (Patty *et al.*, 1930). During the following five decades, studies throughout the world indicated that occupational exposure to vinyl chloride was associated with a wide range of toxic effects in humans.

After the observation of acro-osteolysis in workers exposed to vinyl chloride in 1966, a series of experiments was initiated in an attempt to reproduce the disease in various animal species. In the course of these long-term toxicity studies, vinyl chloride was found to induce tumours in the Zymbal (sebaceous) gland in rats exposed to levels of 30 000 ppm in air (Viola *et al.*, 1971). Exposure by inhalation of a large number of rats resulted in cancers in various organs, with squamous-cell skin carcinomas, lung adenocarcinomas and liver angiocarcinomas being the most common types (Caputo *et al.*, 1974).

Further experimental studies in rats, mice and hamsters have demonstrated that exposure to vinyl chloride by various routes of administration induces cancers of the lung, brain, breast and skin (IARC, 1974a, 1979). The primary types of tumour found in the rat are liver angiosarcomas and hepatocellular carcinomas, but Zymbal-gland and mammary-gland tumours also occur. In the mouse, vinyl chloride produces pulmonary adenomas and liver and mammary cancers following inhalation (Pepelko, 1984).

In humans, the association between vinyl chloride and liver haemangiosarcoma was first discovered in workers in a vinyl-chloride-polymerization factory in the USA (Creech & Johnson, 1974). Since 1974, more than a hundred cases of liver haemangiosarcoma have been reported among workers exposed to vinyl chloride in various countries (Stafford, 1983). The carcinogenic effects of vinyl chloride have been confirmed beyond any doubt by several epidemiological investigations, in which excessively high incidences of cancer of the digestive organs (including liver), lungs, central nervous system, lymphatic system and haematopoietic system have been recurrently reported among workers exposed to vinyl chloride. Based on such evidence, the brain and the lungs have been regarded as plausible target organs (Merletti et al., 1984) in addition to the liver, an established target of vinyl chloride in humans.

These results initiated a plethora of studies on the possible mechanisms of action of vinyl chloride, which has perhaps been studied more intensively than any other human carcinogen. The available evidence shows that vinyl chloride is epoxidized by monooxygenase(s) to form chloroethylene oxide which rearranges to form chloroacetaldehyde. Chloroethylene oxide, in contrast to chloroacetaldehyde, has been found to induce local tumours after repeated subcutaneous administration and to induce skin tumours in a two-stage carcinogenicity experiment in mice (Zajdela et al., 1980). Thus, although chloroacetaldehyde has not been tested in long-term carcinogenicity studies, the existing evidence strongly supports the pivotal role of chloroethylene oxide as the ultimate carcinogen in vinyl-chloride-induced carcinogenicity.

Vinyl bromide

In contrast to vinyl chloride, vinyl bromide is used as a co-monomer in the manufacture of synthetic polymers in the plastics industry; it has flame-retardant properties and is often used in fibres for clothing or for household articles. Commercial production of vinyl bromide only began in the late 1960's and there are no reports on its possible carcinogenicity in humans.

Vinyl bromide is metabolized to an epoxide by microsomal monoxygenase(s); the similarity between this compound and vinyl chloride is evident since they both form the same alkylation products with nucleic acids. Rats exposed by inhalation to 10–1250 ppm vinyl bromide developed angiosarcomas, primarily of the liver, at all exposure levels, and a dose-dependent increase in squamous-cell carcinoma of the Zymbal gland was also observed. Hepatocellular carcinoma and hepatic neoplastic nodules showed significant, but not dose-related, increases in females only (Benya et al., 1982).

Vinyl acetate

Vinyl acetate is a volatile liquid used in films and lacquers; it is also a component of many plastic polymers. In humans, vapour concentrations exceeding 20 ppm are generally irritating to the respiratory tract and eyes. Vinyl acetate is rapidly hydrolysed in biological fluids, with concomitant production of acetaldehyde (Filov, 1959) which has a half-life of a few minutes.

Vinyl acetate gave negative results in mutagenicity studies with *Salmonella typhimurium* (Bartsch et al., 1979); however, in cultured human lymphocytes, it was a potent inducer of sister chromatid exchanges (Norppa et al., 1983). It has been suggested that vinyl acetate is converted either spontaneously or by the action of esterases into acetic acid and, through an unstable vinyl alcohol intermediate, into acetaldehyde.

In a recent study, concentrations of 1000 and 2500 mg/l vinyl acetate were given to groups of 20 male and 20 female rats in their drinking water on five days per week for 100 weeks (Lijinsky & Reuber, 1983). The incidence of most types of neoplasm was similar in the treated and control groups. However, six females in the high-dose, but none in the low-dose or control groups, had neoplastic nodules of the liver; among the males, such nodules were found in two animals in the high-dose group, four in the low-dose group and none in the controls. Five females in the high-dose group had thyroid C-cell adenomas, compared with two in the low-dose group and none in the controls, and five females in the high-dose group developed uterine carcinomas, compared with one in the low-dose group and none in the controls

The authors considered the consistent occurrence of liver and thyroid C-cell neoplasms to be due to the treatment; they concluded that their results were not negative, and suggested that a study using more animals and higher doses should be performed, with daily preparation of the vinyl acetate solutions.

Acrylonitrile (Vinyl cyanide)

Acrylonitrile is used as an intermediate in the manufacture of a wide variety of acrylic fibres, plastics and synthetic rubber. It undergoes reactions at both the nitrile group and the double bond, and its proposed metabolic pathway suggests the formation of a transient epoxide intermediate. Mutagenicity of the epoxide of acrylonitrile (glycidonitrile) has been shown experimentally (Peter *et al.*, 1983). Acrylonitrile is mutagenic in *Salmonella typhimurium* strains that are sensitive to base-pair mutations in the presence of a mammalian metabolic activation system. Urine from rats treated with intraperitoneal injections of acrylonitrile was also mutagenic in *Salmonella typhimurium* (Lambotte-Vandepaer *et al.*, 1980). Acrylonitrile did not induce chromosomal damage or micronuclei in bone-marrow cells of mice exposed *in vivo*, but positive results were obtained in rat bone-marrow cells *in vivo* and in Chinese hamster ovary cells co-cultivated with rat hepatocytes (IARC, 1982).

Acrylonitrile has been shown to be carcinogenic in animals (IARC, 1982). Quast *et al.* (1981) administered dose levels of 35, 100 and 300 ppm acrylonitrile in drinking water to rats for two years. A statistically significant increase in tumour incidence was observed in the central nervous system, Zymbal gland, stomach and gastrointestinal tract in both females and males, and in the mammary glands of female rats. An increased incidence of tumours in the central nervous system and Zymbal gland of rats was also noted in a drinking-water study by Beliles *et al.* (1980).

Maltoni *et al.* (1977) exposed animals to atmospheres containing 5, 10, 20 and 40 ppm acrylonitrile for four hours per day, five days per week, for 12 months. Marginal increases in tumours of the mammary glands in females and the forestomach in males were observed.

Epidemiological follow-up of 1345 workers exposed to acrylonitrile between 1950 and 1966 revealed a significant increase in lung cancer (5 observed *vs* 1.4 expected) among those with at least a moderate exposure and a follow-up time greater than 10 years (O'Berg, 1980). In a similar study, in 1111 workers exposed to acrylonitrile between 1950 and 1968 and followed for 10 years or more, nine cancers of the respiratory tract (7.6 expected), five stomach cancers (1.9 expected) and two brain cancers (0.7 expected) were observed. In a third epidemiological investigation, 9 lung cancer deaths were observed among rubber workers exposed to acrylonitrile, vs 5.9 expected (standardized mortality ratio, 150). Results from four other studies (Kiesselbach et al., 1979; Thiess et al., 1980; Zack, 1980; Nakamura, 1981), one of which indicated an increase in the incidence of lung cancer of approximately two-fold and the other three of which showed no increased incidence, are difficult to evaluate because of weaknesses in design and/or reporting. The IARC Working Group (1982) considered there to be *limited evidence* that acrylonitrile is carcinogenic in humans.

Vinyl carbamate

Ethyl carbamate is a well-known carcinogen (IARC, 1974b) which produces multiple types of tumours in various rodent species and liver tumours in non-human primates (Adamson & Sieber, 1982) and one to which human exposure has historically been widespread due, in large part, to its industrial and medical uses. Studies by Dahl et al. (1980) have suggested that vinyl carbamate might be the proximate carcinogenic metabolite of ethyl carbamate. The carcinogenic activity of vinyl carbamate is qualitatively similar to, but much stronger than, that of ethyl carbamate; this suggests that the two compounds may converge in the formation of an electrophilic reactant (presumably vinyl carbamate epoxide) which binds covalently to nucleic acids (Dahl et al., 1978; Allen et al., 1982).

ALDEHYDES (EXCLUDING FORMALDEHYDE AND ACETALDEHYDE)

Malonaldehyde

Malonaldehyde is found in many foodstuffs and is present at generally high levels in rancid foods. The formation of malonaldehyde in foods seems to be dependent on many factors, including the degree of unsaturation of the fatty acids and the length of cooking (IARC, 1985).

Malonaldehyde is formed endogenously in tissues as an end-product of lipid peroxidation. It is also a by-product of prostaglandin and tromboxane biosynthesis.

Malonaldehyde has been found to be mutagenic in several strains of *Salmonella typhimurium* (Marnett & Tuttle, 1980; Levin et al., 1982; Yamaguchi, 1982) and in *Escherichia coli* (Yonei & Furni, 1981). It induces somatic mutations in *Drosophila melanogaster* (Szabad et al., 1983), and micronuclei, chromosomal aberrations and aneuploidies in cultured rat-skin fibroblasts (Bird & Draper, 1980; Bird et al., 1982).

Malonaldehyde has been tested in mice by administration in the drinking water and by skin application. Because of its instability and high reactivity, malonaldehyde is not available as a free compound and its enolic sodium salt has generally been used in carcinogenicity studies. Oral studies in mice were considered by the IARC Working Group in June 1984 to be inadequate for evaluation, and no skin tumours were observed in two topical application studies. In two-stage mouse-skin bioassays for initiating and promoting activity, malonaldehyde was negative (Fischer et al., 1983); however, in another two-stage experiment using much higher doses (skin applications of 6 and 12 mg), malonaldehyde showed initiating activity (Shamberger et al., 1975).

Crotonaldehyde

Crotonaldehyde is used as a chemical intermediate. It has been identified in automobile exhausts and in tobacco smoke (Hoffmann et al., 1975; Saito et al., 1983).

Crotonaldehyde has been found to be mutagenic in *Salmonella typhimurium*, especially in liquid incubation assays in the absence of liver homogenate (Lijinsky & Andrews, 1980; Lutz et al., 1982). No data are available on the carcinogenicity of crotonaldehyde.

Acrolein

Acrolein is principally employed as a chemical intermediate for the synthesis of acrylic acid and esters; it is also used as a biocide. Acrolein is a pyrolysis product of fats and is also found in tobacco smoke. It is present as an urban air pollutant, arising from a combination of fossil fuels (IARC, 1985).

Acrolein has been tested in mice by skin application, in rats by administration in the drinking water, and in hamsters by inhalation exposure. The studies in mice and rats suffer from several limitations, making them inadequate for evaluation (IARC, 1985). No carcinogenic effect was found in hamsters.

Glycidaldehyde

Glycidaldehyde occurs in vegetable oils; its concentration depends on the rancidity of the oil. It has been used as a vapour-phase desinfectant (IARC, 1976).

Glycidaldehyde has the capacity to form a cyclic $1,N^2$-adduct with deoxyguanosine *in vitro* (Goldsmith et al., 1968) and is mutagenic in various test systems. It is carcinogenic in mice after skin application, and in mice and rats after subcutaneous injection (IARC, 1976).

HALOETHYL NITROSOUREAS

Haloethyl nitrosoureas are strong alkylating agents, used extensively as chemical anticancer drugs (IARC, 1981; Ludlum & Tong, 1981).

During the past decade, the chloroethyl nitrosourea drugs – carmustine (BNCU[1]), lomustine (CCNU[2]) and semustine (methyl-CCNU) – have been used to treat patients with malignant melanoma and cancers of the brain, lung and digestive tract; their antitumour effects are generally due to their alkylating activity. Although the

---

[1] 1,3-bis-(2-chloroethyl)-1-nitrosourea
[2] 1-(2-chloroethyl)-3-cyclohexyl-1-nitrosourea

nitrosoureas display an unusual degree of cumulative bone-marrow toxicity, there have been no data on their carcinogenic effect in human beings, aside from a few case reports of acute nonlymphocytic leukaemia in patients treated with BCNU or CCNU in combination with other cytotoxic drugs and radiation. In a recent report, 2067 patients treated with methyl-CCNU as adjuvant therapy were followed. Leukaemic disorders were observed in 14 patients, compared to only one leukaemic disorder among 1566 patients given other therapies (relative risk, 12.4; 95% confidence intervals, 1.7–250) (Boice et al., 1983).

BCNU is carcinogenic in rats, producing tumours of the lung after intraperitoneal or intravenous administration, and intra-abdominal tumours after intraperitoneal administration. When BCNU was tested in mice by skin application combined with ultraviolet (B) irradiation, skin tumours appeared earlier than in the controls (IARC, 1982). CCNU induces lung carcinomas in rats following intraperitoneal or intravenous injection. It was also tested in mice by intraperitoneal injection: a slight increase in the incidence of lymphomas was observed (IARC, 1982). BCNU and CCNU are directly-acting alkylating agents, which react with DNA. They induce mutations in bacteria and mammalian cells *in vitro*. CCNU induces sister chromatid exchanges in blood lymphocytes from patients treated *in vivo* and BCNU induces chromosomal anomalies in mice exposed *in vivo* (IARC, 1982).

FUROCOUMARINS WITH ULTRAVIOLET (A) LIGHT

Linear furocoumarins (psoralens) occur widely, and in various purities, in nature as constituents of hundreds of plant species (Pathak et al., 1962). Many furocoumarins are potent photosensitizers, and some are used as light-activated drugs for the treatment of certain skin disorders. The furocoumarin derivatives which have long been used in the photochemotherapy of psoriasis, vitiligo and other hyperproliferative skin diseases are 8-methoxypsoralen (8-MOP) and 4,5',8-trimethylpsoralen (TMP) (Anderson & Voorhees, 1980). More recently, 5-methoxypsoralen (5-MOP) has also been studied for the same use (Hönigsmann et al., 1979). 5-MOP is also used in cosmetic preparations as a sun-tan agent.

All these furocoumarins are bifunctional psoralen derivatives, which are able to photobind with DNA to form monoadducts and diadducts (Song & Tapley, 1979).

In addition to exposure through medical treatment, humans are also exposed to furocoumarins by the consumption of certain foods and, dermally, by the use of perfumes, fragrances and sun-tan preparations.

8-Methoxypsoralen and ultraviolet(A)-therapy (PUVA)

The combined application of 8-methoxypsoralen and ultraviolet (A) light (PUVA) is a well-established mutagenic treatment in a wide variety of cellular systems (Bridges et al., 1981; IARC, 1982). There is ample evidence for the carcinogenicity of PUVA on the skin of the mouse (IARC, 1982).

On the basis of data from patients treated with PUVA, an IARC Working Group (1982) concluded that PUVA treatment is carcinogenic in humans, causing cutaneous squamous-cell carcinomas. This conclusion was mainly based on the results of a multicentric prospective study, which has recently been updated to cover an average period of 5.7 years between the final dermatological examination and the initial PUVA treatment of 1286 patients (Stern *et al.*, 1984). After adjustment for exposures to ionizing radiation and topical tar preparations, the risk that cutaneous squamous-cell carcinoma would develop 22 months or more after the first exposure to PUVA was 12.8 times higher in patients exposed to a high dose than in those exposed to a low dose (95 per cent confidence interval, 5.8 to 28.5). No substantial dose-related increase was noted for basal-cell carcinoma. These results confirm that PUVA is carcinogenic in humans.

5-Methoxypsoralen with ultraviolet (A) light

The furocoumarin, 5-methoxypsoralen (5-MOP), is the melanogenic component present in several sun-tan preparations in the form of natural oil of bergamot. It is also present in *Citrus bergamia* extracts frequently used as additives in Eau de Cologne preparations, and as a natural toxin in food, for example in parsnip root (Ivie *et al.*, 1981).

Combined treatment with 5-MOP plus ultraviolet (A) light was found to be carcinogenic on mouse skin, causing rapidly growing squamous-cell carcinomas (Zajdela & Bignami, 1981). No epidemiological studies are available to evaluate the possible effects of 5-MOP in, for example, sun-tan preparations.

DISCUSSION

Metabolism is an important factor in the production of reactive intermediates of vinyl compounds. The metabolic activation and inactivation of vinyl chloride has been studied in detail, and vinyl bromide is assumed to be activated through a similar pathway. Only the epoxide metabolites of vinyl chloride and vinyl bromide bind irreversibly to DNA (Guengerich *et al.*, 1981). Another metabolite of vinyl chloride, chloroacetaldehyde, is able to alkylate protein structures, but not DNA. Haemangiosarcomas of the liver are the marker tumours associated with exposure of experimental animals to vinyl chloride; such tumours have also been induced by vinyl bromide.

Vinyl acetate seems to differ from vinyl chloride and vinyl bromide in that the principal active metabolite is most likely not an epoxide, but an aldehyde, which is the product of either nonenzymatic cleavage or esterase activity. Although it is not mutagenic in bacterial point-mutation assays, vinyl acetate appears to be an effective clastogenic agent, as is acetaldehyde. Vinyl acetate is probably activated by some means other than microsomal monooxygenases; this may also explain why liver haemangiosarcomas are not induced by vinyl acetate.

The brain is repeatedly mentioned as a possible target organ for many of these chemicals (see Table 1); rats inhaling vinyl chloride develop neuroblastomas of the brain, although only at high exposure levels. Brain tumours have also frequently been associated with exposure to vinyl chloride in humans. Animal bioassays have

Table 1. Summary of the epidemiological and experimental findings

Name of chemical (Evaluation of data: H = humans; A = animals)	Humans			Animals			Reference
	Main type of exposure	Main route of exposure	Target organ or tissue	Animal	Route of exposure	Target organ or tissue, or type of tumour	
Acrolein (H. inadequate) (A. inadequate)	food, air	oral, inhalation	?	Mouse	skin	?	IARC 36 (1985), p. 133
				Rat	oral		
				Hamster	inhalation		
Acrylonitrile (H. limited) (A. sufficient)	occupational	inhalation, oral, skin	lung, brain*, stomach*, prostate*, lymphohaemato-poietic tissue*	Rat	inhalation, oral	brain, Zymbal gland, gastro-intestinal tract, mammary gland	IARC 19 (1979), p. 73; IARC Suppl. 4 (1982), p. 25
BCNU (H. inadequate) (A. sufficient)	medicinal (combination)	intravenous	lymphohaemato-poietic tissue*	Mouse	oral	mammary gland	IARC 26 (1981), p. 79; IARC Suppl. 4 (1982), p. 63
				Rat	intravenous, intraperitoneal, oral	lung, mammary gland, intra-peritoneal tumours	
CCNU (H. inadequate) (A. sufficient)	medicinal (combination)	intravenous	lymphohaemato-poietic tissue*	Mouse	intraperitoneal	lymphopoietic tissue, lung	IARC 26 (1981), p. 137; IARC Suppl. 4 (1982), p. 83
				Rat	intravenous, intraperitoneal	lung, mammary gland	
Ethyl carbamate (H. no data) (A. sufficient)	occupational, medicinal (veterinary)	inhalation, skin, intra-venous	?	Mouse	oral, skin, inhalation, intraperitoneal	lung, lympho-poietic tissue, skin	IARC 7 (1974b), p. 111
				Rat	oral, intra-peritoneal	lung, liver, lymphopoietic tissue	
				Hamster	oral, skin, intraperitoneal	forestomach, skin, lung	

Compound	Use	Human exposure	Human target	Animal species	Route	Animal target	Reference
Glycidaldehyde (H. no data) (A. sufficient)	disinfectant, diet	inhalation, oral	?	Mouse	dermal, subcutaneous	skin, (local) sarcomas	IARC 11 (1976), p. 175
				Rat	subcutaneous	(local) sarcomas	
Malonaldehyde (H. no data) (A. inadequate)	food	oral, endogenous	?	Mouse	oral, skin		IARC 36 (1985) p. 163
8-MOP + UVA (H. sufficient) (A. sufficient)	medicinal	oral, skin	skin	Mouse	oral, skin	skin	IARC 24 (1980), p. 101; IARC Suppl. 4 (1982), p. 158
5-MOP + UVA	cosmetic	skin, oral	skin*	Mouse	skin	skin	Zajdela & Bisagni (1981)
Vinyl acetate	occupational	skin, inhalation	?	Rat	oral	thyroid, uterus, liver	Lijinsky & Reuber (1983); Maltoni (1977)
					inhalation	mammary tumours	
Vinyl bromide	occupational	skin, inhalation	?	Rat	inhalation	liver (angiosarcoma), Zymbal gland	Benya et al. (1982)
Vinyl carbamate	no known use			Mouse	intraperitoneal	liver, lung, thymus	Dahl et al. (1980)
				Rat	intraperitoneal	liver, Zymbal gland, mesenchymal tumours	
Vinyl chloride (H. sufficient) (A. sufficient)	occupational	inhalation, skin, oral	liver (angiosarcoma), respiratory tract, brain, lymphatic and haematopoietic system, gastrointestinal tract	Mouse	inhalation	liver (angiosarcoma), mammary gland, lung, brain, nasal cavity	IARC 7 (1974b), p. 291; IARC 19 (1979), p. 377; IARC Suppl. 4 (1982), p. 260

* Suspected target organ or tissue

Name of chemical (Evaluation of data: H = humans; A = animals)	Humans			Animals			Reference
	Main type of exposure	Main route of exposure	Target organ or tissue	Animal	Route of exposure	Target organ or tissue, or type of tumour	
				Rat	inhalation, oral	liver (angio-sarcoma), Zymbal gland, brain, various sites	
				Hamster	inhalation	liver (angio-sarcoma), skin, forestomach	
				Rabbit	inhalation	lung, skin	

demonstrated increases in brain tumours after oral or inhalation exposure to acrylonitrile. Acrylonitrile functions as an ethylating agent (causing cyanoethylation), and could also act as an alkylating agent in the central nervous system.

Surprisingly little carcinogenicity data is available on the aldehydes under consideration; although acrolein, crotonaldehyde and malonaldehyde appear to be mutagenic in bacteria, there are no convincing carcinogenicity studies in animals. Dialdehydes have the capacity to form intramolecular (etheno) cross-links in nucleotides (Moschel & Leonard, 1976). Epidemiological studies are nonexistent, although human exposure to these aldehydes can be extensive.

The strong alkylating agents – chloroethyl nitrosoureas – are carcinogenic in animals, and recent epidemiological studies have provided quantitative evidence that they are probably leukaemogenic in humans. Nitrosoureas decompose in physiological media to form alkylating and carbamoylating moieties (Montgomery, 1976). In addition to the carbonium ion, other alkylating moities can be produced from nitrosoureas. These include vinyl carbonium ions, 2-chloroethylamine, and probably 2-chloroacetaldehyde. Some 2-chloroethanol is produced as well, and this compound can be converted by aldehyde dehydrogenase to 2-chloroacetaldehyde. It has also been purported that vinyl chloride is formed as a decomposition product of BCNU; it has been detected in blood and exhaled air from patients after administration of BCNU (Clemens et al., 1982).

The bifunctional furocoumarins, 8-MOP and 5-MOP, have been shown to be carcinogenic in animals when given in combination with ultraviolet radiation. The combined application of 8-MOP and ultraviolet (A) light is carcinogenic in humans; no epidemiological studies are available on 5-MOP, an agent widely used in sun-tan preparations. But the known experimental evidence suggests the utmost caution in a situation like this where an epidemiological study appears to be a difficult venture.

REFERENCES

Adamson, R.H. & Sieber, S.M. (1982) *Chemical carcinogenesis studies in nonhuman primates.* In: Langenbach, R., Nesnow, S. & Rice, J., eds, *Organ and Species Specificity in Chemical Carcinogenesis,* New York, Plenum Press, pp. 129–156

Allen, J.W., Langenbach, R., Nesnow, S., Sasseville, K., Leavitt, S., Campbell, J., Brock, K. & Sharief, Y. (1982) Comparative genotoxicity studies of ethyl carbamate and related chemicals: further support for vinyl carbamate as a proximate carcinogenic metabolite. *Carcinogenesis,* **3,** 1437–1441

Anderson, T.F. & Voorhees, J.J. (1980) Psoralen photochemotherapy of cutaneous disorders. *Ann. Rev. Pharmacol. Toxicol.,* **20,** 235–257

Bartsch, H., Malaveille, C., Barbin, A. & Planche, G. (1979) Mutagenic and alkylating metabolites of halo-ethylenes, chlorobutadienes and dichlorobutenes produced by rodent or human liver tissue. *Arch. Toxicol.,* **41,** 249–277

Beliles, R.P., Paulin, H.J., Makris, N.G. & Weir, R.J. (1980) *Three-Generation Reproduction Study of Rats Receiving Acrylonitrile in Drinking Water,* Litton Bionetics, Inc., Fredrick, MD

Benya, T.J., Busey, W.M., Darato, M.A. & Berteau, P.E. (1982) Inhalation carcinogenicity bioassay of vinyl bromide in rats. *Toxicol. appl. Pharmacol.*, **64**, 367–379

Bird, R.P. & Draper, H.H. (1980) Effect of malonaldehyde and acetaldehyde on cultured mammalian cells: growth, morphology, and synthesis of macromolecules. *J. Toxicol. environ. Health*, **6**, 811–823

Bird, R.P., Draper, H.H. & Basrur, P.K. (1982) Effect of malonaldehyde and acetaldehyde on cultured mammalian cells: production of micronuclei and chromosomal aberrations. *Mutat. Res.*, **101**, 237–246

Boice, J.D., Jr, Green, M.H., Killen, J.Y., Jr, Ellenberg, S.S., Keehn, R.J., McFadden, E., Chen, T.T. & Fraumeni, J.F. (1983) Leukemia and preleukemia after adjuvant treatment of gastrointestinal cancer with semustine (methyl-CCNU). *New Engl. J. Med.*, **309**, 1079–1084

Bridges, B.A., Greaves, M., Polani, P.E. & Wald, N. (1981) Do treatments available for psoriasis patients carry a genetic or carcinogenic risk? *Mutat. Res.*, **86**, 279–304

Caputo, A., Viola, P.L. & Bigotti, A. (1974) Oncogenicity of vinyl chloride at low concentrations in rats and rabbits. *J. int. Res. Commun.*, **2**, 1582

Clemens, M.R., Frank, H., Remmer, H. & Waller, H.D. (1982) Vinyl chloride: decomposition product of BCNU *in vivo* and *in vitro*. *Cancer Chemother. Pharmacol.*, **10**, 70–71

Creech, J.L. & Johnson, M.N. (1974) Angiosarcoma of liver in the manufacture of polyvinyl chloride. *J. occup. Med.*, **16**, 150–151

Dahl, G.A., Miller, J.A. & Miller, E.C. (1978) Vinyl carbamate as a promutagen and a more carcinogenic analog of ethyl carbamate. *Cancer Res.*, **38**, 3793–3804

Dahl, G.A., Miller, E.C. & Miller, J.A. (1980) Comparative carcinogenicities and mutagenicities of vinyl carbamate, ethyl carbamate and ethyl *N*-hydroxycarbamate. *Cancer Res.*, **40**, 1194–1203

Filov, V.A. (1959) On the fate of complex esters of vinyl alcohol and fatty acids in the organism (Russ.). *Gig. Tr. prof. Zabol.*, **3**, 42–46

Fischer, S.M., Ogle, S., Marnett, L.J., Nesnow, S. & Slaga, T.J. (1983) The lack of initiating and/or promoting activity of sodium malondialdehyde on SENCAR mouse skin. *Cancer Lett.*, **19**, 61–66

Goldsmith, B.M., Biazaj, T.P. & Van Duuren, B.L. (1968) The reaction of guanosine and deoxyguanosine with glycidaldehyde. *Tetrahedron Lett.*, **13**, 1583–1586

Guengerich, F.P., Mason, P.S. Stott, W.T., Fox, T.R. & Watanabe, P.G. (1981) Roles of 2-haloethylene oxides and 2-haloacetaldehydes derived from vinyl bromide and vinyl chloride in irreversible binding to protein and DNA. *Cancer Res.*, **41**, 4391–4398

Hoffmann, D., Brunnemann, K.D., Gori, G.B. & Wynder, E.L. (1975) On the carcinogenicity of marijuana smoke. *Recent Adv. Phytochem.*, **9**, 63–81

Hönigsmann, H., Jashke, E., Gschnait, F., Brenner, W., Fritsch, P. & Wolff, K. (1979) 5-Methoxypsoralen (bergapten) in photochemotherapy of psoriasis. *Br. J. Dermatol.*, **101** 369–377

IARC (1974a) *IARC Monographs on the Evaluation of the Carcinogenic Risk of Chemicals to Humans*, Vol. 4, *Some aromatic amines, hydrazine and related substances, N-nitroso compounds, and miscellaneous alkylating agents,* Lyon, International Agency for Research on Cancer

IARC (1974b) *IARC Monographs on the Evaluation of the Carcinogenic Risk of Chemicals to Humans,* Vol. 7, *Some anti-thyroid and related substances, nitrofurans and industrial chemicals,* Lyon, International Agency for Research on Cancer

IARC (1976) *IARC Monographs on the Evaluation of the Carcinogenic Risk of Chemicals to Humans,* Vol. 11, *Cadmium, nickel, some epoxides, miscellaneous industrial chemicals, and general considerations on volatile anaesthetics,* Lyon, International Agency for Research on Cancer

IARC (1979) *IARC Monographs on the Evaluation of the Carcinogenic Risk of Chemicals to Humans,* Vol. 19, *Some monomers, plastics and synthetic elastomers, and acrolein,* Lyon, International Agency for Research on Cancer

IARC (1980) *IARC Monographs on the Evaluation of the Carcinogenic Risk of Chemicals to Humans,* Vol. 24, *Some pharmaceutical drugs,* Lyon, International Agency for Research on Cancer

IARC (1981) *IARC Monographs on the Evaluation of the Carcinogenic Risk of Chemicals to Humans,* Vol. 26, *Some antineoplastic and immunosuppressive agents,* Lyon, International Agency for Research on Cancer

IARC (1982) *IARC Monographs on the Evaluation of the Carcinogenic Risk of Chemicals to Humans,* Suppl. 4, *Chemicals, industrial processes and industries associated with cancer in humans (IARC Monographs Volumes 1 to 29),* Lyon, International Agency for Research on Cancer

IARC (1985) *IARC Monographs on the Evaluation of the Carcinogenic Risk of Chemicals to Humans,* Vol. 36, *Some allyl and allylic compounds, aldehydes, epoxides and peroxides,* Lyon, International Agency for Research on Cancer

Ivie, G.W., Holt, D.L. & Ivey, M.C. (1981) Natural toxicants in human foods: psoralens in raw and cooked parsnip root. *Science,* **213,** 909–910

Kiesselbach, N., Korallus, V., Lange, H.-J., Niess, A. & Zwinger, T. (1979) Acrylonitrile – Epidemiological study – Bayer 1977. *Zbl. Arbeitsmed.,* **29,** 256–259

Lambotte-Vandepaer, M., Duverger-van Bogaert, M., de Meester, C., Poncelet, F. & Mercier, M. (1980) Mutagenicity of urine from rats and mice treated with acrylonitrile. *Toxicology,* **16,** 67–71

Levin, D.E., Hollstein, M., Christman, M.F., Schwiers, E.A. & Ames, B.N. (1982) A new *Salmonella* tester strain (TA 102) with A.T. base pairs at the site of mutation detects oxidative mutagens. *Proc. natl Acad. Sci. USA,* **79,** 7445–7449

Lijinsky, W. & Andrews, A.W. (1980) Mutagenicity of vinyl compounds in *Salmonella typhimurium. Teratog. Carcinog. Mutag.,* **1,** 259–267

Lijinsky, W. & Reuber, M.D. (1983) Chronic toxicity studies of vinyl acetate in Fischer rats. *Toxicol. appl. Pharmacol.,* **68,** 43–53

Ludlum, D.B. & Tong, W.P. (1981) *Modification of DNA and RNA bases by the nitrosoureas.* In: Serrou, B., Schein, P.S. & Imbach, J.-L., eds, *Nitrosoureas in Cancer Treatment,* Amsterdam, Elsevier, pp. 21–31

Lutz, D., Eder, E., Neudecker, T. & Henschler, D. (1982) Structure-mutagenicity relationship in α,β-unsaturated carbonylic compounds and their corresponding allylic alcohols. *Mutat. Res.,* **93,** 305–315

Maltoni, C., Ciliberti, A. & DiMaio, V. (1977) Carcinogenicity bioassays on rats of acrylonitrile administered by inhalation and by injection. *Med. Lav., 68,* 401–410

Marnett, L.J. & Tuttle, M.A. (1980) Comparison of the mutagenicities of malondialdehyde and the side products formed during its chemical synthesis. *Cancer Res., 40,* 276–282

Merletti, F., Heseltine, E., Saracci, R., Simonato, L., Vainio, H. & Wilbourn, J. (1984) Target organs for carcinogenicity of chemicals and industrial exposures in humans: a review of results in the IARC Monographs on the Evaluation of the Carcinogenic Risk of Chemicals to Humans, *Cancer Res., 44,* 2244–2250

Montgomery, J.A. (1976) Chemistry and structure-activity studies of the nitrosoureas. *Cancer Treat. Rep., 60,* 651–664

Moschel, R.C. & Leonard, N.J. (1976) Fluorescent modification of guanine: reaction with substituted malondialdehyde. *J. org. Chem., 41,* 294–300

Nakamura, K. (1981) *Mortality study of acrylonitrile workers* (Jap.). In: *Annual Report of the National Institute of Industrial Health for 1980,* p. 31

Norppa, H., Tursi, F., Mäki-Paakkanen, J., Järventaus, H. & Sorsa, M. (1983) *Vinyl acetate is a potent inducer of chromosome damage in mammalian cells.* In: *International Seminar on Methods of Monitoring Human Exposure to Carcinogenic and Mutagenic Agents (Abstracts),* Finland, Espoo, p. 57

O'Berg, M.T. (1980) Epidemiologic study of workers exposed to acrylonitrile. *J. occup. Med., 22,* 245–252

Pathak, M.A., Daniels, F. & Fitzpatrik, T.B. (1962) The presently known distribution of furocoumarins (psoralens) in plants. *J. invest. Dermatol., 39,* 225–239

Patty, F.A., Yant, W.P. & Waite, C.P. (1930) Acute response of guinea pigs to vapors of some new commercial organic compounds. *Publ. Health Rep., 45,* 1963–1971

Pepelko, W.E. (1984) Experimental respiratory carcinogenesis in small laboratory animals. *Environ. Res., 33,* 144–188

Peter, H., Schwarz, M., Mathiasch, B., Appel, K.E. & Bolt, H.M. (1983) A note on synthesis and reactivity towards DNA of glycidonitrile, the epoxide of acrylonitrile *Carcinogenesis, 4,* 235–237

Quast, J.F., Humiston, C.G., Wade, C.E., Carreon, R.M., Hermann, E.A., Park, C.N. & Schwetz, B.A. (1981) Results of a chronic toxicity and oncogenicity study in rats maintained on water containing acrylonitrile for 24 months. *Toxicologist, 1,* 129

Saito, T., Takashina, T., Yanagisawa, S. & Shinei, T. (1983) Determination of trace low molecular weight aliphatic carbonyl compounds in auto exhaust by gas chromatography with a glass capillary column. *Bunseki Kaguku, 32,* 33–38

Shamberger, R.J., Tytko, S.A. & Willis, C.E. (1975) Malonaldehyde is a carcinogen. *Fed. Proc., 34,* 827

Song, P.S. & Tapley, H.J. (1979) Photochemistry and photobiology of psoralens. *Photochem. Photobiol., 29,* 1177–1197

Stafford, J. (1983) *Liver Angiosarcoma,* Manchester, UK, Imperial Chemical Industries (limited circulation)

Stern, R.S., Laird, N., Melski, J., Parrish, J.A., Fitzpatrick, T.B. & Bleich, H.L. (1984) Cutaneous squamous-cell carcinoma in patients treated with PUVA. *New Engl. J. Med., 310,* 1156–1161

Szabad, J., Soos, I., Polgar, G. & Hejja, G. (1983) Testing the mutagenicity of malondialdehyde and formaldehyde by the *Drosophila* mosaic and the sex-linked recessive lethal tests. *Mutat. Res., 113,* 117–133

Thiess, A.M., Frentzel-Beyme, R., Link, R. & Wild, H. (1980) Mortality study of workers in the chemical industry in different production operations with concomitant exposure to acrylonitrile (Ger.). *Zbl. Arbeitsmed., 30,* 259–267

Viola, P.L., Bigotti, A. & Caputo, A. (1971) Oncogenic responses of rat skin, lungs and bones to vinyl chloride. *Cancer Res., 31,* 516–522

Yamaguchi, T. (1982) Mutagenicity of trioses and methyl glyoxal on *Salmonella typhimurium. Agric. Biol. Chem., 46,* 849–851

Yonei, S. & Furui, H. (1981) Lethal and mutagenic effects of malondialdehyde, a decomposition product of peroxidized lipids, on *Escherichia coli* with different DNA-repair capacities. *Mutat. Res., 88,* 23–32

Zack, J.A. (1980) The mortality experience of Mousanto workers exposed to acrylonitrile. Unpublished report submitted to the Environmental Protection Agency, June 4, 1980

Zajdela, F. & Bisagni, E. (1981) 5-Methoxypsoralen, the melanogenic additive in suntan preparations, is tumorogenic in mice exposed to 365 nm u.v. radiation. *Carcinogenesis, 2,* 121–127

Zajdela, F., Croisy, A., Barbin, A., Malaveille, C., Tomatis, L. & Bartsch, H. (1980) Carcinogenicity of chloroethylene oxide, an ultimate reactive metabolite of vinyl chloride, and bis(chloromethyl)ether after subcutaneous administration and in initiation-promotion experiments in mice. *Cancer Res., 40,* 352–356

II. CHEMISTRY AND FORMATION OF CYCLIC AND OTHER ADDUCTS

A. Vinyl Halides, Carbamate Esters and Metabolites

SUBSTITUTED ETHENOADENOSINES AND ETHENOCYTIDINES

N.J. LEONARD & K.A. CRUICKSHANK

Department of Chemistry, School of Chemical Sciences, University of Illinois, Urbana, Illinois, USA

Implications concerning the in-vivo formation of etheno-bridged nucleosides upon exposure of rats to certain carcinogens have led us to examine other nucleoside annellating agents (e.g. N-(t-butoxycarbonyl)-2-bromoacetamide and 2-chloroketene diethylacetal) that chemically produce substituted etheno-bridged nucleosides. The resulting 8-(t-butoxycarbonylamino)- and 8-ethoxy-substituted ethenoadenosines are highly fluorescent. The question to be resolved is whether reagents that introduce a substituted etheno bridge onto adenosine or cytidine are suspect mutagens or whether the formation of 1,N^6-ethenoadenosine and 3,N^4-ethenocytidine in rat liver is simply one manifestation of the behavior of an annellating agent.

The importance of etheno-substituted nucleosides and nucleotides is demonstrated in a negative sense by the occurrence of 1,N^6-ethenoadenosine (εAdo) and 3,N^4-ethenocytidine (εCyd) in the enzymatic hydrolysates of hepatic RNA from rats exposed to the carcinogens, vinyl chloride, ethyl carbamate (urethane) and vinyl carbamate, as shown by Laib and Bolt (1977, 1978) and by Ribovich et al. (1982). Vinyl bromide resembles vinyl chloride in its RNA alkylation, according to Ottenwälder et al. (1979), and the reactive metabolites, chloroethylene oxide and chloroacetaldehyde, have been implicated (review by Hemminki, 1983). Chloroacetaldehyde is a known mutagen (McCann et al., 1975) and carcinogen (Van Duuren et al., 1979) that chemically converts adenosine to 1,N^6-ethenoadenosine and cytidine to 3,N^4-ethenocytidine (Barrio et al., 1972) and effects similar conversions in the deoxy series (review by Leonard, 1984).

We are currently investigating the formation of substituted etheno-bridged adenosines and cytidines as a prelude to utilizing these in cross-linking reactions. The annellating reagent, N-(t-butoxycarbonyl)-2-bromoacetamide (BrCH$_2$CONHCO$_2$$t$-Bu) has been found to react with either tri-O-acetyladenosine (Fig. 1A) or with tri-O-acetylcytidine (Fig. 1B). The latent amine that is masked by the carbamate group

Fig. 1 Reaction of N-(t-butoxycarbonyl)-2-bromoacetamide (BrCH$_2$CONHCO$_2$t-Bu) with tri-O-acetyladenosine (A) and with tri-O-acetylcytidine (B). R, tri-O-acetyl-β-D-ribofuranosyl; DMF, dimethylformamide; Py, pyridine; R.T., room temperature. The yield of each reaction is given in percentage.

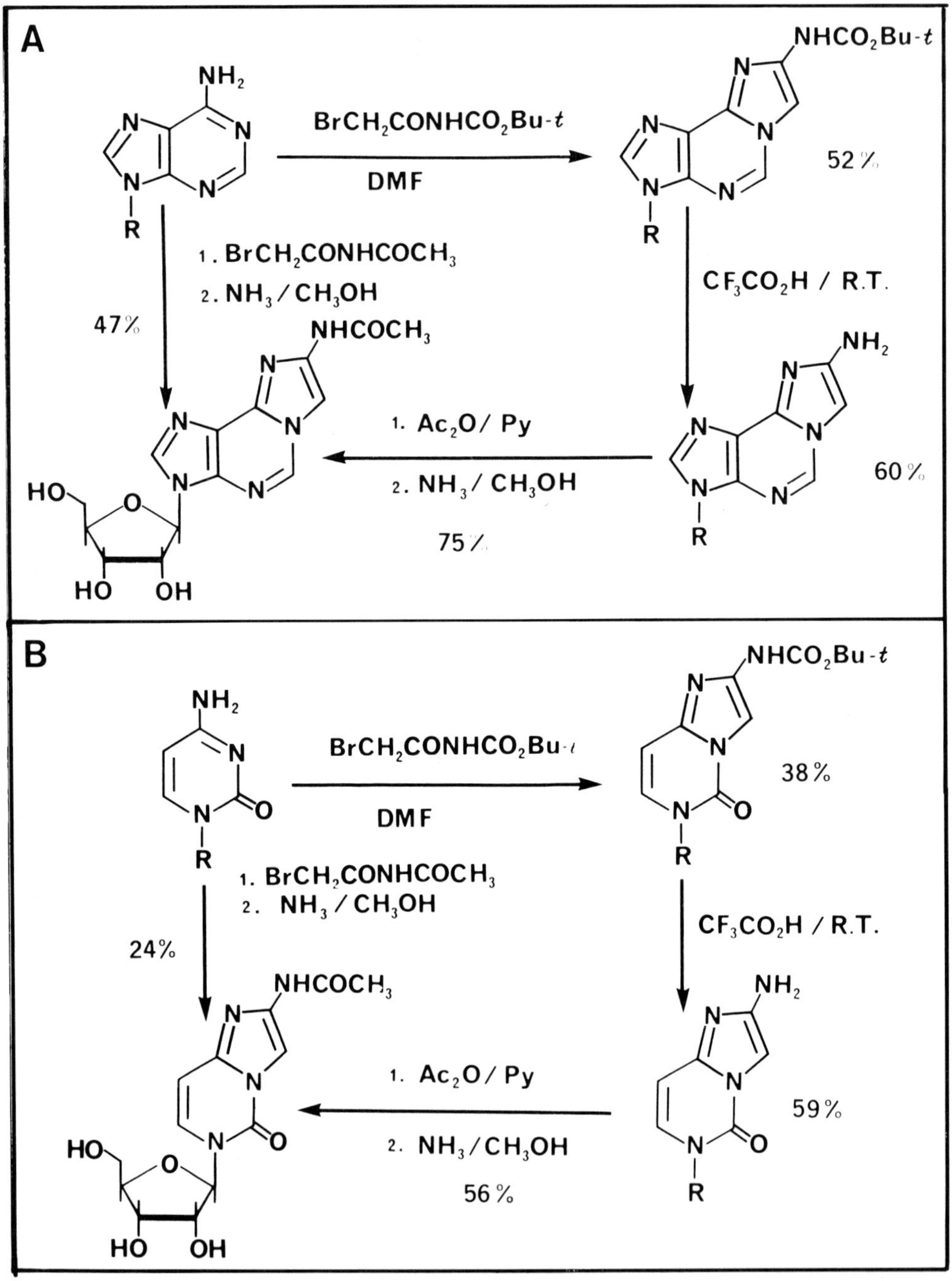

Fig. 2. Reaction of 2-chloroketene diethylacetal with tri-O-acetyladenosine (A) and with tri-O-acetylcytidine (B). DMF, dimethylformamide; MeCN, methyl cyanide; R.T., room temperature. Yields are given in percentages.

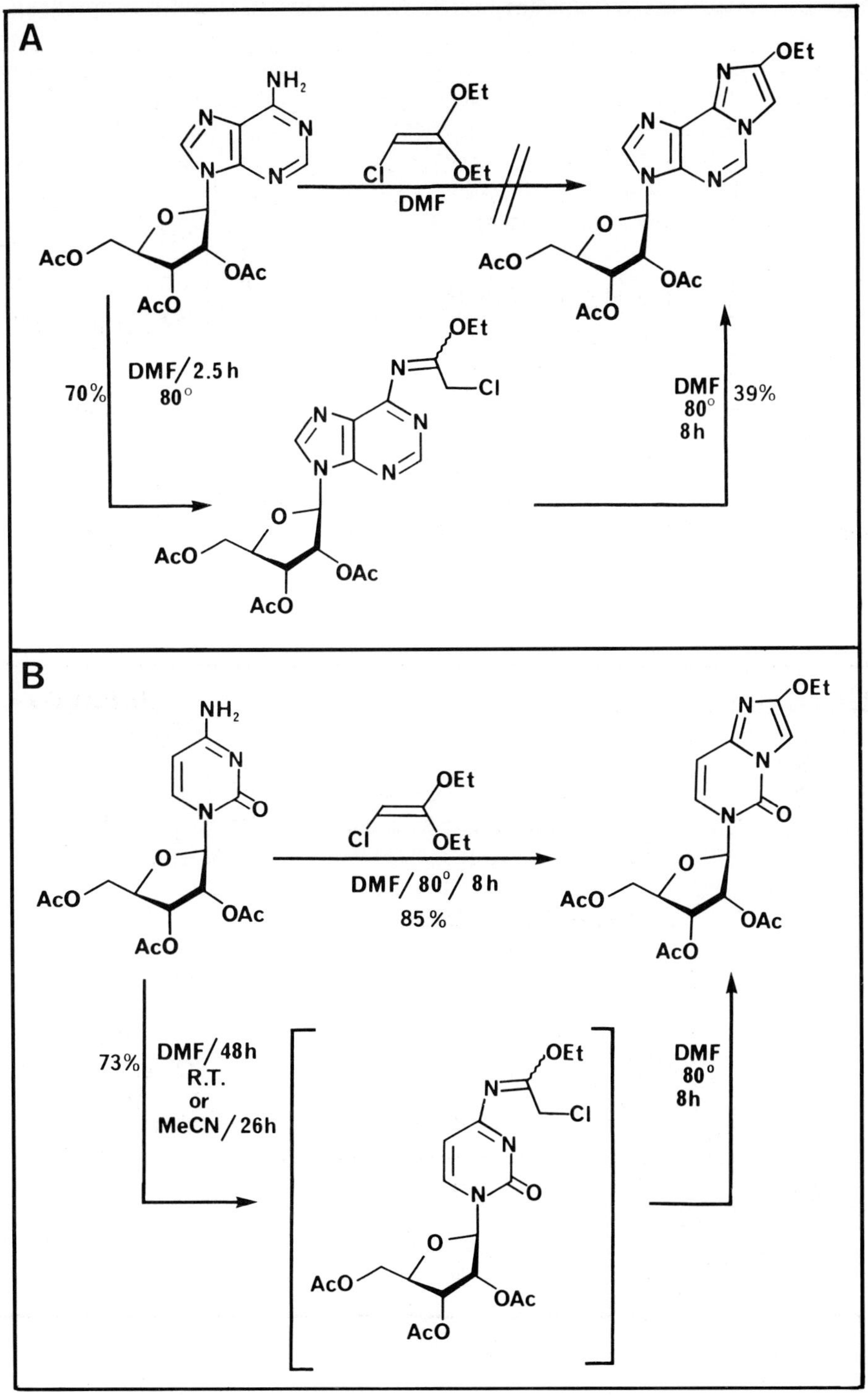

can be unmasked by treatment with acid. The aminoetheno compounds shown in Figure 1 are very labile but can be acetylated to produce compounds identical to the products of adenosine or cytidine, respectively, plus N-acetyl-2-bromoacetamide. Almost all of these compounds have valuable fluorescence properties. Although yields have not yet been optimized, the series has provided excellent models for analysis by nuclear magnetic resonance, ultraviolet absorption and fluorescence spectroscopy, enabling us to make assignments in the next series (Fig. 2) resulting from reactions with 2-chloroketene diethylacetal. This latter compound is a base-substitution mutagen, requiring metabolic activation in the Ames test.

The chloromethyl ethoxyimidate intermediate shown in Figure 2A can be isolated and will be used in cross-linking reactions. The aminoetheno and ethoxyetheno ribonucleosides may provide further chemical, spectroscopic and biological information as to the nature of representative substituted etheno-bridged nucleic acid derivatives.

REFERENCES

Barrio, J.R., Secrist, J.A., III & Leonard, N.J. (1972) Fluorescent adenosine and cytidine derivatives. *Biochem. biophys. Res. Commun., 46,* 597–604

Hemminki, K. (1983) Nucleic acid adducts of chemical carcinogens and mutagens. *Arch. Toxicol., 52,* 249–285

Laib, R.J. & Bolt, H.M. (1977) Alkylation of RNA by vinyl chloride metabolites *in vitro* and *in vivo:* formation of 1,N^6-ethenoadenosine. *Toxicology, 8,* 185–195

Laib, R.J. & Bolt, H.M. (1978) Formation of 3,N^4-ethenocytidine moieties in RNA by vinyl chloride metabolites *in vitro* and *in vivo. Arch. Toxicol., 39,* 235–240

Leonard, N.J. (1984) Etheno-substituted nucleotides and coenzymes: fluorescence and biological activity. *CRC Crit. Rev. Biochem., 15,* 125–199

McCann, J., Simmon, V., Streitwieser, D. & Ames, B.N. (1975) Mutagenicity of chloroacetaldehyde, a possible metabolic product of 1,2-dichloroethane (ethylene chloride), chloroethanol (ethylene chlorohydrin), vinyl chloride, and cyclophosphamide. *Proc. natl Acad. Sci. USA, 72,* 3190–3193

Ottenwälder, H., Laib, R.J. & Bolt, H.M. (1979) Alkylation of RNA by vinyl bromide metabolites *in vitro* and *in vivo. Arch. Toxicol., 41,* 279–286

Ribovich, M.L., Miller, J.A., Miller, E.C. & Timmons, L.G. (1982) Labelled 1,N^6-ethenoadenosine and 3,N^4-ethenocytidine in hepatic RNA of mice given [ethyl-1,2-^{3}H or ethyl-1-^{14}C]ethyl carbamate (urethane). *Carcinogenesis, 3,* 539–546

Van Duuren, B.L., Goldschmidt, B.M., Laemengart, G., Smith, A.C., Melchionne, S., Seidman, I. & Roth, D. (1979) Carcinogenicity of halogenated olefinic and aliphatic hydrocarbons in mice. *J. natl Cancer Inst., 63,* 1433–1439

THE STRUCTURE AND PROPERTIES OF 7-(2-OXOETHYL)GUANINE: A MODEL FOR A KEY DNA ALKYLATION PRODUCT OF VINYL CHLORIDE

P. POLITZER, R. BAR-ADON & B.A. ZILLES

*Department of Chemistry, University of New Orleans,
New Orleans, Louisiana, USA*

SUMMARY

Recent laboratory studies have shown that the major in-vivo DNA alkylation product of vinyl chloride is the 7-(2-oxoethyl) derivative of guanine, and the suggestion has been made that this derivative is responsible for the carcinogenicity of vinyl chloride. Some nuclear magnetic resonance evidence indicates that this alkylation product may be in equilibrium with a cyclic hemiacetal; this would affect the hydrogen bonding between guanine and cytosine. In order to help elucidate the properties of this key DNA alkylation product, we have computed the structure and properties of the same derivative of the isolated guanine molecule, i.e., 7-(2-oxoethyl)guanine. An *ab initio* self-consistent-field molecular orbital procedure was used. In this paper, we present the optimized (STO-3G) structure of 7-(2-oxoethyl)guanine, and the STO-5G energy, atomic charges and dipole moment. We found very little tendency for the formation of a cyclic hemiacetal; the equilibrium constant was estimated to be about 10^{-9}.

INTRODUCTION

The carcinogenicity of vinyl chloride (compound I, Fig. 1) is well established, on the basis of both in-vivo and in-vitro studies (Rannug *et al.*, 1976; Maltoni, 1977; Hong *et al.*, 1981; Woo *et al.*, 1984) as well as epidemiological data (Selikoff & Hammond, 1975; Infante, 1981). The first step in its carcinogenic pathway is believed to be a metabolic epoxidation, producing chlorooxirane (Compound II, Fig. 1) (Rannug *et al.*, 1976; Bartsch *et al.*, 1979; Zajdela *et al.*, 1980).

The epoxide II is an alkylating agent, which forms covalent bonds with nucleophilic sites on nucleic acid bases, DNA, RNA and protein residues (Rannug *et al.*, 1976; Osterman-Golkar *et al.*, 1977; Green & Hathway, 1978; Barbin *et al.*, 1981). The

Fig. 1. Metabolic epoxidation of vinyl chloride (I) to chlorooxirane (II)

Fig. 2. Possible equilibrium between the 7-(2-oxoethyl) derivative of guanine (III) and a cyclic hemiacetal (IV)

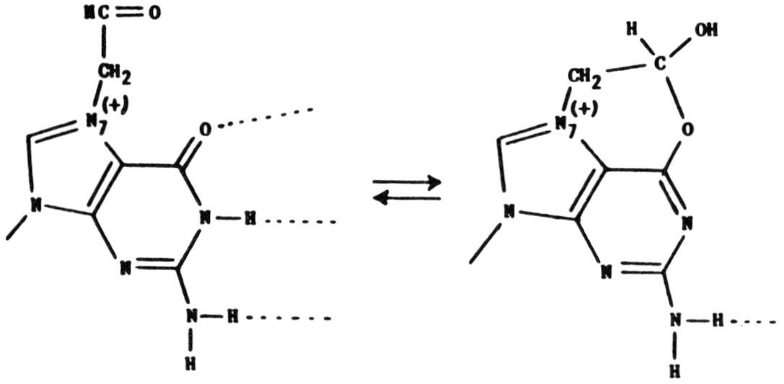

major in-vivo DNA alkylation product of vinyl chloride has recently been shown to be the 7-(2-oxoethyl) derivative of guanine (structure III, Fig. 2) (Osterman-Golkar et al., 1977; Scherer et al., 1981; Laib et al., 1981).

The nuclear magnetic resonance (NMR) spectrum of III has been interpreted as indicating a possible equilibrium with the cyclic hemiacetal IV (Scherer et al., 1981). In this hemiacetal, two of the three base-pairing sites of guanine are affected. It has been suggested that the DNA alkylation product, III, may be responsible for the carcinogenicity of vinyl chloride (Laib et al., 1981).

In our earlier work, we carried out extensive computational studies of the properties and reactive behaviour of vinyl chloride and chlorooxirane, as well as other halogenated alkenes and the corresponding epoxides (Politzer et al., 1981; Politzer & Hedges, 1982; Politzer & Proctor, 1982; Bauer & Politzer, 1984; Laurence & Politzer, 1984; Laurence et al., 1984; Politzer & Laurence, 1984b; Politzer et al., 1985). Based on the results of these investigations, we proposed two possible mechanistic pathways for the formation of the key alkylation product, III, and identified several factors that may help to determine the differing carcinogenic activities of various alkenes (Laurence et al., 1984; Politzer et al., 1985). For example, we were able to show that carcinogenicity appears to be associated with a relatively strong negative electrostatic potential near the epoxide oxygen (Politzer & Laurence, 1984a).

Fig. 3. Atomic charges in 7-(2-oxoethyl)guanine, calculated at the STO-5G level

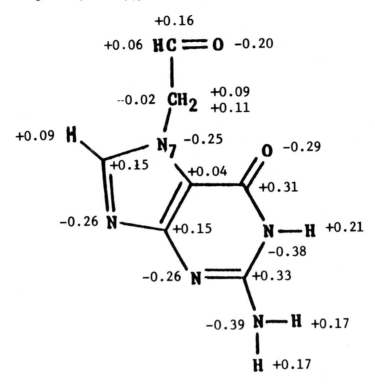

As a first step in elucidating the properties of the DNA alkylation product, III, we report in this paper the results of a computational study of the structure and properties of the 7-(2-oxoethyl) derivative of the isolated guanine molecule. (The structural formula is shown in Table 1 and Fig. 3.) We also examine the likelihood of an equilibrium between this derivative (V) and a cyclic hemiacetal analogous to IV.

METHODS

Our computational procedure is an *ab initio* self-consistent-field molecular orbital approach, using the GAUSSIAN 80 program (Binkley *et al.*, 1981). Optimized geometries were obtained at the STO-3G basis set, and were then used to compute final STO-5G wave functions, energies and other molecular properties. Atomic charges were determined by means of the Mulliken population analysis procedure (Mulliken, 1955).

Table 1. Calculated bond lengths and bond angles in 7-(2-oxoethyl)guanine

Bond lengths (Angstroms)				Bond angles (degrees)			
a:	1.39	h:	1.43	a, b:	114	h, i:	123
b:	1.32	i:	1.22	b, c:	104	j, q:	123
c:	1.41	j:	1.48	q, d:	125	q, k:	106
d:	1.43	k:	1.39	d, e:	113	k, m:	127
e:	1.29	m:	1.47	e, g:	125	m, n:	112
f:	1.46	n:	1.55	f, g:	116	n, p:	124
g:	1.42	p:	1.21	g, h:	125		
		q:	1.36				

RESULTS

We have carried out a thorough STO-3G optimization of the geometry of 7-(2-oxoethyl)guanine. Some of the key bond lengths and bond angles of our final structure are shown in Table 1. (A complete set of the calculated geometrical parameters may be obtained from the authors.) The STO-5G energy of this structure is -688.494 hartrees. The atomic charges, computed at the STO-5G level, are given in Fig. 3. We found the dipole moment to be 5.5 D.

DISCUSSION

When considering the data in Table I, it is helpful to recall some typical interatomic distances (Sutton, 1958): C–C (aliphatic), 1.541 Å; C=C, 1.337 Å; C–C (aromatic), 1.395 Å; C–N (paraffinic), 1.472 Å; C–N (heterocyclic, aromatic), 1.352 Å; C=O, 1.23 Å. A comparison of these values with the bond lengths in Table 1 reveals that the double bonds are rather localized, although both rings do possess some degree of aromatic character.

The atomic charges in Figure 3 show very clearly the readiness of the system to enter into hydrogen bonding such as is depicted for structure III, Fig. 2. The positive charges of the amine and imine hydrogens are approximately double those of the remaining hydrogens, while the oxygen attached to the ring is considerably more negative than a typical carbonyl oxygen, which has a charge in the neighborhood of -0.20.

A point of particular interest is the possibility of an equilibrium between the oxoethyl derivative, V, and a cyclic hemiacetal analogous to IV. The presence of a significant amount of such a hemiacetal in equilibrium with the actual DNA alkylation product, III, would certainly be expected to affect the guanine-cytosine hydrogen bonding in DNA, as mentioned earlier. This could result in miscoding and possible replicational and transcriptional errors.

We have accordingly investigated the stability of the cyclic hemiacetal that would correspond to 7-(2-oxoethyl)guanine, V. Due to the time-consuming nature of the computations, we carried out only a partial optimization of the structure of this hemiacetal; it was sufficient, however, to show the latter to be considerably less stable than 7-(2-oxoethyl)guanine itself. We found an energy difference of about 12 kcal/mol. Using this to estimate an equilibrium constant K for (reaction I)

$$\text{7-(2-oxoethyl)guanine} \rightleftharpoons \text{hemiacetal,}$$

we found $K \sim 10^{-9}$. A more complete optimization of the hemiacetal geometry would somewhat diminish the energy difference and thus tend to increase the value of K. On the other hand, consideration of the entropy term for the formation of the more ordered hemiacetal structure would have the opposite effect upon K. Thus these two errors should at least partially cancel each other. It appears to be fairly safe to conclude that the equilibrium represented by reaction I is very far to the left.

We are presently studying the possibility of a tautomeric equilibrium involving 7-(2-oxoethyl)guanine. This could also significantly affect the hydrogen-bonding characteristics of guanine.

ACKNOWLEDGEMENTS

We thank the U.S. Environmental Protection Agency for partial funding of this work under assistance agreement number CR808866-01-0 to Peter Politzer. The contents do not necessarily reflect the views or policies of the Environmental Protection Agency, nor does mention of trade names or commercial products constitute endorsement or recommendation for use. We are also grateful for financial support provided by the University of New Orleans Computer Research Center.

REFERENCES

Barbin, A., Bartsch, H., Leconte, P. & Radmann, M. (1981) Studies on the miscoding properties of 1,N^6-ethenoadenine, DNA reaction products of vinyl chloride metabolites, during in-vitro DNA synthesis. *Nucleic Acids Res., 9,* 375–387

Bartsch, H., Malaveille, C., Barbin, A. & Planche, G. (1979) Mutagenic and alkylating metabolites. *Arch. Toxicol., 41,* 249–279

Bauer, J. & Politzer, P. (1984) The effect of an epoxide-nucleophile reaction upon hydrogen bonding involving the nucleophile. *Internat. J. quantum Chem., 25,* 869–879

Binkley, J.S., Whiteside, R.A., Krishnan, R., Seeger, R., DeFrees, D.J., Schlegel, H.B., Topiol, S., Kahn, L.R. & Pople, J.A. (1981) GAUSSIAN 80: an *ab initio* molecular orbital program. *Quantum Chem. Prog. Exch., 13,* 406

Green, T. & Hathway, D.E. (1978) Interactions of vinyl chloride with rat liver DNA in vivo. *Chem.-biol. Interactions, 22,* 211–224

Hong, C.B., Winston, J.M., Thornburg, L.P., Lee, C.C. & Woods, J.S. (1981) Follow-up study on the carcinogenicity of vinyl chloride and vinylidene chloride in rats and mice: tumor incidence and mortality subsequent to exposure. *J. Toxicol. environ. Health, 7,* 909–924

Infante, P.F. (1981) Observations of the site-specific carcinogenicity of vinyl chloride to humans. *Environ. Health Perspect., 41,* 89–94

Laib, R.J., Gwinner, L.M. & Bolt, H.M. (1981) DNA alkylation by vinyl chloride metabolites: etheno derivatives or 7-alkylation of guanine? *Chem.-biol. Interactions, 37,* 219–231

Laurence, P.R. & Politzer, P. (1984) Some reactive properties of chlorooxirane, a likely carcinogenic metabolite of vinyl chloride. *Internat. J. quantum Chem., 25,* 493–502

Laurence, P.R., Proctor, T.R. & Politzer, P. (1984) Reactive properties of *trans*-dichlorooxirane in relation to the contrasting carcinogenicities of vinyl chloride and *trans*-dichloroethylene. *Internat. J. quantum Chem., 26,* 425–438

Maltoni, C. (1977) Recent findings in the carcinogenicity of chlorinated olefins. *Environ. Health Perspect., 21,* 1–10

Mulliken, R.S. (1955) Electronic population analysis on LACO-MO molecular wave functions. *J. Chem. Phys., 23* 1833–1840

Osterman-Golkar, S., Hultmark, D., Segerback, D., Calleman, C.J., Gothe, R., Ehrenberg, L. & Wachtmeister, C.A. (1977) Alkylation of DNA and proteins in mice exposed to vinyl chloride. *Biochem. biophys. Res. Commun., 76,* 259–266

Politzer, P. & Hedges, W.L. (1982) A study of the reactive properties of the chlorinated ethylenes. *Internat. J. quantum Chem., quantum Biol. Symp. No. 9,* 307–319

Politzer, P. & Laurence, P.R (1984a) Relationships between the electrostatic potential, epoxide hydrase inhibition and carcinogenicity for some hydrocarbon and halogenated hydrocarbon epoxides. *Carcinogenesis, 5,* 845–848

Politzer, P. & Laurence, P.R. (1984b) Halogenated hydrocarbon epoxides: factors underlying biological activity. *Internat. J. quantum Chem., quantum Biol. Symp. No. 11,* 155–166

Politzer, P. & Proctor, T.R. (1982) Calculated properties of some possible vinyl chloride metabolites. *Internat. J. quantum Chem., 22*, 1271–1279

Politzer, P., Trefonas, P., Politzer, I.R. & Elfman, B. (1981) Molecular properties of the chlorinated ethylenes and their epoxide metabolites. *Ann. N.Y. Acad. Sci., 367*, 478–492

Politzer, P., Laurence, P.R. & Jayasuriya, K. (1985) *Halogenated olefins and their epoxides: factors underlying carcinogenic activity*. In: Rein, R., ed., *The Molecular Basis of Cancer. Part A: Macromolecular Structure, Carcinogens and Oncogenes*, New York, Alan R. Liss, Inc., 227–237

Rannug, V., Gothe, R. & Wachtmeister, C.A. (1976) Mutagenicity of chloroethylene oxide, chloroacetaldehyde, 2-chloroethanol und chloroacetic acid, conceivable metabolites of vinyl chloride. *Chem.-biol. Interactions, 12*, 251–263

Scherer, E., Van der Laken, C.J., Gwinner, L.M., Laib, R.J. & Emmelot, P. (1981) Modification of deoxyguanosine by chloroethylene oxide. *Carcinogenesis, 2*, 671–677

Selikoff, I.J. & Hammond, E.C., eds (1975) *Toxicity of Vinyl Chloride – Polyvinyl Chloride, Ann. N.Y. Acad. Sci., 246*

Sutton, L.E., ed. (1958) *Tables of Interatomic Distances and Configuration in Molecules and Ions*, Spec. Pub. 11, London, Chemical Society

Woo, Y.-T., Lai, D., Arcos, J.C. & Argus, M.F. (1984) *Chemical Induction of Cancer*, Vol. IIIB, New York, Academic Press, section 5.2.2.1

Zajdela, F.A., Croisy, A., Barbin, A., Malaveille, C., Tomatis, L. & Bartsch, H. (1980) Carcinogenicity of chloroethylene oxide, an ultimate reactive metabolite of vinyl chloride, and bis(chloromethyl)ether after subcutaneous administration and in initiation-promotion experiments in mice. *Cancer Res., 40*, 352–356

NEUTRAL REACTIONS OF HALOACETALDEHYDES WITH POLYNUCLEOTIDES: MECHANISMS, MONOMER AND POLYMER PRODUCTS

B. SINGER[1] & S.R. HOLBROOK

Laboratory of Chemical Biodynamics, Laurence Berkeley Laboratory, University of California, Berkeley, CA, USA

H. FRAENKEL-CONRAT

Department of Molecular Biology, University of California, Berkeley, CA, USA

J.T. KUŚMIEREK

Institute of Biochemistry and Biophysics, Polish Academy of Sciences, Warsaw, Poland

SUMMARY

The generally accepted mechanism for the formation of etheno derivatives upon reaction of adenosine or cytidine with haloacetaldehydes involves two intermediates. The first, a primary addition to the exocyclic amino group, has not been experimentally verified. The second, a cyclic form of the first intermediate, has been described in monomers but presumed to be too unstable to exist in polynucleotides since such derivatives would be readily dehydrated to other derivatives at pHs below neutrality. We have found that the cyclic intermediates of adenosine and cytidine are the predominant products in polynucleotides, even upon extensive reaction with chloroacetaldehyde at neutrality. The hydrated compounds have half-lives at pH 7, 37 °C, of 1.4 h and 13 h for adenosine and cytidine, respectively. Two types of evidence are presented for the existence of the first intermediate, a (1-hydroxy-2-chloroethyl)-substituted exocyclic amino group. Firstly, poly d[A–T] cannot form etheno derivatives (except when denatured) and the observed cross-linking is therefore attributed to alkylation by the chlorinated sidechain of the adenine residue (A), acting

---

[1] To whom correspondence should be addressed

on the N^6 of A on the opposite strand. Secondly, our results show that blocking of the acceptor nitrogen, needed for cyclization, leads to the formation of relatively stable derivatives of adenosine and cytidine. Guanosine, as a monomer, is modified extensively, but in synthetic polymers no reaction was detected, possibly due to secondary structure.

INTRODUCTION

The reactions of chloroacetaldehyde (CAA) with nucleosides to yield etheno derivatives of adenosine, cytidine and guanosine have been well-documented and recently reviewed by Leonard (1983) (Fig. 1). Bromoacetaldehyde (BAA) appears to react similarly, Less is known of the reactions with another vinyl halide metabolite, chloroethylene oxide (CEO), but from analogy with other simple epoxides it can be assumed that CEO is a typical alkylating agent (Singer & Grunberger, 1983). This is supported by the isolation of 7-(2-oxoethyl)deoxyguanosine, both *in vitro* (Scherer *et al.*, 1981) (Fig. 2) and *in vivo* (Osterman-Golkar *et al.*, 1977; Laib *et al.*, 1981).

Polynucleotides have also been modified with CAA, but usually these reactions, like those with monomers, are carried out for long times with high concentrations of reagent at pH 4–6 (Steiner *et al.*, 1973). It should be noted that etheno derivatives cannot form unless one of the normal hydrogen-bonded positions is available (Kimura *et al.*, 1977; Lilley, 1983).

In assessing the probability that etheno derivatives are formed *in vivo*, it is therefore necessary to consider the reaction mechanism with single-stranded DNA and RNA under physiological conditions. This mechanism (Fig. 3) postulates a transitory

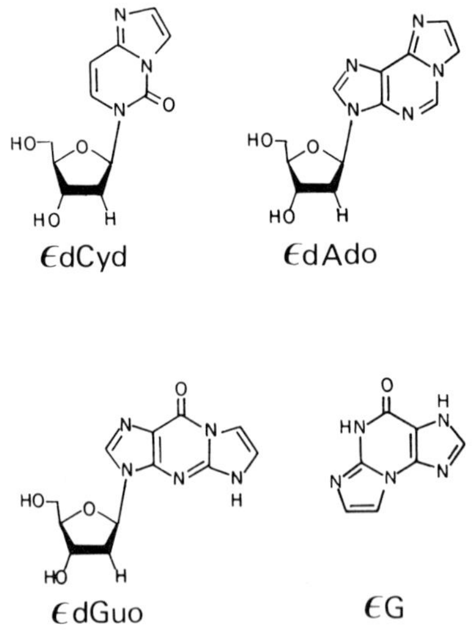

Fig. 1. Etheno derivatives formed by chloroacetaldehyde reaction. εdCyd, 3,N^4-ethenodeoxycytidine; εdAdo, 1,N^6-ethenodeoxyadenosine; εdGuo, 1,N^2-ethenodeoxyguanosine; εG, N^2,3-ethenoguanine

intermediate (*2*) and an initial cyclic adduct (*3*) which is unstable, particularly at low pH, but likely to occur *in vivo*. Wiewiórowski and coworkers (Biernat *et al.*, 1978; Krzyżosiak *et al.*, 1979) did isolate intermediate *3* from reactions with monomers and determined, at pH 3–7, the conditions and rates for dehydration leading to etheno derivatives. These rates of dehydration were much lower at neutrality than at low pH. Although Krzyżosiak *et al.* (1981) indicated that the hydrated form was present in CAA-treated transfer RNA, they did not isolate these compounds but studied their stability in the RNA. Nevertheless, investigators have reacted polynucleotides with CAA, or cells with vinyl chloride (of which CAA is the stable metabolite), and have assumed that only the etheno derivatives are present.

Fig. 2. Reaction product of chloroethylene oxide with deoxyguanosine

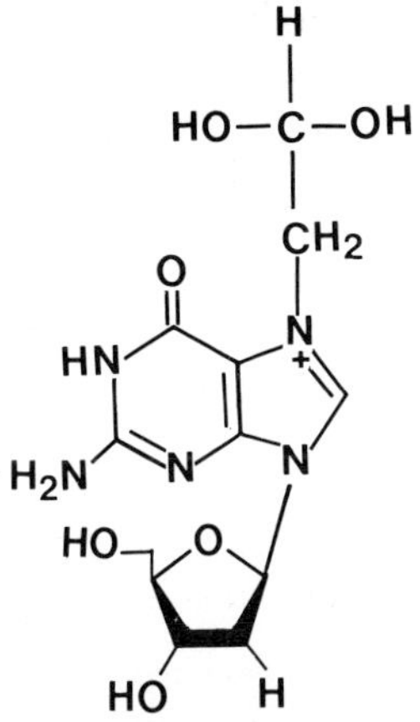

7-(2-oxoethyl)deoxyguanosine
hydrated aldehyde
conformation

Fig. 3. Postulated mechanism for reaction of chloroacetaldehyde with cytidine and adenosine

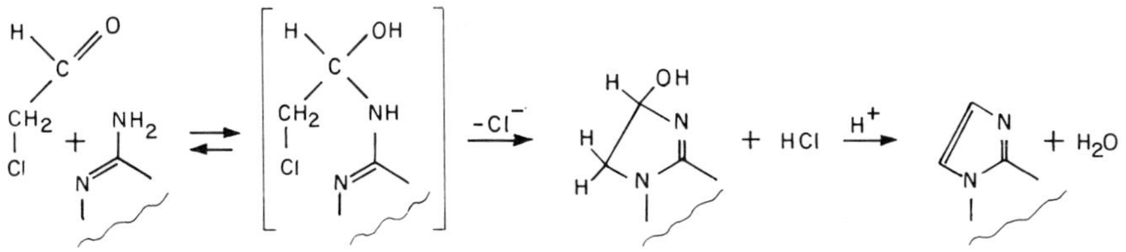

In the case of double-stranded deoxypolynucleotides, a further common assumption is that reaction with CAA only occurs as the result of local denaturation, allowing formation of the etheno ring, while reaction with CEO at the N-7 of guanine residues is not dependent on strandedness.

We now report that after reaction of CAA with poly (C), poly (dC), poly (A) and poly (dA), at neutrality, the hydrate is the predominant product and is only converted to the etheno derivative when heated for several hours at elevated temperatures (Kuśmierek & Singer, 1982). Treatment of alternating poly d[A–T] with CAA or BAA under the same conditions leads to cross-links between adenine (A) residues on opposite strands (Singer et al., 1984). Model experiments suggest that cross-linking results from addition of 1-hydroxy-2-chloroethyl groups to the N^6 of A (intermediate 2, Figure 3), followed by a second reaction, i.e., alkylation of the N^6 of A on the opposite strand. Calculations of surface accessibility of the N^6 of A in DNA (B-form) indicate that this position is available for modification by small reagents (Holbrook & Kim, 1983).

REACTION OF CAA WITH CYTOSINE AND ADENINE RESIDUES IN SINGLE-STRANDED POLYMERS

The etheno derivatives, $3,N^4$-ethenocytidine (εCyd) or $3,N^4$-ethenodeoxycytidine (εdCyd) and $1,N^6$-ethenoadenosine (εAdo) or $1,N^6$-ethenodeoxyadenosine (εdAdo), as well as their hydrated forms (εCyd·H_2O, εdCyd·H_2O, εAdo·H_2O, εdAdo·H_2O) were found as sole products in enzymatic digests of homopolymers reacted with CAA to 65% modification.

Hydrated etheno derivatives have a limited stability as monomers, and can only be detected by mild methods of analysis of modified polymers. Decreasing the time of enzyme digestion to four hours and carrying out high-performance liquid chromatography (HPLC) at 20 °C enabled us to isolate both hydrated and dehydrated derivatives (Kuśmierek & Singer, 1982). Reaction with CAA at pH 7 for up to two hours at 37 °C, resulted in at least 90% of the product in the hydrated form. Figure 4 shows the ultraviolet absorption spectrum of εAdo·H_2O isolated by HPLC and the spectral change upon heating. Similar data were obtained with all treated polymers.

Further proof of the relative stability of the hydrated form of εAdo·H_2O in polymers is shown in Figure 5, where spectral changes were observed for several hours. Using these results, plus similar data for εCyd·H_2O, we found that the half-life ($t_{1/2}$) at 37 °C (pH 7.25) of εAdo·H_2O is 1.4 h and of εCyd·H_2O, 13 h. Poly (dA), reacted with CAA for one hour and four hours at 37 °C, pH 7.25, contains a high proportion of hydrate as shown in Figures 6A and 6B. In Figure 6A, with 15% total modification, there is a clear shift of the λ_{min} characteristic of the ethenoadenine residue upon heating. Surprisingly, even when there is 25% modification after a four-hour reaction, the shift upon heating is dramatic (Fig. 6B).

REACTION OF CAA AND BAA WITH ALTERNATING POLY d[A–T]

When poly d[A–T] was treated with either CAA or BAA at either pH 4.7 or 7.25 for the same times as used to modify poly (dA), as well as more intensively (e.g., 24 h, 37 °C), no εdAdo could be detected by HPLC analysis (Singer *et al.*, 1984). However, the study of the thermal melting profiles of the resulting polymers indicated marked changes.

After treatment with 7 mM CAA for one hour, there was complete loss of the typical hysteresis generally observed upon annealing (Fig. 7). The melting temperature and hyperchromicity were, however, unchanged. With increasing time or molarity of CAA or BAA, poly d[A–T] gradually lost its hyperchromicity and cooperativity (data not shown). These observations indicate that cross-links have been formed between the two chains. The number of cross-links leading to the almost complete loss of helix-coil transition can be calculated to be approximately 1 per 50 base pairs. Only 1 cross-link per molecule, however, is needed for the loss of hysteresis shown in Figure 7.

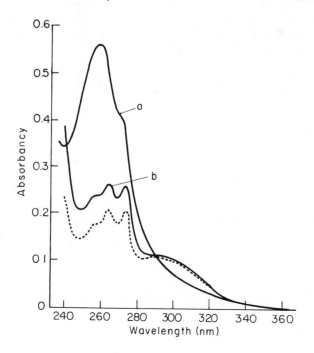

Fig. 4. Ultraviolet absorption spectra of nucleosides isolated by high-performance liquid chromatography of enzymatic digests of chloroacetaldehyde-treated poly (A): (a) hydrated form of 1,N^6-ethenoadenosine (εAdo·H_2O); (b) εAdo·H_2O after heating for 3.5 h at 37 °C. (...), spectrum of authentic εAdo. All spectra were recorded in 0.4 M ammonium formate pH 7.

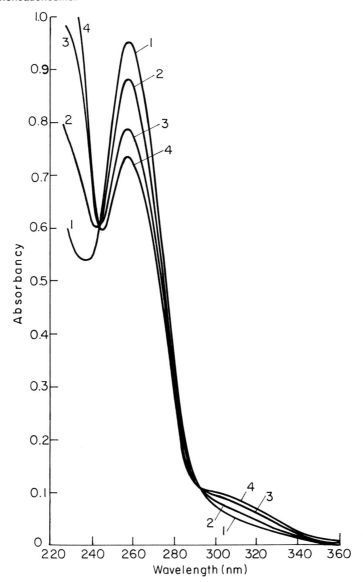

Fig. 5. Spectral changes accompanying dehydration of εAdo·H$_2$O residues in poly (A) at 37°C: (1) 0 time, (2) 1 h, (3) 3 h and (4) 7 and 9 h. The polymer was treated in 0.1 M sodium cacodylate buffer pH 7.25. The extent of modification was 28%. εAdo·H$_2$O is the hydrated form of 1,N^6-ethenoadenosine.

In contrast, CAA treatment of the duplex polymer, poly(dA)·poly(dT), did not affect the normal helix-coil transition, nor did CAA-treated poly(dT) lose its ability to form a normal double-stranded structure with poly(dA).

The conclusion derived from these data is that a reaction occurs between adenine residues on opposite strands. It appears likely that the reaction site is the N^6 of A, since the N-1 is hydrogen-bonded under the reaction conditions. Experiments are in progress to test this possibility by means of model experiments, using 1-methyladenosine and the analogous compound, 3-methylcytidine. These compounds, which cannot form etheno derivatives, give 1-hydroxy-2-chloroethyl derivatives substituted at the N^6 and N^4 positions, respectively. These derivatives might be stable enough to be able to alkylate a suitable receptor group. In the absence of such a group, chlorine is slowly replaced by a hydroxyl group.

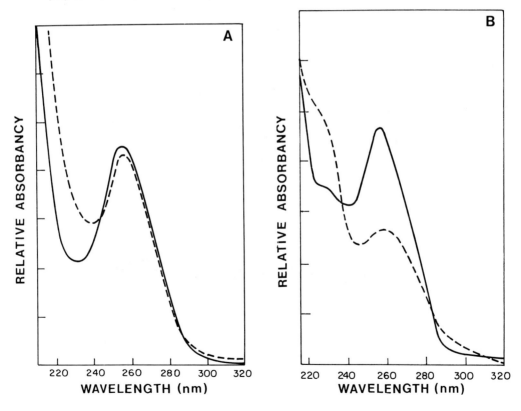

Fig. 6. Ultraviolet absorption spectra of poly (dA) treated with chloro-acetaldehyde at 37 °C, for 1 h (A) or for 4 h (B), and reisolated from a Biogel P-150 column. Spectra were measured in 0.01 M Tris-HCl pH 7.8, directly upon isolation (———), or after heating at 37 °C for 24 h (- - - - -). The treated polymer was 15% modified in A, and 25% modified in B.

REACTIONS OF CAA AND BAA WITH GUANOSINE IN POLYMERS AND AS A MONOMER

No reaction with the guanine residue (G) was detected in CAA-treated poly(dC) · poly(dG), poly [U,G] or poly [A,G] although in the latter polymer over 60% of the A was modified under the conditions used (pH 7.25). This was surprising in view of the report that guanosine was modified by CAA at pH 6.4 (but not at pH 4.5) to yield $1,N^2$-ethenoguanosine (Sattsangi et al., 1977). On the other hand, CAA treatment of calf thymus DNA at pH 7 led to the detection, using fluorescence, of $N^2,3$-ethenoguanine, but not of $1,N^2$-ethenoguanine (Oesch & Doerjer, 1982). However, due to the fact that the latter is only weakly fluorescent, its presence might have been below the level of detection.

Fig. 7. Thermal melting profiles in 0.01 M Tris-HCl pH 7.2. (a) poly d[A–T]: melting temperature, 53.8 °C; hyperchromicity: 46%; (b) poly d[A–T] after treatment with 0.007 M chloroacetaldehyde at pH 7, 37 °C, 1 h: melting temperature, 53.2 °C; hyperchromicity, 45%. The solid lines are absorbancy upon heating and the dashed lines are absorbancy upon cooling. The rate of both heating and cooling was 1 °C/2 min.

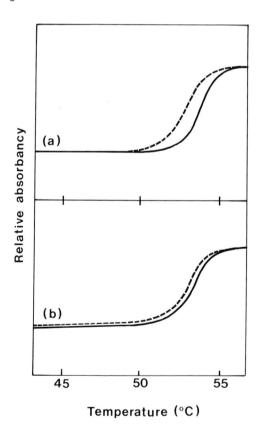

We reinvestigated the reaction of guanosine with CAA and found that after 18–24 h reaction (initial pH 7.25, final pH 5.8), there were several products separable by paper or thin-layer chromatography. Most were fluorescent and alkali-unstable. A nonfluorescent derivative had a spectrum at pH 7 corresponding to $1,N^2$-ethenoguanosine [λ_{max} (H$_2$O), 226 nm, 280 nm (broad)] (Sattsangi et al., 1977). The spectrum changed over a period of 30 min in 0.1 N sodium hydroxide, indicating that this ethenonucleoside was also alkali-unstable.

Reaction at pH 4 led to at least two modified derivatives, one alkali-stable and the other unstable. However, the extent of reaction after 24 h was only approximately 10% at pH 4, compared to approximately 80% at pH 7. Our results for neutral CAA reaction of guanosine are similar to those reported by Sattsangi et al. (1977) for CAA reaction of guanosine and by Kayasuga-Mikado et al. (1980) for BAA reaction of guanylic acid, except for our finding of high amounts of several other products, which are detectable even after 30 min reaction.

DISCUSSION

The mechanism of reaction of haloethylaldehydes with cytosine and adenine residues has not been established unequivocally. Krzyżosiak et al. (1979) favour a primary reaction on the exocyclic amino group, yielding a highly unstable carbinolamine (Fig. 3, intermediate 2) for which they present indirect evidence. Our finding of cross-links being formed in poly d[A–T], in which the ring nitrogen is not available for etheno ring formation, strongly supports this idea.

To confirm the existence of such an initial product, we have used 1-methyladenosine and 3-methylcytidine as model compounds with blocked ring nitrogens, although it is recognized that the quaternary nitrogen of these compounds may affect their reactivity. Both react rapidly with chloroacetaldehyde at pH 6–7. The lack of reaction of 1-methyladenosine reported by Secrist et al. (1972) is probably due to differences in the reaction and isolation conditions.

The aldehyde addition products are reactive alkylating agents which can then react with another exocyclic group on the opposite strand, displacing the halogen. Their presence in BAA- or CAA-treated poly d[A–T] can account for the observed cross-links. This mechanism is similar to that proposed by Tong and Ludlum (1981) for diguanylethane cross-links produced by 1,3-bis-(2-chloroethyl)-1-nitrosourea and by Lown et al. (1978) who found that N^4-(2-chloroethyl)-1-methylcytosine crosslinks DNA efficiently.

In order to test the stereochemical feasibility of interstrand cross-linking by chloroacetaldehyde between the adenine exocyclic amino groups, we constructed a molecular model of such a structure. Starting with the coordinates of a self-complementary canonical B-DNA double helical fragment (Arnott & Hukins, 1972) of sequence 5′-pTpApTpA, methyl groups were attached to the 6-amino groups of the second adenine in each strand in a standard geometry. These methyls were then rotated by approximately equal angles about the C6-N6 bonds so that the distance between the methyls corresponded to a standard carbon-carbon bond distance. This required approximately a 55° torsional rotation out of the base plane and was the only

Fig. 8. Drawing of a double helical DNA fragment with an ethyl cross-link joining N^6-amino groups of adenine residues on opposite strands. The carbons of the ethyl cross-link have been drawn as large circles for emphasis. When this linkage is formed by reaction of DNA with chloroacetaldehyde, a hydroxyl is attached to one ethyl carbon (not shown).

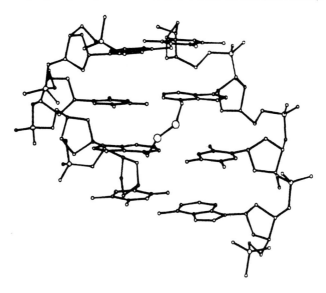

adjustment necessary to form a stereochemically reasonable model. No distortion of the DNA was required, although the N-6 hydrogen-bonding capability was lost. The resulting structure is illustrated in Figure 8. No hydroxyl group is shown on the bridging carbon structure, as its exact location is ambiguous. In any case, there appears to be ample room to position the hydroxyl group in the major groove without any helical distortion. As the bridging ethyl group (actually hydroxyethyl, but hydroxy is not shown) requires a *trans* (fully extended) conformation to span strands, it probably shows little conformational freedom.

In the B-form of DNA, the number of possible cross-linkable sites which could be alkylated on opposite strands is greatly increased since the reactivity of the most nucleophilic site, the N–7 of G, is unaffected by hydrogen bonding.

There is no controversy about the formation of intermediate *3* in Figure 3, although the hydrates have only recently been isolated from CAA-treated polynucleotides. Our finding that the nonfluorescent hydrated intermediates can be isolated in high yield from extensively reacted polynucleotides indicates that these derivatives are almost certainly present *in vivo*, and must be considered in assessing the carcinogenicity of vinyl halides.

The structures of numerous unidentified guanosine derivatives after BAA or CAA reaction have not yet been elucidated except for 1,N^2-ethenoguanosine, which is not the only major product. Further studies are clearly indicated.

ACKNOWLEDGMENT

This work was supported by Grant CA 12316 from the National Cancer Institute, National Institutes of Health, Bethesda, MD, USA.

REFERENCES

Arnott, S. & Hukins D.W.L. (1972) Optimised parameters for A-DNA and B-DNA, *Biochem. biophys. Res. Commun.*, **47**, 1504–1509

Biernat, J., Ciesiołka, J., Górnicki, P., Adamiak, R.W., Krzyżosiak, W.J. & Wiewiórowski, M. (1978) New observations concerning the chloroacetaldehyde reaction with some tRNA constituents. Stable intermediates, kinetics and selectivity of the reaction. *Nucleic Acids Res.*, **5**, 789–804

Holbrook, S.R. & Kim, S.-H. (1983) Correlation between chemical modification and surface accessibility in yeast phenylalanine transfer RNA. *Biopolymers*, **22**, 1146–1166

Kayasuga-Mikado, K., Hashimoto, T., Negishi, T., Negishi, K. & Hayatsu, H. (1980) Modification of adenine and cytosine derivatives with bromoacetaldehyde. *Chem. pharm. Bull.*, **28**, 932–938

Kimura, K., Nakanishi, M., Yamamoto, T. & Tsuboi, M. (1977) A correlation between secondary structure of DNA and reactivity of adenine residues with chloroacetaldehyde. *J. Biochem.*, **81**, 1699–1703

Krzyżosiak, W.J., Biernat, J., Ciesiołka, J., Górnicki, P. & Wiewiórowski, M. (1979) Further studies on adenosine and cytidine reactions with chloroacetaldehyde, a new support for the cyclic carbinolamine structure of the stable reaction intermediate and its relevance to the reaction mechanism and tRNA modification. *Polish J. Chem.*, **53**, 243–252

Krzyżosiak, W.J., Biernat, J., Ciesiołka, J., Gulewicz, K. & Wiewiórowski, M. (1981) The reactions of adenine and cytosine residues in tRNA with chloroacetaldehyde. *Nucleic Acids Res.*, **9**, 2841–2851

Kuśmierek, J.T. & Singer, B. (1982) Chloroacetaldehyde treated ribo- and deoxyribonucleotides. I. Reaction products. *Biochemistry*, **21**, 5717–5722

Laib, R.J., Gwinner, L.M. & Bolt, H.M. (1981) DNA alkylation by vinyl chloride metabolites: etheno derivatives or 7-alkylation of guanine? *Chem.-biol. Interactions*, **37**, 219–231

Leonard, N.J. (1984) Etheno-substituted nucleotides and coenzymes: fluorescence and biological activity. *CRC Crit. Rev. Biochem.*, **15**, 125–199

Lilley, D.M.J. (1983) Structural pertubation in supercoiled DNA: hypersensitivity to modification by a single-strand selective chemical reagent conferred by inverted repeat sequences. *Nucleic Acids Res.*, **11**, 3097–3112

Lown, W.J., McLaughlin, L.W. & Chang, Y.-M. (1978) Mechanism of action of 2-haloethylnitrosoureas on DNA and its relation to their antileukemic properties. *Bioorg. Chem.*, **7**, 97–110

Oesch, F. & Doerjer, G. (1982) Detection of N^6,3-ethenoguanine in DNA after treatment with chloroacetaldehyde *in vitro*. *Carcinogenesis*, **3**, 663–665

Osterman-Golkar, S., Hultmark, D., Segerbäch, D., Calleman, C.J., Göthe, R., Ehrenberg, L. & Wachtmeister, C.A. (1977) Alkylation of DNA and proteins in mice exposed to vinyl chloride. *Biochem. biophys. Res. Commun., 76,* 259–266

Sattsangi, P.D., Leonard, N.J. & Frihart, C.R. (1977) 1,N^2-ethenoguanine and N^2,3-ethenoguanine. Synthesis and comparison of the electronic spectral properties of these linear and angular triheterocycles related to the Y bases. *J. org. Chem., 42,* 3292–3296

Scherer, E., Van Der Laken, C.J., Gwinner, L.M., Laib, R.J. & Emmelot, P. (1981) Modification of deoxyguanosine by chloroethylene oxide. *Carcinogenesis, 2,* 671–677

Secrist, J.A. III, Barrio, J.R., Leonard, N.J. & Weber, G. (1972) Fluorescent modification of adenosine-containing coenzymes. Biological activities and spectroscopic properties. *Biochemistry, 11,* 3499–3506

Singer, B. & Grunberger, D. (1983) *Molecular Biology of Mutagens and Carcinogens,* New York, Plenum Press.

Singer, B., Abbott, L.G. & Spengler, S.J. (1984) Assessment of mutagenic efficiency of two carcinogen-modified nucleosides, 1,N^6-ethenodeoxyadenosine and O^4-methyldeoxythymidine, using polymerases of varying fidelity. *Carcinogenesis, 5,* 1165–1171

Steiner, R.F., Kinnier, W., Lunasin, A. & Delac, J. (1973) Fluorescent derivatives of polyribonucleotides containing ε-adenosine. *Biochim. biophys. Acta, 294,* 24–37

Tong, W.P. & Ludlum, D.B. (1981) Formation of the cross-linked base, diguanylethane, in DNA treated with N,N'-bis-(2-chloroethyl)-N-nitrosourea. *Cancer Res. 41,* 380–382

REACTION KINETICS AND CYTOSINE ADDUCTS OF CHLOROETHYLENE OXIDE AND CHLOROACETALDEHYDE: DIRECT OBSERVATION OF INTERMEDIATES BY FTNMR AND GC-MS

I. O'NEILL[1], A. BARBIN, M. FRIESEN & H. BARTSCH

Unit of Environmental Carcinogens and Host Factors,
International Agency for Research on Cancer,
Lyon, France

SUMMARY

As it is not yet known which are the important miscoding adducts formed in the reaction of the relatively unstable compound chloroethylene oxide (CEO) with double-stranded DNA, proton FTNMR and GC-mass spectroscopy were used to directly detect and characterize reaction intermediates. Reaction of CEO with cytidine gave the (hydrated) 2-oxoethyl derivative at the N-3 position prior to ring closure to $3,N^4$-ethenocytidine; 5-methylcytosine gave an analogous reaction. However, reactions of CEO or chloroacetaldehyde (CAA) with 3-methylcytidine – i.e., with the N-3 blocked as in double-stranded DNA (ds DNA) – were shown by GC-MS of the silylated products to give, at a much slower rate, a pattern of at least 17 adducts all of which contained chlorine. Based on MS fragmentation and considerations of positional, optical and *cis/trans* isomerism, the reaction products of the 3-methylcytosine moiety were assigned as *cis/trans* N^4-(2-chlorovinyl)-3-methylcytosine which may have arisen from the corresponding N^4-(1-hydroxy-2-chloroethyl) adduct. It is postulated that formation of these cytosine- N^4 adducts would be more rapid in double-stranded DNA than in the model compound, and that the N^4-(2-chlorovinyl) group may be a miscoding adduct. The kinetics for CEO rearrangement, hydrolysis and nucleophilic attack have been studied by proton FTNMR and lead to the hypothesis that concerted nucleophilic attack by cytosine-N^4 and CEO rearrangement produce the N^4 adducts.

---

[1] To whom correspondence should be addressed

INTRODUCTION

Reactive metabolites of vinyl chloride are known to be formed through microsomal metabolism (Göthe et al., 1974; Barbin et al., 1975) and chloroethylene oxide (CEO) is considered by the present authors to be the ultimate carcinogenic metabolite (Zadjela et al., 1980). However, it is not yet known which are the important miscoding adducts formed by CEO or its rearrangement product, chloroacetaldehyde (CAA), in double-stranded DNA (dsDNA). Most studies (see Discussion by Barbin & Bartsch, these proceedings[2]) have attempted to identify DNA adducts of significance following reaction with CAA, which is relatively more stable than CEO. The purpose of the present work was to obtain detailed structural and kinetic information on transient and possibly unstable substances formed during in-vitro reactions of CEO, ultilizing high-field Fourier transform nuclear magnetic resonance spectroscopy (FTNMR) to follow reactions, and gas chromatography coupled to mass spectrometry (GC-MS) to examine CEO-nucleoside reaction mixtures *in toto*.

Three separate aspects have been examined: a) CEO-cytidine reactions; b) CEO and CAA reactions with cytidine analogues either to mimic reactions with dsDNA, where the N-3 of cytidine is blocked by hydrogen bonding, or to simplify spectra; c) CEO properties in relation to rearrangement, hydrolysis and nucleophilic attack.

MATERIALS AND METHODS

Proton FTNMR observations were made on a 350 MHz CAMECA spectrometer in 5 mm outer diameter tubes under temperature control of $\pm 0.2\,°C$. GC-MS observations were made with a Perkin-Elmer Sigma 3B gas chromatograph (25m Altex RSL-150 BP capillary column) coupled to a HP 5970A mass selective detector. CEO and CAA were prepared according to published procedures (Gross, 1963; Gross & Freiberg, 1969). Cytidine, 5-methylcytosine, 3-methylcytidine, uridine, 3-methyluridine and 2,4,6-trimethylpyridine were from P.L. Biochemicals, St Goar, FRG of from Sigma, St Louis, MO, USA. Sodium chloride, sulfate and thiosulfate were from Merck, Darmstadt, FRG. Deuterated solvents were from Spectrometrie Spin et Techniques, Paris, France.

For kinetic measurements, samples were mixed in the NMR tube and immediately frozen to $-80\,°C$ until FTNMR observation commenced. Reactions of several-fold molar excess of CEO or CAA with nucleosides or bases were started immediately prior to FTNMR observations, but reactions using nucleophilic salts were maintained for 24 h at ambient temperature prior to observation.

Reactions with a 250-fold molar excess of CEO or CAA were carried out in methanol for 24 h or more (CEO at $4\,°C$, or CAA at room temperature), the CEO/CAA and solvent were then removed in a stream of nitrogen prior to trimethylsilylation with a small excess of N,O-bis(trimethylsilyl)trifluoroacetamide (from Pierce Eurochemie, Beijerland, Netherlands) in pyridine at $70\,°C$ for 10 min.

---

[2] See p. 345.

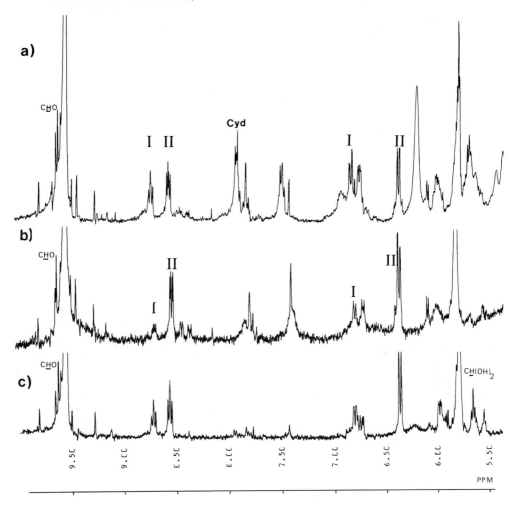

Fig. 1. Fourier transform nuclear magnetic resonance spectra at 350 MHz of the reaction of cytidine (Cyd) with several-fold molar excess of chloroethylene oxide (CEO): a) at 25 °C after a few minutes in deuterated dimethylsulfoxide; b) same as (a), then heated to 50 °C to reveal the reversible I⇌II equilibrium; c) at 25 °C with trace D_2O added to remove NH signals by exchange. Peaks marked I and II correspond to the structures shown in Figure 3.

RESULTS

Reaction of cytidine with CEO

Since parallel work (reported in the section on CEO kinetics below) showed that CEO stability is solvent-dependent, CEO-cytidine reactions were followed in dry deuterated dimethylsulfoxide (DMSO-D_6). Direct observation (Fig. 1a), shortly after

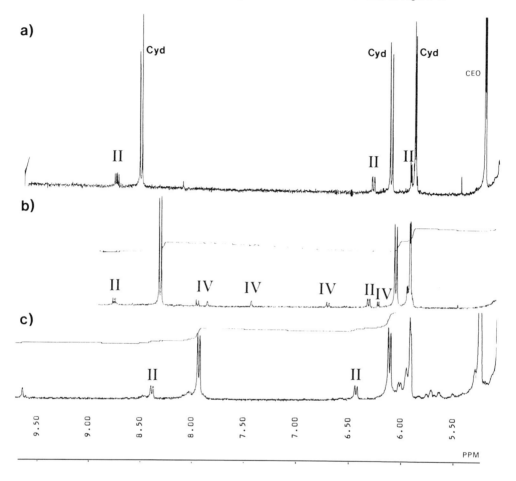

Fig. 2. Fourier transform nuclear magnetic resonance spectra at 350 MHz of the reaction of cytidine (Cyd) with several-fold molar excess of chloroethylene oxide (CEO): a) at 25 °C after a few minutes in deuterated methanol; b) same as (a), but heated at 50 °C for 30 minutes; c) in D_2O at 25 °C. Peaks marked II and IV correspond to the structures shown in Figure 3.

mixing a two-fold molar excess of CEO with cytidine, showed that most of the cytidine had reacted and almost no CEO remained and that cytidine derivatives and CAA were produced. The reaction mixture was probed by spin-spin decoupling, change of temperature (Fig. 1b) and addition of limited amounts of D_2O (Fig. 1c). The main initial intermediates were two pairs of substances in which the hydrogens at positions 5 and 6 of the cytosine base (cytosine H-5 and H-6) were shifted considerably downfield; one pair of intermediates appeared to contain an aldehyde function (9.7 ppm). Depyrimidation had not occurred since the three main products on long storage were nucleoside derivatives.

Fig. 3. Possible mechanism for the reaction of chloroethylene oxide (CEO) with cytidine. I, 3-(2-oxoethyl)cytidine; Ia, hydrated form of 3-(2-oxoethyl)cytidine; II, hydrated form of 3,N^4-ethenocytidine (εCyd); III, εCyd, protonated; IV, εCyd; R, ribosyl

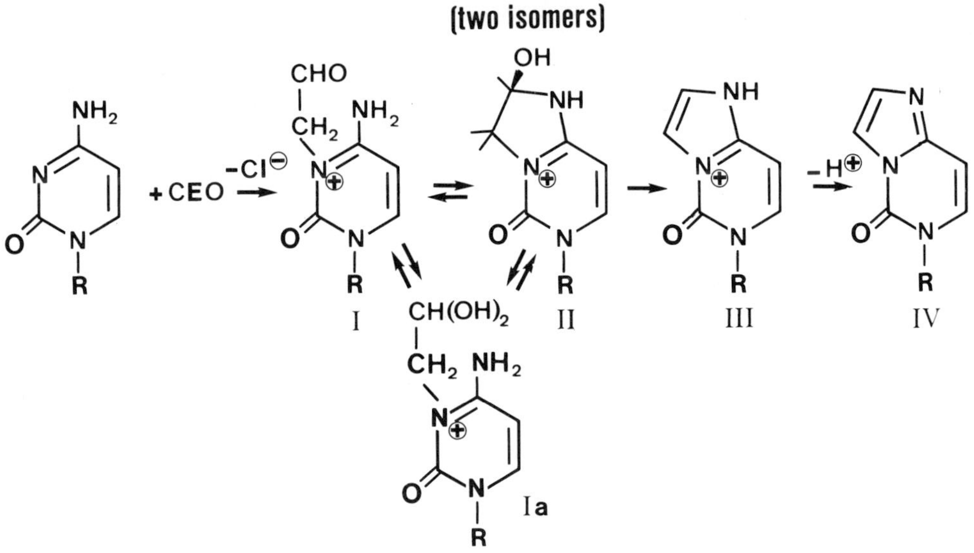

NMR-spectra (Fig. 1a–1c) characterized the principal initial products formed in DMSO-D_6 as pairs of compounds in equilibrium, apparently following an irreversible reaction of cytidine in such a manner as to deshield the cytosine H-5 and H-6 and form an aldehyde. In contrast, reaction in deuterated methanol (CD_3OD) gave a much simpler spectrum (Fig. 2a), revealing a small amount of one pair of intermediates; subsequent heating of CEO and cytidine produced some 3,N^4-ethenocytidine (εCyd) (IV; Fig. 2b). In D_2O, some reaction of CEO with cytidine was also observed (Fig. 2c); in both CD_3OD and D_2O, CEO was still present after several minutes. As the same pair of intermediates was apparently formed in all three solvents, it is possible that these intermediates have the structures I, Ia and II, shown in Figure 3, with II accumulating in CD_3OD or D_2O since the great excess of water or methanol would not favour retention of the free aldehyde group. The extent of reaction by CEO was clearly far greater in DMSO-D_6 than in either CD_3OD or D_2O. Storage of the CEO and cytidine mixture (having the spectrum shown in Figure 1a) for 15 days at ambient temperature yielded εCyd (IV) as the main product, with possibly protonated εCyd (III) as another compound present (spectra shown in Fig. 4; structures of III and IV, in Fig. 3). The relative downfield shifts for cytosine H-5 and H-6 and for the etheno-bridge hydrogens, as well as the small upfield shift for one anomeric hydrogen (Kozerski et al., 1984), are in accord with this assignment. Identities of the third nucleoside product and the many minor signals (Fig. 4b) are unknown.

Fig. 4. Fourier transform nuclear magnetic resonance spectra of the sample from Figure 1a stored at ambient temperature for 15 days: a) full spectrum; b) aromatic region. Small letters signify coupled pairs of aromatic hydrogens either from positions 5 and 6 of the cytosine base or from the etheno bridge. Peaks marked III and IV correspond to the structures shown in Figure 3. CAA, chloroacetaldehyde; CAAh, chloroacetaldehyde hydrate

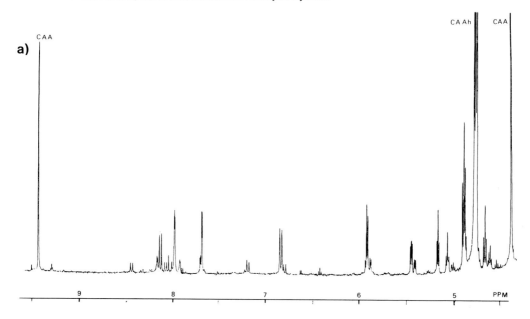

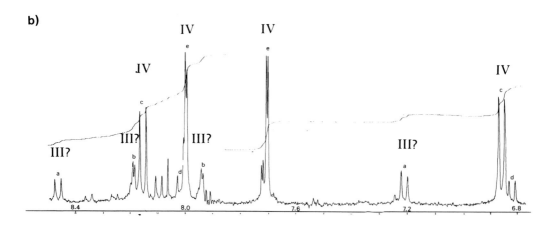

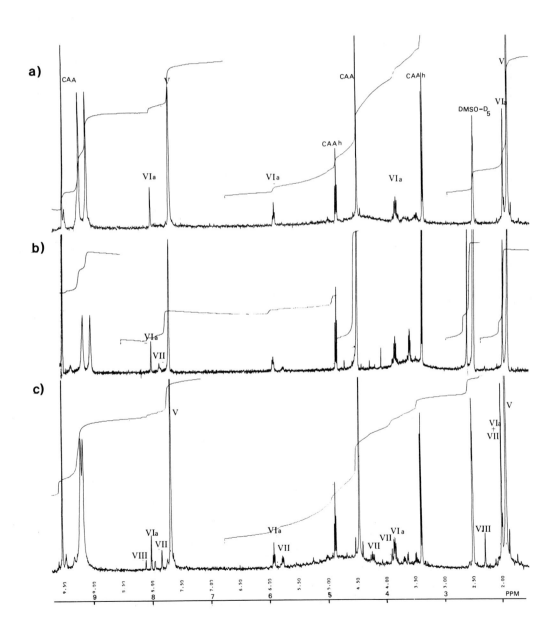

Fig. 5. Fourier transform nuclear magnetic resonance spectra at 350 MHz of the reaction of 5-methylcytosine with several-fold excess of chloroethylene oxide (CEO) or chloroacetaldehyde (CAA) in deuterated dimethylsulfoxide (DMSO-d_6): a) reaction at 25 °C with CAA, spectrum recorded after a few minutes; b) reaction at 25 °C with CEO, spectrum recorded after a few minutes; c) same as (a), but heated at 50 °C for 45 min. Peaks marked V, VIa, VII and VIII correspond to the structures shown in Figure 6.

From these observations and the experiments with 5-methylcytosine (following section), it was concluded that the major observable reaction of cytidine with CEO occurs at N-3 to yield 2-oxoethyl derivatives (Fig. 3) as initial products.

Reaction of 5-methylcytosine, 3-methyl(deoxy)cytidine and 3-methyluridine with CEO and CAA

The reaction of CEO and CAA with 5-methylcytosine was studied since the lack of a ribose moiety and the substitution of cytosine H-5 by a methyl function facilitate interpretation of NMR spectra. Both CEO and CAA appeared to give the same initial intermediates; the spectra given in Figure 5 show the initial products (VI, VIa, VII) and the final product assigned as $(3,N^4$-etheno)-5-methylcytosine (VIII) (structures shown in Fig. 6). When the reactions were followed with time, it was apparent that two intermediates existed for which signals for the cytosine H-6, the 5-methyl group and the CEO-derived moieties could all be detected (Table 1). Minor signals in the aldehyde region, and the $-CH-CH_2-$ moiety detected at 5.92 and 3.8/3.9 ppm, appear to be consistent with the formation of a 2-oxoethyl(solvate) intermediate, prior to conversion to the hydrated $3,N^4$-etheno adduct (VII) (Fig. 6). The temperature-dependent sharpening of signals for cytosine H-6 in VII but not in VI are consistent with these intermediates being cyclic and noncyclic, respectively; the chemical shifts for VII are consistent with reported data (Biernat et al., 1978) for hydrated $3,N^4$-ethenocytidine.

Fig. 6. Possible mechanism for the reaction of chloroethylene oxide (CEO) with 5-methylcytosine (V). VI, 3-(2-oxoethyl)-5-methylcytosine; VIa, hydrated form of VI; VII, hydrated form of $(3,N^4$-etheno)-5-methylcytosine; VIII, $(3,N^4$-etheno)-5-methylcytosine

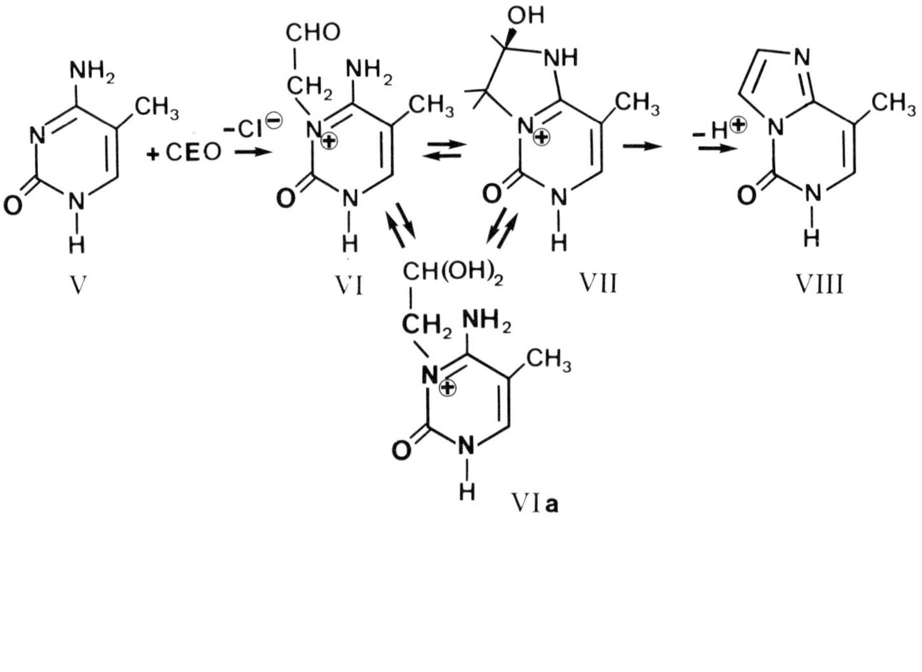

Table 1. Chemical shifts in the Fourier transform nuclear magnetic resonance spectra for reactions of 5-methylcytosine with chloroacetaldehyde in deuterated dimethylsulfoxide at 25 °C

Compound	Chemical shift (in ppm) for		Alkyl substituent	
	H-5	CH$_3$		
5 meC[a]	7.71	1.91	–	–
VIa[b]	8.03	2.00	CH 5.92 (4 & 4 Hz);	CH$_2$ 3.8 & 3.9
VII[b]	7.83	2.00	CH 5.77 (3 & 9 Hz);	CH$_2$ 3.9 & 4.2
VIII[b]	7.64	2.31	CH 8.10	CH 7.95

[a] 5-methylcytosine
[b] Structures as shown in Figure 6

When 3-methylcytidine was treated with a two-fold molar excess of CEO in DMSO-D$_6$, no apparent reaction could be detected by FTNMR spectroscopy under the same conditions in which cytidine had shown rapid reaction. Under forcing conditions (250-fold molar excess of CEO for 24 h at 4 °C or of CAA at ambient temperature), a reaction could be detected using either DMSO or CH$_3$OH as solvent; analysis by GC-MS was possible after trimethylsilylation of the reaction mixture. At least 17 adducts were detected (Fig. 7a), with either [ClC$_2$H$_2$], [ClC$_2$H$_3$OTMS] or [ClC$_2$H$_2$ + ClC$_2$H$_3$OTMS] substitutions (TMS = trimethylsilyl). Both CEO and CAA gave the same pattern of adducts. The structures assigned in Figure 7b are based on MS fragmentation and the multiple isomers are consistent with positional, optical or *cis/trans* isomerism. Fragmentation patterns of the two major adducts with molecular weights of 534 and 536 showed that a ClC$_2$H$_2$ group was attached to the cytosine moiety. Natural chlorine isotopic distribution confirmed that mono- and diadducts were present.

A greater reactivity of the cytosine base moiety compared to the carbohydrate moiety was also apparent with 3-methyldeoxycytidine, but with 3-methyluridine, the ribose moiety was more reactive than the base. The formation of four ClC$_2$H$_2$ adducts to the uracil base can be explained by *cis/trans* isomerism of an adduct attached at either of the two uracil oxygen atoms. Since hydroxy-chloroethyl moieties are observed with the OH groups of ribose, a similar adduct is believed to be the precursor of the chlorovinyl groups attached to the bases. This mechanism is further supported by the report by Singer *et al.* (these proceedings[3]) of the formation of N^4-(1-hydroxy-2-chloroethyl)-3-methylcytidine following the reaction of 3-methylcytidine with CAA.

The presence of two chlorovinyl isomers for 3-methylcytidine, following reaction with CEO, can be explained by assuming these to be *cis/trans* isomers attached to the N^4 of the cytosine base, which in turn requires that they have a 2-chlorovinyl structure. Proton FTNMR observation of the product of 3-methyldeoxycytidine treated with a 250-fold excess of CEO showed signals suggestive of chlorovinyl groups which were predominantly *cis* isomers [coupled (~5Hz) signals at 6.1 and 6.9 ppm].

---

[3] See p. 45.

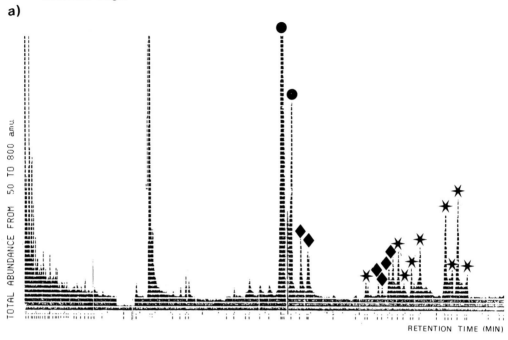

Fig. 7. Gas chromatography / mass spectrometry trace (a) of the total trimethylsilylated product of the reaction of 3-methylcytidine with a 250-fold excess of chloroethylene oxide. Peaks represent total ion current and are keyed by the symbols ♦, ● and ★ to the three structural types (b) revealed by the mass spectrum of each peak. TMS, trimethylsilyl; c/t, *cis/trans*; M.W., molecular weight

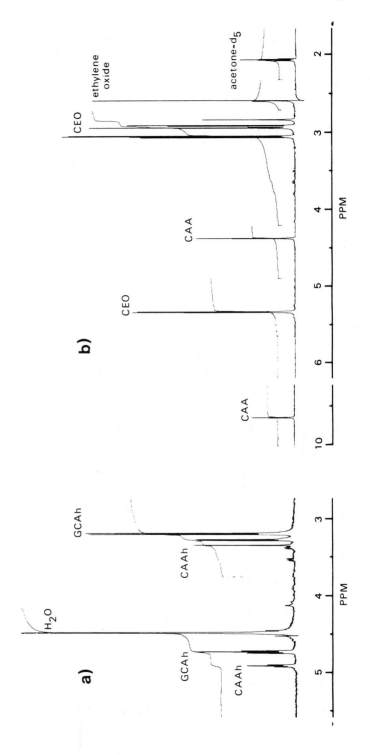

Fig. 8. Fourier transform nuclear magnetic resonance spectra (at 350 MHz) typical of the kinetic observation of (a) chloroethylene oxide hydrolysis/rearrangement in D_2O (b) chloroethylene oxide rearrangement in deuterated acetone. CEO, chloroethylene oxide; CAA, chloroacetaldehyde; CAAh, chloroacetaldehyde hydrate; GCAh, glycolaldehyde hydrate

CEO kinetics for rearrangement, hydrolysis and nucleophilic attack

Proton FTNMR observations were made in deuterated acetone:D_2O (1:1) at several temperatures to monitor the disappearance of CEO by rearrangement and hydrolysis, or by reaction with various nucleophiles. Typical spectra for hydrolysis/rearrangement are shown in Figure 8, and the relevant reactions are summarized in Figure 9.

In kinetic studies, proton FTNMR was a particularly effective tool for simultaneous quantitation of several compounds. The concentrations of CAA and GCA (glycolaldehyde) thus measured were plotted against CEO concentration and K_R and K_W were determined from the slopes (equations 1 to 3). K_H' was determined from the plot $\ln[CEO] = f(t)$.

$$[CEO] = [CEO]_o e^{-K_H' t} \qquad \text{equation 1}$$

$$K_H' = K_R + K_w[D_2O]$$

$$[CAA] = \frac{[CEO]_o \times K_R}{K_H'} (1 - e^{-K_H' t}) \qquad \text{equation 2}$$

$$[GCA] = \frac{[CEO]_o \times K_w[D_2O]}{K_H'} (1 - e^{-K_H' t}) \qquad \text{equation 3}$$

In these equations, $[CEO]_o$ is the initial CEO concentration, K_R is the first-order rate constant for rearrangement, K_w, the second-order rate constant for hydrolysis and K_H', the pseudo first-order rate constant for the overall reaction of CEO. Results are summarized in Table 2; a typical plot of CAA and GCA formation in D_2O from CEO is shown in Figure 10.

In the presence of 0.22 M sodium chloride in acetone: D_2O (1:1) at 7 °C, K_R showed an increase from 0.84 to 4.83 h^{-1}. If this is considered to be due to a nucleophilic

Table 2. Rate constants for the hydrolysis and rearrangement of chloroethylene oxide in D_2O at various temperatures

Temperature (°C)	K_w[a] ($M^{-1} h^{-1}$)	K_R[b] (h^{-1})
4	0.0149	0.48
14	0.0362	2.27
24	0.0635	6.31
E_α (Kcal × mole^{-1})	12[c]	21[c]
ln A	17.4[c]	37.7[c]

[a] Second-order rate constant for hydrolysis
[b] First-order rate constant for rearrangement
[c] Calculated from $\ln K = -\frac{E_\alpha}{RT} + \ln A$

Fig. 9. Summary of the reactions involved in the disappearance of chloroethylene oxide by rearrangement and hydrolysis, or by reaction with nucleophiles (Nu)

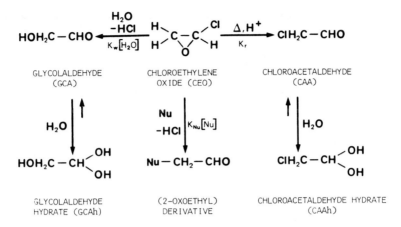

Table 3. Chemical shifts of the products, NuCH$_2$CH(OH)$_2$, formed from attack by various nucleophiles (Nu) on chloroethylene oxide

Nucleophile	Chemical shift (in ppm) for	
	CH$_2$	CH(OD)$_2$
D$_2$O	3.4	5.0
Cl$^-$	3.5	5.1
SO$_4^=$	3.9	5.1
S$_2$O$_3^=$	3.1 & 3.8	5.0
Pyridine	3.7 & 3.8	5.5
5 meC[a]	3.8 & 3.9	5.92
Cytidine	Obscured	5.9

[a] 5-methylcytosine

attack by Cl$^-$ on CEO rather than an effect of the medium, it can be calculated from the increased rate of appearance of CAA that the second-order rate constant, K_{Cl^-}, is 18 M^{-1}h^{-1}. This is comparable to the result of 14 M^{-1}h^{-1} that can be calculated using the Swain-Scott relationship, log (K_{Cl^-}/K_w) = s × n_{Cl^-} (Swain & Scott, 1953), where s is the nucleophilic selectivity of CEO (s = 0.71; unpublished data), n_{Cl^-}, the nucleophilic constant of Cl$^-$ (n_{Cl^-} = 2.7, Swain & Scott, 1953), and K_w = 0.17 M^{-1}h^{-1} in H$_2$O (extrapolated at 7°C from data by Osterman-Golkar, 1984).

Similar studies were carried out on the attack by various nucleophiles (Nu) on CEO. Probable products were, amongst others, hydrated aldehydes of the type Nu-CH$_2$(OH)$_2$ (Table 3). The amount of product formed relative to GCA (hydrate) increased with increasing nucleophilicity of Nu. The postulated 2-oxoethyl hydrates of 5-methylcytosine and cytidine are included for comparison.

Fig. 10. Plot of glycolaldehyde (GCA) and chloroacetaldehyde (CAA) concentrations *versus* chloroethylene oxide (CEO) concentration from the nuclear magnetic resonance study of the hydrolysis and rearrangement of CEO in D_2O (GCA and CAA observed as hydrates)

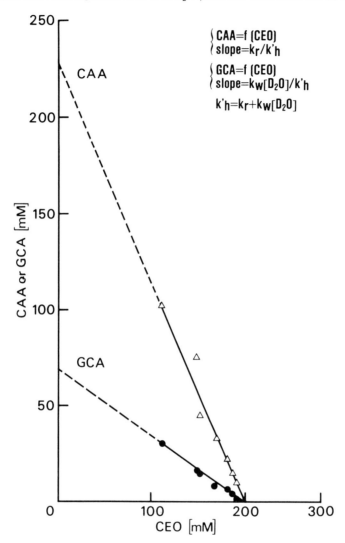

DISCUSSION

Examination by FTNMR of the decomposition of CEO allowed for the direct measurement of the rearrangement and hydrolysis constants for CEO. Similarly, the study of products obtained from reaction with various nucleophiles showed that hydrated 2-oxoethyl groups are formed, amongst other products, which is compatible

Fig. 11. Possible mechanisms for rearrangement of chloroethylene oxide (CEO) and for nucleophilic attack on CEO. CAA, chloroacetaldehyde; Nu, nucleophile

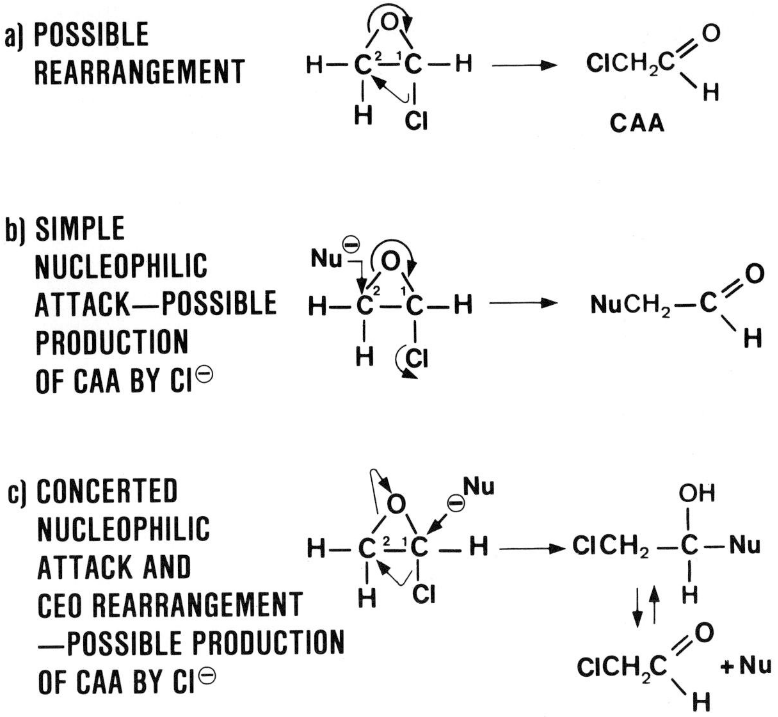

a) POSSIBLE REARRANGEMENT

b) SIMPLE NUCLEOPHILIC ATTACK—POSSIBLE PRODUCTION OF CAA BY Cl$^\ominus$

c) CONCERTED NUCLEOPHILIC ATTACK AND CEO REARRANGEMENT —POSSIBLE PRODUCTION OF CAA BY Cl$^\ominus$

with the reported formation of a hydrated 2-oxoethyl adduct at N-7 of deoxyguanosine (Scherer et al., 1981). FTNMR kinetic measurements provided evidence that CEO can yield CAA not only by thermal rearrangement but also by reaction with chloride ion.

It was further shown that FTNMR allows the observation and characterization of intermediates from the reactions of nucleosides or their bases with carcinogens. Cytidine was found to react rapidly with CEO, when N-3 was not blocked, yielding several different intermediates; however εCyd was the major final product. A possible reaction sequence (Fig. 3) involves attachment of a 2-oxoethyl group at N-3 of cytidine (N-3 being the most nucleophilic site), followed by reversible ring closure at N^4. An etheno derivative was also formed by reaction of 5-methylcytosine with CEO; two intermediates in equilibrium were characterized as the hydrated 2-oxoethyl adduct at N-3 and the hydrated 3, N^4-etheno adduct (Fig. 6). However, under similar conditions, no reaction was detected by FTNMR between CEO and 3-methylcytidine; this is consistent with the rapid reaction of CEO with cytosine being primarily at N-3.

Nevertheless, to test the hypothesis that specific CEO adducts could be formed in dsDNA (due to the reduced accessibility or nucleophilicity of nucleophilic centres such as N-3 of cytosine residues and N-1 of adenine residues),

3-methyl(deoxy)cytidine was reacted with a large excess of CEO and the adducts were observed by GC-MS. The data showed that the base adduct is probably a 2-chlorovinyl group attached to N^4, possibly arising from the intermediate, N^4-(1-hydroxy-2-chloroethyl)-3-methyl(deoxy)cytidine. Considering the likely deactivating effect of the 3-methyl group in the substrate, deoxycytidine in dsDNA would probably form such an N^4 adduct more readily than these in-vitro experiments would suggest. In dsDNA, irreversible ring closure at N-3 of the intermediate to yield 3,N^4-ethenodeoxycytidine would be essentially precluded; instead formation of the N^4-(2-chlorovinyl) group would stabilize adduct formation and could be a miscoding adduct analogous to N^4-methoxycytosine (Singer et al., 1984). Therefore, CEO may not only produce 2-oxoethyl adducts at nucleophilic ring nitrogens but may also yield 1-hydroxy-2-chloroethyl adducts on free amino groups. Utilizing the chlorine shift postulated for rearrangement of other α-chloroepoxides (Bonse & Henschler, 1976), and noting that CEO/CAA rearrangement occurs in D_2O without deuterium incorporation, a mechanism for such a reaction is suggested in Figure 11c.

ACKNOWLEDGEMENTS

The collaboration of CNRS (Solaise) and Centre d'Etudes Nucleaires (Grenoble) for FTNMR facilities and Dr A. Croisy (Institut Curie, Orsay) for CEO preparation are acknowledged, as is the secretarial assistance of Mrs M. Wrisez.

REFERENCES

Barbin, A., Brésil, H., Croisy, A., Jacquignon, P., Malaveille, C., Montesano, R. & Bartsch, H. (1975) Liver microsome-mediated formation of alkylating agents from vinyl bromide and vinyl chloride. *Biochem. biophys. Res. Commun.*, **67**, 596–603

Biernat, J., Ciesiołka, J., Górnicki, P., Adamiak, R.W., Krzyżosiak, W.J. & Wiewiórowski, M. (1978) New observations concerning the chloroacetaldehyde reaction with some tRNA constituents. Stable intermediates, kinetics and selectivity of the reaction. *Nucleic Acids Res.*, **5**, 789–804

Bonse, G. & Henschler, D. (1976) Chemical reactivity, biotransformation and toxicity of polychlorinated aliphatic compounds. *CRC Crit. Rev. in Toxicol.*, **4**, 395–409

Göthe, R., Calleman, C.J., Ehrenberg, L. & Wachtmeister, C.A. (1974) Trapping with 3,4-dichlorobenzenethiol of reactive metabolites formed *in vitro* from the carcinogen vinyl chloride. *Ambio*, **3**, 234–236

Gross, H. (1963) Monochloroacetaldehyde or derivatives of glycolaldehyde and glyoxal from α-halo-ethers (German). *J. prakt. Chem.*, **21**, 99–102

Gross, H. & Freiberg, J. (1969) Concerning the existence of chloroethylene oxide (German). *J. prakt. Chem.*, **311**, 506–510

Kozerski, L., Sierpputowska-Gracz, H., Krzyżosiak, W., Bratek-Wiewiórowska, M., Jaskólski, M. & Wiewiórowski, M. (1984) Comparative structural analysis of cytidine, ethenocytidine and their protonated salts. III. ^1H, ^{13}C and ^{15}N NMR studies at natural isotope abundance. *Nucleic Acids Res.*, **12**, 2205–2223

Osterman-Golkar, S. (1984) Reaction kinetics in water of chloroethylene oxide, chloroacetaldehyde, and chloroacetone. *Hereditas, 101,* 65–68

Scherer, E., Van Der Laken, C.J., Gwinner, L.M., Laib, R.J. & Emmelot, P. (1981) Modification of deoxyguanosine by chloroethylene oxide. *Carcinogenesis, 2,* 671–677

Singer, B., Fraenkel-Conrat, H., Abbott, L.G. & Spengler, S.J. (1984) N^4-Methoxydeoxycytidine triphosphate is in the imino tautomeric form and substitutes for deoxythymidine triphosphate in primed poly d[A–T] synthesis with *E. coli* DNA polymerase I. *Nucleic Acids Res. 12,* 4609–4618

Swain, C.G. & Scott, C.G. (1953) Quantitative correlation of relative rates. Comparison of hydroxide ion with other nucleophilic reagents toward alkyl halides, esters, epoxides, and aryl halides. *J. Am. chem. Soc., 75,* 141–147

Zajdela, F., Croisy, A., Barbin, A., Malaveille, C., Tomatis, L. & Bartsch, H. (1980) Carcinogenicity of chloroethylene oxide, an ultimate reactive metabolite of vinyl chloride, and bis(chloromethyl)ether after subcutaneous administration and in initiation-promotion experiments in mice. *Cancer Res., 40,* 352–356

CHEMICAL MODIFICATION OF ADENINE AND CYTOSINE RESIDUES WITH CHLOROACETALDEHYDE AT THE NUCLEOSIDE AND THE tRNA LEVEL: THE STRUCTURAL EFFECT OF CHLOROACETALDEHYDE MODIFICATION

W.J. KRZYŻOSIAK & M. WIEWIÓROWSKI[1]

Institute of Bioorganic Chemistry, Polish Academy of Sciences, Poznan, Poland

M. JASKÓLSKI

Institute of Chemistry, A. Mickiewicz University, Poznan, Poland

INTRODUCTION

Chemical modification is a widely used approach in studying the structure-function relationship of nucleic acids. In cases in which the original biological properties of the studied system are changed, the changes are correlated with modification sites and functionally important regions of nucleic acids are identified. Chemical modification is also used to probe the structural properties of folded macromolecules, the studies on transfer RNA tertiary structure being the best example (Rich & RajBhandary, 1976).

In both kinds of application the goal is to establish or at least to gain some insight into the structural consequences of modification reactions. However, on the polymer level, the precise identification of the changes in conformational parameters of modified units, as well as of the changes in the base electronic structure which modifies the ability to undergo intra- and intermolecular interactions, is still a difficult goal to achieve. Nevertheless, indirect approaches can shed some light on this problem. Combined X-ray and solution studies carried out at the monomer level can reveal many modification-dependent changes in structural properties of nucleosides or nucleotides and some of them can safely be extrapolated to the level of polymers. On the other hand, the chemical modification pattern in a transfer RNA molecule of known tertiary structure can be considered as a valuable test capable of determining whether a particular modification used only detects accessible bases or whether it produces some structural transitions, as well.

---

[1] To whom correspondence should be addressed

Fig. 1. Formation of etheno derivatives by reaction of adenosine or cytidine with chloroacetaldehyde (CAA). Ado, adenosine; Cyd, cytidine; εAdo, 1,N^6-ethenoadenosine; εCyd, 3,N^4-ethenocytidine; εAdo·H_2O and εCyd·H_2O, cyclic carbinolamine intermediates

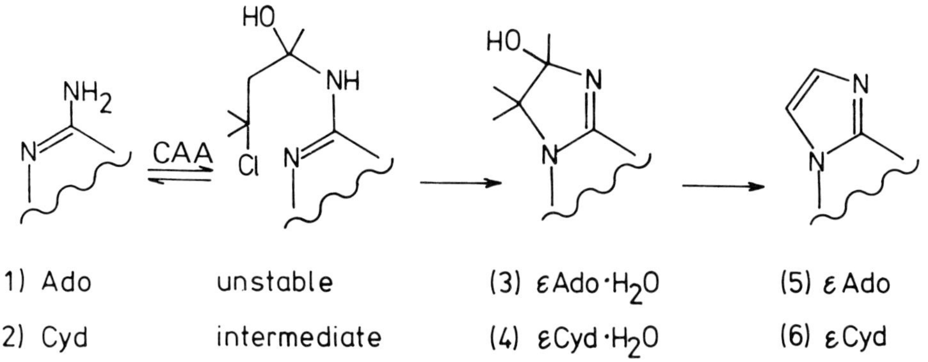

(1) Ado   unstable   (3) εAdo·H_2O (5) εAdo
(2) Cyd   intermediate   (4) εCyd·H_2O (6) εCyd

The above problems will be discussed here briefly using as an example the modification of adenine and cytosine residues with chloroacetaldehyde (Kochetkov et al., 1971).

REACTION INTERMEDIATES

The kinetic and spectroscopic studies on reactions of adenosine (Ado) and cytidine (Cyd) with chloroacetaldehyde (CAA) enabled us to detect, isolate and characterize the stable reaction intermediates (Biernat et al., 1978a, b; Krzyżosiak et al., 1979). For present structural considerations, the cyclic carbinoloamine intermediates (3) and (4) shown in Figure 1 are of importance due to their relatively long life-times. Unfortunately, only a little is known about their electronic structure (Krzyżosiak et al., 1979) and practically nothing about their conformational properties. One serious limitation which makes rapid progress in structural characterization of these intermediates difficult is their chemical instability, which makes it impossible to obtain crystals suitable for X-ray analysis.

According to our recent results, cyclic intermediates which are considerably more stable can be obtained in the reactions of Ado and Cyd with α-bromopropionaldehyde (Krzyżosiak et al., 1983). It is also possible to stabilize completely the hydroxyethano bridge on adenine and cytosine moieties using 6-methyladenosine and 4-methylcytidine as substrates for CAA reaction (Biernat et al., 1978a, b). However, it seems unlikely that such approaches for obtaining more manageable analogues which would mimic the structures of intermediates (3) and (4) could be fully fruitful even in the case of a successful X-ray analysis. The presence of an extra methyl group at, or in close proximity to, the hydroxyethano bridge could be the source of additional effects and the structures obtained for the intermediate analogues would not necessarily closely resemble the structures of (3) and (4). Till now, even the detailed solution structures of the intermediates (3) and (4) have not been determined.

Fig. 2. Electronic structures and hydrogen schemes (dotted lines) for cytidine (Cyd), 3,N^4-ethenocytidine (εCyd), adenosine (Ado) and 1,N^6-ethenoadenosine (εAdo). The bond distances (in Å) are shown for Cyd and Ado and the dimensional differences (in Å, × 10^3) for εCyd and εAdo. The ribosyl groups (R) were not considered.

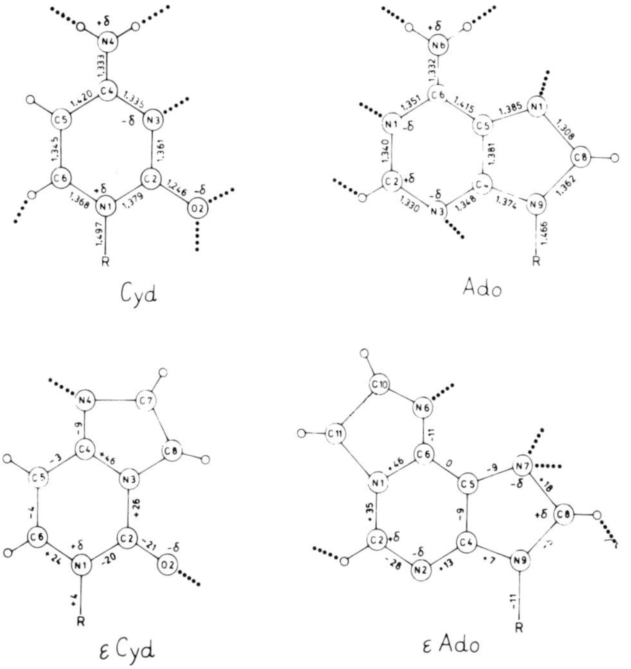

ETHENO DERIVATIVES

In contrast to the reaction intermediates, the final ethenonucleosides have been satisfactorily characterized, in terms of their structures, in recent years (Jaskólsky *et al.*, 1981; Jaskólski, 1982; Krzyżosiak *et al.*, 1982; Kozerski *et al.*, 1984). This applies to both the neutral species and their hydrochlorides. The pKa values of 1,N^6-ethenoadenosine (εAdo) and 3,N^4-ethenocytidine (εCyd) show that only neutral species are present in the vicinity of pH 7 – the pH at which most biochemical tests are carried out – so the present discussion will deal with these structures only.

Etheno-bridging of Ado and Cyd causes drastic changes in the electronic structure of the base, as shown in Figure 2. The pyrimidine ring of Ado is a mesomeric system to which the polar forms contribute significantly. A partial positive charge is present at the C-2 atom, while the N^6/C-6/N-1 region is a mesomeric system with partial negative and partial positive charges at N-1 and N^6, respectively. Conjugation also exists between N-9 and the C-8/N-7 double bond. After etheno bridging, the mesomerism within the pyrimidine ring is practically disabled and the C-6/N^6 bond becomes typically double-conjugated with the etheno bridge. Conjugation still exists

Table 1. Conformational parameters of adenosine (Ado), 1,N^6-ethenoadenosine (εAdo), cytidine (Cyd) and 3,N^4-ethenocytidine (εCyd)

Nucleoside	Glycosyl χ^a	Ribose pucker	Pseudorotation parameters		Exocyclic CH_2OH ψ^d
			P^b	τ_m^c	
Ado	9.9	$^3T2$	7.0	36.8	177.0
εAdo	26.2	^2E	163.6	44.3	56.7
Cyd	18.4	$^3T2$	9.1	37.9	47.1
εCyd	9.9	$^3T2$	10.6	40.8	47.6

^{a} Glycosidic bond torsion angle, in degrees
^{b} Phase angle of pseudorotation, in degrees
^{c} Ribose ring puckering amplitude, in degrees
^{d} Torsion angle along the C5'–C4' bond, in degrees

between N-9 and C-8 and again there is evidence of charge separation in the C-2 region resulting in a partial positive charge at C-2. Some charge separation also exists in the N-7/C-8 region (Jaskólski, 1982).

In the case of Cyd, partial charges are scattered all over the pyrimidine ring, with partial positive charges at N-1 and C-6 and negative charges concentrated mainly at the carbonyl oxygen and to a lesser extent at N-3. Introduction of the etheno bridge limits charge separation to the lactam group only and consequently stabilizes the system of three conjugated double bonds (Jaskólski et al., 1981). The results of solution studies (Kozerski et al; 1984) are in perfect agreement with this interpretation.

The conformational parameters of the ribose and of the whole molecule are completely different for Ado and εAdo (Table 1). Ado is 3'-endo puckered and has the exocyclic -CH_2OH group in the *trans* orientation while εAdo has 2'-endo pucker and the exocyclic -CH_2OH group in the g^+ orientation. Although both nucleosides have the glycosyl bond in the *anti* region, the χ values are very different from each other (Jaskólski, 1982). On the other hand, the conformational parameters of Cyd and εCyd are more similar despite the presence of the short intramolecular hydrogen bond between the H at position 6 and the O at the 5' position in the Cyd structure (Jaskólski et al., 1981). It is difficult to decide if the conformational changes are a consequence of different electronic structures or of different systems of intermolecular interactions in the crystal structures.

The practical conclusion emerging from these studies is that εCyd should mimic Cyd units in polynucleotides better than εAdo would mimic the Ado ones. In other words, εAdo should be considered as a potentially more effective inducer of conformational changes than εCyd.

TRANSFER RNA MODIFICATION

CAA, like many other base-modifying reagents, shows specificity for single-stranded regions of nucleic acids (Shulman & Pelka, 1976) and is sensitive to the

Fig. 3. Secondary structure model of yeast phenylalanine transfer RNA. Arrows indicate sites of chloroacetaldehyde modification at pH 5.5.

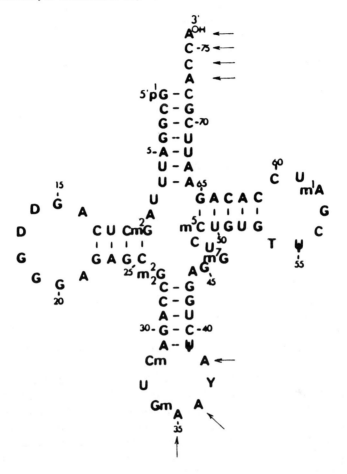

strength of stacking interactions (Kimura et al., 1976). In order to be attacked by the reagent, two reaction centres (the N^6 amino group and the N-1 of adenine or the N^4 amino group and the N-3 of cytosine) have to be, in the first place, free of significant hydrogen bonding interactions. Yeast phenylalanine transfer RNA (tRNA^{Phe}) is the only native RNA molecule for which the tertiary structure has been resolved at a high level of accuracy by the X-ray diffraction method (Hingerty et al., 1978; Sussman et al., 1978). The hydrogen bonding scheme, the system and strength of stacking interactions, as well as the strong magnesium binding sites have been established for this molecule.

Recently, the CAA modification pattern in yeast tRNA^{Phe} has been determined (Krzyżosiak & Ciesiołka, 1983) and it is shown in Figure 3. Despite extensive modification of the adenine and cytosine residues which are accessible in the tertiary structure, all other residues remain unaffected by the reagent. This means that in the case of highly-folded structures like tRNAs, which are stabilized by many intramolecular interactions, the introduction of the etheno bridge into a few bases (located at the CCA-end and in the anticodon loop, in the case of tRNA^{Phe}) is unable to break the existing system of interactions. However, on the basis of the chemical modification experiment, we cannot say anything about possible local conformational perturbations which could take place in the vicinity of modified units.

In conclusion, the reactions of CAA with adenine and cytosine derivatives are now relatively well-defined with respect to both mechanistic and structural aspects. We hope that improved understanding of the structural consequences of these modifications will prove to be an important element in the link between chemical reactions and biological consequences. Further studies are being undertaken to investigate such consequences.

REFERENCES

Biernat, J., Ciesiołka, J., Górnicki, P., Adamiak, R.W., Krzyżosiak, W.J. & Wiewiórowski, M. (1978a) New observations concerning the chloroacetaldehyde reaction with some tRNA constituents. Stable intermediates, kinetics and selectivity of the reaction. *Nucleic Acids Res.*, **5,** 789–804

Biernat, J., Ciesiołka, J., Górnicki, P., Krzyżosiak, W.J. & Wiewiórowski, M. (1978b) Further studies on the reaction of adenine, cytosine, and corresponding nucleosides with chloroacetaldehyde. *Nucleic Acids Res., Special Publ. No. 4,* pp. 203–206

Hingerty, B., Brown, R.S. & Jack, A. (1978) Further refinement of the structure of yeast tRNA^{Phe}. *J. mol. Biol.*, **124,** 523–534

Jaskólski, M. (1982) *Comparison of molecular dimensions and intermolecular interactions of adenosine, ethenoadenosine and their hydrochlorides.* In: Katuski, Z., ed., *Proceedings of the Fourth Symposium on Organic Crystal Chemistry,* Poznań-Kiekrz, UAM, pp. 221–245

Jaskólski, M., Krzyżosiak, W.J., Sierzputowska-Gracz, H. & Wiewiórowski, M. (1981) Comparative structural analysis of cytidine, ethenocytidine and their protonated salts. I. Crystal and molecular structure of ethenocytidine. *Nucleic Acids Res.*, **9,** 5423–5442

Kimura, K., Nakanishi, M., Yamamoto, T. & Tuboi, M. (1976) Detection of adenine residue free from base-pairing by means of fluorescence measurement of ethenoadenosine. *Nucleic Acids Res., Special Publ. No. 2,* 125–128

Kochetkov, N.K., Shibaev, V.N. & Kost A.A. (1971) New reaction of adenine and cytosine derivatives, potentially useful for nucleic acids modification. *Tetrahedron Lett.*, **22,** 1993–1996

Kozerski, L., Sierzputowska-Gracz, H., Krzyżosiak, W.J., Bratek-Wiewiórowska, M., Jaskólski, M. & Wiewiórowski, M. (1984) Comparative structural analysis of cytidine, ethenocytidine and their protonated salts. III. ^{1}H, ^{13}C and ^{15}N NMR studies at natural isotope abundance, *Nucleic Acids Res.*, **12,** 2205–2223

Krzyżosiak, W.J. & Ciesiołka, J. (1983) Long-range conformational transition in yeast tRNA^{Phe}, induced by the Y-base removal and detected by chloroacetaldehyde modification. *Nucleic Acids Res.*, **11,** 6913–6921

Krzyżosiak, W.J., Biernat, J., Ciesiołka, J., Górnicki, P. & Wiewiórowski, M. (1979) Further studies on adenosine and cytidine reactions with chloroacetaldehyde. A new support for the cyclic carbinolamine structure of the stable reaction intermediate and its relevance to the reaction mechanism and tRNA modification. *Polish J. Chem.*, **53**, 243-252

Krzyżosiak, W.J., Jaskólski, M., Sierzputowska-Gracz, H. & Wiewiórowski, M. (1982) Comparative structural analysis of cytidine, ethenocytidine and their protonated salts. II. IR spectral studies. *Nucleic Acids Res.*, **10**, 2741-2753

Krzyżosiak, W.J., Biernat, J., Ciesiołka, J., Górnicki, P. & Wiewiórowski, M. (1983) Comparative studies on the reactions of adenosine and cytidine with chloroacetaldehyde, α-bromopropionaldehyde and chloroacetone. Synthesis of ethenoadenosine and ethenocytidine derivatives methylated at the etheno bridge. *Polish J. Chem.*, **57**, 779

Rich, A. & RajBhandary, U.L. (1976) Transfer RNA: molecular structure, sequence and properties. *Ann. Rev. Biochem.*, **45**, 805-860

Schulman, L.H. & Pelka, H. (1976) Location of accessible bases in *E. coli* formylmethionine transfer RNA as determined by chemical modification. *Biochemistry*, **15**, 5769-5775

Sussman, J.L., Holbrook, S.R., Wade, W.R., Church, G.M. & Kim, S.H. (1978) Crystal structure of yeast phenylalanine tRNA. I. Crystalographic refinement. *J. mol. Biol.*, **123**, 607-630

CYCLIC ADDUCT FORMATION AT STRUCTURAL PERTURBATIONS IN SUPERCOILED DNA MOLECULES

D.M.J. LILLEY

Department of Biochemistry, Medical Sciences Institute, The University, Dundee, United Kingdom

SUMMARY

Supercoiled DNA can stabilize the existence of micropolymorphic structures, which may be sequence dependent. A good example is cruciform formation from inverted repeat sequences. DNA sequences exhibiting a high degree of two-fold symmetry, for instance, the 13-bp inverted repeat of ColE1, can adopt cruciform geometry when supercoiled, as judged by sensitivity to single-strand-specific nucleases and to resolvase and by topological changes revealed by gel electrophoresis. The present report shows that these structures are also excellent substrates for the local formation of cyclic adducts by bromoacetaldehyde, glyoxal or osmium tetroxide. Individual reagents may show pronounced differences in sequence selectivity, although all appear to react strongly at the ColE1 sequence. The reactions initially occur at the centre of the inverted repeats; this is followed by a conformational change such that the modification propagates outwards from the centre. These compounds constitute a valuable addition to the list of probes available for the investigation of local perturbations in DNA structure.

INTRODUCTION

Fibre diffraction studies of DNA structure have revealed two principal types of structure for double-helical DNA, the A and the B structure (Langridge *et al.*, 1960; Fuller *et al.*, 1965). Recent studies using a variety of experimental approaches, the most significant being single crystal X-ray diffraction, have shown that at a level of detail corresponding to individual torsion angles, the picture is very much more complex. To take the example which has been studied the most, i.e., the dodecamer analysed by Dickerson and co-workers (Wing *et al.*, 1980; Dickerson & Drew, 1981), individual torsion angles may vary from nucleotide to nucleotide by 100° or more, and even the ribose pucker is not immune. Out of the twelve ribose rings, only four are found to be C_2'-endo, long held to be the definitive pointer to the B-conformation. Perhaps this should not be so surprising; what it really tells us is that there is no completely general DNA structure, and that at the level of the single nucleotide the structure is decidedly polymorphic; i.e., the three-dimensional structure of the DNA reflects the underlying nucleotide base sequence. This is rather satisfying from a biological point of view since it implies that particular sequences may well have self-advertising DNA conformations which can be recognized by specific proteins.

Sequence-dependent DNA structure may be taken to greater extremes. Under certain conditions, alternating purine-pyrimidine sequences may change handedness to adopt the now-familiar Z conformation (Wang *et al.*, 1979). Inverted repeat sequences, sometimes referred to as palindromes, may form an even greater perturbation in which the base-pairing is reorganized. Within the sequences related by the two-fold symmetry, intrastrand base-pairing is possible to form what is most frequently called a cruciform (Platt, 1955; Gierer, 1966; Gellert *et al.*, 1979; Lilley, 1980; Panayotatos & Wells, 1981; Lyamichev *et al.*, 1983). Structural variants which cause local alteration in the DNA twist will be affected by an additional factor in DNA structure, namely, DNA supercoiling. Most natural DNA molecules are negatively supercoiled; i.e., the DNA is underwound (Vinograd & Lebowitz, 1966). This torsional stress in the molecule may be relieved, totally or partially, by a structural transformation resulting in a reduction of twist, such as the adoption of Z conformation by a $(C-G)_n$ stretch or cruciform extrusion by an inverted repeat. The relaxation of the stress of supercoiling leads to a lowering of free energy and hence such structures are favoured in negatively supercoiled DNA.

The potential biological significance of both structural microheterogeneity and DNA supercoiling has prompted a detailed investigation of cruciform structures in this laboratory. A variety of probes have been developed including single-strand-specific nucleases such as S1 nuclease (Lilley, 1980, 1981; Panayotatos & Wells, 1981), micrococcal nuclease (Dingwall *et al.*, 1981; Lilley 1983a) and BAL-31 nuclease (L. R. Hallam & D.M.J. Lilley, in preparation), restriction nucleases (Mizuuchi *et al.*, 1982b; Lilley & Hallam, 1983) and resolving enzyme (Mizuuchi *et al.*, 1982a; Lilley & Kemper, 1984). Probing methods coupled with topological studies (Lilley & Hallam, 1984) have revealed a consistent picture of the cruciform extruded as a well-defined structure having moderate free energy of formation. Enzyme probes, despite their invaluable selectivity, do suffer from some disadvantages. Firstly, they are bulky molecules which may experience steric problems which reduce their accessibility to certain structural features. Secondly, there is a potential theoretical problem which applies to enzyme probes in general, and particularly to single-strand-specific nucleases. Since the enzyme must necessarily recognize a given structural feature, it must possess a binding domain with an affinity for the feature in question. The binding process may influence the formation of the structure either thermodynamically or kinetically. We have therefore sought alternative structure-specific probes, which should be small molecules. Such low-molecular-weight chemical probes would not have a binding domain and could only react opportunistically with available nucleic acid reactive centres. We have studied three such compounds, bromoacetaldehyde (Lilley, 1983b), osmium tetroxide (Lilley & Paleček, 1984) and glyoxal, each of which reacts selectively with DNA bases possessing single-stranded character. We find that all these probes give rise to selective modification of nucleotides in and around cruciform structures, and the use of these compounds both confirms and extends results obtained with enzymes. These reagents are valuable additions to the battery of probes now available for exploring new structural perturbations in DNA.

MATERIALS AND METHODS

Plasmids

pColIR215 is a 4427-bp (base-pair) recombinant molecule derived (Lilley, 1981) by cloning a 440-bp fragment of ColE1, containing the ColE1 13-bp inverted repeat known to have high cruciform potential, between the *Eco*R1 and *Bam*H1 sites of pBR322. pColIR515 was derived (Lilley & Hallam, 1983) from pColIR215 by deletion of sequence between the *Bam*H1 and *Pvu*II sites to give a 2735-bp plasmid containing the ColE1 inverted repeat. pIRbke8 was constructed (Lilley & Markham, 1983) by cloning two synthetic 13-base oligonucleotides into the *Bam*H1 site of pAT153. All plasmids were prepared from chloramphenicol-amplified log phase *Escherichia coli* K12 HB101 cells by lysozyme/EDTA/Triton X-100 lysis and caesium chloride/ethidium bromide density banding (Clewell & Helinski, 1969).

Pure topoisomers

Individual topoisomers of pColIR515 were prepared by partial relaxation of supercoiled plasmid by chicken reticulocyte topoisomerase I in the presence of ethidium bromide, agarose gel electrophoresis in the presence of chloroquine, and excision and electroelution of topoisomer bands (Lilley & Hallam, 1984).

Bromoacetaldehyde

Bromoacetaldehyde was prepared according to Secrist *et al.* (1972) and Kohwi-Shigematsu *et al.* (1983) and used as a 1M stock solution. Supercoiled DNA was reacted with bromoacetaldehyde in 50 mM sodium acetate pH 4.5 at 37 °C.

Glyoxal

Glyoxal (Trimeric, dihydrate) reactions were performed in 18 mM Tris-HCl pH 7.8, 18 mM boric acid, 0.5 mM EDTA at 37 °C.

Osmium tetroxide

Osmium tetroxide was dissolved in distilled water to give an 8 mM stock solution. Supercoiled DNA was reacted with osmium tetroxide in 10 mM Tris-HCl pH 7.8, 1 mM EDTA, 1% (v/v) pyridine at 25 °C.

Enzyme reactions

S1 nuclease reactions were performed in 50 mM sodium acetate pH 4.6, 50 mM sodium chloride, 1 mM $ZnCl_2$. For excision of chemically modified regions of DNA, the DNA was incubated with 5 units S1 nuclease at 37 °C for 60 min. For S1 cleavage of unmodified supercoiled DNA, lower temperatures (typically 15 °C) were used in general. 5'-termini of DNA molecules were radioactively labelled with ^{32}P using T_4 polynucleotide kinase and either the exchange-labelling procedure (Berkner & Folk, 1977) or direct labelling of dephosphorylated termini (Maxam & Gilbert, 1980). Topoisomerase reactions were performed in 10 mM Tris-HCl pH 8.0, 200 mM NaCl, 10 mM EDTA at 25 °C. Restriction enzymes (New England Biolabs or Bethesda Research Labs) were used in accordance with manufacturers' instructions.

Electrophoresis

Agarose and polyacrylamide gel electrophoresis was performed in TBE buffer (90 mM Tris-HCl pH 8.3, 90 mM boric acid, 10 mM EDTA) at ambient temperature. When required for autoradiography, gels were dried on 3 mM paper (Whatman) and exposed to Kodak X-omat S film at −70 °C.

RESULTS

The most well-studied inverted repeat in this laboratory is the ColE1 sequence, present in pColIR215 and derivatives thereof. We have also made considerable use of a synthetic inverted repeat present in pIRbke8. The maps of these plasmids, and the sequences of the inverted repeats, are shown in Figure 1. Except where otherwise stated, the following studies have been performed on plasmid DNA at its native superhelix density, i.e., $\sigma = -0.06$.

Bromoacetaldehyde

The haloacetaldehydes react with the exocyclic amino group and the N-1 or N-3 of adenine and cytosine, respectively, to generate $1,N^6$-ethenoadenine or $3,N^4$-ethenocytosine derivatives (Kochetkov et al., 1971; Secrist et al., 1972). Bromoacetaldehyde reacts faster than the chloro analogue (Kayasuga-Mikado et al., 1980) and was therefore employed in these studies. Since the reaction proceeds at the sites of hydrogen bonding between base pairs, the chemistry would be expected to be single-strand-selective, and this has proved to be the case experimentally (Kayasuga-Mikado et al., 1980). As a corollary to this, the DNA is left permanently unpaired following the modification reaction. The fixed 'bubble' may be recognized in some manner.

We have used the following three-stage protocol to explore site-selective modification by this and all other compounds:

(a) The supercoiled DNA is reacted with the single-strand-selective reagent.

Fig. 1. Circular restriction maps of pColIR215 and pIRbke8, with the positions of the ColE1 and synthetic inverted repeats, respectively, represented by filled boxes. The corresponding sequences are written below the maps.

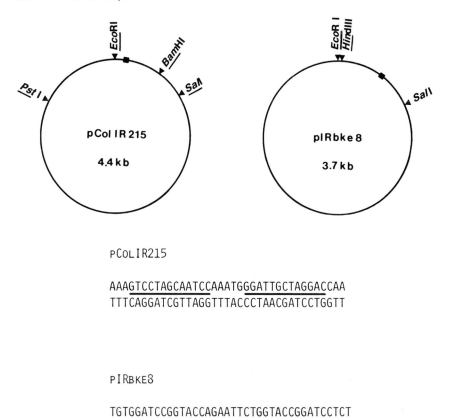

PCOLIR215

AAAGTCCTAGCAATCCAAATGGGATTGCTAGGACCAA
TTTCAGGATCGTTAGGTTTACCCTAACGATCCTGGTT

PIRBKE8

TGTGGATCCGGTACCAGAATTCTGGTACCGGATCCTCT
ACACCTAGGCCATGGTCTTAAGACCATGGCCTAGGAGA

(b) The modified DNA is cleaved by a restriction enzyme. The result of this reaction should be a linear (i.e., torsionally unstressed) molecule with an unpaired modified region(s).

(c) The linear DNA is incubated with S1 nuclease, which should digest the DNA rendered permanently unpaired.

The DNA is then examined by gel electrophoresis. If site-selective modification has occurred, then S1 nuclease should remove this sequence selectively, thereby generating discrete fragments which are smaller than full-length linear molecules; i.e., discrete bands should be visible on the gel. Figure 2 shows such an experiment. pColIR215 was reacted with bromoacetaldehyde at concentrations between 5 and 500 mM for 30 min at 37 °C. The DNA was completely cleaved by the restriction enzyme *Sal*I, and then digested by S1 nuclease. At the lowest bromoacetaldehyde concentration, no site-selective modification is apparent; only full-length linear plasmid is evident. At higher reagent concentrations, a new 4090-bp band appears,

Fig. 2. Site-selective modification of supercoiled pColIR215 by bromoacetaldehyde. Plasmid was incubated with the reagent at 37 °C, linearized with SalI and then incubated with S1 nuclease. The resulting DNA fragments were electrophoresed in a 1% agarose gel. Track 1, supercoiled pColIR215; Track 2 to 6, pColIR215 modified by 5, 15, 50. 150 and 500 mM bromoacetaldehyde, respectively; Track 7, phage PM2 DNA digested by HindIII as marker fragments (lengths in base pairs given on the right-hand side. L and S denote full-length linear and supercoiled pColIR215, respectively. Bands corresponding to site-selective modification of the ColE1 cruciform are indicated by the large arrows on the right-hand side.

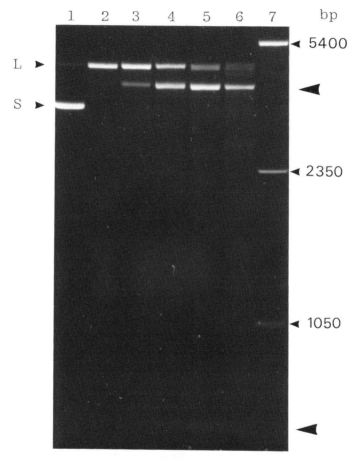

due to modification and subsequent excision of DNA at the ColE1 inverted repeat. This site has been mapped unambiguously using several restriction enzymes of different target sequence (Lilley, 1983b). At the highest bromoacetaldehyde concentration, some over-reaction occurred and some smearing is visible in this track. However, at intermediate concentrations, the modification reaction is extremely site-selective. Since this same sequence is hypersensitive to S1 nuclease (Lilley, 1981), micrococcal nuclease (Lilley, 1983a), BAL 31 nuclease (Lilley & Hallam, 1984), and T_4 endonuclease VII (Lilley & Kemper, 1984) and exhibits reduced sensitivity to

Fig. 3. Distribution of modification around the synthetic inverted repeat of pIRbke8. Supercoiled pIRbke8 was modified by 50 mM bromoacetaldehyde and nuclease-digested before labelling 5′-termini with ^{32}P. The resulting 5% polyacrylamide gel was autoradiographed. Track 1, phage φX174 replicative form DNA cleaved by HinfI to give marker fragments (lengths given in base pairs); Track 2, pIRbke8 incubated with S1 nuclease, followed by HindIII cleavage; Track 3, bromoacetaldehyde-modified pIRbke8.

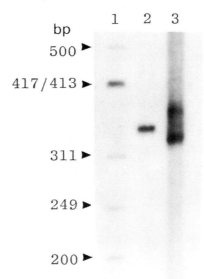

AvaII (Lilley & Hallam, 1983) when supercoiled, we conclude that it is cruciform formation by the inverted repeat which causes the elevated sensitivity to bromoacetaldehyde. This hypersensitivity is not present in either linear or relaxed pColIR215, implying a supercoil-dependence. The synthetic inverted repeat of pIRbke8 is similarly selectively modified by bromoacetaldehyde, consistent with cruciform extrusion by this sequence. It should be noted that there is no similarity between these two sequences, beyond the two-fold symmetry.

At the level of resolution of an agarose gel, the bands resulting from modification appear very narrow, comparable in fact to an S1 nuclease experiment on the same plasmid. In order to study this in finer detail, short restriction-site/modification-site fragments were visualized by [5′-^{32}P]-labelling and autoradiography. An example is shown in Figure 3 for the synthetic inverted repeat of pIRbke8. At this level of resolution, it is clear that the S1-nuclease-sensitive modification extends over a wider region than the S1-nuclease-hypersensitivity alone; in fact, it extends over the entire region of the inverted repeat and 10 or 20 nucleotides into the flanking DNA. Thus, there is a clear contrast between the precision of S1 nuclease cleavage in supercoiled DNA, and the broader target for bromoacetaldehyde modification. Similar results have been obtained for the ColE1 inverted repeat. When the S1 nuclease cleavage of supercoiled pColIR215 was performed at 37 °C, however, a small but significant change was found in the pattern of fragments generated. This is best seen in the

Fig. 4. Comparison of S1 nuclease and bromoacetaldehyde patterns of attack in supercoiled pColIR215 at 37 °C. Supercoiled pColIR215 was cleaved by S1 nuclease, or modified with 50 mM bromoacetaldehyde and then cleaved by *Sal*I. Samples were electrophoresed on a 5% polyacrylamide gel and the photographic negative was scanned densitometrically. Upper scan (S1), S1 nuclease; lower scan (BA), bromoacetaldehyde. The scale of the schematic is matched to that of the scans, and the position of the ColE1 inverted repeat is indicated by the filled box.

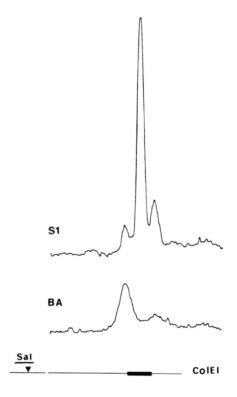

densitometric scans shown in Figure 4. Small satellite bands have appeared flanking the major central cleavage by S1 nuclease, and these closely correspond to the most strongly sensitive modified positions in the sequence. This will be discussed further below.

Glyoxal

Glyoxal is another compound which reacts with nucleic acids in a single-strand-selective manner. Reaction occurs selectively at guanine bases, forming a cyclic adduct between N-1 and the exocyclic nitrogen (Broude & Budowsky, 1971). As with bromoacetaldehyde, the reaction is single-strand-selective since hydrogen-bonding functions are modified. The actual target nucleotide is, of course, different. Moreover, the reaction may be performed at pH 7.8, rather than the low pH at which the bromoacetaldehyde reaction was carried out. The protocol of modification, restriction enzyme cleavage and S1 nuclease cleavage remains the same. An example

Fig. 5. Site-selective modification of supercoiled pColIR215 by glyoxal. Plasmid was incubated with 220 mM glyoxal at 37 °C for 50 min, cleaved with restriction enzyme and incubated with S1 nuclease, before electrophoresis in 1% agarose. Track 1, supercoiled pColIR215; Tracks 2 and 3, phage λ DNA incubated with HindIII and Ava1, respectively (lengths of resulting marker fragments are given in base pairs); Tracks 4 to 6, glyoxal-modified pColIR215 cleaved with BamH1, SalI or NruI, respectively, before S1 nuclease digestion. Full-length linear and supercoiled pColIR215 are denoted by L and S, respectively.

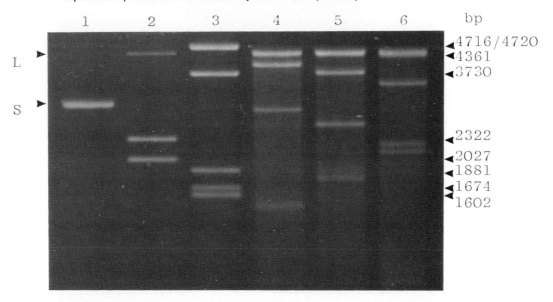

is shown in Figure 5. Two sites of hypersensitivity to glyoxal modification are present in supercoiled pColIR215. One is the ColE1 inverted repeat, confirming the abnormal structure of this sequence in the supercoiled plasmid. The second site has not yet been identified precisely, but is most likely to be the minor cruciform of pBR322. Close examination of the shorter fragments arising from modification at the ColE1 inverted repeat shows that they migrate as doublets. This is reminiscent of the behaviour of the fragments from bromoacetaldehyde reactions and, as we shall see, from osmium tetroxide modification as well.

Osmium tetroxide

Osmium tetroxide reacts selectively with single-stranded nucleic acids (Hodgson, 1977; Marzilli, 1977; Lukasova *et al.*, 1984). The chemistry is, however, quite different from that discussed above for bromoacetaldehyde and glyoxal. Osmium tetroxide adds across the unsaturated 5,6 bond of thymine or cytosine to generate a *cis* ester. There is no reaction with the hydrogen-bonding functions of the bases and it is not obvious at first sight why single-strand-selectivity should be manifested. This probably arises from a combination of steric and electronic effects, since the osmium cyclic adduct is not coplanar with the pyrimidine ring (Neidle & Stuart, 1976) and the latter loses aromaticity. Reaction of pColIR215 with 1 mM osmium tetroxide at

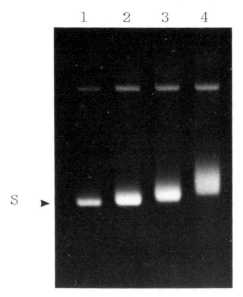

Fig. 6. The mobility of supercoiled pColIR215 in agarose is retarded upon reaction with osmium tetroxide. pColIR215 was incubated with 1 mM osmium tetroxide at 25 °C for various times and electrophoresed in 1% agarose. Tracks 1 to 4 correspond to 0, 3, 10 and 20 min reaction, respectively, with osmium tetroxide. S denotes the position of unreacted supercoiled plasmid.

25 °C leads to a progressive lowering in the mobility of the supercoiled plasmid in agarose, as shown in Figure 6. This suggests that a 'bubble' of unpaired DNA is being generated and fixed by the reaction with osmium tetroxide, leading to a topological relaxation of the supercoiled DNA. Using a modification, restriction and S1 nuclease protocol as before, we tested for site-selective modification. Figure 7 shows such an experiment employing *Ava*1, *Bam*H1 and *Sal*1 restriction enzymes. Only one major site of modification is revealed, corresponding to modification at the ColE1 inverted repeat. Just as with the other reagents, the structure of the shorter fragment bands indicates doublet formation – this is well illustrated by the ~1390-bp band arising from *Ava*1 cleavage. To examine this point in greater detail, supercoiled pColIR215 was reacted with 3.2 mM osmium tetroxide and, following the usual cleavage reactions, labelled with ^{32}P; the ~340-bp *Bam*H1 modification site fragments were then examined by electrophoresis and autoradiography. The results of this experiment are shown in Figure 8. It may be seen that the initial reaction is primarily at the centre of the ColE1 inverted repeat, but that longer reaction time leads to a shortening of this fragment.

The degree of supercoiling

We know that supercoiling is essential to the stability of cruciform structures, and that linear or totally relaxed DNA is not modified by bromoacetaldehyde or osmium tetroxide. To define the limit more precisely, we have purified individual topoisomers

Fig. 7. Site-selective modification of pColIR215 by osmium tetroxide. Supercoiled pColIR215 was reacted with 1.6 mM osmium tetroxide at 25°C before restriction cleavage and S1 nuclease digestion and agarose gel electrophoresis. Track 1, supercoiled pColIR215; Tracks 2 to 4, osmium-tetroxide-modified pColIR215 cleaved by AvaI, BamHI and SalI, respectively; Track 5, S1-nuclease-incubated pColIR215 cleaved with SalI; Track 6, phage PM2 DNA digested by HindIII (lengths of resulting marker fragments are given in base pairs). S and L denote supercoiled and full-length linear pColIR215, respectively.

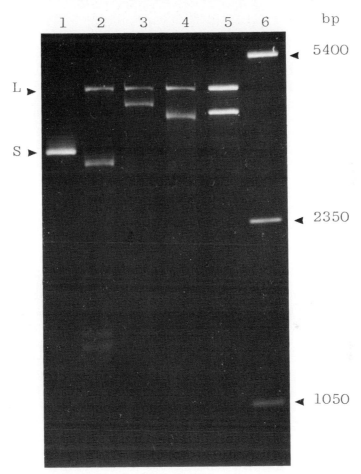

of plasmid pColIR515 bearing a small ColE1 inverted repeat. When these were separately tested for site-selective modification by osmium tetroxide, the results shown in Figure 9 were obtained. Topoisomers having a linkage difference (ΔLk) of -13 or greater (i.e., lower supercoiling; $\sigma < -0.050$) were completely unmodified by the reagent but, as the linkage was progressively reduced, the hypersensitivity rapidly rose and was maximal at $\Delta Lk = -15$ ($\sigma > -0.057$). This behaviour was exactly paralleled by sensitivity to S1 nuclease, BAL 31 and T4 endonuclease VII and by writhing changes revealed by polyacrylamide-agarose gel electrophoresis (Lilley &

Fig. 8. Pattern of osmium-tetroxide modification about the ColE1 inverted repeat of pColIR215 as a function of time. Supercoiled pColIR215 was incubated with 3.2 mM osmium tetroxide at 25 °C, cleaved with BamHI, and digested with S1 nuclease. The DNA fragments were [5'-^{32}P]-labelled and electrophoresed on a 5% polyacrylamide gel; the short fragments were located by autoradiography. Track 1, phage φX174 replicative form DNA cleaved by HinfI (lengths of resulting marker fragments are given in base pairs); Track 2, S1-nuclease-incubated pColIR215 cleaved by BamHI; Tracks 3 and 4, pColIR215 modified by osmium tetroxide for 3 and 20 min, respectively, before BamHI and S1 nuclease cleavage reactions.

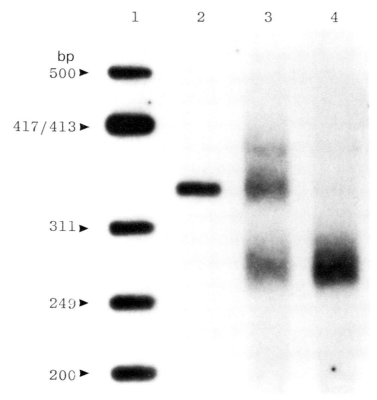

Hallam, 1984). The critical ΔLk observed was corrected for temperature effects to −13.2 at 37 °C and this value, in conjunction with a helical twist change (ΔTw) measured by gel electrophoresis, allows a calculation of the free energy of formation of the ColE1 cruciform of $\Delta G = 18.4 \pm 0.5$ kcal mole^{-1}.

DISCUSSION

It is clear that the ColE1 cruciform may serve as a selective target for adduct formation by bromoacetaldehyde, glyoxal and osmium tetroxide, when stabilized by negative supercoiling. Indeed, these reagents provide valuable confirmatory evidence

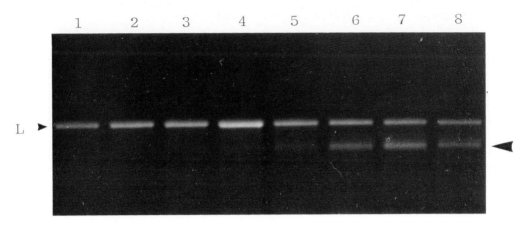

Fig. 9. Osmium-tetroxide modification of single topoisomers of pColIR515. Single topoisomer preparations were reacted with 1.6 mM osmium tetroxide at 25 °C for 10 min, and then cleaved by *Bam*HI and S1 nuclease. Site-selective modification results in the generation of the fragment the migration of which in 1% agarose is indicated by the arrow on the right-hand side. Full-length linear pColIR515 is denoted by L. Tracks 1 to 8 correspond to analogous experiments on topoisomers having linking differences of −10 through to −17, respectively.

for the existence of this cruciform, in a manner which is free of potential artefacts arising from the binding of enzyme probes. The conditions under which the chemical probes are employed differ widely in terms of buffer composition and pH, and the mechanisms of reaction of bromoacetaldehyde and osmium tetroxide are totally unrelated.

A common feature observed for all three chemical probes is the generation of double bands on gel electrophoresis – what could be their origin? Two explanations are possible. First, the structure of the DNA flanking the cruciform may be perturbed in some manner, leading to elevated reactivity towards chemical modification. This possibility gains some support for the ColE1 inverted repeat in view of the coincidence of bromoacetaldehyde modification peaks and S1 nuclease satellite cleavages. Second, initial modification of the bases in the inverted repeat may lead to conformational changes which create new sites of enhanced reactivity. It should be borne in mind that there is a fundamental distinction to be made between nuclease cleavage experiments (with, for example, S1 nuclease or T_4 endonuclease VII) and chemical modification experiments. The former are self-limiting in that the first cleavage introduced releases the torsional constraint, and thus destabilizes the cruciform. By contrast, chemical modification leaves the backbone of the supercoiled DNA intact; i.e., the molecule remains torsionally stressed and still subject to chemical reaction. Indeed, modification of residues may predispose neighbouring bases to adduct formation. It is therefore quite possible to envisage a mechanism whereby the initial modification event is in the loop of the cruciform, as the results using osmium tetroxide suggest; this would induce a change in the conformation of the molecule to a local 'bubble', which may then expand, driven by the supercoiling, as further modification to contiguous nucleotides occurs. To explain the observed results, it is necessary to propose that this 'bubble expansion' is not a smooth process, but that specific pause points are encountered, thereby generating the distinctive pattern of bands which are seen. Of course, the central factor remains; namely, it is the cruciform structure which is responsible for hypersensitivity in the local region of DNA.

How sequence-dependent are these chemical probes? We have observed differences in the reactivity of these reagents towards different inverted repeats. For instance, whilst all three react strongly with the ColE1 inverted repeat, bromoacetaldehyde reacts with pIRbke8 whereas osmium tetroxide does not. These differences require further investigation, but may to some extent be reflections of the kinetic properties of different cruciforms. The ColE1 cruciform extrudes at lower ionic strength than that of pIRbke8 (Lilley, 1985), and hence the differences may be partially due to different buffer conditions between reactions.

How general are these studies? Clearly these probes are of considerable utility for structural studies of cruciforms, but what of other structures? The B–Z conformation is sensitive to nucleases (Singleton *et al.*, 1984) and might therefore be worthy of investigation by chemical probes. Many enzymes and chemical probes have been applied to the study of eukaryotic genes both in chromatin and DNA (Wu, 1980; Cartwright & Elgin, 1982; Hentschel, 1982; Jesse *et al.*, 1982; Larsen & Weintraub, 1982; Goding & Russell, 1983; Nickol & Felsenfeld, 1983; Mace *et al.*, 1983; Weintraub, 1983; McKeon *et al.*, 1984). Kohwi-Shigematsu *et al.* (1983) have shown that chicken globin genes possess regions of enhanced reactivity towards bromoacetaldehyde, and Glikin *et al.* (1984) have shown that a supercoiled plasmid clone of *Drosophila* histone genes contains a site of strong reactivity towards osmium tetroxide in the spacer between the H1 and H3 genes. We have recently studied a sequence – $(A-T)_{34}$ – present in the first intron of a *Xenopus* globin gene (Greaves *et al.*, 1985). When supercoiled to a moderate extent, this sequence adopts a cruciform structure, as revealed by S1 nuclease, BAL 31, T_4 endonuclease VII and band shift methods. The sequence is extremely reactive towards bromoacetaldehyde and osmium tetroxide.

In summary, the application of single-strand-selective chemical probes has been very useful in the study of cruciform structures present in supercoiled DNA. Experience gained with these relatively well-understood systems will be valuable as the methods are applied to systems of lower structural definition but perhaps greater biological interest.

ACKNOWLEDGEMENTS

I thank Gerald Gough for performing the glyoxal modification experiments, and Emil Paleček (Biophysics Institute, Brno, Czechoslovakia) for valuable collaboration on osmium tetroxide modification studies. I also thank Terumi Kohwi-Shigematsu and Hal Weintraub for provision of a bromoacetaldehyde protocol before publication. The financial support of the Medical Research Council and the Royal Society is gratefully acknowledged.

REFERENCES

Berkner, K.L. & Folk, W.R. (1977) Polynucleotide kinase exchange reaction. *J. biol. Chem., 252,* 3176–3184

Broude, N.E. & Bodowsky, E.I. (1971) The reaction of glyoxal with nucleic acid components. III. Kinetics of the reaction with monomers. *Biochem. biophys. Acta., 254,* 380–388

Cartwright, I.L. & Elgin, S.C.R. (1982) Analysis of chromatin structure and DNA sequence organisation: use of the 1,10-phenanthroline-cuprous complex. *Nucleic Acids Res., 10,* 5835–5852

Clewell, D.B. & Helinski, D. (1969) Supercoiled circular DNA-protein complex in *Escherichia coli:* purification and induced conversion to an open circular DNA form. *Proc. natl Acad. Sci. USA, 62,* 1159–1166

Dickerson, R.E. & Drew, H.R. (1981) Structure of B-DNA dodecamer. II. Influence of base sequence on helix structure. *J. mol. Biol., 149,* 761–786

Dingwall, C., Lomonossoff, G.P. & Laskey, R.A. (1981) High sequence specificity of micrococcal nuclease. *Nucleic Acids Res., 9,* 2659–2673

Fuller, W., Wilkins, M.F.H., Wilson, H.R. & Hamilton, L.D (1965) The molecular configuration of deoxyribonucleic acid. IV. X-ray diffraction study of the A form. *J. mol. Biol., 12,* 60–80

Gellert, M., Mizuuchi, K., O'Dea, M.H., Ohmori, H. & Tomizawa, J. (1979) DNA gyrase and DNA supercoiling. *Cold Spring Harbor Symp. quant. Biol., 43,* 35–40

Gierer, A. (1966) A model for DNA-protein interactions and the function of the operator. *Nature, 212,* 1480–1481

Glikin, G.C., Vojtišková, M., Rena-Descalzi, L. & Paleček, E. (1984) Osmium tetroxide: a new probe for site-specific distortions in supercoiled DNA. *Nucleic Acids Res., 12,* 1725–1735

Goding, C.R. & Russell, W.C. (1983) S1 sensitive sites in adenovirus DNA. *Nucleic Acids Res., 11,* 21–36

Greaves, D.R., Patient, R.K. & Lilley, D.M.J. (1985) Facile cruciform formation by an (A–T)$_{34}$ sequence from a *Xenopus* globin gene. *J. mol. Biol., 185,* 461–478

Hentschel, C.C. (1982) Homopolymer sequences in the spacer of a sea urchin histone gene repeat are sensitive to S1 nuclease. *Nature, 295,* 714–716

Hodgson, D.J. (1977) Stereochemistry of metal complexes of nucleic acid constitutents. *Proc. inorg. Chem., 23,* 211–254

Jesse, B., Garguilo, G., Ravzi, F. & Worcel, A. (1982) Analogous cleavage of DNA by micrococcal nuclease and a 1,10-phenanthroline-cuprous complex. *Nucleic Acids Res., 10,* 5823–5834

Kayasuga-Mikado, K., Hashimoto, T., Negishi, T., Negishi, K. & Kayatsu, H. (1980) Modification of adenine and cytosine derivatives with bromoacetaldehyde. *Chem. pharm. Bull., 28,* 932–938

Kochetkov, N.K., Shabaev, V.N. & Kost, A.A. (1971) New reaction of adenine and cytosine derivatives, potentially useful for nucleic acid modification. *Tetrahedron Lett., 22,* 1993–1996

Kohwi-Shigematsu, T., Gelinas, R. & Weintraub, H. (1983) Detection of an altered DNA conformation at specific sites in chromatin and supercoiled DNA. *Proc. natl Acad. Sci. USA,* **80,** 4389–4393

Langridge, R., Marvin, D.A., Seeds, W.E., Wilson, H.R., Hooper, C.W., Wilkins, M.F.H. & Hamilton, L.D. (1960) The molecular configuration of deoxyribonucleic acid. II. Molecular models and their Fourier transforms. *J. mol. Biol.,* **2,** 38–64

Larsen, A. & Weintraub, H. (1982) An altered DNA conformation detected by S1 nuclease occurs at specific regions in active chick globin chromatin. *Cell,* **29,** 609–622

Lilley, D.M.J. (1980) The inverted repeat as a recognisable structural feature in supercoiled DNA molecules. *Proc. natl Acad. Sci. USA,* **77,** 6468–6472

Lilley, D.M.J. (1981) Hairpin-loop formation by inverted repeats in supercoiled DNA is a local and transmissible property. *Nucleic Acids Res.,* **9,** 1271–1289

Lilley, D.M.J. (1983a) Dynamic, sequence-dependent DNA structure as exemplified by cruciform extrusion from inverted repeats in negatively supercoiled DNA. *Cold Spring Harbor Symp. quant. Biol.,* **47,** 101–112

Lilley, D.M.J. (1983b) Structural perturbation in supercoiled DNA: hypersensitivity to modification by a single-strand-selective chemical reagent conferred by inverted repeat sequences. *Nucleic Acids Res.,* **11,** 3097–3112

Lilley, D.M.J. (1985) The kinetic properties of cruciform extrusion are determined by DNA base-sequence. *Nucleic Acids Res.,* **13,** 1443–1465

Lilley, D.M.J. & Hallam, L.R. (1983) The interactions of enzyme and chemical probes with inverted repeats in supercoiled DNA. *J. biomol. Struct. Dyn.,* **1,** 169–182

Lilley, D.M.J. & Hallam, L.R. (1984) Thermodynamics of the ColE1 cruciform. Comparisons between probing and topological experiments using single topoisomers. *J. mol. Biol.,* **180,** 179–200

Lilley, D.M.J. & Kemper, B. (1984) Cruciform-resolvase interactions in supercoiled DNA. *Cell,* **36,** 413–422

Lilley, D.M.J. & Markham, A.F. (1983) Dynamics of cruciform extrusion in supercoiled DNA: use of a synthetic inverted repeat to study conformational populations. *EMBO J.,* **2,** 527–533

Lilley, D.M.J. & Paleček, E. (1984) The supercoil-stabilised cruciform of ColE1 is hyper-reactive to osmium tetroxide. *EMBO J.,* **3,** 1187–1192

Lukasova, E., Vojtiskova, M., Jelen, F., Sticzay, T. & Paleček, E. (1984) Osmium-induced alteration in DNA structure. *Gen. Physiol. Biophys.* **3,** 175–191

Lyamichev, V.I., Panyutin, I.G. & Frank-Kamenetskii, M.D. (1983) Evidence of cruciform structures in superhelical DNA provided by two-dimensional gel electrophoresis. *FEBS Lett.,* **153,** 298–302

Mace, H.A.F., Pelham, H.R.B. & Travers, A.A. (1983) Association of an S1 nuclease-sensitive structure with short direct repeats 5′ of *Drosophila* heat shock genes. *Nature,* **304,** 555–557

Marzilli, L.G. (1977) Metal ion interactions with nucleic acid and nucleic acid derivatives. *Progr. inorg. Chem.,* **23,** 255–378

Maxam, A.M. & Gilbert, W. (1980) Sequencing end-labelled DNA with base-specific chemical cleavages. *Methods Enzymol.,* **65,** 499–560

McKeon, C., Schmidt, A. & deCrombrugghe, B. (1984) A sequence conserved in both chicken and mouse α 2(I) collagen promoter contains sites sensitive to S1 nuclease. *J. biol. Chem.*, **259,** 6636–6640

Mizuuchi, K., Kemper, B., Hays, J. & Weisberg, R.A. (1982a) T4 endonuclease VII cleaves Holliday structures. *Cell, 29,* 357–365

Mizuuchi, K., Mizuuchi, M. & Gellert, M. (1982b) Cruciform structures in palindromic DNA are favored by DNA supercoiling. *J. mol. Biol., 156,* 229–243

Neidle, S. & Stuart, D.I. (1976) The crystal and molecular structure of an osmium bispyridine adduct of thymine. *Biochim. biophys. Acta., 418,* 226–231

Nickol, J.M. & Felsenfeld, G. (1983) DNA conformation at the 5' end of the chicken adult β-globin gene. *Cell, 35,* 467–477

Panayotatos, N. & Wells, R.D. (1981) Cruciform structures in supercoiled DNA *Nature, 289,* 466–470

Platt, J.R. (1955) Possible separation of inter-twined nucleic acid chains by transfer-twist. *Proc. natl. Acad. Sci. USA, 41,* 181–183

Secrist, J.A., Barrio, J.R., Leonard, N.J., Villar-Palasi, C. & Gilman, A.G. (1972) Fluorescent modification of adenosine 3'5' monophosphate: spectroscopic properties and activity in enzyme systems. *Science, 177,* 279–280

Singleton, C.K., Kilpatrick, M.W. & Wells, R.D. (1984) S1 nuclease recognises DNA conformational junctions between left-handed helical (dT-dG)n, (dC-dA)n and contiguous right-handed sequences. *J. biol. Chem., 259* 1963–1967

Vinograd, J. & Lebowitz, J. (1966) Physical and topological properties of circular DNA. *J. gen. Physiol., 49,* 103–125

Wang, A.H.-J., Quigley, G.J., Kolpak, F.J., Crawford, J.L. van Boom, J.H. van der Marel, G. & Rich, A. (1979) Molecular structure of a left-handed double helical DNA fragment at atomic resolution. *Nature, 282,* 680–686

Weintraub, H. (1983) A dominant role for DNA secondary structure in forming hypersensitive structures in chromatin. *Cell, 32,* 1191–1203

Wing, R., Drew, H., Takano, T., Broka, C., Tanaka, S., Itakura, K. & Dickerson, R.E. (1980) Crystal structure analysis of a complete turn of B-DNA. *Nature, 287,* 755–758

Wu, C. (1980) The 5' ends of *Drosophila* heat shock genes in chromatin are sensitive to DNaseI. *Nature, 286,* 854–860

THE ROLE OF CYCLIC BASE ADDUCTS IN VINYL-CHLORIDE-INDUCED CARCINOGENESIS: STUDIES ON NUCLEIC ACID ALKYLATION *IN VIVO*

R.J. LAIB

*Institut für Arbeitsphysiologie, University of Dortmund,
Dortmund 1, Federal Republic of Germany*

INTRODUCTION

Although there is now a considerable amount of evidence for the alkylation of nucleic acid moieties by reactive metabolites of vinyl chloride (VC), the role of these alkylated moieties in VC-induced carcinogenesis has not yet been clearly defined.

Metabolic pathways for vinyl chloride in man and in laboratory animals (IARC, 1979; Laib, 1982) have been intensively investigated. Chloroethylene oxide (CEO) and chloroacetaldehyde (CAA), both reactive intermediates in VC metabolism, have been considered responsible for the toxicity and carcinogenicity observed. Both electrophilic metabolites react with nucleic acid bases *in vitro* and *in vivo*, and defined reaction products have been characterized. On the basis of the somatic mutation theory of carcinogenesis, attempts have been made to establish a link between these chemical reactions and the biological consequences.

ADDUCTS FORMED AFTER CHEMICAL REACTION OF NUCLEIC ACID BASES WITH CAA AND CEO

Targets for alkylation in nucleic acids are the adenine, cytosine, and guanine residues in synthetic polynucleotides, RNA, and DNA. Treatment of ribo- and deoxyribopolynucleotides with CAA results in formation of $3,N^4$-ethenocytosine and $1,N^6$-ethenoadenine, as well as their hydrated intermediates, $3,N^4$-(N^4-hydroxyethano)cytosine and $1,N^6$-(N^6-hydroxyethano)adenine (Barbin et al., 1981; Hall et al., 1981; Kuśmierek & Singer, 1982).

When DNA was treated with CAA, formation of $3,N^4$-ethenodeoxycytidine, $1,N^6$-ethenodeoxyadenosine, and N^2,3-ethenodeoxyguanosine could be shown (Green & Hathway, 1978; Oesch & Doerjer, 1982). It should be emphasized, however, that in double-stranded, helical DNA, prepared by hydroxyapatite chromatography and/or

S1-nuclease digestion, CAA did not react with adenine residues to yield $1,N^6$-ethenoadenine (Kimura et al., 1977). That positions involved in base-pairing are not reactive towards haloacetaldehydes is also supported by a study on nucleic acid alkylation by bromoacetaldehyde (Kayasuga-Mikado et al., 1980).

Chemical reaction of CEO with deoxyguanosine resulted in the formation of 7-(2-oxoethyl)deoxyguanosine as the single alkylation product (Scherer et al., 1981); in contrast, in the reaction of CAA with guanosine, $1,N^2$-ethenoguanosine and/or $N^2,3$-ethenoguanosine were formed (Sattsangi et al., 1977; Oesch & Doerjer, 1982). These results led to the formulation of a reaction scheme for the formation of $N^2,3$-ethenoguanine in which the condensation of the aldehyde function of CAA with the exocyclic amino group of guanine is the critical step (Oesch & Doerjer, 1982). The formation of 7-(2-oxoethyl)guanine is consistent with a bimolecular S_N2 mechanism of alkylation between CEO and the N-7 of guanine, one of the most nucleophilic centres in DNA (Roe et al., 1973). On this basis, it was suggested that the analysis of both adducts ($N^2,3$-ethenoguanine and 7-(2-oxoethyl)guanine) could provide information on the ratios of both metabolites (CAA and CEO) present and on their relative roles in VC-induced carcinogenicity (Oesch & Doerjer, 1982).

ADDUCTS FORMED AFTER INCUBATION OF NUCLEIC ACIDS WITH ^{14}C-VC AND RAT LIVER MICROSOMES

Principal alkylation products after incubation of RNA with ^{14}C-VC in a rat liver microsomal system are $1,N^6$-ethenoadenosine and $3,N^4$-ethenocytidine (Laib & Bolt, 1977, 1978). $1,N^6$-ethenodeoxyadenosine, $3,N^4$-ethenodeoxycitidine, and 7-(2-oxoethyl)guanine are readily formed upon in-vitro incubation of DNA with labelled VC and rat liver microsomes, with the guanine derivative being the major alkylation product (Laib et al., 1981). In accordance with the study of Kimura et al. (1977), only 7-(2-oxoethyl)guanine, but not the etheno derivatives, are formed when the DNA is protected from the denaturing effects of the microsomal suspension by a membrane permeable to CAA and CEO (Laib et al., 1981).

ADDUCTS FORMED AFTER EXPOSURE OF LABORATORY ANIMALS TO VC

Radioactive $1,N^6$-ethenoadenosine, $3,N^4$-ethenocytidine, and 7-(2-oxoethyl)-guanine could be characterized in RNA hydrolysates from rat liver after exposure of the animals to ^{14}C-VC (Laib & Bolt, 1977, 1978; Laib et al., 1981). Furthermore, analysis of RNA from mouse liver or kidney after intraperitoneal injection of the animals with ^{14}C-VC revealed the formation of $1,N^6$-ethenoadenosine and $3,N^4$-ethenocytidine (Bergman, 1982).

Mass fragmentographic evidence (by single ion detection) for the presence of $3,N^4$-ethenodeoxycytidine in the liver DNA of rats which had received 250 ppm VC in their drinking water for two years has been reported by Green and Hathway (1978). Unfortunately, no supporting quantitative data have been presented by these authors.

After a single intraperitoneal injection of mice with ^{14}C-VC and subsequent chromatographic analysis of the liver DNA hydrolysate, a minor peak of radioactivity coeluted with added 1,N^6-ethenodeoxyadenosine but could not be definitely identified (Bergman, 1982). The major product of DNA alkylation was also not identified in this study, but was most probably 7-(2-oxoethyl)guanine.

A single exposure of rats or mice to ^{14}C-VC resulted in formation of 7-(2-oxoethyl)guanine as the major product of alkylation of liver DNA (Osterman-Golkar et al., 1977; Laib et al., 1981). Neither 1,N^6-ethenodeoxyadenosine nor 3,N^4-ethenodeoxycytidine could be detected in the study with rats (Laib & Bolt, 1981).

Young rats (Maltoni, 1977), particularly in the second and third week of life (Laib et al., 1985a), are very sensitive to the carcinogenic stimulus of VC. After acute exposure of 12-day old rats and adult animals to ^{14}C-VC on two subsequent days, analysis of liver DNA revealed 7-(2-oxoethyl)guanine as the major alkylation product. Even in liver DNA of the young animals, in which the level of oxoethylguanine was 5-fold higher than in adult liver, radioactive ethenodeoxyadenosine or ethenodeoxycytidine could not be detected (Laib et al., 1983). However, in more recent studies, chromatographic analysis of liver DNA hydrolysates from young animals have revealed a minor peak of radioactivity comigrating with authentic N^2,3-ethenoguanine, indicating that small amounts of this compound are formed (Laib et al., 1985b).

POSSIBLE BIOLOGICAL CONSEQUENCES OF DNA ADDUCTS REPORTED IN LABORATORY RODENTS AFTER EXPOSURE OF THE ANIMALS TO VC

1,N^6-Ethenoadenine and 3,N^4-ethenocytosine residues and their hydrated intermediates

Experiments in various systems were carried out to determine whether these adducts had miscoding properties or would alter the transcription process. Transcription by DNA polymerase or DNA-dependent RNA polymerase of various synthetic templates containing 1,N^6-ethenoadenine, 3,N^4-ethenocytosine or their hydrated intermediates provided evidence for miscoding properties of these alkylation products (Spengler & Singer, 1980; Barbin et al., 1981; Hall et al., 1981; Kuśmierek & Singer, 1982), and it was concluded that such cyclic base adducts may represent "promutagenic lesions" responsible for VC-induced carcinogenesis. However, from recent studies on the replication of deoxyribopolynucleotides containing etheno derivatives, it has been reported that neither the hydrated form of 3,N^4-ethenocytosine nor 1,N^6-ethenoadenine or its hydrate are mutagenic, but that 3,N^4-ethenocytosine is (Singer & Spengler, these proceedings[1]).

---

[1] See p. 359.

Table 1. Alkylation of DNA bases by vinyl chloride (VC) in laboratory rodents *in vivo*

Species/Organ/Treatment	Alkylated bases[a]				Reference
	OetG	EtG	EtA	EtC	
	(mol modified DNA base / 10^6 mol unmodified base)				
mouse/liver/single intra-peritoneal injection of ^{14}C-VC	8.3	–	(1.0)[b]	n.d.	Bergman (1982)
rat/liver/250 ppm VC in drinking water for 2 years	–	–	traces detected, no quantitative data		Green & Hathway (1978)
rat/liver/single exposure to ^{14}C-VC	4.5	–	n.d.	n.d.	Laib & Bolt (1981)
rat/liver/2 subsequent exposures to ^{14}C-VC	5.2	n.d.	n.d.	n.d.	Laib *et al.* (1983)
12-day old rats/liver/ 2 subsequent exposures	25.7	0.25	n.d.	n.d.	Laib *et al.* (1985b)

[a] Abbreviations used: OetG, 7-(2-oxoethyl)guanine; EtG, N^2,3-ethenoguanine; EtA, 1,N^6-ethenoadenine; EtC, 3,N^4-ethenocytosine; –, not investigated; n.d., not detected
[b] This value was calculated from a measurement of 8 dpm above background.

N^2,3-Ethenoguanine

Miscoding properties have been suggested for this cyclic adduct on the basis of its molecular structure (Doerjer *et al.*, 1984). Only recently, evidence has been obtained for the formation of minor amounts of N^2,3-ethenoguanine in the liver DNA of 12-day old rats after repeated exposure of the animals to ^{14}C-VC (Laib *et al.*, 1985b).

7-(2-Oxoethyl)guanine

This adduct is quantitatively by far the major DNA alkylation product of VC. The amount formed upon exposure of laboratory rodents to VC has been found to exceed that of etheno adducts by two orders of magnitude (see Table 1). This adduct is characterized by the introduction of a reactive oxoethyl group at the N-7 position of guanine. Physico-chemical investigations have provided evidence that the attachment of this free aldehyde group may lead, after subsequent formation of a hemiacetal, to a modification at the O^6 position which is involved in the base-pairing of DNA (see Fig. 1, Scherer *et al.*, 1981). Faulty base-pairing during replication of DNA containing O^6,7-(1'-hydroxyethano)guanine has been suggested to occur after exposure to VC (Scherer *et al.*, 1981). However, essential miscoding properties could not be attributed to this adduct in a replication-fidelity assay using DNA polymerase and CEO-modified poly(dG-dC) (Barbin *et al.*, 1985; Barbin & Bartsch, these proceedings[2]).

---

[2] See p. 345.

Fig. 1. Base adducts derived from covalent binding of reactive vinyl chloride metabolites to DNA and/or RNA

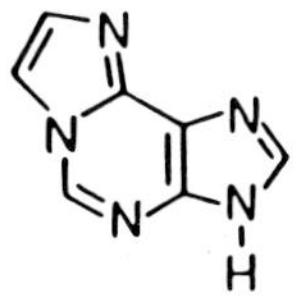

1, N^6-ethenoadenine

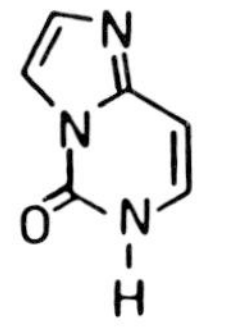

3, N^4-ethenocytosine

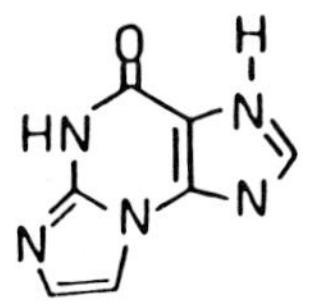

N^2,3-ethenoguanine

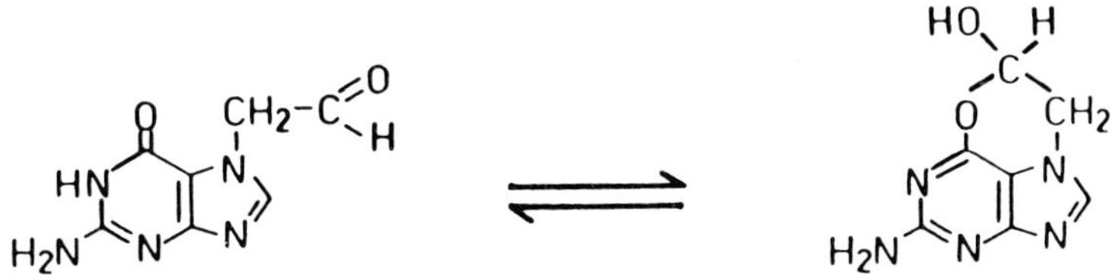

7-(2-oxoethyl)guanine ⇌ O^6, 7-(1'-hydroxyethano)guanine

When discussing possible biological consequences of 7-(2-oxoethyl)guanine in DNA, the chemical reactivity of the aldehyde function should also be considered. It is well-accepted that DNA adducts of chemical carcinogens may exert their biological activity by the formation of DNA cross-links (Doerjer et al., 1984). In this respect, the formation of cross-links in DNA, and/or from DNA to transcriptionally active proteins, may be mediated by the reactive oxoethyl residue; this would have biological consequences. This possibility has been discussed with respect to intrastrand cross-linking between 7-(2-oxoethyl)guanine and an adjacent cytosine residue (Scherer et al., these proceedings[3]).

CONCLUSIONS AS TO THE ROLE OF BASE ADDUCTS IN VC-INDUCED CARCINOGENESIS: FUTURE REASEARCH NEEDS

Studies on possible mechanisms of VC-induced carcinogenicity have clearly shown that metabolic features of reactive metabolites, structural aspects of target molecules,

---

[3] See p. 109

reactivity of the ultimate carcinogenic metabolite, and molecular properties of the adducts are factors to be considered.

The high amount (compared to the etheno adducts) of 7-(2-oxoethyl)guanine detected in rat liver DNA after exposure of the animals to VC supports the principal role of CEO in VC-induced carcinogenesis. Other studies (Gwinner *et al.*, 1983; Bolt, these proceedings[4]) have shown that CAA generated metabolically within liver cells *in vivo* does not alkylate nucleic acid bases to form etheno compounds. Therefore, we can conclude that the DNA alkylation products observed *in vivo* are formed exclusively by CEO.

A critical evaluation of the literature shows that the roles of the different base adducts in the mechanism of VC-induced carcinogenesis have not yet been clearly defined. On a quantitative basis, 7-(2-oxoethyl)guanine, with its reactive aldehyde group, is by far the major DNA adduct of VC. However, no measurable miscoding activity could be attributed to this adduct in a replication-fidelity assay *in vitro* and its cross-linking potential remains to be investigated. Very weak quantitative evidence is available for the formation in liver DNA of $1,N^6$-ethenoadenine, $3,N^4$-ethenocytosine, and their hydrated intermediates; the miscoding potential of these has, however, been reasonably well demonstrated. The detection of minor amounts of $N^2,3$-ethenoguanine in the DNA of young rats after repeated exposure of the animals to VC stresses the need to further investigate the influence of this potential adduct on replication fidelity. Furthermore, the importance of the O^6 position of guanine or the O^4 position of thymine as targets for alkylation by CEO remains to be elucidated.

In summary, DNA alkylation and the role of base adducts in VC-induced carcinogenesis have been intensively investigated. Further research, however, is needed for a comprehensive understanding of the mechanisms involved in the action of this human carcinogen.

ACKNOWLEDGEMENTS

The author wishes to thank Dr H.M. Bolt for critically reading this manuscript and Dr G. Doerjer for providing $N^2,3$-ethenoguanine.

REFERENCES

Barbin, A., Bartsch, H., Leconte, P. & Radman, M. (1981) Studies on the miscoding properties of $1,N^6$-ethenoadenine and $3,N^4$-ethenocytosine, DNA reaction products of vinyl chloride metabolites, during in-vitro DNA synthesis. *Nucleic Acids Res.*, **9,** 375–387

Barbin, A., Laib, R.J. & Bartsch, H. (1985) Lack of miscoding properties of 7-(2-oxoethyl)guanine, the major vinyl chloride DNA adduct. *Cancer Res.*, **45,** 2440–2444

---

[4] See p. 261.

Bergman, K. (1982) Reactions of vinyl chloride with RNA and DNA of various mouse tissues *in vivo*. *Arch. Toxicol.,* **49,** 117–129

Doerjer, G., Bedell, M.A. & Oesch, F. (1984) *DNA adducts and their biological relevance.* In: Obe, G., ed., *Mutations in Man,* Berlin, Springer Verlag, pp. 20–34

Green, T. & Hathway, D.E. (1978) Interactions of vinyl chloride with rat-liver DNA *in vivo*. *Chem.-biol. Interactions,* **22,** 211–224

Gwinner, L.M., Laib, R.J., Filser, J.G. & Bolt, H.M. (1983) Evidence of chloroethylene oxide being the reactive metabolite of vinyl chloride towards DNA: comparative studies with 2,2′-dichlorodiethylether. *Carcinogenesis,* **4,** 1483–1486

Hall, J.A., Saffhill, R., Green, T. & Hathway, D.E. (1981) The induction of errors during in-vitro DNA synthesis following chloroacetaldehyde-treatment of poly(dA-dT) and poly(dC-dG) templates. *Carcinogenesis,* **2,** 141–146

IARC (1979) *IARC Monographs on the Evaluation of the Carcinogenic Risk of Chemicals to Humans.* Vol. 19, *Some monomers, plastics and synthetic elastomers, and acrolein,* Lyon, International Agency for Research on Cancer

Kayasuga-Mikado, K., Hashimoto, T., Negeshi, T., Negeshi, K. & Hayatsu, H. (1980) Modification of adenine and cytosine derivatives with bromoacetaldehyde. *Chem. pharm. Bull.,* **28,** 932–936

Kimura, K., Nakanishi, M., Yamamoto, T. & Tsuboi, M. (1977) A correlation between the secondary structure of DNA and the reactivity of adenine residues with cloroacetaldehyde. *J. Biochem.,* **81,** 1699–1703

Kuśmierek, J.T. & Singer, B. (1982) Chloroacetaldehyde treated ribo- and deoxyribonucleotides. 2. Errors in transcription by different polymerases resulting from ethenocytosine and its hydrated intermediate. *Biochemistry,* **21,** 5723–5728

Laib, R.J. (1982) *Specific covalent binding and toxicity of aliphatic halogenated xenobiotics.* In: Beckett, A.H. & Gorrod, J.W., eds, *Reviews on Drug Metabolism and Drug Interactions, Vol. 4,* London, Freund Publishing House Limited, pp. 1–48

Laib, R.J. & Bolt, H.M. (1977) Alkylation of RNA by vinyl chloride metabolites *in vitro* and *in vivo*: formation of 1,N^6-ethenoadenosine. *Toxicology,* **8,** 185–195

Laib, R.J. & Bolt, H.M. (1978) Formation of 3,N^4-ethenocytidine moieties in RNA by vinyl chloride metabolites *in vitro* and *in vivo*. *Arch. Toxicol.,* **39,** 235–240

Laib, R.J., Gwinner, L.M. & Bolt, H.M. (1981) DNA alkylation by vinyl chloride metabolites: etheno derivatives or 7-alkylation of guanine? *Chem.-biol. Interactions,* **37,** 219–231

Laib, R.J., Cartier, R., Bartsch, H. & Bolt, H.M. (1983) Influence of age on induction of preoplastic foci and on alkylation of rat liver DNA by vinyl chloride. *J. Cancer Res. clin. Oncol.,* **105,** A21

Laib, R.J., Klein, K.-P. & Bolt, H.M. (1985a) The rat liver foci bioassay. I. Age-dependence of induction by vinyl chloride of ATPase-deficient foci. *Carcinogenesis,* **6,** 65–68

Laib, R.J., Doerjer, G. & Bolt, H.M. (1985b) Detection of N^2,3-ethenoguanine in liver DNA hydrolysates of young rats after exposure of the animals to ^{14}C-vinyl chloride. *J. Cancer Res. clin. Oncol.,* **109,** A7

Maltoni, C. (1977) Recent findings on the carcinogenicity of chlorinated olefins. *Environ. Health Perspect.,* **21,** 1–5

Oesch, F. & Doerjer, G. (1982) Detection of N^2,3-ethenoguanine in DNA after treatment with chloroacetaldehyde *in vitro*. *Carcinogenesis, 3,* 663–665

Osterman-Golkar, S., Hultmark, D., Segerbäck, D., Calleman, C.J., Göthe, R., Ehrenberg, L. & Wachtmeister, C.A. (1977) Alkylation of DNA and proteins in mice exposed to vinyl chloride. *Biochem. biophys. Res. Commun., 76,* 259–266

Roe, R., Jr., Paul, J.S. & Montgomery, P.O'B., Jr. (1973) Synthesis and PMR spectra of 7-hydroxyalkylguanosinium acetates. *J. heterocyclic Chem., 10,* 849–857

Sattsangi, P.D., Leonard, N.J. & Frihart, C.R. (1977) 1,N^2-ethenoguanine and N^2,3-ethenoguanine: synthesis and comparison of the electronic spectral properties of these linear and angular triheterocycles related to the Y bases. *J. org. Chem., 42,* 3292–3296

Scherer, E., van der Laken, C.J., Gwinner, L.M., Laib, R.J. & Emmelot, P. (1981) Modification of deoxyguanosine by chloroethylene oxide. *Carcinogenesis, 2,* 671–677

Spengler, S. & Singer, B. (1980) Transcriptional errors and ambiguity resulting from the presence of 1,N^6-ethenoadenosine or 3,N^4-ethenocytidine in polyribonucleotides. *Nucleic Acids Res., 9,* 365–373

MODIFICATION OF DNA AND METABOLISM OF ETHYL CARBAMATE *IN VIVO:* FORMATION OF 7-(2-OXOETHYL)GUANINE AND ITS SENSITIVE DETERMINATION BY REDUCTIVE TRITIATION USING ^{3}H-SODIUM BOROHYDRIDE

E. SCHERER[1], H. WINTERWERP & P. EMMELOT

Division of Chemical Carcinogenesis, The Netherlands Cancer Institute, Amsterdam, The Netherlands

SUMMARY

The modification of liver DNA of mice and rats by ethyl carbamate and its putative proximate metabolite, vinyl carbamate, has been investigated. Following treatment with [ethyl-1-^{14}C]-ethyl carbamate, the main radioactive DNA adduct was identified as 7-(2-oxoethyl)guanine by cochromatography with the authentic marker in several separation systems. After reduction by sodium borohydride ($NaBH_4$) to 7-(2-hydroxyethyl)guanine, the radioactive material again cochromatographed with the respective marker. Reduction of modified liver DNA by ^{3}H-$NaBH_4$, following administration of unlabelled ethyl carbamate or vinyl carbamate, allowed the quantitation of 7-(2-oxoethyl)guanine (as 7-(2-hydroxy-2-[^{3}H]-ethyl)guanine). Vinyl carbamate led to about 100 times as much 7-(2-oxoethyl)guanine (on a molar basis) as did ethyl carbamate. Both the formation of 7-(2-oxoethyl)guanine by ethyl carbamate and vinyl carbamate, and the much higher activity of the latter compound, strongly support the existence of the metabolic activation pathway, ethyl carbamate → vinyl carbamate → epoxyethyl carbamate, as proposed by Dahl *et al.* (1978, 1980). The possible role of 7-(2-oxoethyl)guanine in the initiation of the carcinogenic process is discussed in view of the structural equilibrium with its hemiacetal conformation, O^6,7-(1'-hydroxyethano)guanine. In the latter conformation, it is assumed to represent a promutagenic lesion. In addition, intrastrand cross-links between modified guanine and adjacent cytosine or adenine seem possible and may have promutagenic consequences. Replication of DNA containing such lesions may lead

---

[1] To whom correspondence should be addressed

Fig. 1. Multi-Hit model of chemical carcinogenesis, illustrated by a three-hit process. (1) normal cells; (2) initiated cells; (3) more advanced precancerous cells; (4) tumour cells. A (genetic) rare event, or a hit, is indicated by a thunderbolt (wavy arrow).

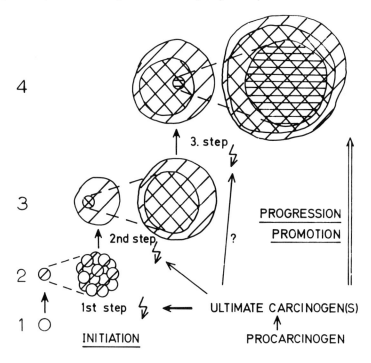

to the induction of mutations. This may be a critical event in the initiation, and eventually progression, of the carcinogenic process as determined by ethyl carbamate and other carcinogens, such as vinyl chloride, which lead to the same DNA modification.

INTRODUCTION

Most chemical carcinogens must be activated *in vivo* to form electrophilic metabolites which can react with macromolecules. DNA is considered as the most important target, since its modification can lead to discrete hereditable changes, mutations, which may underly the initiation step of the carcinogenic process. There is an increasing amount of evidence that the conversion of a normal cell into a tumour cell requires a certain number of such discrete changes, or hits (Fig. 1, Emmelot & Scherer, 1980; Scherer, 1984). These (genetic) changes can be variously expressed phenotypically and may or may not be discernible by the detection methods employed. Carcinogenesis can thus be described as a protracted developmental process leading step-by-step, in a sequence of increasingly neoplastic cell stages, to the malignant tumour cell stage. The activation of specific oncogenes may be related

Fig. 2. Metabolic activation of ethyl carbamate (A), (as proposed by Dahl et al. (1978), compared to that of vinyl chloride (B) (from Scherer & Emmelot, 1981). X refers to unknown reactive intermediate(s) which may also be formed.

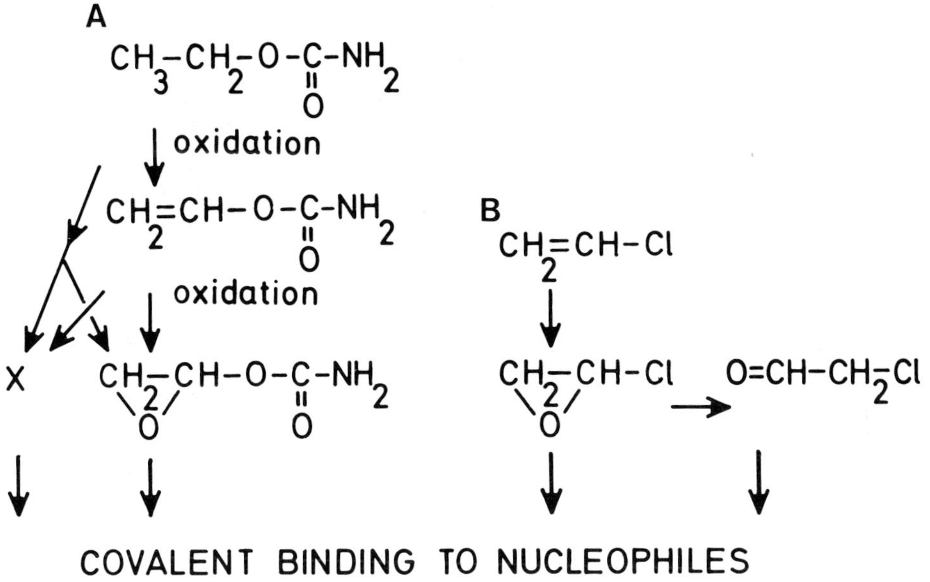

A

$$CH_3-CH_2-O-\underset{\underset{O}{\|}}{C}-NH_2$$

↓ oxidation

$$CH_2=CH-O-\underset{\underset{O}{\|}}{C}-NH_2 \qquad \textbf{B} \quad CH_2=CH-Cl$$

↓ oxidation

$$X \qquad \underset{\diagdown O \diagup}{CH_2-CH}-O-\underset{\underset{O}{\|}}{C}-NH_2 \qquad \underset{\diagdown O \diagup}{CH_2-CH}-Cl \quad \longrightarrow \quad O=CH-CH_2Cl$$

↓ ↓ ↓ ↓

COVALENT BINDING TO NUCLEOPHILES

to some of these step-like changes (Land et al., 1983; Klein & Klein, 1984; Schimke, 1984). It is obvious that such a sequence of stochastic events can operate only if the participating cells, initiated or more advanced, have acquired a certain growth autonomy, or if they can be stimulated to proliferate, thus forming clones of cells and an increasing cell population at risk to be further converted. Tumour promoters seem to act, at least partially, by stimulating the proliferation of initiated cells (Scherer, 1984). Recent experiments performed on the mouse skin (Hennings et al., 1983) and in rat liver systems (Scherer et al., 1984) indicate that not only the initiation step, but also the second rare event – leading in mouse skin from papillomas to carcinomas and in rat liver from precancerous foci to neoplastic nodules – depends more on (mutagenic) carcinogens than on (nonmutagenic) tumour promoters. It is thus conceivable that mutational events are involved in the whole step-by-step progression from the initiated to the frank tumour cell. Chemical carcinogens, their metabolic activation and DNA modification, and the repair of DNA adducts may thus be important not only for the initiation phase but for the whole process of carcinogenic evolution.

For a great number of carcinogens, metabolic activation to electrophilic species, as well as covalent binding to nucleophilic sites present in DNA, RNA and protein, has been firmly established (Miller & Miller, 1981). In contrast, for ethyl carbamate – a versatile carcinogen in many tissues of various species (IARC, 1974) – the search for the metabolic pathway leading to more reactive compounds and to the modifications of macromolecules is still in progress. Very recently, this search was newly inspired by the suggestion that the metabolic activation of ethyl carbamate might be similar to that of vinyl chloride (Fig. 2), thus leading to vinyl carbamate and epoxyethyl carbamate as intermediate and ultimate metabolites (Dahl et al., 1978, 1980). This hypothesis is supported by the identification of 7-(2-oxoethyl)guanine in DNA (Scherer & Emmelot, 1981, 1982; this paper) and of cyclic base modifications in RNA (Ribovich et al., 1982) from ethyl-carbamate-treated rats and mice; the same modifications are also found in nucleic acids of vinyl-chloride-treated animals (Osterman-Golkar et al., 1977; Laib et al., 1981; Laib, these proceedings[2]). The DNA modification, 7-(2-oxoethyl)guanine, may be promutagenic in its hemiacetal ring conformation, $O^6,7$-(1'-hydroxyethano)guanine (Scherer et al., 1981), or it may confer mispairing properties after formation of guanine-cytosine (G-C) or guanine-adenine (G-A) intrastrand cross-links, and thus be responsible for the carcinogenic and mutagenic activity of ethyl carbamate and of other carcinogens which lead to the same DNA adduct.

MATERIALS AND METHODS

Specific pathogen-free Sprague-Dawley rats (random-bred from Zentral-Institut für Versuchstiere, Hannover, FRG), weighing 250–300 g, and male C3HF/A mice (our own inbred breeding colony), weighing 30 g, had free access to standard food pellets (Hope Farms, Woerden, The Netherlands) and water. [Ethyl-1-^{14}C]-ethyl carbamate (4.8 mCi/mmol), [carbonyl-^{14}C]-ethyl carbamate (2.21 mCi/mmol) and ^{3}H-NaBH$_4$ (10 Ci/mmol) were from NEN (Dreieichenhein, FRG). Ethyl carbamate (Merck, Darmstadt, FRG) and vinyl carbamate (a gift from Dr J.A. Miller) were administered in saline by intraperitoneal injection. DL-Ethionine (Sigma, St. Louis, Mo., USA) was applied in 0.1 N hydrochloric acid by gastric instillation.

Liver DNA was isolated according to the method of Kirby and Cook (1967). DNA was hydrolysed enzymatically using DNase I, snake venom phosphodiesterase and alkaline phosphatase (Millipore Corp., Freehold, NJ, USA). The enzyme digest was adjusted to pH 8.0 and eluted from a 30 × 1 cm Dowex 50W-X4 column (Bio-Rad, Richmond, Ca. USA; −400 mesh, NH$_4^+$ form) by 0.1 M ammonium formate pH 8.0, 70 fractions of 3 ml, followed by 0.3 M ammonium formate pH 9.0, 20 fractions of 10 ml. 7-(2-Oxoethyl)guanine, prepared according to the method of Scherer et al. (1981) by the reaction of deoxyguanosine with chloroethylene oxide in acetic acid, was added as marker to the hydrolysate. 7-(2-Hydroxyethyl)guanine was obtained by reduction (NaBH$_4$) of 7-(2-oxoethyl)guanine. DNA modified by chloroacetaldehyde (Green & Hathway, 1978) was added in some cases to the ethyl-carbamate-modified DNA before hydrolysis as a source of 1,N^6-ethenodeoxyadenosine and 3,N^4-ethenodeoxycytidine. Hydrolysis was carried out overnight at 37 °C, since shorter treatment was found to lead to only partial hydrolysis (approx. 30%) of the bound

---

[2] See p. 101.

radioactivity, ([ethyl-1-^{14}C]-ethyl carbamate), while unmodified DNA and DNA containing etheno compounds were almost completely hydrolysed after three hours.

Absolute radioactivity of column fractions and of fractions from thin-layer plates was measured by liquid scintillation counting, using a Triton X100-toluene-Omnifluor scintillation mixture (Scherer et al., 1977). Thin-layer chromatograms were run over a distance of 10 cm on silica gel plates (Merck, Darmstadt, FRG) in ethanol: n-butanol: 5% acetic acid(10:80:25, by volume). Samples were applied in methanol.

Reduction of 7-(2-oxoethyl)guanine in DNA modified in vitro by chloroethylene oxide (1 mg DNA/ml, 0.1 M Tris-HCl pH 7.5, 5 μl chloroethylene oxide/ml, reaction for 2 min at room temperature, precipitation by approx. 0.1 vol of 2% Cetavlon (n-hexadecyltrimethylammonium bromide)), or in vivo by unlabelled ethyl carbamate or vinyl carbamate, was performed in dry isopropanol (distilled from isopropanol/NaBH$_4$). The Cetavlon salt of modified DNA was dissolved in dry ethanol at about 1 mg/ml. For reductive tritiation, samples of about 1.5 mg DNA were evaporated under a stream of nitrogen, dried in vacuo and dissolved in 2 ml distilled isopropanol. One millilitre of ^{3}H-NaBH$_4$(2.5 mCi, 0,1 mg), freshly dissolved in isopropanol, was added and reduction performed at room temperature for seven days. The DNA was precipitated by the addition of three drops of saturated sodium acetate in water, and the precipitate washed out onto a glass-fibre filter (Whatman GF/A) with four volumes of 10 ml ethanol (70%, containing 5 mM sodium acetate). The filter was put back into the original reaction vial, and again washed with three changes of chloroform:methanol (1:1, by volume). After drying, the DNA was partially depurinated by boiling in 2 ml water for 30 min. Chromatography on Dowex NH$_4^+$ was followed by rechromatography of the 7-(2-hydroxyethyl)guanine-containing fractions on Sephadex G10 (Pharmacia, Uppsala, Sweden; 1 × 30 cm, eluted with H$_2$O), and on silica gel thin-layer plates.

RESULTS

It has been suggested by Dahl et al. (1978, 1980) that ethyl carbamate Fig. 2), can be activated metabolically to vinyl carbamate and epoxyethyl carbamate (Fig. 2), as proximate and ultimate carcinogens, respectively, of which the latter is structurally very similar to a proposed ultimate metabolite of vinyl chloride, chloroethylene oxide (Zajdela et al., 1980). Therefore, it seemed important to test whether modifications of nucleic acids as described for vinyl chloride can be identified in DNA of ethyl-carbamate-treated animals. The presence of 1,N^6-ethenoadenine was indicated in DNA of rats treated for two years with vinyl chloride (Green & Hathway, 1978). Recent evidence from Laib et al. (1981), however, suggests that etheno compounds are formed mainly in RNA. A modified guanine has been reported in liver DNA after vinyl chloride treatment in vivo (Osterman-Golkar et al., 1977) and tentatively identified as 7-(2-oxoethyl)guanine. This assignment has been confirmed by Laib et al. (1981). Reaction of deoxyguanosine with chloroethylene oxide in vitro led to the same modification (Scherer et al., 1981). In the present study, the main DNA adduct observed in DNA modified by ethyl carbamate in vivo (Scherer et al., 1980) was identified as 7-(2-oxoethyl)guanine.

Fig. 3. Representative Dowex 50W-X4 (NH_4^+-form) column chromatogram of enzymatically hydrolysed rat liver DNA modified *in vivo* by [ethyl-1-^{14}C]-ethyl carbamate (top), or [carbonyl-^{14}C]-ethyl carbamate (bottom). DNA was isolated 3 or 4.5 hours, respectively, after carcinogen application. No ^{14}C-labelled material was detected in the fractions in which 3,N^4-ethenodeoxycytidine and 1,N^6-ethenodeoxyadenosine eluted from the column: fraction 27 and fraction 55, respectively. Deoxynucleosides (dThd, deoxythymidine; dCyd, deoxycytidine; dAdo, deoxyadenosine; dGuo, deoxyguanosine) were detected by absorption at 260 nm (dotted line). (from Scherer & Emmelot, 1981)

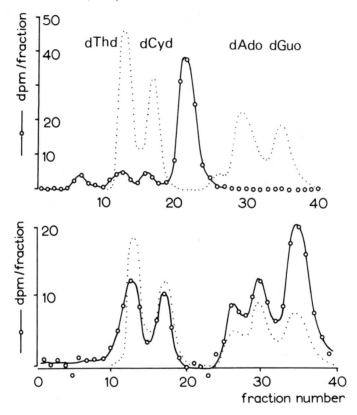

Enzymatic hydrolysis, or partial depurination by neutral heating of liver DNA modified *in vivo* by [ethyl-1-^{14}C]-ethyl carbamate followed by Dowex cation-exchange chromatography, led to a clear separation of a single main peak of radioactivity eluting between the positions of the parent pyrimidine and purine deoxyribonucleosides (Fig. 3, top; Scherer *et al.*, 1980). 7-(2-Oxoethyl)guanine added as a marker cochromatographed with the radioactive modification, indicating that 7-(2-oxoethyl)deoxyguanosine depurinated under the conditions of enzymatic hydrolysis. If ethyl carbamate labelled in the carbonyl-carbon was used, no radioactivity was observed in the fractions between deoxycytidine and deoxyadenosine; only the parent deoxyribonucleosides were labelled (Fig. 3, bottom). Upon rechromatography, the radioactive modification comigrated with the 7-(2-oxoethyl)guanine marker in the

Fig. 4. Thin-layer rechromatography of the ^{14}C-labelled DNA modification (Fig. 3, fractions 18–24) of [ethyl-1-^{14}-C]-ethyl carbamate (top), and of the product obtained after reduction by an excess of sodium borohydride (bottom), with authentic markers 7-(2-oxoethyl)guanine and 7-(2-hydroxyethyl)guanine, respectively. The positions of the markers, indicated by the shaded areas, was detected by decreased fluorescence at 254 nm of the indicator dye. The thin-layer system used was silica gel/ethanol:n-butanol:acetic acid (5%), 10:80:25.

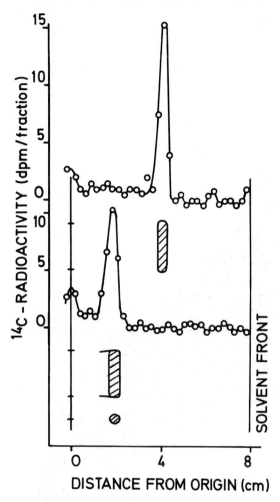

thin-layer systems used (silanized silica gel/chloroform:methanol:ammonia (5%), 90:30:10, and silica gel/ethanol:n-butanol:acetic acid (5%), 80:10:25; Scherer & Emmelot, 1981) (Fig. 4, top). Reduction in methanol of the radioactive modification and the 7-(2-oxoethyl)guanine marker by an excess of NaBH$_4$ led to a decrease in the mobility of both compounds in the latter thin-layer system, consistent with the expected formation of 7-(2-hydroxyethyl)guanine (Fig. 4, bottom).

These results strongly indicate that ethyl carbamate leads to the same major modification of DNA as vinyl chloride. In contrast to reported results on RNA modified by ethyl carbamate (Ribovich et al., 1982), neither $1,N^6$-ethenodeoxyadenosine nor $3,N^4$-ethenodeoxycytidine were detected in ethyl-carbamate-modified liver DNA of either rat or mouse (sacrificed between 3 and 24 hours after administration of [ethyl-1-^{14}C]-ethyl carbamate). These compounds elute from the Dowex NH_4^+ column (Fig. 3) at about fraction 27 (ethenodeoxycytidine), fraction 45 (ethenoadenine) and fraction 55 (ethenodeoxyadenosine). In none of these fractions were significant amounts of radioactivity detected. Since both etheno compounds are formed *via* a hydrated intermediate form, which under physiological conditions slowly changes to the respective etheno derivative (Biernat et al., 1978; Krzyżosiak et al., 1981; Kuśmierek & Singer, 1981; authors' own unpublished observations), it could be argued that only the hydrated intermediates were present shortly after the modification of DNA by ethyl carbamate *in vivo*. The hydrated form of $1,N^6$-ethenodeoxyadenosine (obtained by modification of deoxyadenosine with chloroethylene oxide in acetic acid) is strongly bound to the Dowex NH_4^+ column, and elutes with 0.1 M ammonium formate pH 8 only after conversion to the etheno derivative. In the hydrated form, it can be eluted at elevated pH and higher salt concentration (0.3 M ammonium formate pH 9). Since radioactivity has never been observed in the high salt high pH eluate of Dowex columns run for the separation of the enzymatic hydrolysate of ethyl-carbamate-modified DNA, both $1,N^6$-ethenodeoxyadenosine and its hydrated form appear not to be present in detectable amounts in DNA modified *in vivo* by ethyl carbamate. This is consistent with the failure to detect etheno compounds in DNA modified *in vivo* by vinyl chloride (Laib et al., 1981; Laib, these proceedings[3]).

The reduction of 7-(2-oxoethyl)guanine to 7-(2-hydroxyethyl)guanine introduces a stable proton at the 2-carbon of the ethyl group. We therefore investigated whether reduction by tritiated $NaBH_4$ could furnish a sensitive method for the detection and quantitation of 7-(2-oxoethyl)guanine in DNA modified by unlabelled carcinogen.

In order to avoid as far as possible reactions of tritiated $NaBH_4$ with water, which would lead to rapid inactivation and a decrease of the specific activity due to isotope exchange, reduction should take place under nonaqueous conditions, and in a solvent system in which $NaBH_4$ is stable and both DNA and $NaBH_4$ are soluble. Since the cetyltrimethylammonium (Cetavlon) salt of DNA is soluble in ethanol, experiments were performed with the Cetavlon salt of modified DNA. Several organic solvents were tested. Isopropanol proved to be a good solvent for both compounds, and $NaBH_4$ was stable for prolonged periods of time, at concentrations of about 10 μM. Reduction by ^3H-$NaBH_4$ of DNA modified *in vitro* by chloroethylene oxide was followed by thorough washing, partial depurination by neutral heating in order to release 7-(2-hydroxyethyl)guanine from the DNA polymer, and cation-exchange chromatography. A peak of high radioactivity was obtained at the position where 7-(2-hydroxyethyl)guanine was expected to elute from the column (Fig. 5, left).

---

[3] See p. 101.

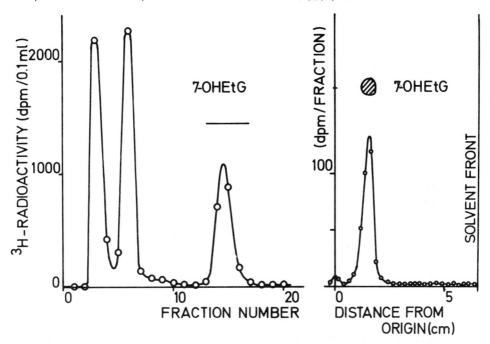

Fig. 5. Dowex NH_4^+ column chromatography (left), following partial depurination of DNA modified *in vitro* by chloroethylene oxide and reduced by ^{3}H-sodium borohydride in isopropanol (5 µg/ml, 5 days), and thin-layer rechromatography (right) of the Dowex fractions containing 7-(2-hydroxyethyl)guanine (7-OHEtG) on silica gel (solvent system: see Fig. 4). (----), expected position of 7-OHEtG upon elution from Dowex column; (○), position of authentic marker

Rechromatography of the radioactive peak fractions with authentic marker in a thin-layer system (Fig. 5, right), proved that 7-(2-hydroxy-2-[^{3}H]-ethyl)guanine had been formed. The amount of radioactive 7-(2-hydroxyethyl)guanine obtained upon reduction by various concentrations of ^{3}H-NaBH$_4$, and for 2, 5 or 7 days, is shown in Figure 6. Concentrations of 5–40 µg/ml NaBH$_4$ led to a high amount of radioactivity after only two days, with only a slight further increase at longer reaction times, while at lower concentrations, the yield of radioactivity was proportionally lower.

The reduction, by 40 µg/ml ^{3}H-NaBH$_4$, of liver DNA from rats treated with a single administration of ethyl carbamate (300 mg/kg) or vinyl carbamate (10 mg/kg) is illustrated for vinyl carbamate in Figure 7. After the first column chromatography on Dowex (Fig. 7, A), only one high peak of radioactivity was obtained. Rechromatography on Sephadex G10 of the radioactive material eluting in the fractions in which 7-(2-hydroxyethyl)guanine was expected, together with marker, led to a small peak cochromatographing with the marker (Fig. 7, B); after a further (thin-layer) chromatography on silica gel, the radioactive 7-(2-hydroxyethyl)guanine peak was separated from other contaminating material (Fig. 7, C).

Fig. 6. ^{3}H-radioactivity obtained in the 7-(2-hydroxyethyl)guanine fractions (Dowex NH_4^+) after reduction in isopropanol of DNA modified *in vitro* by chloroethylene oxide, by various concentrations of ^{3}H-sodium borohydride, for 2, 5 or 7 days. △, 40; ▽, 20; ○, 10; x, 5; ●, 2.5; ▲, 1.2; ▼, 0.6, µg/ml ^{3}H-sodium borohydride

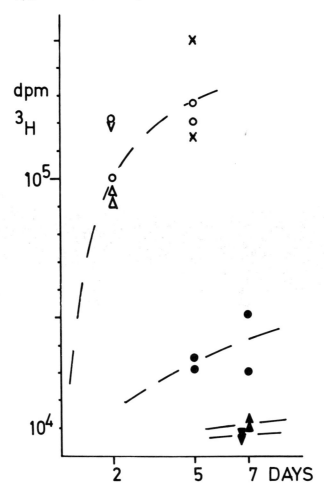

Radioactive reduction was performed with liver DNA of rats given a single administration of ethyl carbamate, vinyl carbamate, *N*-nitrosodiethylamine, DL-ethionine, or saline. The amount of radioactive 7-(2-hydroxyethyl)guanine obtained was related to the degree of modification of DNA as measured by the use of [ethyl-1-^{14}C]-ethyl carbamate, and to the number of precancerous liver foci/cm^3 (Scherer *et al.*, 1980) induced in regenerating rat liver (Table 1). On a molar basis, vinyl carbamate was about 90 times more potent than ethyl carbamate. This again indicates that vinyl carbamate is a proximate carcinogen in the course of metabolic activation of ethyl carbamate. The number of precancerous foci induced by both substances

Fig. 7. Chromatographic separation of 7-(2-hydroxy-2-[^{3}H]-ethyl)guanine obtained after reduction by ^{3}H-sodium borohydride in isopropanol (40 µg/ml, 5 days) of rat liver DNA modified *in vivo* by vinyl carbamate (10 mg/kg). Dowex fractions containing 7-(2-hydroxyethyl)guanine, indicated by (———) in (A), were first rechromatographed with authentic marker on a Sephadex G10 column (B), then by thin-layer chromatography on silica gel (C) (see legend to Fig. 4).

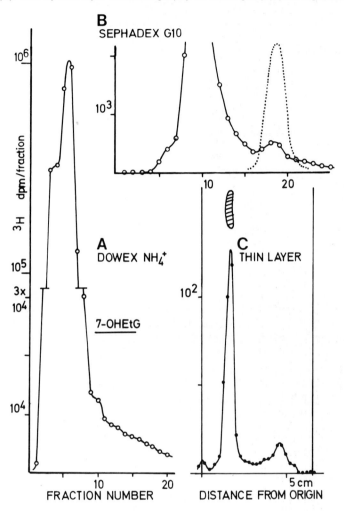

correlated well with the amount of 7-(2-oxoethyl)guanine formed in liver DNA. Neither DL-ethionine nor *N*-nitrosodiethylamine led to the comigration of radioactivity with the marker 7-(2-hydroxyethyl)guanine. For *N*-nitrosodiethylamine, this result was expected; for DL-ethionine, it points against the proposed metabolic activation of ethionine (Leopold *et al.*, 1979) by dehydrogenation of the ethyl moiety to form *S*-vinyl homocysteine and *S*-epoxyethyl homocysteine, as the latter compound would be expected to lead again to the formation of 7-(2-oxoethyl)guanine in DNA.

Table 1. Formation of 7-(2-oxoethyl)guanine (7-OEtG) in rat liver DNA *in vivo* by DL-ethionine, *N*-nitrosodiethylamine, ethyl carbamate and vinyl carbamate, compared with the formation of precancerous foci in regenerating rat liver

Compound	Dose mg/kg	7-OHEtG[a] ^{3}H-dpm	7-OEtG[b] µmol/mol DNA-phosphate	Precancerous liver foci/cm^3 ± SE
Saline	–	<15		
N-Nitrosodiethylamine	10	<15		
DL-Ethionine	300	<15		
Ethyl carbamate	300	490	4.4 ± 1	73 ± 15
Vinyl carbamate	10	1500		170 ± 15

[a] Amount of 7-OEtG was determined by reduction to 7-(2-hydroxyethyl)guanine (7-OHEtG) using ^{3}H-sodium borohydride in isopropanol
[b] Determined using [ethyl-1-^{14}C]-ethyl carbamate

DISCUSSION

An extensive review of the metabolism of ethyl carbamate was given by Mirvish (1968). Most of the ethyl carbamate administered to mice is metabolized within 24 hours to CO_2, NH_3 and ethanol. The metabolism leading to *N*-substituted compounds possibly occurs by way of oxidation to *N*-hydroxyethyl carbamate, since *N*-acetyl-*N*-hydroxyethyl carbamate and *N*-acetoxyethyl carbamate could be identified as urinary metabolites (Boyland & Nery, 1965). Reaction of metabolites of ethyl carbamate with glutathione was indicated by the excretion in the urine of *N*-acetyl-*S*-carboxyethylcysteine and other possible metabolites of conjugation products with glutathione. None of these *N*-substituted carbamates was more carcinogenic than ethyl carbamate itself (Pound, 1967; Mirvish, 1968). In view of the carcinogenicity of vinyl chloride (IARC, 1979), for which metabolic activation to chloroethylene oxide and chloroacetaldehyde is strongly indicated (Gwinner *et al.*, 1983), Dahl *et al.* (1978, 1980) proposed that the metabolic activation of ethyl carbamate involves oxidation of the ethyl group. The similarity to the mode of activation of vinyl chloride (Fig. 2), safrole and aflatoxin B_2 (Swenson *et al.*, 1977) makes this a very attractive hypothesis.

In its support, Dahl *et al.* (1978, 1980) reported that vinyl carbamate was up to 100 times more efficient in inducing lung adenomas and in initiating skin papillomas in mice than ethyl carbamate, and that it was mutagenic in *Salmonella typhimurium* in the presence of a microsomal activation system. Using the induction of precancerous liver foci in regenerating rat liver as a biological endpoint for initiation, we found a similar difference between ethyl and vinyl carbamate (Scherer *et al.*, 1980; Table 1). The induction of sister chromatid exchange was also reported to be much more efficient by vinyl carbamate than by ethyl carbamate (Csukás *et al.*, 1981; Allen *et al.*, 1982; Conner, these proceedings[4]). The ^{3}H/^{14}C ratio of modified hepatic

---

[4] See p. 313.

Fig. 8. Base-pairing scheme between the cyclic hemiacetal, O^6-7-(1'-hydroxyethano)deoxyguanosine and thymidine, indicating possible miscoding properties of the 7-(2-oxoethyl)guanine residue (from Scherer & Emmelot, 1981)

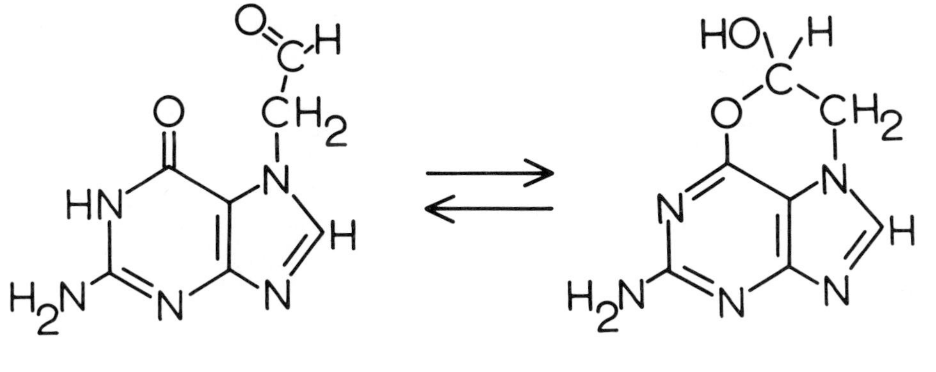

7-(2-OXOETHYL)GUANINE O^6,7-(1'-HYDROXYETHANO)GUANINE

macromolecules was found to be much lower than the ratio of the double-labelled [ethyl-1-^{14}C; 1,2-^{3}H]-ethyl carbamate applied (Dahl et al., 1978). Such a decrease is consistent with the proposed metabolic activation which involves the loss of protons of the ethyl group of ethyl carbamate. The direct attempt to detect vinyl carbamate as a metabolite of ethyl carbamate in mice in vivo failed, however, despite the use of sensitive methods (Dahl et al., 1978, 1980; Miller & Miller, 1983). The metabolism of ethyl carbamate and vinyl carbamate to a common ultimate carcinogen has been proposed to account for this failure.

The similarity of the modification of DNA (Scherer et al., 1980; Scherer & Emmelot, 1981, 1982) and of RNA (Ribovich et al., 1982) observed for ethyl carbamate and vinyl chloride, also strongly supports the enzymatic activation pathway proposed by Dahl et al. (1978, 1980).

Early work in various laboratories on the modification of DNA by ethyl carbamate in vivo, as discussed recently by Scherer and Emmelot (1981), did not lead to the identification of specific sites of modification. Thus, 7-(2-oxoethyl)guanine is the first and major DNA modification product identified until now for ethyl carbamate (Scherer et al., 1980; Scherer & Emmelot, 1981, 1982). In contrast to 7-alkylguanine or 7-hydroxyalkylguanine, the aldehyde group of 7-(2-oxoethyl)guanine confers chemical reactivity on this derivative. The β-carbon of the oxoethyl group may react with amino groups of other bases or of proteins under formation of a Schiff's base, and thus result in cross-links leading eventually to sister chromatid exchange (Csukás et al., 1981; Conner & Cheng, 1983), or to mispairing during DNA replication. Hemiacetal ring closure between O^6 and the β-carbon of the oxoethyl group would be expected to confer promutagenic potential on 7-(2-oxoethyl)guanine (Fig. 8). The actual conformation of 7-(2-oxoethyl)guanine in DNA is, however, not known. From

Fig. 9. Hypothetical intrastrand cross-linking between 7-(2-oxoethyl)guanine residues and adjacent cytosine residues. In double-stranded polymers this may lead to the illustrated C=A mispairing.

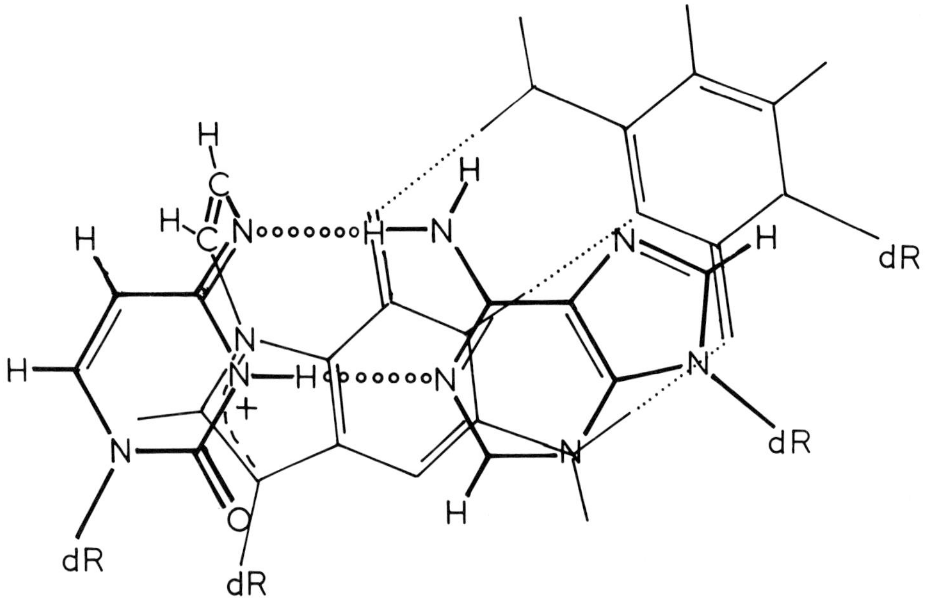

the calculations of Politzer *et al.* (these proceedings[5]) it follows that, for the 7-(2-oxoethyl)guanine base, the hemiacetal conformation is highly unlikely. In the case of the deoxyribonucleoside, however, the positive charge introduced by the alkylation of N-7 may increase the reactivity of O^6 by the zwitterionic charge distribution (Rose *et al.*, 1973), and 7-(2-oxoethyl)guanine residues in DNA or synthetic polymers may actually be present in the promutagenic O^6,7-(1'-hydroxyethano) form.

The promutagenic properties of 7-(2-oxoethyl)guanine have been investigated by Barbin & Bartsch (these proceedings[6]). They found that chloroethylene-oxide-modified poly(dG-dC) containing up to 27% 7-(2-oxoethyl)guanine did not lead to misincorporation opposite the (modified) guanine residues, but rather, opposite the cytosine residues. This argues against the presence of the hemiacetal conformation in the modified polymer. The involvement of modified guanine in the miscoding of cytosine can, however, not be excluded since – according to model building – an intrastrand cross-link between modified guanine and adjacent cytosine seems possible and could favour the incorporation of adenine opposite the cross-linked cytosine (Fig. 9). A similar intrastrand cross-link seems possible in a G–A sequence resulting

---

[5] See p. 37.
[6] See p. 345.

eventually in miscoding of cross-linked adenine. The relative persistence of 7-(2-oxoethyl)guanine in rat liver DNA (Laib et al., 1981) should favour the induction of mutations during replication of modified DNA, and this seems to be a basic requirement for the initiation and possibly the progression of the carcinogenic process.

REFERENCES

Allen, J.W., Langenbach, R., Nesnow, S., Sasseville, K., Leavitt, S., Campbell, J., Brock, K. and Sharief, Y. (1982) Comparative genotoxicity studies of ethyl carbamate and related chemicals: further support for vinyl carbamate as a proximate carcinogenic metabolite. *Carcinogenesis, 3,* 1437–1441

Biernat, J., Ciesiołka, J., Górnicki, P., Adamiak, R.W., Krzyżosiak W.J., & Wiewiórowski, M. (1978) New observations concerning the chloroacetaldehyde reaction with some tRNA constituents. Stable intermediates, kinetics and selectivity of the reaction. *Nucleic Acids Res., 5,* 789–804

Boyland, E. & Nery, R. (1965) The metabolism of urethane and related compounds. *Biochem. J., 94,* 198–208

Conner, M.K. & Cheng, M. (1983) Persistence of ethyl carbamate-induced DNA damage *in vivo* as indicated by sister chromatid exchange analysis *Cancer Res., 43,* 965–971

Csukás, I., Gungl, E., Antoni, F., Vida, G. & Solymosy, F. (1981) Role of metabolic activation in the sister chromatid exchange-inducing activity of ethyl carbamate (urethane) and vinyl carbamate. *Mutat. Res., 89,* 75–82

Dahl, G.A., Miller, J.A. & Miller, E.C. (1978) Vinyl carbamate as a promutagen and a more carcinogenic analog of ethyl carbamate. *Cancer Res., 38,* 3793–3804

Dahl, G.A., Miller, E.C. & Miller, J.A. (1980) Comparative carcinogenicities and mutagenicities of vinyl carbamate, ethyl carbamate, and ethyl *N*-hydroxycarbamate. *Cancer Res., 40,* 1194–1203

Emmelot, P. & Scherer, E. (1980) The first relevant cell stage in rat liver carcinogenesis. A quantitative approach. *Biochim. biophys. Acta, 605,* 247–304

Green, T. & Hathway, D.E. (1978) Interactions of vinyl chloride with rat liver DNA *in vivo. Chem.-biol. Interactions, 22,* 211–224

Gwinner, L.M., Laib, R.J., Filser, J.G. & Bolt, H.M. (1983) Evidence of chloroethylene oxide being the reactive metabolite of vinyl chloride towards DNA: comparative studies with 2,2'-dichlorodiethylether. *Carcinogenesis, 4,* 1483–1486

Hennings, H., Shores, R., Wenk, M.L., Spangler, E.F., Tarone, R. & Yuspa, S.H. (1983) Malignant conversion of mouse skin tumours is increased by tumour initiators and unaffected by tumour promoters. *Nature, 304,* 67–69

IARC (1974) *IARC Monographs on the Evaluation of the Carcinogenic Risk of Chemicals to Humans,* Vol. 7, *Some anti-thyroid and related substances, nitrofurans and industrial chemicals,* Lyon, International Agency for Research on Cancer, pp. 111–140

IARC (1979) *IARC Monographs on the Evaluation of the Carcinogenic Risk of Chemicals to Humans,* Vol. 19, *Some monomers, plastics and synthetic elastomers, and acrolein,* Lyon, International Agency for Research on Cancer, pp. 377–438

Kirby, K.S. & Cook, E.A. (1967) Isolation of deoxyribonucleic acid from mammalian tissues. *Biochem. J., 104* 254–257

Klein, G. & Klein, E. (1984) Oncogene activation and tumor progression. *Carcinogenesis, 5,* 429–435

Krzyżosiak, W.J., Biernat, J., Ciesiołka, J., Gulewicz, K. & Wiewiórowski, M. (1981) The reactions of adenine and cytosine residues in tRNA with chloroacetaldehyde. *Nucleic Acids Res., 9,* 2841–2851

Kuśmierek, J.T. & Singer, B. (1982) Chloroacetaldehyde-treated ribo- and deoxyribopolynucleotides. I. Reaction products. *Biochemistry, 21,* 5717–5722

Laib, R.J., Gwinner, L.M. & Bolt, H.M. (1981) DNA alkylation by vinyl chloride metabolites: etheno derivatives or 7-alkylation of guanine? *Chem.-biol. Interactions, 37,* 219–231

Land, H., Paradat, L.F. & Weinberg, R.A. (1983) Cellular oncogenes and multistep carcinogenesis. *Science, 222,* 771–778

Leopold, W.R., Miller, J.A. & Miller, E.C. (1979) S-vinyl homocysteine (vinthionine): a highly mutagenic analog of ethionine. *AACR Abstracts, 20,* 110

Miller, E.C. & Miller, J.A. (1981) Mechanisms of chemical carcinogenesis. *Cancer, 47,* 1055–1064

Miller, J.A. & Miller, E.C. (1983) The metabolic activation and nucleic acid adducts of naturally-occurring carcinogens: recent results with ethyl carbamate and the spice flavors safrole and estragole. *Br. J. Cancer, 48,* 1–15

Mirvish, S.S. (1968) The carcinogenic action and metabolism of urethane and *N*-hydroxyurethane. *Adv. Cancer Res., 11,* 1–42

Osterman-Golkar, S., Hultmark, D., Segerbäck, D., Calleman, C.J., Göthe, R., Ehrenberg, L. & Wachtmeister, C.A. (1977) Alkylation of DNA and proteins in mice exposed to vinyl chloride. *Biochem. biophys. Res. Commun., 76,* 259–266

Pound, A.W. (1967) The initiation of skin tumours in mice by homologues and *N*-substituted derivatives of ethyl carbamate. *Aust. J. exptl. Biol. med. Sci., 45,* 507–516

Ribovich, M.L., Miller, J.A., Miller, E.C. & Timmins, L.G. (1982) Labeled $1,N^6$-ethenoadenosine and $3,N^4$-ethenocytidine in hepatic RNA of mice given [ethyl-1,2-^{3}H or ethyl-1-^{14}C]-ethyl carbamate (urethan). *Carcinogenesis, 3,* 539–546

Roe, R., Jr., Paul, J.S. & Montgomery, P.O'B., Jr. (1973) Synthesis and PMR spectra of 7-hydroxyalkylguanosinium acetates. *J. heterocycl. Chem., 10,* 849–857

Scherer, E. (1984) Neoplastic progression in experimental hepatocarcinogenesis. *Biochim. biophys. Acta, 738,* 219–236

Scherer, E. & Emmelot, P. (1981) *Modification of DNA by ethyl carbamate* in vivo: *implications for its metabolic activation.* In: Pani, P., Feo, F. & Columbano, A., eds, *Recent Trends in Chemical Carcinogenesis,* Vol. 1, Cagliari, Italy, ESA, pp. 17–34

Scherer, E. & Emmelot, P. (1982) 7-(2-Oxoethyl)guanine, a possibly promutagenic base, in DNA modified *in vivo* by ethyl carbamate (urethane). *Proceedings of the 13th International Cancer Congress,* UICC, A550

Scherer, E., Steward, A.P. & Emmelot, P. (1977) Kinetics of formation of O^6-ethylguanine in, and its removal from, liver DNA of rats receiving diethylnitrosamine. *Chem.-biol. Interactions, 19,* 1–11

Scherer, E., Steward, A.P. & Emmelot, P. (1980) *Formation of precancerous islands in rat liver and modification of DNA by ethyl carbamate: implications for its metabolism.* In: Holmstedt, B., Lauwerys, R., Mercier, M. & Roberfroid, M., eds, *Mechanisms of Toxicity and Hazard Evaluation,* Amsterdam, Elsevier/North-Holland Biomedical Press, pp. 249–254

Scherer, E., Van der Laken, C.J., Gwinner, L.M., Laib, R.J. & Emmelot, P. (1981) Modification of deoxyguanosine by chloroethylene oxide. *Carcinogenesis, 2,* 671–677

Scherer, E., Feringa, A.W. & Emmelot, P. (1984) *Initiation-promotion-initiation. Induction of neoplastic foci within islands of precancerous liver cells in the rat.* In: Börzsönyi, M., Day, N.E., Lapis, K. & Yamasaki, H., eds, *Models, Mechanisms and Etiology of Tumour Promotion (IARC Scientific Publications No. 56),* Lyon, International Agency for Research on Cancer, pp. 57–68

Schimke, R.T. (1984) Gene amplification, drug resistance, and cancer. *Cancer Res., 44,* 1735–1742

Swenson, D.H., Lin, J.K., Miller, E.C. & Miller, J.A. (1977) Aflatoxin B_1-2,3-oxide as a probable intermediate in the covalent binding of aflatoxins B_1 and B_2 to rat liver DNA and ribosomal RNA *in vivo. Cancer Res., 37,* 172–181

Zajdela, F., Croisy, A., Barbin, A., Malaveille, C., Tomatis, L. & Bartsch, H. (1980) Carcinogenicity of chloroethylene oxide, an ultimate reactive metabolite of vinyl chloride, and bis(chloromethyl)ether after subcutaneous administration and in initiation-promotion experiments in mice. *Cancer Res., 40,* 352–356

II. CHEMISTRY AND FORMATION OF CYCLIC AND OTHER ADDUCTS

B. Haloalkylnitrosoureas

ISOLATION AND CHARACTERIZATION OF ELECTROPHILES FROM 2-HALOETHYLNITROSOUREAS FORMING CYTOTOXIC DNA CROSS-LINKS AND CYCLIC NUCLEOTIDE ADDUCTS AND THE ANALYSIS OF BASE SITE-SELECTIVITY BY *AB INITIO* CALCULATIONS

J.W. LOWN, R.R. KOGANTY, U.G. BHAT & S.M.S. CHAUHAN

Department of Chemistry, University of Alberta, Edmonton, Alberta, Canada

A.-M. SAPSE & E.B. ALLEN

City University of New York, New York, NY, USA

SUMMARY

E- and *Z*-2-haloethyldiazotates – electrophilic species hitherto suggested as intermediates in the reactions of 2-haloethylnitrosoureas (HENUs) under physiological conditions – were synthesized and characterized by ^{1}H-, ^{15}N- and ^{13}C-NMR (nuclear magnetic resonance). They were stabilized and solubilized in organic solvents as their 18-crown-6 ether complexes. Characterization of the Z-2-fluoroethyldiazotate by ^{19}F- and ^{13}C-NMR, and comparison with the Z-2-chloroethyl compound, confirmed facile cyclization to the 1,2,3-oxadiazoline and subsequent decomposition to nitrogen and ethylene oxide. The *E*-2-haloethyldiazotates form DNA interstrand cross-links at a rate, and to an extent, and with a DNA base dependence, which parallels the behaviour of the parent HENUs, while the Z isomers alkylate DNA but show minimal cross-linking. Both *E*-and Z-(2′-chloroethyl)thioethyldiazotates, neither of which can undergo cyclization, cross-link DNA efficiently.

Self-consistent-field (SCF) *ab initio* calculations provided optimized geometries, atomic charges and LUMO (Lowest Unoccupied Molecular Orbital) atom contributions for the *E*- and Z-2-haloethyldiazohydroxides. The HSAB (Hard and Soft Acids and Bases) theory, in conjunction with HOMO (Highest Occupied Molecular Orbital) values on key DNA base sites, accounted for the observed site-selectivity in the formation of identified cross-links produced by 1,3-bis-(2-chloroethyl)-1-nitrosourea. Independent chemical studies on cytosine derivatives corroborated the predicted site selectivity of attack by electrophiles and the formation of ethanocytidine cyclic adducts.

INTRODUCTION

Clinically active 2-haloethylnitrosoureas (HENUs) decompose spontaneously under physiological conditions to form electrophiles which react with cellular macromolecules, principally DNA (Wheeler, 1976). The lesions which have been correlated with biological activity include interstrand cross-linking (Kohn, 1977; Lown *et al.*, 1978; Kohn & Gibson, these proceedings[1]) and base alkylation including cyclic nucleotide adduct formation (Singer & Kuśmierek, 1982). The structures of several key intermediates were either identified by isolation and synthesis (in the case of N-nitrosooxazolidine, carbamates, and 2-haloethanol) or inferred from ^{1}H-, ^{15}N- and ^{13}C-NMR (in the case of E- and Z-2-(haloethyl)- and -2(hydroxyethyl) diazohydroxides (Lown & Chauhan, 1981)). Specific N-^{18}O-labelling of 1,3-bis-(2-chloroethyl)-1-nitrosourea (BCNU) in conjunction with in-situ reduction with alcohol dehydrogenase confirmed that a 1,2,3-oxadiazoline, plausibly arising from the Z-2-chloroethyldiazohydroxide, was the key intermediate and ruled out a diazonium ion (Lown & Chauhan, 1982). We now report the synthesis and characterization by ^{1}H-, ^{15}N-, ^{13}C- and ^{19}F-NMR of the key electrophiles, E- and Z-2-haloethyldiazotates, and the investigation of their characteristic reactions. *Ab initio* calculations in conjunction with the Hard and Soft Acids and Bases (HSAB) theory (Pearson, 1966) permit an interpretation of the site-selectivity of base alkylation and the formation of lesions, including cross-links and cyclic nucleotide adducts, which are of direct relevance to the in-vivo effects of N-nitroso compounds.

MATERIALS AND METHODS

Sodium nitrite (^{15}N-labelled; 99% enrichment) and potassium phthalimide (^{15}N-labelled; 95–99% enrichment) were obtained from Merck, Sharpe and Dohme (Montreal, QUE). E- and Z-2-haloethyldiazotates specifically labelled with ^{15}N were prepared by reaction of ^{15}N-labelled nitrosohydrazines and nitrosoureas under anhydrous conditions with potassium tertiary butoxide in tetrahydrofuran. The diazotate potassium salts were stabilized and rendered soluble in organic solvents as the 18-crown-6 ether complexes (Fig. 1).

The ^{15}N-NMR spectra were recorded on 0.01 M solutions in 0.01–0.1 M chromium acetyl acetonate with a Bruker WH-200 spectrometer operating at 20.283 MHz, using dimethylformamide as an external reference. Interstrand cross-linking of λ-DNA was detected and quantified by the method previously described (Lown *et al.*, 1978).

---

[1] See p. 155.

Fig. 1. Preparation of E-2-haloethyldiazotates. KOtBu, potassium tertiary butoxide; THF, tetrahydrofuran

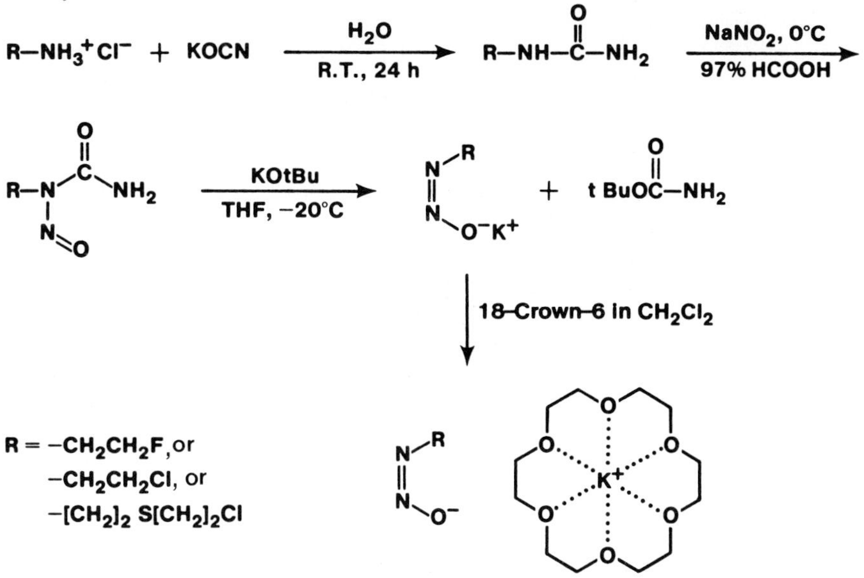

R = $-CH_2CH_2F$, or
$-CH_2CH_2Cl$, or
$-[CH_2]_2 S[CH_2]_2Cl$

RESULTS

The E- and Z-2-haloethyldiazotates are reasonably stable at 0 °C or below and both the stability and solubility in organic solvents were improved by preparation of the 18-crown-6-ether complexes. The E isomers decompose under physiological conditions to give 2-chloroethanol and ethylene glycol, while the Z isomers give ethylene oxide. The E isomer of chloroethyldiazotate reacts with deoxycytidine to form N^4-(2-chloroethyl)deoxycytidine which cyclizes to the cyclic nucleoside adduct, 3,N^4-ethanodeoxycytidine; this adduct has also been isolated from the reaction of BCNU with DNA (Gombar et al., 1980; Ludlum, these proceedings[2]). The model compound, 1-methyl-3-(2-chloroethyl)cytosine, is converted to 1-methyl-3,N^4-ethanocytosine at 37 °C, pH 7.0 with a $t_{1/2}$ of 53 min, while the isomeric 1-methyl-N^4-(2-chloroethyl)cytosine forms the identical product under these conditions with a $t_{1/2}$ of 16 min (Lown et al., 1978).

Analysis of Z-2-fluoroethyldiazotate by ^{13}C-NMR at -40 °C to -20 °C showed C-1 at δ 39.69 ($J^{13}C - ^{19}F$ = 20.3 Hz) and C-2 at δ 80.57 ($J^{13}C - ^{19}F$ = 166 Hz). At -10 °C, new signals, lacking ^{19}F coupling, appeared at δ 55.48 (C-4) and δ 70.47 (C-5). Since identical signals were obtained from the Z-2-chloroethyldiazotate, they are ascribed to 1,2,3-oxadiazoline, which subsequently loses N_2 and forms ethylene oxide.

E-2-chloroethyldiazotate (1 mM) cross-links λ-DNA to 25% in 15 min at 37 °C and pH 7.4, whereas the Z isomer at 20 mM results in 45% DNA-cross-linking only after

---

[2] See p. 137.

Fig. 2. Formation of interstrand cross-link by 2-chloroethyldiazohydroxide (E). I, O^6-(2-chloroethyl)deoxyguanosine; II, 1,O^6-oxazoliniumdeoxyguanosine; III, 1-(3-deoxycytidyl),2-(1-deoxyguanosinyl)ethane; IV, O^6-(2-hydroxyethyl)deoxyguanosine; dR, deoxyribose

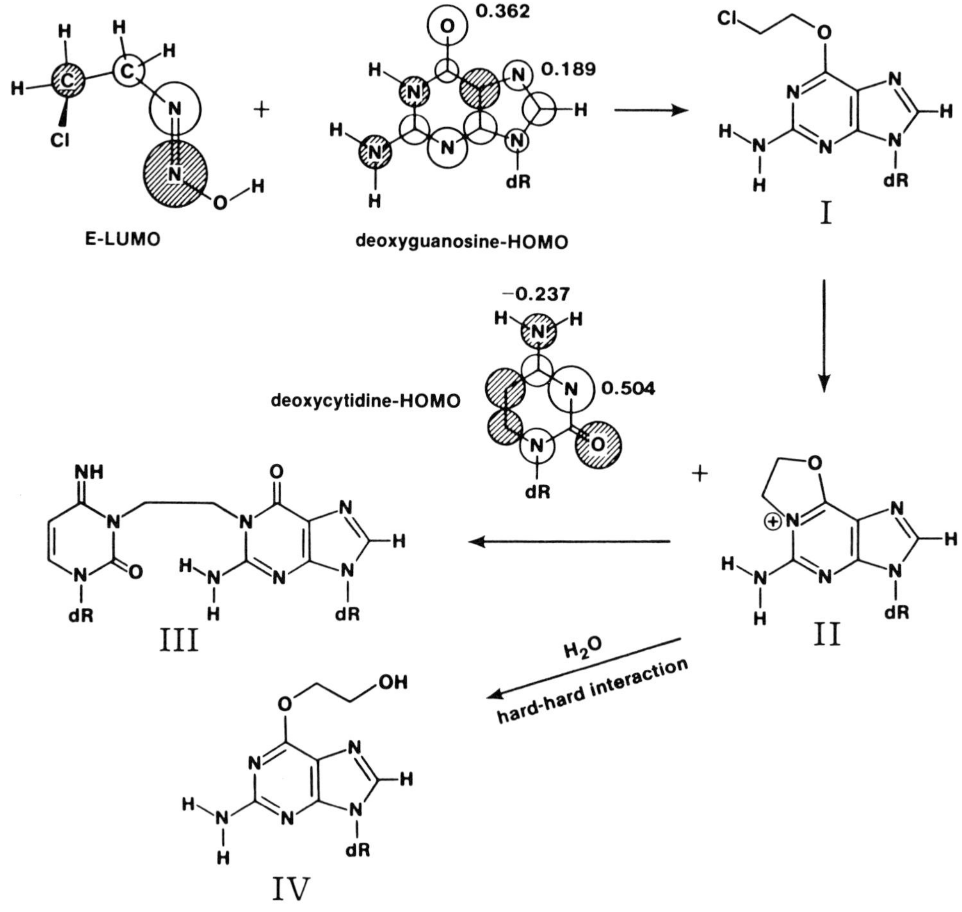

20 h under comparable conditions. *E*-2-Fluoroethyldiazotate (1 mM) cross-links λ-DNA to 24% in 75 min under physiological conditions, whereas the corresponding Z isomer generated *in situ* gives no evidence of interstrand cross-linking, only of alkylation.

The *E*-(2'-chloroethyl)thioethyldiazotate generated *in situ* gives 92% cross-linking of λ-DNA in 10 min when present at a concentration of 2 mM, whereas the corresponding Z isomer at the same concentration gives 91% λ-DNA cross-linking in 8 min. Conversion of either *E*- or *Z*-thiodiazotates to the corresponding sulfoxide derivatives removes their ability to crosslink but permits alkylation of DNA. The 2-hydroxyethyldiazotate alkylates, but does not crosslink, DNA. SCF-3-21G calculations of optimized geometry, atomic charges and LUMO contributions (see Table 1) indicate that the structure shown in Figure 2 is the most stable of the *E*

ELECTROPHILES FROM 2-HALOETHYLNITROSOUREAS 133

Table 1. LUMO atom contributions and energies for E- and Z-2-haloethyldiazohydroxides obtained by SCF 3-21G

Atom[a]	2-Fluoroethyldiazohydroxides				2-Chloroethyldiazohydroxides			
	E-I	E-II	Z-I	Z-II	E-I	E-II	Z-I	Z-II
N-1	−0.842	0.895	0.888	0.895	0.802	0.915	−0.853	−0.946
N-2	0.960	−1.023	0.955	0.987	−0.898	−1.009	0.927	0.987
O	−0.301	0.349	−0.313	−0.291	0.279	0.309	−0.316	−0.272
C-1	−0.310	0.218	−0.261	−0.229	0.331	0.164	−0.268	−0.154
(H¹)′	0.478	0.379	−0.424	0.376	−0.517	0.327	−0.466	0.316
(H¹)″	−0.506	0.361	0.456	−0.376	0.490	−0.307	0.462	−0.316
C-2	−0.070	−0.093	−0.109	0.241	1.064	−0.100	−0.271	0.210
(H²)′	−0.097		0.150	0.129	0.122		0.128	0.147
(H²)″				−0.129				−0.147
F/Cl	0.034		0.150	0.129				
Energy (au)	0.12147	0.15359	0.13677	0.13624	0.11956	0.14149	0.13219	0.12462

[a] Orbital is 1s for hydrogen and 2py for other atoms.

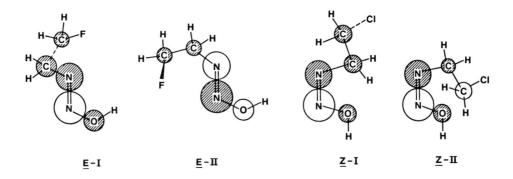

E-I   E-II   Z-I   Z-II

configurations in each case. The LUMO contributions for C-1 in the 2-haloethyldiazohydroxides are significantly greater than for the methyl group in methyldiazohydroxide (Lown *et al.*, 1984). In addition, in the more reactive form of the Z isomer, the oxygen and the C-2 atoms are only 1.8 Å apart, virtually within bonding distance, which is in accord with the experimentally observed facile cyclization of this isomer. The calculated energy difference between the E and Z isomers of 31.5 kcal. mol^{-1} is in accord with their observed lack of interconversion.

DISCUSSION

The results demonstrate that the previously suggested ultimate electrophiles generated under physiological conditions from HENUs, namely the E- and Z-2-

Table 2. HOMO atom contributions and energies for guanosine[a] and cytidine[a]

Guanosine		Cytidine	
Atom	HOMO Contribution	Atom	HOMO Contribution
N-1	−0.229	N-1	0.425
C-2	0.248	C-2	—
N-3	0.453	N-3	0.504
C-4	−0.323	C-4	0.173
C-5	−0.457	C-5	−0.403
C-6	0.025	C-6	−0.248
O^6	0.362	O^2	−0.520
N-7	0.189	N^4	−0.237
C-8	0.391		
N^2	−0.279		
Energy (au)	−0.21550		−0.33635[b]

[a] Obtained by STO-3G calculations
[b] Obtained by SCF 3-21G calculations

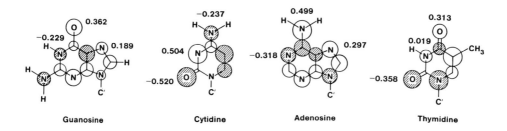

Guanosine   Cytidine   Adenosine   Thymidine

haloethyldiazohydroxides (Wheeler, 1976; Lown et al., 1978) can be isolated and characterized. Moreover, they may be stabilized by formation of their 18-crown-6-ether complexes which permits their dissolution in organic solvents. These electrophilic species reproduce the characteristic features of the parent HENUs including interstrand cross-linking (Kohn, 1977; Lown et al., 1978) and alkylation of DNA, as well as giving rise to the spectrum of products isolated from the HENUs under physiological conditions (Wheeler, 1976). The E and Z configurational isomers do not interconvert (Lown et al., 1984) and, whereas the Z isomers undergo facile cyclization to 1,2,3-oxadiazoline (Lown & Chauhan, 1981), the E isomers are capable of forming interstrand cross-links in DNA. In contrast, both the E and Z isomers of the thioether-diazohydroxides crosslink DNA and they do so by an order of magnitude more efficiently, because neither can cyclize.

A site-specific interaction between the more nucleophilic O^6 HOMO of guanine (see Table 2) and the C-1 LUMO of E-2-chloroethyldiazohydroxide predicts formation of O^6-(2-chloroethyl)deoxyguanosine (I, Fig. 2). The latter forms the cyclic adduct 1,O^6-oxazoliniumdeoxyguanosine (II, Fig. 2) (Tong et al., 1982). A subsequent soft-soft interaction between the more nucleophilic N-3 HOMO of deoxycytidine and the softer carbon adjacent to O^6 in the adduct accounts for the selective formation of the observed ethano-bridged cross-link (III, Fig. 2) between the N-3 of deoxycytidine and

Fig. 3. Reaction of 1-methylcytosine (C) with 2-chloroethyldiazohydroxide (E). I, 1-methyl-3-(2-chloroethyl)cytosine; II, 1-methyl-N^4-(2-chloroethyl)cytosine; III, 1-methyl-3,N^4-ethanocytosine; R, CH_3

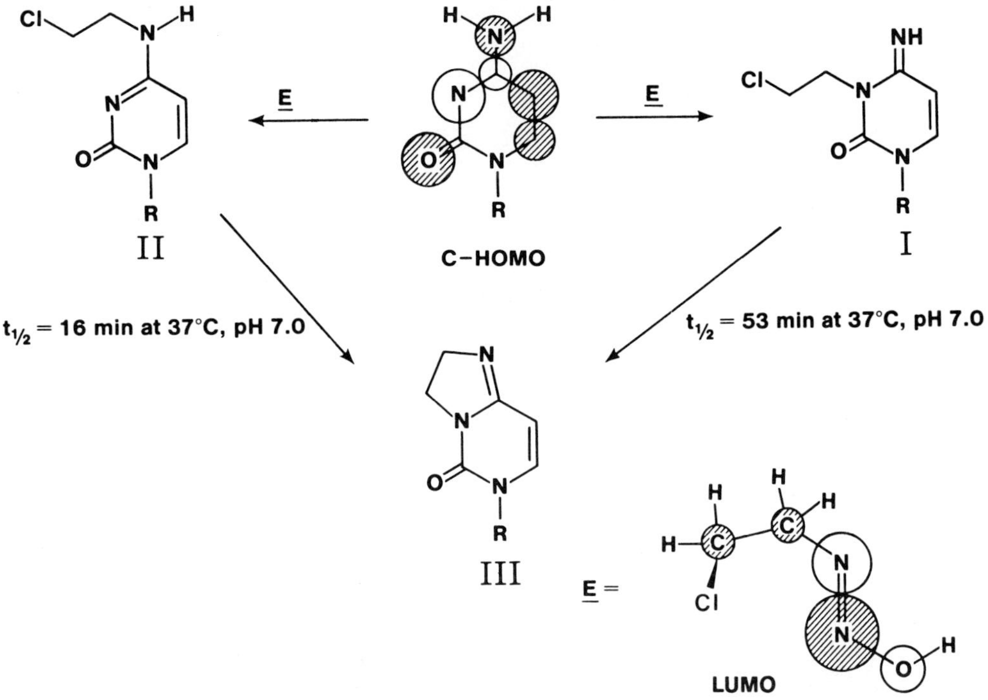

the N-1 of deoxyguanosine (Tong et al., 1982). In contrast, a hard-hard HSAB interaction demanded by the harder base, water, predicts attack at the hard carbon adjacent to the positive nitrogen in the 1-O^6-oxazoliniumdeoxyguanosine and this accounts for the concomitant formation of O^6-(2-hydroxyethyl)deoxyguanosine (IV, Fig. 2) (Tong et al., 1981).

The greater calculated HOMO component on the N-3 of cytidine of 0.504 (Table 2) reflects the greater nucleophilicity of this centre. Selective interaction of this site with the C-1 LUMO of the E-2-chloroethyldiazohydroxide results in the formation of 3-(2-chloroethyl)deoxycytidine from the reaction of BCNU with DNA. Subsequent intramolecular chloride displacement in this adduct gives rise to the isolated 3,N^4-ethanodeoxycytidine (Gombar et al., 1980). The model reaction with 1-methylcytosine (Fig. 3) demonstrates that this cyclization proceeds with a $t_{1/2}$ of 53 min under physiological conditions, whereas the isomeric N^4-(2-chloroethyl) adduct cyclizes to the same 3,N^4-ethano adduct with a $t_{1/2}$ of 16 min under these conditions (Lown et al., 1978).

A hard-hard HSAB interaction between the C-1 LUMO (0.183) of E-2-hydroxyethyldiazohydroxide and, preferentially, the N-7 HOMO (0.189) of deoxyguanosine accounts for the isolation of 7-(2-hydroxyethyl)deoxyguanosine from the reaction of BCNU with DNA (Gombar et al., 1980).

Although the calculated HOMO contribution for the O^6 of guanosine is greater than that for the N-7 (Table 2), this does not necessarily mean that the O^6 will always be the preferred site of alkylation. The fact that the N-7 is often selected (Singer & Kuśmierek, 1982) implies that additional factors such as hydrogen bonding and steric hindrance operate. Our future work is designed to assess these additional influences.

ACKNOWLEDGEMENTS

This work was supported by National Cancer Institute DHHS grant 1 RO1 CA21488-01 (to JWL), by the Alberta Cancer Board, and by a contract with the National Foundation for Cancer Research.

REFERENCES

Gombar, C.T., Tong, W.P. & Ludlum, D.B. (1980) Reactions of bischloroethyl nitrosourea and chloroethyl cyclohexyl nitrosourea with DNA. *Biochem. Pharmacol., 29,* 2639–43

Kohn, K.W. (1977) Interstrand cross-linking of DNA by BCNU and other 1-(2-haloethyl)-1-nitrosoureas. *Cancer Res., 37,* 1450–1454

Lown, J.W. & Chauhan, S.M.S. (1981) Synthesis of 2-(alkylimino)-3-nitrosooxazolidines and other intermediates in CENU aqueous decomposition. *J. org. Chem., 46,* 2479–2489

Lown, J.W. & Chauhan, S.M.S (1982) Discrimination between alternative pathways of aqueous decomposition of antitumor (2-chloroethyl)nitrosoureas using specific ^{18}O-labeling. *J. org. Chem., 47,* 851–856

Lown, J.W., McLaughlin, L.W. & Chang, Y.M. (1978) Mechanism of action of 2-haloethylnitrosoureas on DNA and its relation to antileukemic properties. *Bioorg. Chem., 7,* 97–110

Lown, J.W., Chauhan, S.M.S., Koganty, R.R. & Sapse, A.-M. (1984) Alkyldinitrogen species implicated in the carcinogenic, mutagenic and anticancer activities of N-nitroso compounds. Characterization by ^{15}N-NMR of ^{15}N-enriched compounds and analysis of DNA base site-selectivity by *ab initio* calculations. *J. Am. chem. Soc., 106,* 6401–6408

Pearson, R.G. (1966) Acids and Bases. *Science, 151,* 171–179

Singer, B. & Kuśmierek, J.T. (1982) Chemical mutagenesis. *Ann. Rev. Biochem., 51,* 655–693

Tong, W.P., Kirk, M.C. & Ludlum, D.B. (1981) Formation of 6-hydroxyethylguanosine in DNA treated with BCNU (N,N'-bis[2-chloroethyl]-N-nitrosourea). *Biochem. biophys. Res. Commun., 100,* 351–357

Tong, W.P., Kirk, M.C. & Ludlum, D.B. (1982) Formation of the cross-link 1-[N^3-deoxycytidyl], 2-(N^1-deoxyguanosinyl]ethane in DNA treated with N,N'-bis(2-chloroethyl)-N-nitrosourea. *Cancer Res., 42,* 3102–5

Wheeler, G.D. (1977) A review of studies on the mechanism of action of nitrosoureas. *Am. chem. Soc. Symp. Ser., 30,* 87–119

FORMATION OF CYCLIC ADDUCTS IN NUCLEIC ACIDS BY THE HALOETHYLNITROSOUREAS

D.B. LUDLUM

Division of Oncology, Department of Medicine, Albany Medical College, Albany, NY, USA

SUMMARY

The haloethylnitrosoureas react with adenine and cytosine nucleosides in aqueous solution at neutral pH to form $1,N^6$-ethanoadenine and $3,N^4$-ethanocytosine nucleosides, respectively. These cyclic nucleosides probably result from a two-step reaction in which a haloethyl group first attaches to the ring nitrogen and then reacts with the exocyclic amino group. They are relatively stable at neutral pH, and can be formed in both DNA and synthetic polynucleotides. Since the base-pairing positions are affected, the presence of these modified nucleosides in DNA would presumably change the informational content of this molecule, leading to mutagenic or cytotoxic effects.

Haloethyl groups can also attack the O^6 position of guanine nucleosides. Attempts to synthesize O^6-chloroethylguanine nucleosides have not yet been successful, but O^6-fluoroethylguanosine has been fully characterized. This compound undergoes a base-catalysed hydrolysis to 1-hydroxyethylguanosine in aqueous solution at 37 °C, evidently through the cyclic intermediate, $1,O^6$-ethanoguanosine. This intermediate can be isolated by high-performance liquid chromatography and has been partially characterized by ultraviolet spectrometry. We have hypothesized that the corresponding $1,O^6$-ethanoguanine deoxynucleoside is formed in DNA by rearrangement of O^6-haloethyldeoxyguanosine, and that this cyclic nucleoside is an intermediate in the formation of the DNA cross-link, 1-(3-deoxycytidyl),2-(1-deoxyguanosinyl) ethane. The formation of this cross-link in DNA probably explains some of the antitumour activity of the haloethylnitrosoureas. A similar mechanism may lead to cross-links between deoxyadenosine and thymidine.

INTRODUCTION

The formation of cyclic nucleosides by the haloethylnitrosoureas evidently plays a key role in the expression of their biological activities. Historically, the isolation of one such nucleoside, $3,N^4$-ethanocytidine, from polycytidylic acid which had been reacted with bis-chloroethylnitrosourea focused attention on the reactions of these agents with the nucleic acids as a basis for their cytotoxicity (Kramer et al., 1974; Ludlum et al., 1975).

The following report describes the isolation and characterization of cyclic nucleosides of adenine and cytosine and provides evidence for the formation of $1,O^6$-ethanoguanine-containing cyclic nucleosides. The significance of these reactions in explaining the mutagenic, carcinogenic and cytotoxic actions of the haloethylnitrosoureas is also discussed.

MATERIALS AND METHODS

Cyclic nucleoside adducts have been isolated from the reaction of haloethylnitrosoureas with both synthetic polyribonucleotides and DNA. Polyribonucleotides for these studies were obtained from Miles Laboratory (Naperville, IL) or from Schwarz/Mann (Orangeburg, NY). 1,3-Bis-(2-chloroethyl)-l-nitrosourea (BCNU), 1-(2-chloroethyl)-3-cyclohexyl-1-nitrosourea (CCNU), 1,3-bis-(2-fluoroethyl)-1-nitrosourea (BFNU), and 1-fluoroethyl-3-cyclohexyl-1-nitrosourea (FCNU) were obtained from the Division of Cancer Treatment, National Cancer Institute (Bethesda, MD); enzymes for digesting polyribonucleotides and DNA came from Worthington Biochemical Corp. (Freehold, NJ); all other reagents were obtained from standard commercial sources.

Typically, polyribonucleotide or DNA was incubated at a concentration of 2 mg/ml in 50 mM cacodylate buffer pH 7, at 37 °C with BCNU or CCNU at 1 mg/ml for six hours. (This reaction time corresponds to several half-lives of these reagents.) The polyribonucleotide or DNA was then precipitated with ethanol, redissolved and reprecipitated several times to wash it free of unreacted nitrosourea. Modified nucleosides were released from polyribonucleotides by digestion with venom phosphodiesterase and bacterial alkaline phosphatase for 14 hours at 37 °C. DNA was digested to its component nucleosides with a combination of deoxyribonuclease I, venom phosphodiesterase, spleen phosphodiesterase and bacterial alkaline phosphatase.

In some cases, separations of modified nucleosides including the cyclic adducts were performed on a cation-exchange resin (SP-Sephadex C-25), but high-performance liquid chromatography (HPLC), with its superior resolving power, was used to isolate and purify most of the modified nucleosides. Most HPLC separations were performed on a C_{18} reverse-phase column eluted with 50 mM KH_2PO_4, pH 4.5 or 6, containing varying amounts of acetonitrile. Nucleosides can also be separated on a strong cation-exchange resin (SCX) and their purity verified in this way. Eluents were monitored with a Perkin-Elmer LC-55B variable wavelength detector, or more recently with a Hewlett-Packard 1040A fixed diode array detector, interfaced with a Perkin-Elmer Sigma 10 Data System.

In most cases, structures of modified nucleosides were determined from ultraviolet and mass spectral data. Additional quantities of ethanocytosine and ethanoadenine

Fig. 1. Formation of 3,N^4-ethanocytosine nucleosides from haloethylnitrosoureas. R', the 3-substituent in the haloethylnitrosourea (a chloroethyl group in BCNU and a cyclohexyl group in CCNU); R, ribosyl; dR, deoxyribosyl; X, halogen atom; Cyd, cytidine; dCyd, deoxycytidine

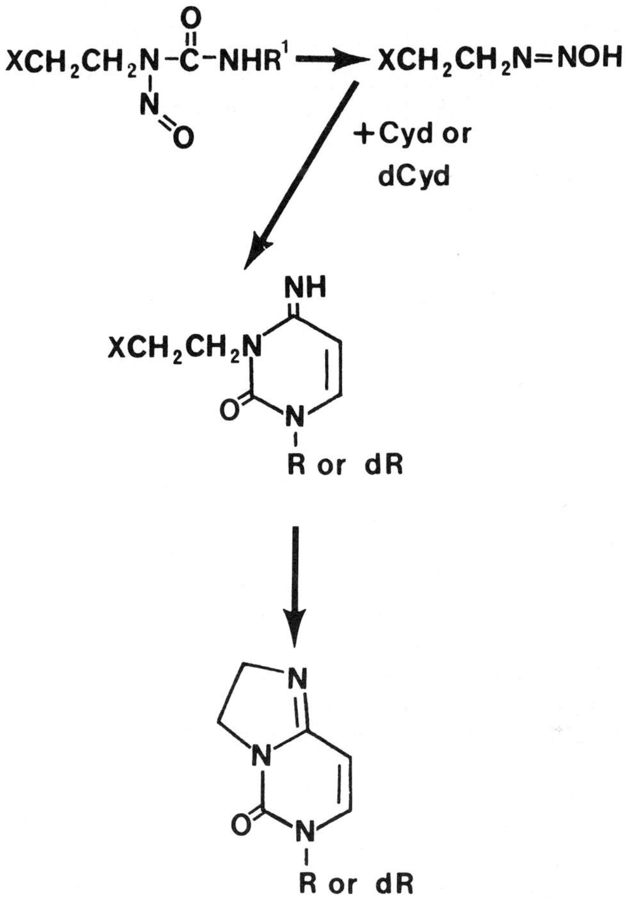

nucleosides were synthesized from 1,2-dibromoethane or 1-bromo-2-chloroethane and the parent nucleosides as described previously (Tong & Ludlum, 1979; Gombar et al., 1980).

Ultraviolet spectra were recorded on a Beckman Model 35 spectrophotometer in 0.1 N HCl, 0.1 N sodium cacodylate buffer, pH 7, as well as in 0.1 M NaOH. Early in these investigations, mass spectrometry was performed on trimethylsilylated nucleosides using an electron impact technique. More recently, through the courtesy of M. Kirk of the Southern Research Institute, we have obtained mass spectral data directly on nonsilylated nucleosides by field desorption and fast atom bombardment techniques.

Fig. 2. Formation of 1,N^6-ethanoadenine nucleosides from haloethylnitrosoureas. Ado, adenosine; dAdo, deoxyadenosine; other symbols as in Figure 1

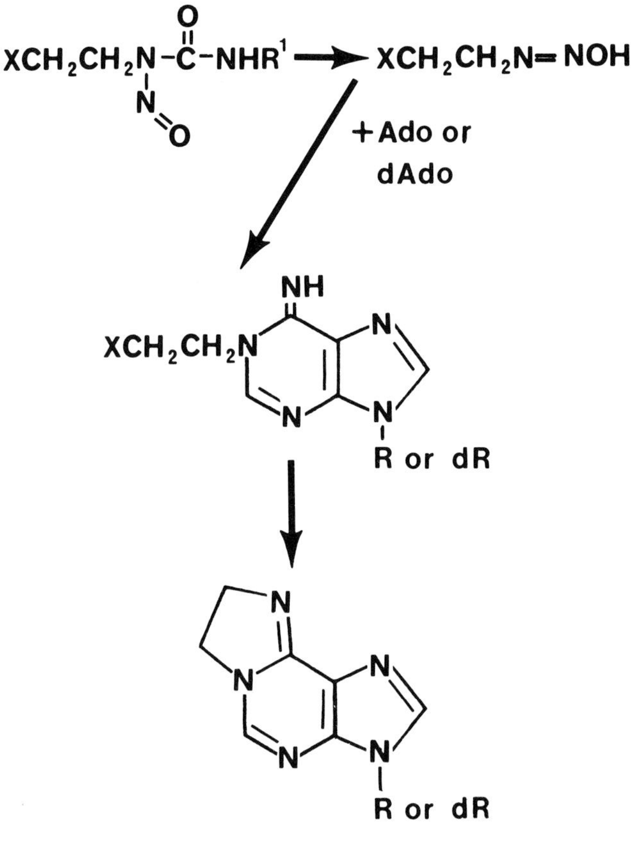

RESULTS

The first cyclic nucleoside adduct identified as a product of reaction with the haloethylnitrosoureas was 3,N^4-ethanocytidine. This product and 3-hydroxyethylcytidine were isolated from the hydrolysate of a polycytidylic acid which had been reacted with BCNU (Kramer et al., 1974). The work of Lijinsky et al. (1972), which showed that a CD_3– group was transferred intact from deuterated methylnitrosourea to nucleophilic sites in DNA, suggested to us that ethanocytidine was formed by the initial transfer of a $ClCH_2CH_2$– group from BCNU to the 3 position of cytidine. Ring closure with the exocyclic amino group would then occur with the elimination of chlorine. This pathway for the formation of ethanocytidine is shown in Figure 1.

Initial attachment of the chloroethyl group to the N-3 of cytidine is in agreement with the observation that the other modified cytidine identified by Kramer et al.

(1974), 3-hydroxyethylcytidine, also bears a substituent in that position. Alternatively, as considered by Lown et al. (1978), 3,N^4-ethanocytidine could be formed following initial attachment of the chloroethyl group to the N^4 position of cytidine.

The next step in demonstrating the mechanism of ethanocytidine formation was to isolate the intermediate haloethylcytidine. This was accomplished by reacting the fluorine analogue of BCNU, i.e. 1,3-bis-(2-fluoroethyl)-1-nitrosourea (BFNU), with cytidine. Since the fluoroethyl group is less susceptible to nucleophilic displacement than the chloroethyl group, it was possible to isolate 3-fluoroethylcytidine from the reaction mixture. This compound has a half-life of approximately 80 minutes in potassium phosphate buffer pH 7.4, at 37 °C, cyclizing to 3,N^4-ethanocytidine (Tong & Ludlum, 1978). The absence of any N^4-fluorethylcytidine confirmed that, in the reaction with BFNU, the initial attack was at the 3 position as shown in Figure 1. However, it is possible that the site selectivity for the chloroethyl group is different than for the fluoroethyl group.

An entirely analogous reaction occurs with the adenine nucleosides (Tong & Ludlum, 1979). When adenosine was incubated with BFNU in aqueous solution at pH 7, 1-fluoroethyladenosine and 1,N^6-ethanoadenosine, as well as 1-hydroxyethyladenosine, were identified as products. Similarly, reaction with BCNU produced 1-chloroethyladenosine, 1,N^6-ethanoadenosine, and 1-hydroxyethyladenosine. When 1-fluoroethyladenosine was isolated and incubated in neutral buffer at 37 °C, it had a half-life of approximately 28 hours, cyclizing to form 1,N^6-ethanoadenosine. Thus, the pathway shown in Figure 2 for the formation of ethanoadenosine is in all respects equivalent to that for the formation of ethanocytidine shown in Figure 1.

Equivalent reactions occur with the deoxynucleosides, and both ethanodeoxycytidine and ethanodeoxyadenosine have been identified in hydrolysates of DNA which was reacted with BCNU or with CCNU. The use of HPLC has been of great importance in identifying these derivatives; retention times on a C_{18} reverse-phase column are shown in Table 1. These retention times, together with the ultraviolet spectral data shown in Table 2, can be used to verify the presence of these derivatives in DNA hydrolysates.

The final cyclic nucleoside adduct, 1,O^6-ethanoguanosine, has been less fully characterized but, in many ways, is the most interesting of the cyclic adducts formed by the haloethylnitrosoureas. This is because of its apparent role as an intermediate in DNA cross-linking.

DISCUSSION

The formation of cyclic nucleoside adducts in DNA by the haloethylnitrosoureas is unique to this group of antitumour agents. Since the biological activities of the haloethylnitrosoureas are somewhat different from other agents which modify DNA covalently, it is reasonable to assume that their unique biological properties are related to the formation of cyclic adducts.

Table 1. High-performance liquid chromatographic retention times (minutes) of cyclic adducts of nucleosides and deoxynucleosides

Compound	System[a]	
	A	B
Cytidine	4.8	4.8
Deoxycytidine	6.1	6.2
3,N^4-Ethanocytidine	5.7	7.9
3,N^4-Ethanodeoxycytidine	8.3	11.5
Adenosine	23.2	24.5
Deoxyadenosine	35.2	40.0
1,N^6-Ethanoadenosine	8.9	11.5
1,N^6-Ethanodeoxyadenosine	12.8	16.5
Guanosine	9.4	9.5
Deoxyguanosine	12.8	13.3
1,O^6-Ethanoguanosine	8.8	–

[a] System A: Spherisorb ODS 5μ (4.6 × 250 mm) column eluted isocratically with 3% acetonitrile in 50 mM KH_2PO_4, pH 4.5, at 1 ml/min. System B: Same column eluted isocratically with 3% acetonitrile in 50 mM KH_2PO_4, pH 6, at 1 ml/min.

Table 2. Ultraviolet spectral analysis of cyclic adducts of nucleosides and deoxynucleosides[a]

Compound	Acid		pH 7		Base	
	max	min	max	min	max	min
3,N^4-Ethanocytidine	286	244	283	251	279	256
3,N^4-Ethanodeoxycytidine	285	243	284	250	279	255
1,N^6-Ethanoadenosine	262	235	262	235	269	239
1,N^6-Ethanodeoxyadenosine	263	239	263	237	269	238
1,O^6-Ethanoguanosine	250, 290	237, 267	unstable			

[a] The wavelength of the maximum (max) or minimum (min) absorption is given in each case.

Following the suggestion of Kramer et al. (1974) and Ludlum et al. (1975) that the transfer of the haloethyl group to a DNA base could lead to cross-linking, Kohn (1977) and Lown et al. (1978) demonstrated by physical methods that interstrand cross-links were indeed formed in DNA. Later, Erickson et al. (1980) showed that cell lines which have the Mer$^+$ phenotype and which are able to remove alkyl groups from the O^6 position of guanine are resistant to cross-linking. Finally, Tong et al. (1982) demonstrated the existence of the cross-link, 1-(3-deoxycytidyl),2-(1-deoxyguanosinyl)ethane (dCydCH$_2$CH$_2$dGuo), in DNA which had been treated with BCNU, and proposed that the pathway for the formation of this cross-link is as shown in Figure 3.

Fig. 3. Formation of the intermediate, 1,O^6-ethanodeoxyguanosine, and its reaction with deoxycytidine to form a DNA cross-link. Guo, guanosine; dGuo, deoxyguanosine; other symbols as in Figure 1

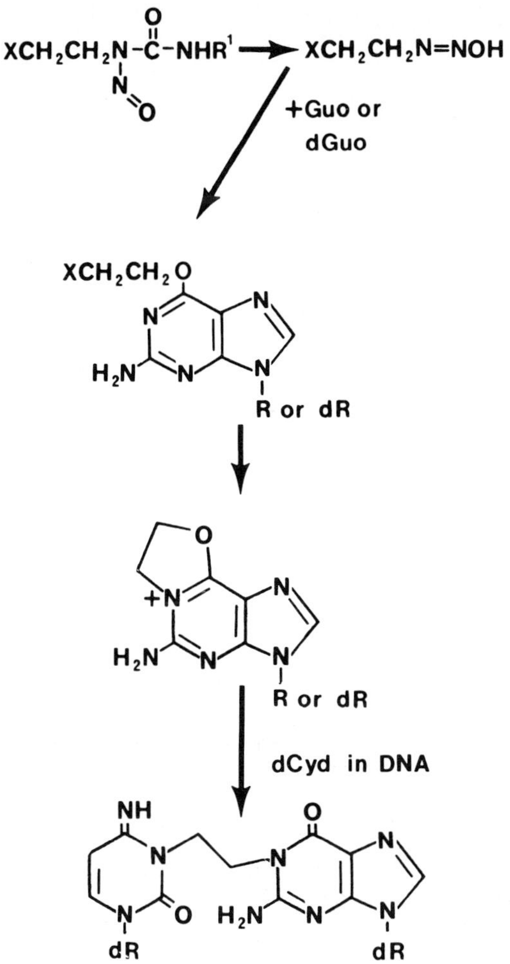

Later, Tong et al. (1983) provided support for this mechanism by showing that O^6-fluoroethylguanine was formed in DNA that had been treated with FCNU. At the same time, they showed that O^6-fluoroethylguanosine hydrolysed in aqueous solution at 37 °C to 1-hydroxyethylguanosine. They also observed an unstable intermediate, with the chromatographic properties shown in Table 1 and the ultraviolet spectral characteristics shown in Table 2, which hydrolysed to 1-hydroxyethylguanosine. Since this intermediate is formed from O^6-fluoroethylguanosine and decomposes to 1-hydroxyethylguanosine, it probably has the cyclic structure 1,O^6-ethanoguanosine. The corresponding deoxy adduct, 1,O^6-ethanodeoxyguanosine, is probably formed as an intermediate in the cross-linking of DNA.

Fig. 4. Postulated formation of 3,O^4-ethanothymidine and its reaction with deoxyadenosine to form DNA cross-link. Thd, thymidine; dThd, deoxythymidine; dAdo, deoxyadenosine; other symbols as in Figure 1

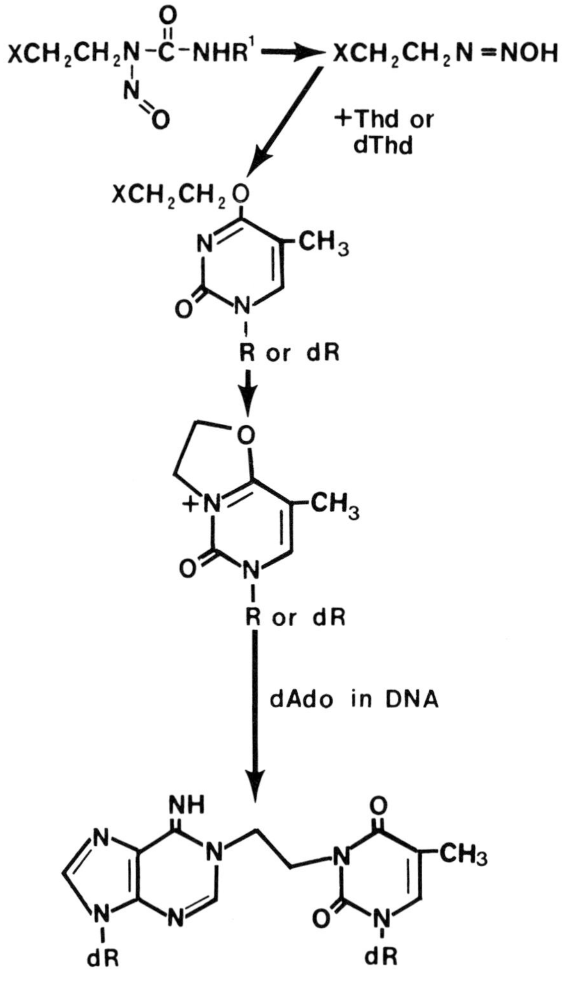

The presence of these cyclic adducts would certainly disrupt the base-pairing relationships in the DNA helix; ethanocytosine formation would disrupt GC (guanine · cytosine) base pairs, and ethanoadenine formation would disrupt AT (adenine · thymine) pairs. Presumably, this would lead to alterations in the informational content of DNA containing these modifications. The presence of ethanocytosine and ethanoadenine could also lead to either chemical or enzymatic degradation of DNA.

The evidence linking 1,O^6-ethanoguanine formation to DNA cross-linking suggests that this intermediate plays a key role in the cytotoxic action of the haloethylnitrosoureas. The observation that Mer$^+$ cells, which have normal levels of

O^6-methylguanine-DNA methyltransferase are resistant to the cytotoxic action of the haloethylnitrosoureas suggests that O^6-alkylation of guanine is involved in cytotoxicity. However, neither of the two cross-linked dinucleosides which have been isolated, 1,2-(diguanosin-7-yl) ethane nor dCydCH$_2$CH$_2$dGuo, involves a linkage through the O^6 position. Since the rearrangement shown in Figure 3 has been demonstrated to occur, removal of a haloethyl group from the O^6 position of guanine before rearrangement would prevent formation of the dCydCH$_2$CH$_2$dGuo cross-link and would explain the importance of the Mer$^+$ trait.

Recently, Robins et al. (1983) and Brent (1984) have shown that the addition of purified O^6-methylguanine-DNA methyltransferase to incubation mixtures containing DNA treated with the haloethylnitrosoureas suppresses cross-link formation. Since DNA cross-linking was measured by physical methods in both cases, however, these results do not prove that it was dCydCH$_2$CH$_2$dGuo cross-link formation which was suppressed. An additional pathway for cross-linking, suggested by Robins et al. (1983), involves an initial attack of the haloethyl group at the O^4 position of thymine followed by cross-link formation with the N^6 position of adenine. This suggestion is in accordance with their observation that O^6-methyltransferase can remove alkyl groups from the O^4 position of thymine so that cross-linking by this mechanism would also be suppressed by O^6-methyltransferase.

One further possibility which we would like to suggest would involve the formation of an intermediate 3,O^4-ethanothymine which could react with the N-1 of adenine, forming a dAdoCH$_2$CH$_2$dThd cross-link (Fig. 4). There is no direct evidence for this mechanism at the present time, but it would extend the parallelism that exists in the way in which the haloethylnitrosoureas modify the DNA bases. Thus, AT base pairs would be disrupted by ethanoadenine formation in the same way as GC base pairs are disrupted by ethanocytidine formation. This mechanism would lead to AT cross-linking by a route analogous to that by which GC cross-linking evidently occurs.

ACKNOWLEDGEMENTS

This work was supported by grant CA32171 from the National Cancer Institute, National Institutes of Health, Department of Health, Education, and Welfare. We thank Suzanne Wissel for editorial assistance.

REFERENCES

Brent, T.P. (1984) Suppression of cross-link formation in chloroethylnitrosourea-treated DNA by an activity in extracts of human leukemic lymphoblasts. *Cancer Res.*, **44**, 1887–1892

Erickson, L.C., Laurent, G., Sharkey, N.A. & Kohn, K.W. (1980) DNA cross-linking and monoadduct repair in nitrosourea-treated human tumour cells. *Nature*, **288**, 727–729

Gombar, C.T., Tong, W.P. & Ludlum, D.B. (1980) Mechanism of action of the nitrosoureas. IV. Reactions of BCNU and CCNU with DNA. *Biochem. Pharmacol., 29,* 2639–2643

Kohn, K.W. (1977) Interstrand cross-linking of DNA by 1,3-bis-(2-chloroethyl)-1-nitrosourea and other 1-(2-haloethyl)-1-nitrosoureas. *Cancer Res., 37,* 1450–1454

Kramer, B.S., Fenselau, C.C. & Ludlum, D.B. (1974) Reaction of BCNU (1,3-bis[2-chloroethyl]-1-nitrosourea) with polycytidylic acid. Substitution of the cytosine ring. *Biochem. biophys. Res. Commun., 56,* 783–788

Lijinsky, W., Garcia, H., Keefer, L., Loo, J. & Ross, A.E. (1972) Carcinogenesis and alkylation of rat liver nucleic acids by nitrosomethylurea and nitrosoethylurea administered by intraportal injection. *Cancer Res., 32,* 893–897

Lown, J.W., McLaughlin, L.W. & Chang, Y.-M. (1978) Mechanism of action of 2-haloethylnitrosoureas on DNA and its relation to their antileukemic properties. *Bioorg. Chem., 7,* 97–110

Ludlum, D.B., Kramer, B.S., Wang, J. & Fenselau, C. (1975) Reaction of 1,3-bis(2-chloroethyl)-1-nitrosourea with synthetic polynucleotides. *Biochemistry, 14,* 5480–5485

Robins, P., Harris, H.L., Goldsmith, I. & Lindahl, T. (1983) Cross-linking of DNA induced by chloroethylnitrosourea is prevented by O^6-methylguanine-DNA transferase. *Nucleic Acids Res., 11,* 7743–7758

Tong, W.P. & Ludlum, D.B. (1978) Mechanism of action of the nitrosoureas. I. Role of fluoroethyl cytidine in the reaction of BFNU with nucleic acids. *Biochem. Pharmacol., 27,* 77–81

Tong, W.P. & Ludlum, D.B. (1979) Mechanism of action of the nitrosoureas. III. Reaction of bis-chloroethyl nitrosourea and bis-fluoroethyl nitrosourea with adenosine. *Biochem. Pharmacol., 28,* 1175–1179

Tong, W.P., Kirk, M.C. & Ludlum, D.B. (1982) Formation of the cross-link, 1-[N^3-deoxycytidyl],2-[N^1-deoxyguanosinyl]ethane, in DNA treated with N,N'-bis(2-chloroethyl)-N-nitrosourea (BCNU). *Cancer Res., 42,* 3102–3105

Tong, W.P., Kirk, M.C. & Ludlum, D.B. (1983) Mechanism of action of the nitrosoureas. V. The formation of O^6-(2-fluoroethyl)guanine and its probable role in the cross-linking of deoxyribonucleic acid. *Biochem. Pharmacol., 32,* 2011–2015

INVESTIGATION OF 6-THIODEOXYGUANOSINE ALKYLATION PRODUCTS AND THEIR ROLE IN THE POTENTIATION OF BCNU CYTOTOXICITY

W.J. BODELL

The Brain Tumor Research Center of the Department of Neurological Surgery, School of Medicine, University of California, San Francisco, CA, USA

SUMMARY

The principal products of the reaction of 6-thio-2′-deoxyguanosine (6-TdGuo) with dimethylsulfate are S^6-methyl-6-thiodeoxyguanosine and 7-methyl-6-thiodeoxyguanosine, identified by ultraviolet and mass spectrometry. To study the reactions of 6-TdGuo in DNA, cells were treated with 6-thioguanine, which is incorporated into DNA during S-phase; DNA was purified from cell lysates and reacted with ^{3}H-methylnitrosourea. In addition to the expected methylated purines (O^6-methylguanine, 7-methylguanine and 3-methyladenine), 0.6% of the total product was S^6-methyl-6-thioguanine. On the basis of thioguanine content, the formation of S^6-methyl-6-thioguanine occurs 70-fold more efficiently than O^6-methylguanine, which indicates that 6-thioguanine incorporated into DNA is very susceptible to chemical modification by alkylating agents. The ultraviolet and mass spectra of two of the major products of the reaction between 6-TdGuo and 2-chloroethyl methanesulfonate suggest that the structures are $(1,S^6$-ethano)-6-thiodeoxyguanosine and $(S^6,7$-ethano)-6-thiodeoxyguanosine, which are presumably formed by an internal cyclization reaction that proceeds through a sulfonium ion intermediate. In cells, this intermediate could react with DNA nucleophiles to form both DNA intra- and interstrand cross-links.

INTRODUCTION

The guanine analogue, 6-thioguanine (6-TG), is used in the treatment of cancer (Paterson & Tidd, 1975). We have shown that treatment of 9L rat brain tumour cells for 48 h with 0.2 µM 6-TG potentiates the cytotoxicity of subsequently administered 1,3-bis-(2-chloroethyl)-1-nitrosourea (BCNU) by 10-fold, and causes a 30% increase in the number of BCNU-induced sister chromatid exchanges and a 50% increase in the number of DNA interstrand cross-links formed (Bodell *et al.*, 1985). Similar observations have been made by Fujimoto *et al.* (1982). Because BCNU cytotoxicity is thought to be related to the number of DNA cross-links formed, we have suggested that the increase in DNA cross-links may be the result of the formation of S^6-(2-chloroethyl)-6-thioguanine, which subsequently reacts to form DNA cross-links (Bodell *et al.*, 1985). The studies reported here were conducted to identify the alkylation products of 6-thio-2'-deoxyguanosine (6-TdGuo), both as the free nucleoside and after incorporation into DNA, with various alkylating agents.

MATERIALS AND METHODS

Equipment

High-performance liquid chromatography (HPLC) was performed using a Chromatronix 3500 pump coupled to a Rheodyne 7120 sample injector valve with a 500 µl sample loop. A 5-micron C-18 reverse-phase column (Alltech 605-RP) was used for separation. Ultraviolet (UV) absorbance at 315 nm was monitored with a Perkin Elmer Model 55 detector. The mobile phase consisted of 16% methanol in 0.01 M ammonium acetate (pH 5.1) eluted at a flow rate of 1.2 ml/min.

UV-absorbing peaks were collected and their UV spectra were obtained at pH 1, 5.1 and 12 on a Cary 219 spectrophotometer. Mass spectra were obtained on a Kratos MS-9 spectrometer with a direct insertion probe, source temperature of 250–270 °C, and an ionization potential of 70 ev.

Reaction of 6-dTGuo with dimethylsulfate

Ten mg of 6-TdGuo (35 µmoles) (Drug Synthesis Branch, National Cancer Institute, Bethesda, MD) was dissolved in 1.0 ml of anhydrous dimethylformamide (DMF), 10 µl of dimethylsulfate (DMS, 105 µmoles, Aldrich, Milwaukee, WI) was added, and the mixture was incubated at 37 °C for two hours. A portion of the reaction mixture was acid-hydrolysed to convert deoxyribonucleoside to free base using the following procedure: 30 µl of the reaction mixture was added to 25 µl of 1 N HCL and 195 µl of water and heated at 70 °C for 60 min. The pH was adjusted to 6–7 with NH$_4$OH, and products were analysed by HPLC.

Reaction of 6-TdGuo with 2-chloroethyl methanesulfonate

One hundred µl of 2-chloroethyl methanesulfonate (ClEMS) (876 µmol, Aldrich, Milwaukee, WI) was added to 10 mg of 6-TdGuo dissolved in 1.0 ml of DMF, and the reaction mixture was heated at 70 °C for two hours. After the reaction period, 30 µl of the sample was acid-hydrolysed as described above, and the products were separated by HPLC.

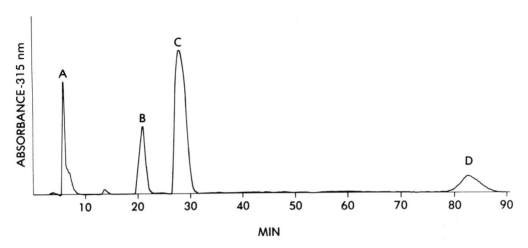

Fig. 1. High-performance liquid chromatographic separation of the reaction products of dimethylsulfate and 6-thio-2'-deoxyguanosine, after acid hydrolysis to free base. A, starting material; B, 7-methyl-6-thioguanine; C, S^6-methyl-6-thioguanine; D, structure not determined

Reaction of 6-TG residues in DNA with ^{3}-H-methylnitrosourea

9L cells were grown with 0.4 μM 6-TG (Sigma, St. Louis, MO) for 24 h, collected, and DNA was isolated and purified as described (Bodell & Banerjee, 1976). Purified DNA (550 μg in 500 μl of cacodylic buffer, pH 7.0) was reacted with 50 μg of ^{3}H-methylnitrosourea (^{3}H-MNU, 1 Ci/mmol, Amersham-Searle, Arlington Heights, IL) for one hour at 37 °C. After the reaction, DNA was repeatedly precipitated with sodium acetate and ethanol until a constant specific activity (^{3}H cpm/ug DNA) was obtained. An aliquot of the alkylated DNA was hydrolysed with 0.1 N HCl for 45 min at 70 °C. The pH of the solution was adjusted to 6.5 with NH$_4$OH, internal standards (7-methylguanine, 3-methyladenine, O^6-methylguanine and S^6-methyl-6-thioguanine) were added, and the components of the mixture were analysed by HPLC. Fractions (1.2 ml each) were collected at 1 min intervals and mixed with 10 ml of Aquasol (New England Nuclear Corp., Boston, MA). The radioactivity of each fraction was determined on a Beckman LS-250 liquid scintillation counter.

RESULTS AND DISCUSSION

Reaction of 6-TdGuo with DMS

The HPLC chromatogram of the reaction products is shown in Figure 1. The first compound which eluted (Peak A) was identified as starting material on the basis of its UV spectrum. Structures of products eluted in Peaks B and C were determined by mass spectrometry. The compound eluted in Peak B had a parent ion at m/e 181 with a fragmentation ion at 149, equivalent to [M$^+$-S], which is consistent with 7-methyl-6-

Table 1. Ultraviolet absorption maxima of thiopurines at various pH values

Compound	pH	Absorption maxima			
6-thioguanine	1			(257)	(347)
	5.1			(257)	(341)
	12				(320)
1-methyl-6-thioguanine	1			(257)	(347)
	5.1		(225)		(342)
	12		(235)		(338)
7-methyl-6-thioguanine	1			(259)	(348)
	5.1				(352)
	12			(272)	(326)
S^6-methyl-6-thioguanine	1		(241)	(274)	(318)
	5.1	(216)	(242)		(310)
	12				(314)
Product A, Figure 2	1				(328)
[(S^6,7-ethano)-6-thioguanine] [a]	5.1				(328)
	12			(281)	(331)
Product C, Figure 2	1			(276)	(324)
	5.1				(325)
	12				(323)
Product D, Figure 2	1			(260)	(324)
[(1,S^6-ethano)-6-thioguanine] [a]	5.1			(255)	(341)
	12			(253)	(320)

[a] Tentative structural assignments

Table 2. Reaction of ^{3}H-methylnitrosourea with 9L rat brain tumour DNA containing 6-thioguanine residues

Alkylation product	% of total radioactivity
7-methylguanine	59.8
3-methyladenine	10.2
O^6-methylguanine	7.4
S^6-methyl-6-thioguanine	0.6

thioguanine. The structure was confirmed by independent synthesis of the compound, by reaction of 7-methylguanine with phosphorous pentasulfide (Elion, 1962). UV spectra for the compound in Peak B and authentic 7-methyl-6-thioguanine were identical. UV absorption maxima for 7-methyl-6-thioguanine and all other compounds are summarized in Table 1. The compound in peak C had a parent ion at m/e 181, with a fragmentation ion at 134 consistent with [M$^+$-SCH$_3$], which suggests that the structure is S^6-methyl-6 thioguanine. The structure of the compound eluted in Peak D has not been determined.

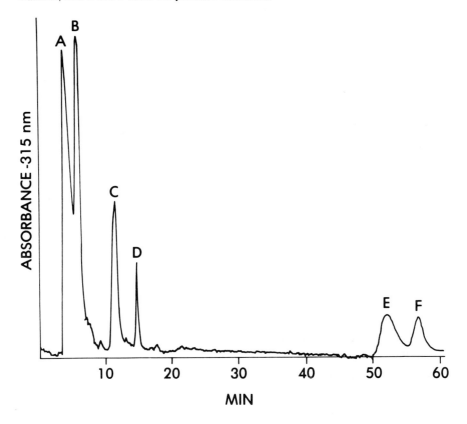

Fig. 2. High-performance liquid chromatographic separation of the reaction products of 2-chloroethyl methanesulfonate and 6-thio-2′-deoxyguanosine, after acid hydrolysis to free base. Tentative structural assignments, based on ultraviolet and mass spectrometry, are: A, (S^6,7-ethano)-6-thioguanine; C, S^6-methyl-X-methyl-6-thioguanine (where X indicates that the position of the second methyl group is not known); D, (1,S^6-ethano)-6-thioguanine. B corresponds to starting material, and E and F have not yet been identified.

Reaction of 6-TG residues in DNA with ^{3}H-MNU

The major alkylation products obtained by reacting 9L DNA containing 6-TG residues with ^{3}H-MNU were 3-methyladenine, 7-methylguanine and O^6-methylguanine (Table 2). The relative percentages of these alkylation products are in good agreement with published data (Lawley et al., 1973; Singer, 1975; Beranek et al., 1980). In addition to these alkylation products, S^6-methyl-6-thioguanine was detected as a minor product (Table 2). From the specific activity of the alkylated DNA, it can be calculated that 36.6 μmol O^6-methylguanine/mol guanine was formed. From the results of previous experiments (Bodell et al., 1985), we can estimate that 9L cells treated with 0.4 μM 6-TG would incorporate approximately 221 μmol 6-TG/mol DNA. With this information, we calculated the specific activity of S^6-methyl-6-thioguanine as 0.571 μmol/221 μmol 6-TG, or 2.58 mmol/mol 6-TG.

Fig. 3. Proposed mechanism for the formation, via a sulfonium ion intermediate (II), of (1,S^6-ethano)-6-thiodeoxyguanosine (III) and (S^6,7-ethano)-6-thiodeoxyguanosine (IV). Compound I is the proposed initial product, S^6-(2-chloroethyl)-6-thiodeoxyguanosine, in the alkylation of 6-thiodeoxyguanosine by chloroethyl methanesulfonate.

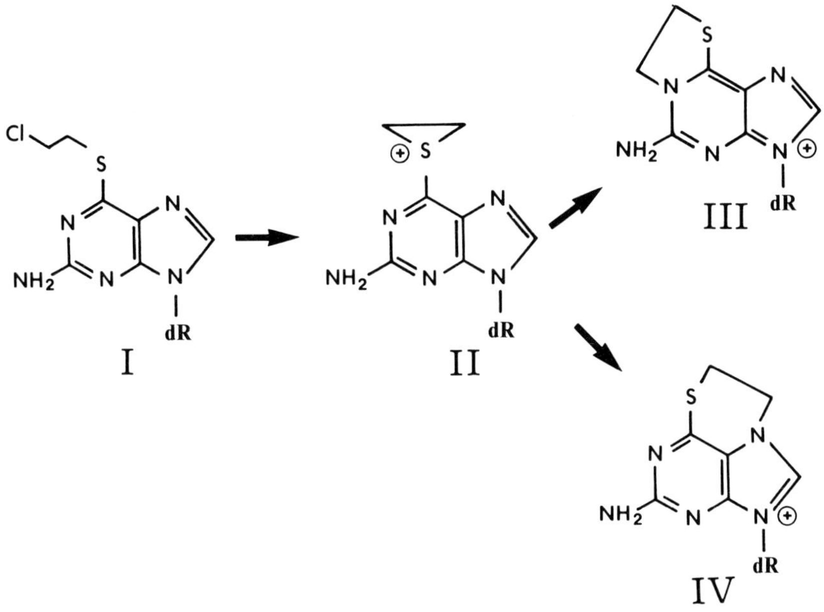

Therefore, although the absolute amount of S^6-methyl-6-thioguanine formed is low, the efficiency of its formation, calculated on the basis of 6-TG content in the DNA, is approximately 70 times higher than that of O^6-methylguanine.

Reaction of 6-TdGuo with ClEMS

The HPLC chromatogram of the reaction mixture is shown in Figure 2. The compound in Peak B was identified as starting material. Mass spectra of both products A and C gave a parent ion at m/e 193, consistent with the addition of –CH$_2$CH$_2$– to 6-TG. Neither product had a fragmentation peak at m/e 161 ([M$^+$-S]), indicating that in both compounds the thiol was alkylated. Product A had a fragmentation peak at m/e 133, corresponding to [M$^+$-CH$_2$CH$_2$].

The UV spectra of the major reaction products (Peaks A, C and D) were not consistent with monoalkylation at either the 1, 6 or 7 positions of 6-TG (Table 1). Product A has a UV spectrum that is similar to S^6-methyl-7-methyl-6-thioguanine, but with a bathochromic shift of approximately 20 nm. Therefore, product A may be (S^6,7-ethano)-6-thioguanine. Product C has a UV spectrum nearly identical to S^6-methyl-X-methyl-6-thioguanine, a dimethylated product of the reaction of 6-TdGuo, with dimethylsulfate. Product D has a UV spectrum similar to 1-methyl-S^6-methyl-6-thioguanine, also with a bathochromic shift of approximately 20 nm. On the basis of the similarities in the UV spectra of products A and D, product D has been assigned the structure (1,S^6-ethano)-6-thioguanine. Products E and F have not yet been identified. These assignments of structure are tentative, and the products are being more fully characterized by standard analytical methods.

A possible mechanism for the formation of these products is given in Figure 3. Based on the results obtained with DMS, we propose that the initial product of the alkylation of 6-TdGuo by ClEMS would be S^6-(2-chloroethyl)-6-thiodeoxyguanosine. Because this compound is related to sulfur mustards, it is possible that a sulfonium intermediate would form (Bartlett & Swain, 1949) and then undergo an internal cyclization reaction to give the ethano derivatives, as shown in Figure 3. A similar mechanism has been proposed to account for the formation of 3,N^4-ethanodeoxycytidine (Lown & McLaughlin, 1979).

After incorporation of 6-TG into cellular DNA, the alkylation product formed by reaction of BCNU with 6-TG might further react to form the sulfonium ion intermediate shown in Figure 3. This intermediate might react with a nucleophile on adjacent bases in DNA to form intra- and interstrand DNA cross-links. This is consistent with our finding that 6-TG pretreatment potentiates DNA cross-linking in cells treated with BCNU (Bodell *et al.*, 1985).

ACKNOWLEDGMENTS

This work was supported by NIH Program Project Grant CA-23525 and the Phi Beta Psi Sorority. The author thanks Jytte Rasmussen for technical assistance, Dr. A. Burlingame of the UCSF mass spectrometry facility for the mass spectra, and Neil Buckley for helpful discussions and editorial assistance.

REFERENCES

Bartlett, P.D. & Swain, C.G. (1949) Kinetics of hydrolysis and displacement reactions of β,β′-dichlorodiethylsulfide (mustard gas) and of β-chloro-β′-hydroxydiethylsulfide (mustard chlorohydrin). *J. Am. chem. Soc.*, **71**, 1406–1415

Beranek, D.T., Weis, C.C. & Swenson, D.H. (1980) A comprehensive quantitative analysis of methylated and ethylated DNA using high-pressure liquid chromatography. *Carcinogenesis*, **1**, 595–606

Bodell, W.J. & Banerjee, M.R. (1976) Reduced DNA repair in mouse satellite DNA after treatment with methyl-methanesulfonate and *N*-methyl-*N*-nitrosourea. *Nucleic Acids Res.*, **3**, 1689–1701

Bodell, W.J., Morgan, W.F., Rasmussen, J., Williams, M.E. & Deen, D.F. (1985) Potentiation of BCNU-induced cytotoxicity in 9L cells by pretreatment with 6-thioguanine. *Biochem. Pharmacol.*, **34**, 515–520

Elion, G.B. (1962) Condensed pyrimidine systems. XXII. *N*-methylpurines. *J. org. Chem.*, **27**, 2478–2491

Fujimoto, S., Ogawa, M. & Sakurai, Y. (1982) Hypothetical mechanism of therapeutic synergism induced by the combination of 6-thioguanine and 3-[(4-

amino-2-methyl-5-pyrimidinyl)methyl]-1-(2-chloroethyl)-1-nitrosourea hydrochloride. *Cancer Res., **42**,* 4079–4085

Lawley, P.D., Orr, D.J., Shah, S.A., Farmer, P.B. & Jarman, M. (1973) Reaction products from *N*-methyl-*N*-nitrosoureas and DNA containing thymidine residues. *Biochem. J., **135**,* 193–201

Lown, J.W. & McLaughlin, C.W. (1979) Mechanism of action of 2-haloethylnitrosoureas on deoxyribonucleic acid: nature of the chemical reactions with deoxyribonucleic acid. *Biochem. Pharmacol., **28**,* 2123–2128

Paterson, A.R.P. & Tidd, M.M. (1975) *6-Thiopurines.* In: Sartorelli, A.C., Johns, D.G. eds, *Antineoplastic and Immunosuppressive Agents,* Berlin, Springer-Verlag, pp. 384–403

Singer, B. (1975) *The chemical effects of nucleic acid alkylation and their relation to mutagenesis and carcinogenesis.* In: W.E. Cohn, ed., *Prog. Nucleic Acids Res. Mol. Biol.,* Vol. 15, New York, Academic Press, pp. 219–284

DNA CROSS-LINKING BY CHLOROETHYLATING AGENTS

K.W. KOHN & N.W. GIBSON

Laboratory of Molecular Pharmacology, Division of Cancer Treatment, National Institutes of Health, Bethesda, MD, USA

SUMMARY

The mechanism of DNA cross-linking and cell killing by DNA chloroethylating agents is reviewed. DNA cross-links probably arise by reactions which occur after initial chloroethylation at the O^6 position of guanine. Evidence indicates that the DNA repair enzyme, O^6-methylguanine-DNA methyltransferase, can remove such chloroethyl groups from guanine. The formation of DNA interstrand cross-links is thereby prevented and the survival of cells containing this enzyme and treated with chloroethylating agents is enhanced. There are now four chemical classes of DNA chloroethylating agents, all of which are effective antitumour drugs in animal test systems. The killing of cells by these chloroethylating agents is quantitatively dependent on the extent of DNA interstrand cross-linking.

Bifunctional alkylating agents which can form cross-links between nucleophilic sites are among the most active and most useful anticancer drugs. Cross-linking capability occurs not only in typical alkylating agents, such as bis-(2-chloroethyl)amines, but also in the 1-(2-chloroethyl)-1-nitrosoureas (ClEtNUs), which are among the most effective antitumour drugs in animal test systems. The ClEtNUs produce interstrand cross-links (ISCs) in purified DNA (Kohn, 1977; Lown *et al.*, 1978) and both ISCs and DNA-protein cross-links (DPCs) in intact mammalian cells (Ewig & Kohn, 1977, 1978). The mechanism by which ClEtNUs produce ISCs has been hypothesized to involve, firstly, chloroethylation at the O^6 position of guanine in DNA, followed by a slow further reaction with the paired cytosine to form an ethano bridge between the two bases on opposite DNA strands (Kohn, 1977).

Tong *et al.* (1982) have isolated and chemically identified an ethano-bridge cross-link from DNA that had been reacted with a ClEtNU. The sites of linkage of the ethano bridge were, however, the N-1 of guanine and the N-3 of cytosine, neither of which would normally be accessible to outside attack in the DNA helix. As discussed by Ludlum (these proceedings[1]), chemical evidence is consistent with an

---
[1] See p. 137.

initial chloroethylation of the O^6 of guanine, followed by a cyclic rearrangement that leads to the observed cross-link.

Evidence to support chloroethylation of the O^6 of guanine as the initial reaction that leads to ISCs and cell death has come from studies involving the DNA repair enzyme, O^6-methylguanine-DNA methyltransferase (MGMT). MGMT rapidly removes methyl groups from the O^6 of guanine in DNA; the methyl group is transferred to an SH group on the enzyme protein, and the enzyme is thereby inactivated (Harris et al., 1983; Pegg et al., 1983). MGMT can also remove ethyl and hydroxyethyl groups, although at slower rates (Pegg et al., 1984). Purified bacterial MGMT has been found to prevent ISC formation in ClEtNU-treated DNA (Robins et al., 1983). Since the bacterial enzyme can remove alkyl groups from the O^4 of thymine, as well as from the O^6 of guanine, it may be inferred that ISC formation arises from chloroethylation of either or both of these positions. ISCs once formed are not reversed by the enzyme. MGMT activity in mammalian cell extracts has been found to copurify with an activity that suppresses ISC formation in DNA treated with ClEtNUs (Brent, 1984). The ISC suppression activity has the same heat sensitivity as the MGMT activity. This strongly indicates that MGMT can remove DNA chloroethyl groups that otherwise could go on to form ICSs.

The above findings confirm the earlier inference that MGMT can prevent ISC formation by this mechanism in normal human cells treated with ClEtNUs (Erickson et al., 1980a,b). This inference was derived from studies of certain human tumour cells that lack MGMT activity; such cells are said to have the Mer$^-$ or Mex$^-$ phenotype (Day & Ziolkowski, 1979; Day et al., 1980a,b; Harris et al., 1983; Yarosh et al., 1983). Mer$^-$ cells are highly sensitive to killing by ClEtNUs and exhibit the formation of dose-dependent ISCs. Mer$^+$ cells, from normal as well as from some neoplastic cell lines, have normal levels of MGMT activity; they are less sensitive to the ClEtNUs and show no detectable ISCs when treated with these drugs.

In contrast to ISCs, the formation of DPCs in cells treated with ClEtNUs is unaffected by the Mer phenotype (Erickson et al., 1980b). This may be because initial chloroethylation in this case occurs at protein sulfhydryl or amino groups which could then rapidly alkylate and cross-link to any accessible nucleophilic site in DNA.

Mer$^+$ cells can, however, be made to respond to ClEtNUs, in the sense of exhibiting the formation of ISCs, by pretreating the cells with DNA methylating agents (1-methyl-1-nitroso-3-nitroguanidine or 1-methyl-1-nitrosourea) (Zlotogorski & Erickson, 1983, 1984). The interpretation of this finding is that the preloading of methyl groups onto the DNA depletes the cell of active MGMT molecules, thereby permitting the subsequently introduced chloroethyl groups to survive on the DNA long enough to allow them to be converted to ISCs.

The evidence suggests that chloroethylation of the O^6 of guanine (or of other oxygen atoms of DNA bases) is responsible for the selective killing of certain tumour cell types by the ClEtNUs. However, as reviewed by Lown et al. (these proceedings[2]), the ClEtNUs react via several alkylation pathways to yield not only chloroethylations, but also hydroxyethylations, and possibly other alkylation

---

[2] See p. 129.

Fig. 1. Outline of reaction paths leading to chloroethylation at the O^6 position of guanine in DNA. ClEtNU, 1-(2-chloroethyl)-1-nitrosourea; ClEtSoSo, 2-chloroethylmethylsulfonylmethanesulfonate; Mitozolomide, 8-carbamyl-3-(2-chloroethyl)-imidazo[5,1-d]-1,2,3,5-tetrazin-4(3H)-one; ClEtTI, chloroethyltriazenoimidazolecarboxamide

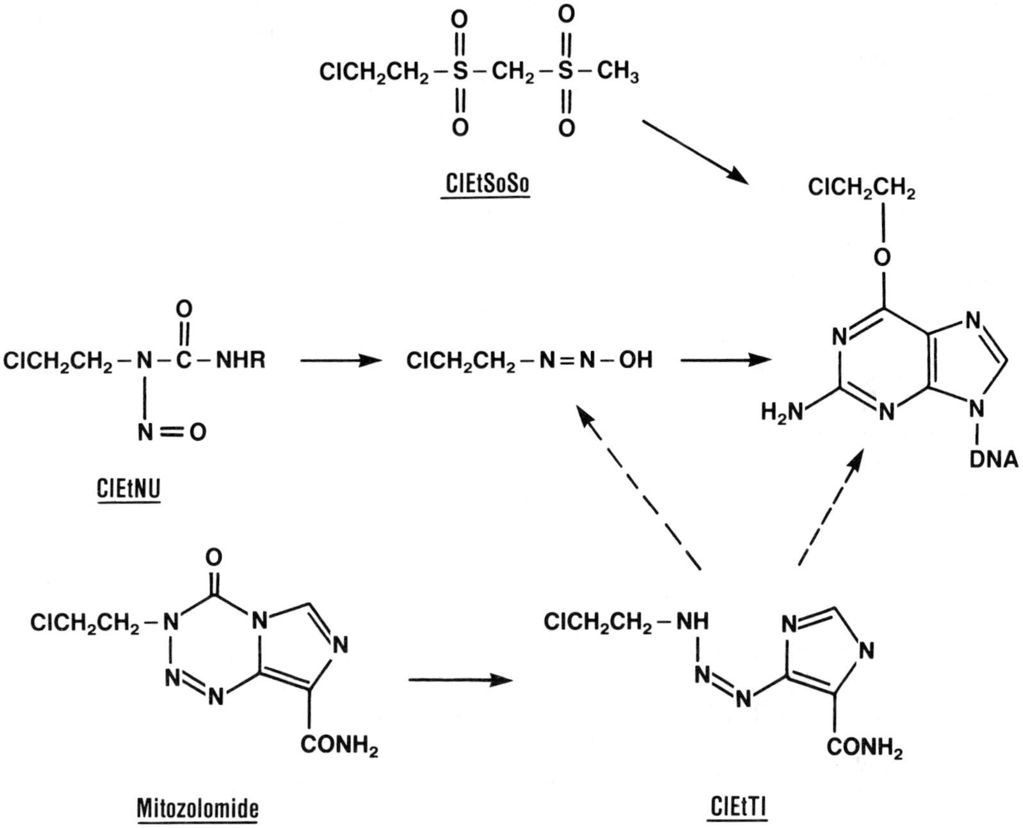

products. Chloroethylations may indeed constitute only a minority of the alkylation products. It would be desirable, therefore, to develop a chloroethylating agent that does not follow other alkylation routes.

To this end, we have studied the compound, 2-chloroethylmethylsulfonyl-methanesulfonate (ClEtSoSo) (Fig. 1) which was recently prepared by Shealy et al. (1984). ClEtSoSo has been found to be highly effective against several murine tumour systems: P388 and L1210 leukaemias, B16 melanoma and Lewis lung carcinoma. We have studied the effects of ClEtSoSo on DNA in normal human embryo cells (IMR-90, Mer$^+$) and transformed human embryo cells (VA-13, Mer$^-$) (Gibson et al., 1985). The effects of ClEtSoSo were essentially the same as those of the ClEtNUs in the following respects: (1) ISCs were produced in VA-13 (Mer$^-$) cells but not in IMR-90 (Mer$^+$) cells (Fig. 2); (2) the ISCs in VA-13 cells formed after a delay of 6–12 h and then their numbers declined (Fig. 2), perhaps because of the action of a DNA repair

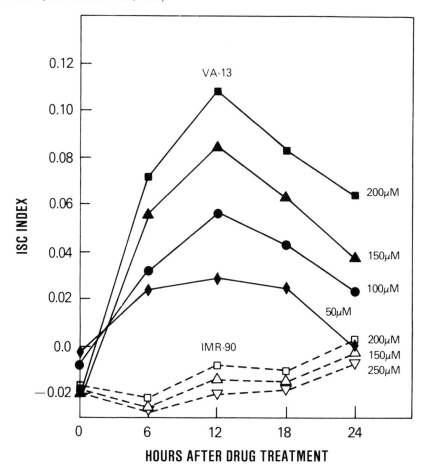

Fig. 2. DNA interstrand cross-linking (ISC) in IMR-90 (Mer$^+$) and VA-13 (Mer$^-$) human cells treated for two hours with the indicated concentrations of 2-chloroethylmethylsulfonylmethanesulfonate and then incubated in drug-free medium for various times. DNA was analysed by alkaline elution (from Gibson et al., 1985)

process; (3) DPCs formed without delay in both cell types and appeared to undergo repair (Fig. 3); (4) VA-13 cells were more sensitive than were IMR-90 cells (dose modification factor = 5) (Fig. 4). In view of the similarity of the effects on DNA, it is highly likely that the mechanism of action of ClEtSoSo is basically the same as that of the ClEtNUs, the initial step almost certainly being the chloroethylation of DNA.

Recent work in our laboratory has shown that upon reaction of DNA with ClEtSoSo, fewer major products are detected by high-performance liquid chromatographic than upon reaction with the ClEtNUs. This is in accord with the expectation that ClEtSoSo is not involved in as great a variety of alkylation pathways as are the ClEtNUs.

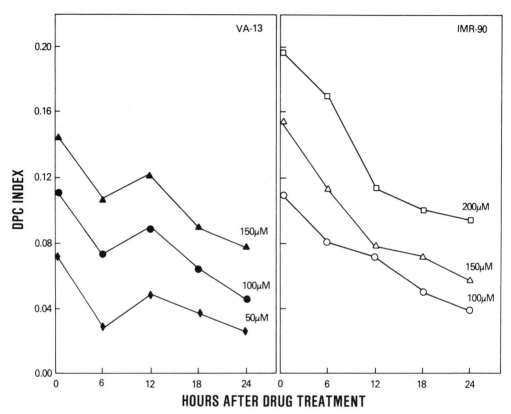

Fig. 3. DNA-protein cross-linking (DPC) in IMR-90 (Mer$^+$) and VA-13 (Mer$^-$) human cells treated with 2-chloroethylmethylsulfonylmethanesulfonate for two hours and then incubated in drug-free medium for various times. DNA was analysed by alkaline elution (from Gibson et al., 1985)

Another direction in the further development of chloroethylating agents as antitumour drugs would be to modify the selectivity for reaction with various sites, either with respect to particular positions on various bases or with respect to dependence on nucleotide sequence. The various ClEtNUs all react *via* the same chloroethylating intermediate, 2-chloroethyldiazohydroxide (Lown *et al.*, these proceedings[3]), and therefore all should yield the same relative frequencies of products.

A compound that may overcome this limitation is 8-carbamoyl-3-(2-chloroethyl)-imidazo[5,1-d]-1,2,3,5-tetrazin-4(3H)-one (Mitozolomide) (Fig. 1). In the proposed decomposition of Mitozolomide, the carbonyl group is lost, leaving chloroethyltriazenoimidazolecarboxamide (ClEtTI) (Stevens *et al.*, 1984). It is not known whether ClEtTI decomposes further to the chloroethyldiazohydroxide, which would be the same chloroethylating intermediate as that produced by the ClEtNUs

---

[3] See p. 129.

Fig. 4. Inhibition of the colony-forming ability of IMR-90 and VA-13 cells by 2-chloroethylmethylsulfonylmethanesulfonate (two-hour treatment). (from Gibson et al., 1985)

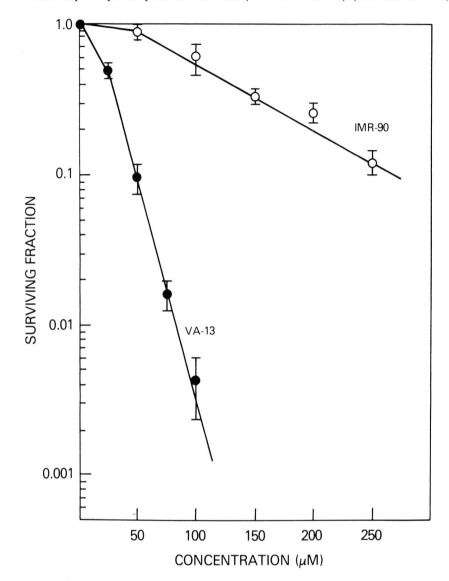

(Fig. 1). If this is the case, then the sites of chloroethylation should be the same as with the ClEtNUs. On the other hand, if ClEtTI reacts directly with DNA, the sites of reaction should be altered, and perhaps affected by the nature of the substituent on the imidazole ring. By altering this substituent, it might then be possible to affect the relative intensity of reaction at various DNA sites.

Like the ClEtNUs and ClEtSoSo, Mitozolomide and ClEtTI produce delayed ISCs selectively in Mer$^-$ cells and selectively kill Mer$^-$ cells (Gibson et al., 1984a,b). The importance of DNA chloroethylation and ISC formation in cell killing is further indicated by the finding that cell killing as a function of ISCs is quantitatively similar for the four chloroethylating agents that have been studied: ClEtNU, ClEtSoSo, Mitozolomide and ClEtTI.

REFERENCES

Brent, T.P. (1984) Suppression of cross-link formation in chloroethylnitrosourea-treated DNA by an activity in extracts of human leukaemic lymphoblasts. *Cancer Res., 44,* 1887–1892

Day, R.S. & Ziolkowski, C.H. (1979) Human brain tumour cell strains with deficient host-cell reactivation of *N*-methyl-*N'*-nitro-*N*-nitrosoguanidine-damaged adenovirus 5. *Nature, 279,* 797–799

Day, R.S., Ziolkowski, C.H., Scudiero, D.A., Meyer, S.A., Lubiniecki, A.S., Girardi, A.J., Galloway, S.M. & Bynum, G.D. (1980) Defective repair of alkylated DNA by human tumour and SV-40 transformed human cell strains. *Nature, 288,* 724–727

Day, R.S., Ziolkowski, C.H., Scudiero, D.A., Meyer, S.A. & Mattern, M.R. (1980) Human tumour cell strains defective in the repair of alkylation damage. *Carcinogenesis, 1,* 21–32

Erickson, L.C., Bradley, M.O., Ducore, J.M., Ewig, R.A. & Kohn, K.W. (1980a) DNA cross-linking and cytotoxicity in normal and transformed human cells treated with antitumour nitrosoureas. *Proc. natl Acad. Sci. USA, 77,* 467–471

Erickson, L.C., Laurent, G., Sharkey, N.A. & Kohn, K.W. (1980b) DNA cross-linking and monoadduct repair in nitrosourea-treated human tumour cells. *Nature, 288,* 727–729

Ewig, R.A. & Kohn, K.W. (1977) DNA damage and repair in mouse leukaemia L1210 cells treated with nitrogen mustard, 1,3-bis (2-chloroethyl)-1-nitrosourea, and other nitrosoureas. *Cancer Res., 37,* 2114–2122

Ewig, R.A. & Kohn, K.W. (1978) DNA cross-linking and DNA interstrand cross-linking by haloethylnitrosoureas in L1210 cells. *Cancer Res., 38,* 3197–3203

Gibson, N.W., Erickson, L.C. & Hickman, J.A. (1984a) Effects of the antitumour agent 8-carbamoyl-3-(2-chloroethyl)imidazo[5,1-d]-1,2,3,5-tetrazin-4(3H)-one on the DNA of mouse L1210 cells. *Cancer Res., 44,* 1767–1771

Gibson, N.W., Hickman, J.A. & Erickson, L.C. (1984b) DNA cross-linking and cytotoxicity in normal and transformed human cells treated *in vitro* with 8-carbamoyl-3-(2-chloroethyl)imidazo[5,1-d]-1,2,3,5-tetrazin-4(3H)-one. *Cancer Res., 44,* 1772–1775

Gibson, N.W., Erickson, L.C. & Kohn, K.W. (1985) 2-chloroethyl-methylsulfonylmethanesulfonate (NSC 338947, 'ClEtSoSo'), a new DNA chloroethylating agent: DNA damage and differential cytotoxicity in human cells. *Cancer Res., 45,* 1674–1679

Harris, A.L., Karran, P. & Lindahl, T. (1983) O^6-methylguanine-DNA methyltransferse of human lymphoid cells: structural and kinetic properties and absence in repair-deficient cells. *Cancer Res., 43,* 3247–3252

Kohn, K.W. (1977) Interstrand cross-links of DNA by 1,3-bis(2-chloroethyl)-1-nitrosourea and other 1-(2-haloethyl)-1-nitrosoureas. *Cancer Res., 37,* 1450–1454

Lown, J.W., McLaughlin, L.W. & Chang, Y.M. (1978) Mechanism of action of 2-haloethylnitrosoureas on DNA and its relation to their antileukaemic properties. *Bioorg. Chem., 7,* 97–110

Pegg, A.E., Wiest, L., Foote, R.S., Mitra, S. & Perry, W. (1983) Purification and properties of O^6-methylguanine-DNA transmethylase from rat liver. *J. biol. Chem., 258,* 2327–2333

Pegg, A.E., Scicchitano, D. & Dolan, M.E. (1984) Comparison of rates of repair of O^6-alkylguanines in DNA by rat liver and bacterial O^6-alkylguanine-DNA alkyltransferase. *Cancer Res., 44,* 3806–3811

Robins, P., Harris, A.L., Goldsmith, I. & Lindahl, T. (1983) Cross-linking of DNA induced by chloroethylnitrosoureas is prevented by O^6-methylguanine-DNA methyltransferase. *Nucleic Acids Res., 11,* 7743–7758

Shealy, Y.F., Krauth, C.A. & Laster, W.R., Jr. (1984) 2-Chloroethyl-(methylsulfonyl)methanesulfonate and related (methylsulfonyl)methanesulfonates, antineoplastic activity *in vivo*. *J. med. Chem., 27,* 664–670

Stevens, M.F.G., Hickman, J.A., Stone, R., Gibson, N.W., Baig, G.U., Lunt, E. & Newton, C.G. (1984) Antitumour imidazotetrazines. I. Synthesis and chemistry of 8-carbamoyl-3-(2-chloroethyl)imidazo[5,1-D]-1,2,3,5-tetrazin-4(3H)-one, a novel broad-spectrum antitumour agent. *J. med. Chem., 27,* 196

Tong, W.P., Kirk, M.C. & Ludlum, D.B. (1982) Formation of the cross-link 1-[N^3-deoxycytidyl],2-[-N^1-deoxyguanosinyl]ethane in DNA treated with N,N'-bis(2-chloroethyl)-N-nitrosourea. *Cancer Res., 42,* 3102–3105

Yarosh, D.B., Foote, R.S., Mitra, S. & Day, R.S. (1983) Repair of O^6-methylguanine in DNA by demethylation is lacking in Mer$^-$ human tumour cell strains. *Carcinogenesis, 4,* 199–205

Zlotosorski, C. & Erickson, L.C. (1983) Pretreatment of normal human fibroblasts and human colon carcinoma cells with MNNG allows chloroethylnitrosourea to produce DNA interstrand cross-links not observed in cells treated with chloroethylnitrosourea alone. *Carcinogenesis, 4,* 759–763

Zlotosorski, C. & Erickson, L.C. (1984) Pretreatment of human colon tumour cells with DNA methylating agents inhibits their ability to repair chloroethyl monoadducts. *Carcinogenesis, 5,* 83–87

II. CHEMISTRY AND FORMATION OF CYCLIC AND OTHER ADDUCTS

C. Bifunctional Aldehydes

REACTIONS OF NUCLEOSIDES WITH GLYOXAL AND ACROLEIN

R. SHAPIRO, R.S. SODUM, D.W. EVERETT S.K. KUNDU

Department of Chemistry, New York University, New York, N.Y., USA

SUMMARY

The structural determination of the products formed by the reaction of derivatives of glyoxal with guanine are reviewed, as are the applications of these reactions. Conditions have now been defined for the use of glyoxal as a probe for microdetermination of the structure of unidentified nucleic acid components. Unexpectedly, 3-methylguanine shows a positive reaction with glyoxal. The adduct produced has been isolated and characterized. A new ring is formed by substitution at the 1- and N^2-positions of 3-methylguanine in this product.

Acrolein reacts smoothly with 1-methylcytosine, cytidine and deoxycytidine at pH 4. The structures of the adducts have been determined. The amino group of cytosine adds to the double bond of acrolein, while the aldehyde binds to the 3-position, forming a new ring. Adenosine, deoxyadenosine and 9-methyladenine react in an analogous manner with acrolein, with ring formation involving the amino group and the N-1 of adenine. The orientation of addition is similar to that in the cytosine series.

Guanosine reacts with excess acrolein at pH 4 to form a bis-adduct. Two new rings are fused to the guanine structure, one at N-1 and the amino group, and the other involving N-7 and C-8. A number of derivatives of this product have been obtained by glycosyl cleavage and imidazole-ring-opening reactions. However, the assignment of the orientation of addition for both acroleins remains tentative in this series.

INTRODUCTION

This conference takes place exactly on the sixteenth anniversary of an earlier one concerned with the biological effects of alkylating agents, held in New York. Cyclic adducts of nucleic acids had been reported only for glyoxal derivatives and glycidaldehyde at that time, and only two papers at that meeting were concerned with the topic. One of them, my own contribution (Shapiro, 1969), concluded with this statement: "As additional reactions of related compounds with nucleic acids are

Fig. 1. Reactions of guanine derivatives glyoxal and substituted glyoxals. I, guanine derivative where R is methyl, ribose or deoxyribose; II, anion of the guanine derivative; III, glyoxal or substituted-glyoxal adduct of the guanine derivative; IV, anion of adduct III; V, N^2-acylguanine derivative; VI, cyclic borate ester of adduct III

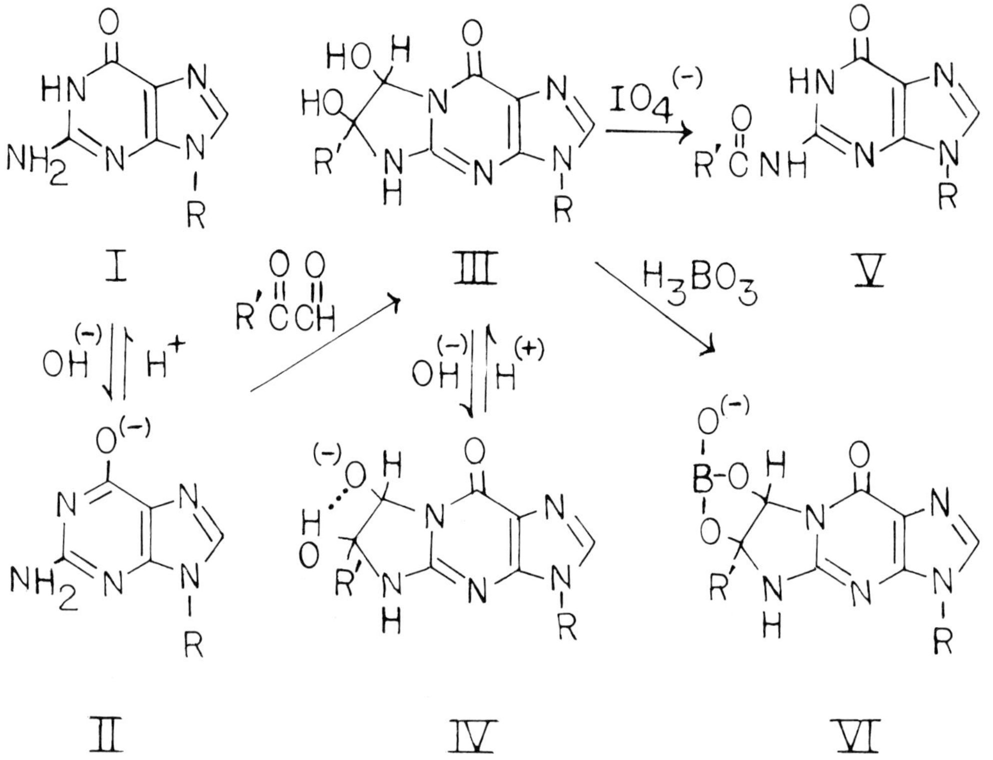

discovered, the biological importance of this class of compounds will undoubtedly grow". The present symposium certainly marks the fulfillment of that prophecy.

At this time, I wish to review unpublished results from our laboratory in the context of earlier developments. Two separate types of bifunctional reagents, glyoxal derivatives and acrolein, will be considered. The discussion will be limited to chemistry, as biological effects are considered elsewhere in these proceedings (Chung et al.,[1] Kasai et al.,[2] Marnett et al.,[3] Nagao et al.,[4] and Whitmore et al.[5]).

---

[1] See p. 207.
[2] See p. 413.
[3] See p. 175.
[4] See p. 283.
[5] See p. 185.

Fig. 2. Reaction of 3-methylguanine (VII) with glyoxal to form a cyclic adduct (VIII)

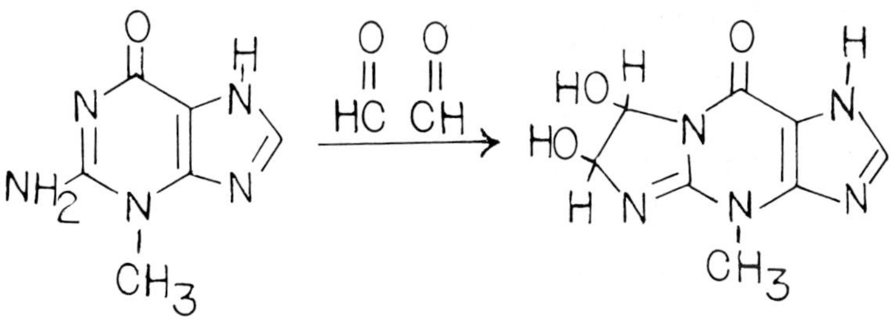

REACTIONS OF GLYOXAL DERIVATIVES

The fundamental chemistry of the reaction of glyoxals with guanine nucleosides is summarized in Figure 1. The orientation of addition in the case of unsymmetrically substituted glyoxal derivatives is illustrated by the structure of compound III. This structure was established by periodate cleavage of III to give the acylguanine, V (Shapiro & Hachmann, 1966; Shapiro et al., 1970). The possibility of N-3 substitution in compound III was not supported by its ultraviolet spectra, and was ruled out conclusively by unambiguous synthesis (Czarnik & Leonard, 1980).

Deprotonation of the N-1 of guanine (I) to yield the anion (II) is obligatory for reaction at that position. For this reason, the reaction proceeds more rapidly in base than in acid. The equilibrium constant for adduct formation is high; approximately 12 000 in the case of glyoxal and 9-methylguanine at pH 7. This equilibrium, I + glyoxal ------→ III, is adversely affected by anion formation, which withdraws guanine from the equilibrium. However, the effect is compensated in part by the conversion of the adduct to its own anion, IV (pK = 10.4) (Everett, 1980).

Adenine and cytosine nucleosides also react with glyoxal, but at much higher concentrations, with several molar glyoxal needed to effect complete conversion (Broude & Budowsky, 1971). The interactions can be demonstrated by ultraviolet spectroscopy. Alternatively, addition of periodate converts the labile adenine adduct in part to N^6-formyladenine (Shapiro et al., 1969). These observations, however, do not establish the structure of the adduct; in particular, it is not known whether the adduct has a cyclic or noncyclic structure. The instability of these adducts in the absence of concentrated glyoxal raises problems for their isolation. We have, however, been able to isolate and purify the glyoxal adduct of 3-methylguanine (VII, Fig. 2). This analogue, like adenine and cytosine, contains an amino group adjacent to a pyridine-type ring nitrogen. The adduct, isolated by precipitation from glyoxal solution, has been demonstrated to have a cyclic structure (VIII, Fig. 2) by nuclear magnetic resonance spectroscopy (Everett, 1980).

The glyoxal-guanine reaction has been of use for the microidentification of unknown nucleosides. For this purpose, it is desirable to work at a glyoxal concentration at which adenine and cytosine derivatives do not react appreciably. We have found 0.01 M glyoxal to be suitable. Under these conditions, 1-methyl- and O^6-methylguanine do not react while 3-methylguanine shows a positive response. Guanines which are mono- (but not di-) substituted in the N^2 position react if the substituent is alkyl, but not if it is acyl. A positive response can usually be detected by a shift in the ultraviolet spectrum, but the upward shift in the pK of a proton at the reactive site is a more sensitive determinant.

Selectivity for guanine and for single-stranded regions of nucleic acids have made glyoxal derivatives desirable as structural probes for transfer RNA and ribosomal RNA. The adducts (III, Fig. 1) revert slowly to guanine in the absence of glyoxal at pH 7, and more rapidly at alkaline pH. They can be stabilized by the addition of borate buffer, which converts them to compound VI (Fig. 1) (Litt, 1969). Structures III and VI interfere with chain cleavage by ribonuclease T1. Upon reversal of adduct formation, the sensitivity to T1 is restored. If a more permanent modification of guanine is desired, the adduct can be converted to the acylguanine with periodate (III ------→ V, Fig. 1).

More recent developments in this area have featured the synthesis of elaborate reagents containing two glyoxal units; for example, ethylene glycol bis[3-(2-ketobutyraldehyde) ether] (Brewer et al., 1983). Such reagents have found application in the cross-linking of RNA chains and the formation of RNA-protein cross-links.

REACTIONS WITH ACROLEIN

The mutagenic and toxic effects of acrolein, as well as its chemical reactivity, have made it a likely candidate for reaction with the nucleic acids and their components. Nelsestuen (1980) reported the reaction of acrolein with adenine, cytosine, uracil and hypoxanthine. The products were not characterized, however. They were assumed to result from simple Michael addition at the N-1 of the pyrimidines and the N-9 of the purines. Recently, Chung et al. (1984) reported the reaction of acrolein, in a 20-fold excess, with deoxyguanosine at pH 7. The products were identified as IX and X (Fig. 3; R = deoxyribose) by elegant spectroscopic and chemical modification studies. Product X was shown to be a mixture of two rapidly equilibrating diastereomers. The orientation of the two types of cyclic adducts was established by borohydride reduction of IX (R = H), with ring opening.

We have also studied nucleoside reactions of acrolein in our own laboratory (Sodum, 1984), but have chosen to work at a more acidic pH. In acid, self-polymerization of acrolein is avoided. Under our conditions (pH 4.2, 25 °C, 24-48 hours, with a 1.1- to 1.5-fold excess of acrolein), adenine and cytosine derivatives reacted smoothly with acrolein. Only a single product was observed upon thin-layer chromatography for the reactions studied (Fig. 3, XI and XIV: cytosine and adenine derivatives, respectively, in which R = methyl, ribose, or deoxyribose). The purification method varied from case to case. Preparative thin-layer chromatography in 2-propanol : ammonia : water (7 : 1 : 2) was used for the 1-methylcytosine derivative, Amerlite CG-50 chromatography for the 9-methyladenine adduct, and Dowex I-X8 chromatography for the four nucleoside products. Preparative yields

Fig. 3. Reactions of guanine, cytosine and adenine derivatives with acroleins: monoadduct formation. I, guanine derivative; IX and X, cyclic 1,N^2-adducts of acrolein with guanine derivatives; XI, cytosine derivative; XII, cyclic 3,N^4-adduct of acrolein with the cytosine derivative, cationic form; XIII, N^4-(2-carboxyethyl)cytosine derivative; XIV, adenine derivative; XV, cyclic 1,N^6-adduct of acrolein with the adenine derivative; XVI, N^6-(2-carboxyethyl)adenine derivative; R, methyl, ribose or deoxyribose

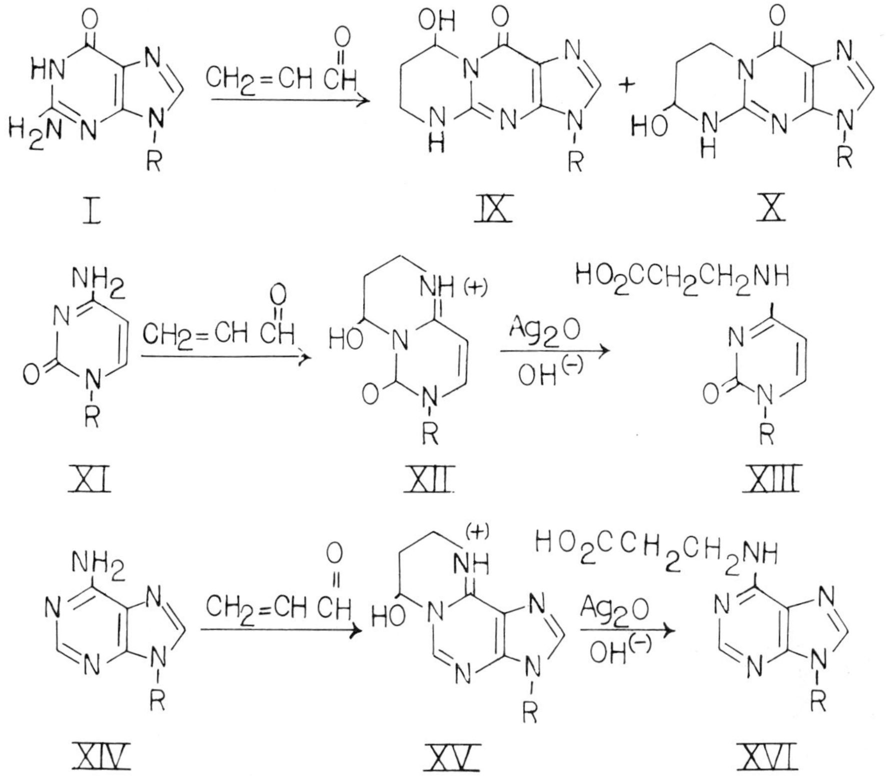

varied from 22% to 54%. The same products were also observed at pH 7, but the reactions proceeded more slowly and the yields were poorer.

We have assigned structures XII and XV (Fig. 3) to the products on the basis of the following evidence. Elemental analyses indicated that they were monoadducts of acrolein. The pK values (between 8 and 9) in the cytosine series were consistent with 3-substitution, and the ultraviolet spectra of the adducts were also in accord with this structure. Similarly, in the adenine series, N-1 substitution was indicated by pK values between 9 and 9.5 and by a resemblance to the ultraviolet spectra of 1-methyladenosine. The protein magnetic resonance spectra of the products revealed no peak due to a free aldehyde group, and were in accord with structures involving ring formation. These data did not suffice to distinguish the orientation shown in XII and XV from that resulting from the reverse mode of addition. As noted above, both possibilities occurred in the reaction of acrolein with deoxyguanosine. This problem was resolved

Fig. 4. Reaction of guanosine with acrolein: Diadduct formation. I, guanosine; XVII, bicyclic 1,N^2,7,8-diadduct of acrolein and guanosine, cationic form; XVIII, bicyclic 1,N^2,7,8-diadduct of acrolein and guanine; XIX, imidazole ring-opened derivative of XVII; XX, dehydration product of XVIII, structure unassigned; XXI, imidazole ring-opened derivative of XVIII

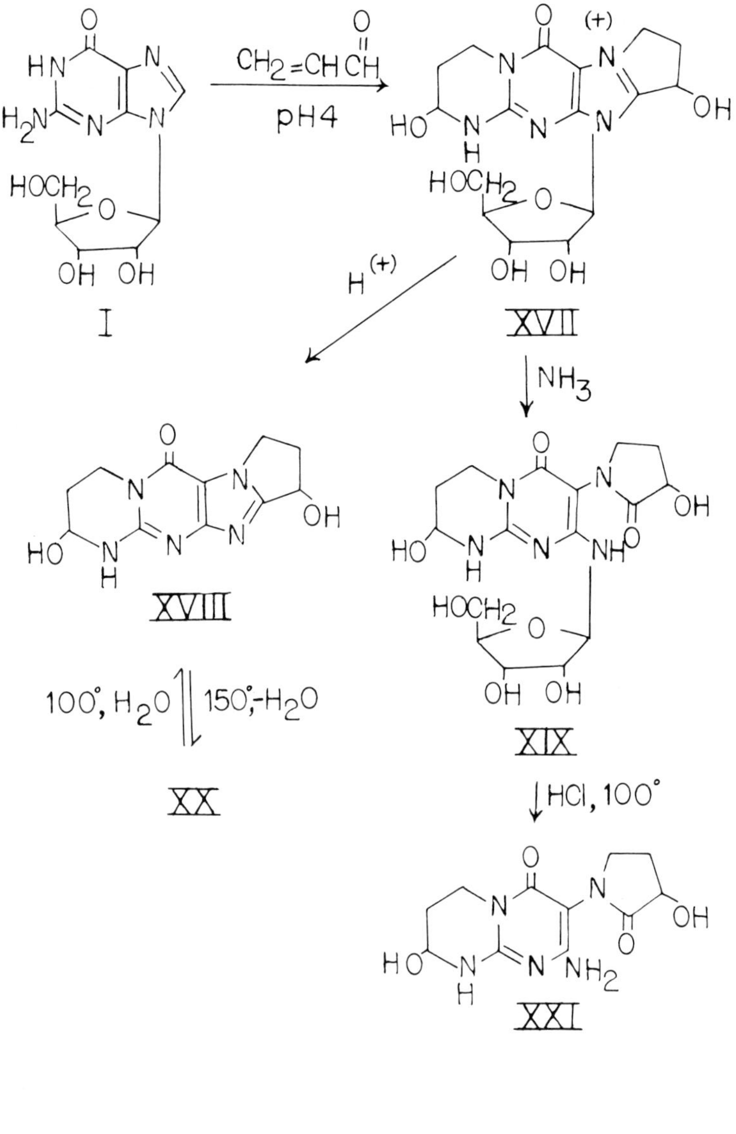

through oxidation of the 1-methylcytosine and 9-methyladenine adducts (silver oxide, 2% KOH, 15 minutes, 60 °C) to yield compounds XIII and XVI (Fig. 3). The yields obtained were 37% and 65%, respectively. Product XIII has been reported in the literature (Ueda & Fox, 1964), while product XVI was readily identified on the basis of its spectroscopic properties. A note of caution must be added with respect to the structure assigned in the case of the 9-methyladenine adduct, however. Adenines substituted in the 1-position are known to rearrange in base to yield products with substitution on the amino group (Singer et al., 1974). No intermediate was observed by thin-layer chromatography during the course of the reaction, however, which renders this possibility less likely.

At an earlier time, another co-worker and I had also investigated the reaction of guanosine with acrolein (S.K. Kundu & R. Shapiro, unpublished results). The conditions used (pH 4, 55 °C, 8 hours, with a seven-fold excess of acrolein) and the products obtained (Fig. 4) differed from those reported by Chung et al. (1984) for the deoxyguanosine reaction. The initial workup involved adsorption on Amberlite CG-120, acidic form, and elution with ammonia. Under these conditions, two products, XVIII and XIX, were isolated in yields of 36% and 46%, respectively. Subsequently we found that another substance, XVII, was formed initially. It could be isolated using a gentler procedure, i.e., chromatography on weakly acidic Amberlite CG-50 resin. Treatment of XVII with 1% ammonia (25 °C, 16 hours) converted it to XIX, while hydrolysis of XVII with 10% HCl produced XVIII.

Further transformation products could be obtained from XVIII and XIX. Compound XVIII, upon heating in dimethylformamide at 150 °C (two hours), yielded a dehydration product, XX. Vinylic protons were observed in the nuclear magnetic resonance spectrum of this substance. It reverted to XVIII upon boiling in water for three hours. Acidic treatment of XIX (0.1 N HCl, 100 °C, 2.5 hours) converted it to the base, XXI.

The various compounds indicated in Figure 4 have been characterized primarily by satisfactory elemental analyses (except for the unstable compound, XVII), homogeniety on thin-layer chromatography in several solvent systems, and ultraviolet spectral analysis at several pH values. The spectra resembled, in each case, those of the appropriate control compounds, and served to exclude such alternatives as N-3 or O^6 substitution. Nuclear magnetic resonance indicated the absence of any free aldehyde group, and of H-8 of guanine, but was otherwise uninformative. Most peaks appeared as broad multiplets, which precluded exact assignments to individual protons, and suggested that the products were mixtures of diastereomers.

The above information indicated that guanosine had formed a bis-adduct with acrolein, with the closure of two new rings. The positions of ring fusion were likely to be those indicated in Figure 4. Little evidence was at hand concerning the orientation involved in forming each ring, however, and the work was not published. At the present time, we are willing to make the assignments indicated in Figure 4, but these must be considered quite tentative. In the new data of Chung et al. (1984), adducts having the structure IX show a peak at 6.3δ in their proton magnetic resonance spectra. This feature is absent in our spectra, so we assume the opposite orientation for the left-most ring in the structures in Figure 4. Even less evidence is available about the right-most ring. Jones & Young (1970) reported the reaction of acrylic acid with guanosine, which involved Michael addition at the N-7 of guanosine to give a monoadduct. If this occurred in the acrolein reaction as well, then ring closure between the aldehyde group and the C-8 of guanine might follow, perhaps through deprotonation at position 8 to form a zwitterion. This mechanistic argument alone supports the orientation shown in Figure 4 for the various products.

Although further studies are needed for a full elucidation of the structures of this series of compounds, their novel nature is apparent. As interest concerning the biological role of polyfunctional carbonyl compounds grows and the study of their chemistry increases, many additional unexpected reactions are likely to be uncovered. Hopefully, such discoveries will also pave the way for new applications, not only as tools for biochemical research, but also in drug design.

ACKNOWLEDGEMENTS

This research was supported by U.S. Public Health Service research grant GM 20176 from the National Institute of General Medical Sciences. We wish to thank Dr. J. Weisgras, Dr. S. Dubelman and Ms. B. Lipton for performing preliminary experiments.

REFERENCES

Brewer, L.A., Goelz, S. & Noller, H.F. (1983) Ribonucleic acid-protein cross-linking within the intact *Eschericia coli* ribosome, utilizing ethylene glycol bis[3-(2-ketobutyraldehyde) ether], a reversible, bifunctional reagent: synthesis and cross-linking within 30S and 50S subunits. *Biochemistry,* **22,** 4303–4309

Broude, N.E. & Budowsky, E.I. (1971) The reactions of glyoxal with nucleic acid components. III. Kinetics of the reaction with monomers. *Biochim. biophys. Acta,* **254,** 380–388

Chung, F.-L., Young, R. & Hecht, S.S. (1984) Formation of cyclic $1,N^2$-propanodeoxyguanosine adducts in DNA upon reaction with acrolein or crotonaldehyde. *Cancer Res.,* **44,** 990–995

Czarnik, A.W. & Leonard, N.J. (1980) Unequivocal assignment of the skeletal structure of the guanine-glyoxal adduct. *J. org. Chem.,* **45,** 3514–3517

Everett, D.W. (1980). I. Reaction of pseudouridine with bisulfite. II. Reaction of glyoxal with guanine derivatives: a spectrophotometric probe of molecular structure. Ph.D. thesis, New York University

Jones J.B. & Young, J.M. (1970) Carcinogenicity of lactones. IV. Alkylation of analogues of DNA guanine groups such as imidazole, *N*-methylimidazole, and guanosine by α,β-unsaturated acids. *Can. J. Chem.,* **48,** 1566–1573

Litt, M. (1969) Structural studies on transfer RNA. I. Labeling of exposed sites in yeast phenylalanine transfer ribonucleic acid with kethoxal. *Biochemistry,* **8,** 3249–3253

Nelsestuen, G.L. (1980) Origin of life: consideration of alternatives to proteins and nucleic acids. *J. mol. Evol.,* **15,** 59–72

Shapiro, R. (1969) Reactions with purines and pyrimidines. *Ann. N.Y. Acad. Sci.,* **163,** 624–630

Shapiro, R. & Hachmann, J. (1966) The reaction of guanine derivatives with 1,2-dicarbonyl compounds. *Biochemistry,* **5,** 2799–2807

Shapiro, R., Cohen, B.I., Shiuey, S.J. & Maurer, H. (1969) On the reaction of guanine with glyoxal, pyruvaldehyde and kethoxal, and the structure of the acylguanines. A new synthesis of N^2-alkylguanines. *Biochemistry, 8,* 238–245

Shapiro, R., Cohen, B.I. & Clagett, D.C. (1970) Specific acylation of the guanine residues of ribonucleic acid. *J. biol. Chem., 245,* 2633–2639

Singer, B., Sun, L. & Fraenkel-Conrat, H. (1974) Reaction of adenosine with ethylating agents. *Biochemistry, 13,* 1913–1920

Sodum, R.S. (1984) The reactions of acrolein and diepoxybutane with nucleic acid derivatives. Ph.D. thesis, New York University

Ueda, T. & Fox, J.J. (1964) Pyrimidines. III. A novel rearrangement in the syntheses of imidazo- or pyrimido[1,2-c]pyrimidines. *J. org. Chem., 29,* 1762–1769

THE ROLE OF CYCLIC NUCLEIC ACID ADDUCTS IN THE MUTATIONAL SPECIFICITY OF MALONDIALDEHYDE AND β-SUBSTITUTED ACROLEINS IN *SALMONELLA*

L.J. MARNETT, A.K. BASU & S.M. O'HARA

Department of Chemistry, Wayne State University, Detroit, MI, USA

INTRODUCTION

Malondialdehyde (MDA) is a product of prostaglandin endoperoxide metabolism and of lipid peroxidation that has been detected in virtually all mammalian tissues (Bernheim *et al.*, 1948; Diczfalusy *et al.*, 1977). It is mutagenic in *Salmonella typhimurium* and mammalian cells but is not a tumour initiator or promoter or a complete carcinogen on mouse skin (Basu & Marnett. 1983; Yau, 1979; Fischer *et al.*, 1983). MDA is the prototype β-dicarbonyl compound but it exists as its α,β-unsaturated aldehyde tautomer, β-hydroxyacrolein (Fig. 1A) (George & Mansell, 1968). Because it has a pK_a of 4.4, MDA exists as an enolate at physiological pH (Osman, 1972). The tautomeric equilibrium in Fig. 1A indicates that MDA is structurally related to dicarbonyl compounds and enals such as methylglyoxal and acrolein (Fig. 1B). Methylglyoxal is a major mutagen in freshly roasted coffee and 4-hydroxyalkenals are important products of lipid peroxidation that have recently been found to be mutagenic (Kasai *et al.*, 1982; Marnett *et al.*, 1984).

MUTAGENESIS IN *SALMONELLA TYPHIMURIUM*

In spite of their close structural homology, MDA, methylglyoxal and acrolein exhibit strikingly different patterns of mutagenicity in *Salmonella typhimurium* (Table 1). MDA reverts strains carrying the frameshift mutation *hisD3052* but not strains carrying the base-pair substitution mutation *hisG46* (Mukai & Goldstein, 1976). In contrast, methylglyoxal and acrolein are mutagenic in *hisG46*-carrying strains but not in *hisD3052*-carrying strains (Kasai *et al.*, 1982; Lutz *et al.*, 1982). Thus, MDA acts as a frameshift mutagen and methylglyoxal and acrolein as base-pair substitution mutagens in *Salmonella*. MDA can induce base-pair substitutions in strains carrying the *hisG428* mutation (Levin *et al*; 1982; Marnett *et al.*, 1984). The compounds also differ as mutagens in their sensitivity to the genetic background of the host.

Fig. 1. Tautomeric equilibrium of malondialdehyde (MDA) with β-hydroxyacrolein, and its structural relationship with dicarbonyl compounds and enals

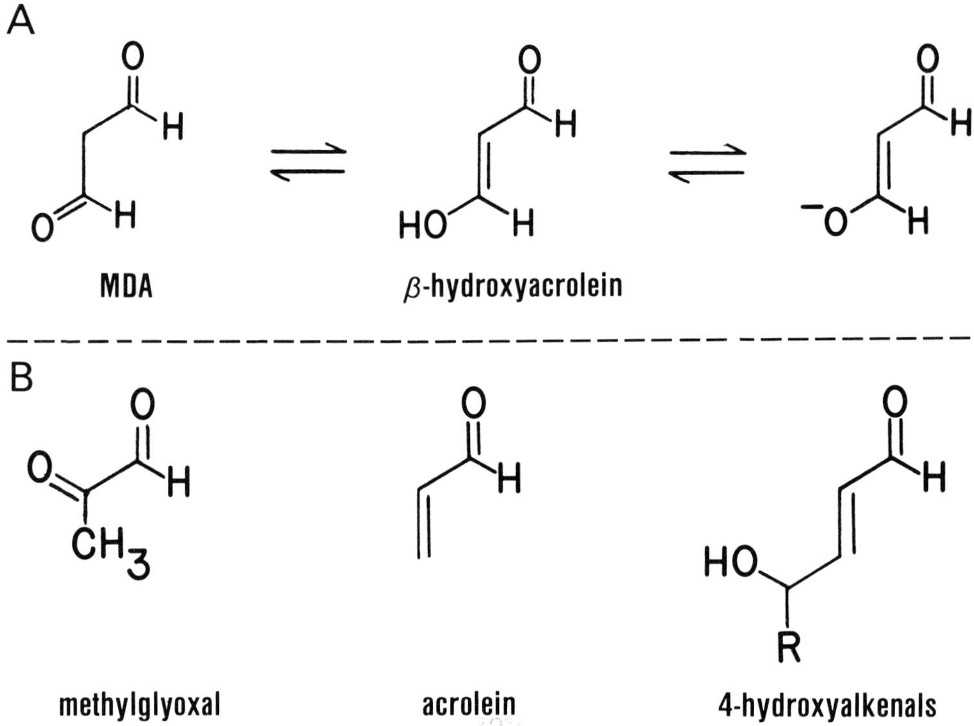

A

MDA ⇌ β-hydroxyacrolein ⇌

B

methylglyoxal   acrolein   4-hydroxyalkenals

Mutagenesis by MDA is unaffected or decreased by deletions of the *uvrB* (ultraviolet resistance) gene whereas mutagenesis by methylglyoxal and enals is enhanced (Mukai & Goldstein, 1976; Marnett *et al.*, 1984). Introduction of plasmid pKM101, which contains an analogue of the *umuC* (ultraviolet sensitivity gene) of *Escherichia coli* (Perry & Walker, 1982), to *Salmonella* strains has no effect on MDA mutagenesis but dramatically stimulates methylglyoxal and acrolein mutagenesis ($\geq$ 100-fold) (Marnett *et al.*, 1984). In fact, acrolein cannot be detected as a mutagen in strains that lack pKM101 (Marnett *et al.*, 1984). Although methylglyoxal is several orders of magnitude more potent as a mutagen than MDA in strain TA104 (pKM101$^+$, *uvrB*$^-$), it exhibits comparable mutagenicity in *hisG428*-carrying strains (pKM101$^-$, *uvrB*$^+$) (Marnett *et al.*, 1984). A lack of effect of pKM101 or *uvrB* deletions on MDA mutagenesis is observed in *Salmonella* strains that detect frameshift or base-pair substitution mutagens.

MOLECULAR REQUIREMENTS FOR MDA MUTAGENESIS

Several pieces of information are required before we can attempt to understand how a small molecule like MDA induces frameshift mutations and why its sensitivity

Fig. 2. Chemical steps required for frameshift mutagenesis by malondialdehyde and by acroleins substituted with a leaving group (X) in the β-position. Nu, nucleic acid base

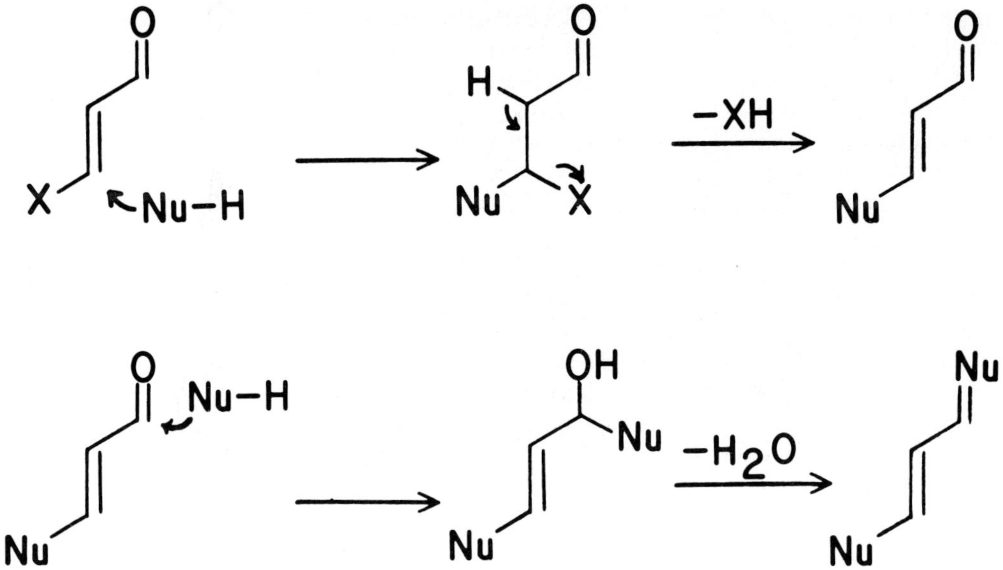

Table 1. Reversion of *Salmonella typhimurium* strains by malondialdehyde, dicarbonyls and enals

Compound	Reverted Mutation[a]		
	hisD3052 (frameshift)	hisG46	hisG428
		(base-pair substitution)	
Malondialdehyde	+	−	+
Methylglyoxal	−	+	+
Acrolein	−	+	+
4-Hydroxynonenal	−	N.T.[b]	+

[a] Deletion of the *uvrB* gene and introduction of plasmid pKM101 generates strains TA98 (*hisD3052*), TA100 (*hisG46*), and TA104 (*hisG428*). This *hisG428* mutation is also contained on plasmid pAQ1 in strain TA102. Data taken from Marnett & Tuttle, 1980; Kasai et al., 1982; Lutz et al., 1982; Marnett et al., 1984
[b] Not tested

to host factors differs so dramatically from methylglyoxal and enals. In order to define the chemical steps required for MDA reversion of the frameshift mutation, *hisD3052*, we have analysed the effect of systematic structural modifications on mutagenesis (Basu & Marnett, 1984). The results of our study indicate that the first step in MDA mutagenesis is Michael addition of a nucleic acid base to the β-carbon of MDA (Fig. 2). The intermediate eliminates a leaving group (H_2O in the case of MDA) to generate MDA substituted at the β-carbon with a nucleic acid base. As expected, analogues of MDA that have better leaving groups at the β-position are more potent mutagens than MDA. Addition of the nucleic acid base followed by elimination of the leaving group *is necessary but not sufficient to cause mutagenesis.* Our studies indicate that addition of a *second* nucleophile to the aldehyde carbon is required to induce mutation (Fig. 2). When functional groups that retain the high reactivity of the conjugated double bond are substituted for the aldehyde, mutagenicity is eliminated. Interestingly, these substitutions do not substantially reduce toxicity. It seems that toxicity only requires addition of a single nucleophile, whereas mutagenicity requires addition of two. This suggests that the mechanisms of toxicity and of mutagenicity by MDA are different. The best combination of high mutagenicity and low toxicity is exhibited by β-methoxyacrolein (Marnett & Tuttle, 1980).

The requirement for two nucleophilic additions to MDA can be fulfilled by monoadditions of two separate bases or by diaddition of a single base. Two monoadditions will result in interstrand or intrastrand cross-links of a DNA molecule and a single diaddition will form a cyclic adduct. No correlation has been found between the ability of a series of substituted MDAs to induce interstrand cross-links between complementary strands of DNA molecules and mutagenicity (Basu *et al.*, 1984). Furthermore, MDA and β-methoxyacrolein are mutagenic in TA104 and TA98, respectively, i.e., in strains that contain *uvrB* deletions (Levin *et al.*, 1982; Basu *et al.*, 1984). Agents such as mitomycin C that induce cross-linking of complementary DNA strands are toxic but not mutagenic in $uvrB^-$ strains (Kondo *et al.*, 1970; Murayama & Otsuji, 1973). This suggests that base-pair substitution and frameshift mutations induced by MDA and β-substituted MDAs are not the result of interstrand cross-links.

It is unlikely that mutation arises as a result of intrastrand cross-links because of the different mutagenic profiles observed with MDA and acrolein. Cross-links induced by both molecules would only differ by the presence of a single double bond, which seems an insufficient structural basis for induction of frameshift mutations, as opposed to base-pair substitution mutations. This argument is not compelling, though, so it is difficult to absolutely rule out the involvement of intrastrand cross-links.

MDA ADDUCTS TO DEOXYNUCLEOSIDES, NUCLEOSIDES, AND BASES

Cyclic adducts to a single nucleic acid base appear to be the best candidates for establishing a chemical basis for MDA mutagenesis. We have compared the extent of reaction of MDA and a series of β-substituted MDAs with the four common deoxynucleosides. Only deoxyguanosine and deoxyadenosine react to any significant extent. The profile obtained from high-performance liquid chromotatography of the products of the reaction of β-(*p*-nitrophenoxy)acrolein with guanosine in dimethylsulfoxide is given in Figure 3. A similar profile is observed with deoxyguanosine. Compound A is unreacted guanosine and compounds B, C, and D are adducts. When radiolabelled guanosine is used in the reaction, only peaks B, C

Fig. 3. Reverse-phase high-performance liquid chromatographic profile of the products of reaction of guanosine with β-(p-nitrophenoxy)acrolein. A, unreacted guanosine; B, C and D, adducts. The structure of adduct D is shown in the inset.

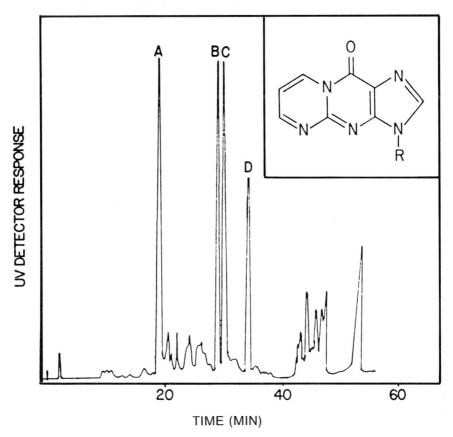

and D are labelled. The ultraviolet (UV), fluorescence, nuclear magnetic resonance (NMR) and mass spectra of peak D indicate that it is 3-(β-D-erythropentofuranosyl)-pyrimido-[1,2-a]-purin-10(3H)-one (Moschel & Leonard, 1976; Seto et al., 1981), the structure of which is shown in the inset to Figure 3.

Peaks B and C have identical UV and NMR spectra and their circular dichroism spectra are mirror images. This suggests that they are diastereomers formed by attachment of enantiomeric guanine adducts to ribose. They do not interconvert at ambient temperature but do at elevated temperature. Neither adduct is fluorescent. The proton and ^{13}C NMR spectra of the ribosides or deglycosylated adducts indicate the presence of seven non-exchangeable protons and eleven carbons. This is consistent with a structure in which *two* molecules of MDA react with one molecule of guanosine with elimination of two molecules of water. The chemical shifts and coupling patterns of the NMR spectra indicate the presence of an aldehyde, a trisubstituted alkene

Fig. 4. Reverse-phase high-performance liquid chromatographic profile of the products of reaction of deoxyadenosine with β-(p-nitrophenoxy)acrolein. Arrows indicate the position of the adducts formed, and their structures are shown in the inset.

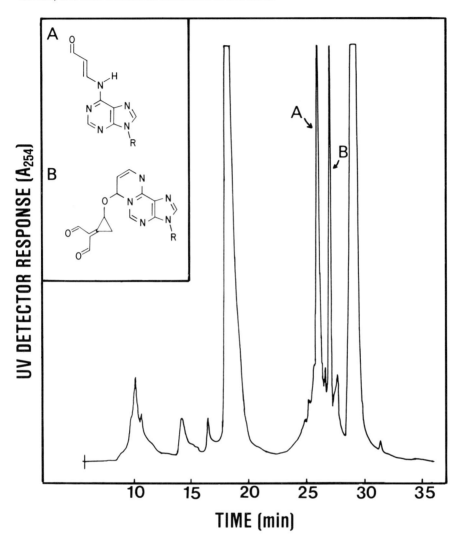

conjugated to the aldehyde, a carbinolamine, a methylene group, and perhaps an enol. Decoupling experiments are in progress to better define the structure.

Reaction of β-(p-nitrophenoxy)acrolein with deoxyadenosine generates two adducts that are indicated by the arrows in the high-performance liquid chromatographic profile of Figure 4. The major adduct (A) has spectral properties identical to β-(N^6-adenosyl)acrolein and the minor adduct (B), to a novel 3 : 1 adduct of MDA to adenosine that contains an alkylidene-cyclopropyl group. Both adducts have been identified recently by Nair *et al.* (1984), and their structures are shown in the inset to Figure 4.

RELATIONSHIP OF ADDUCT STRUCTURE TO MUTAGENICITY

Which MDA-deoxynucleoside adducts are important in induction of mutagenesis? This question is somewhat preliminary because methodology does not yet exist for the sensitive detection of MDA adducts to DNA so it is not certain which adducts are formed in *Salmonella*. Nevertheless, it is clear from the results of our structure-mutagenicity studies that monoaddition products such as β-(N^6-adenosyl)acrolein cannot be important (Basu & Marnett, 1984). Furthermore, addition of one or two additional molecules of MDA to an MDA-adducted base in preference to unadducted bases of a DNA molecule seems unlikely. It is possible that a dimer or trimer of MDA forms in solution and then adds to deoxyguanosine or deoxyadenosine. Polymerization of MDA decreases the ability of MDA solutions to induce frameshift mutations in *hisD3052* (Marnett & Tuttle, 1980). This suggests that the polymeric MDA-DNA adducts may not be responsible for induction of frameshift mutations.

The adduct that seems the most likely choice at present to explain frameshift mutagenesis by MDA is the pyrimido-purinone adduct to deoxyguanosine. This molecule is relatively flat and nonpolar and should lie at the position of the modified guanine base with minimal distortion of the double-helix aside from some displacement of the complementary cytosine base. The pyrimido-purinone adduct is structurally analogous to acridine dyes that intercalate in DNA and induce frameshift mutations (Orgel & Brenner, 1961). Enals (acrolein or crotonaldehyde) and α-dicarbonyl compounds (methylglyoxal) cannot form pyrimido-purinone adducts but form mono- or dihydroxy adducts to deoxyguanosine. These adducts are not flat and have one or more hydroxyl groups that project up or down into adjacent base pairs of the double helix. This may explain why α-dicarbonyl compounds and enals do not induce frameshift mutations. These adduct structures may also explain the differential effects of *uvrB* deletions and pKM101 on the frequency of mutations induced by MDA, α-dicarbonyls and enals. If the pyrimido-purinone adduct intercalates within the stacked bases and does not cause significant local distortions, its presence may not be efficiently detected by repair proteins. In contrast, methylglyoxal and acrolein adducts cannot remain cryptic and may be easily detected and processed by repair proteins. This hypothesis, although speculative, can be tested and experiments to do so are in progress.

CONCLUSION

MDA, α-dicarbonyls, and enals are ubiquitous components of the human environment (Marnett *et al.*, 1984). The dramatic differences in mutational specificity and potency that arise from small structural alterations in this series of compounds are interesting and represent stimulating experimental challenges. We believe that the structures of the cyclic nucleic acid adducts formed from them play a major role in determining their biological activity.

ACKNOWLEDGEMENTS

This work was supported by a research grant from the National Cancer Institute (CA 22206). L.J.M. is a recipient of a Faculty Research Award from the American Cancer Society (FRA 243).

REFERENCES

Basu, A.K. & Marnett, L.J. (1983) Unequivocal demonstration that malondialdehyde is a mutagen. *Carcinogenesis, 4,* 331–333

Basu, A.K. & Marnett, L.J. (1984) Molecular requirements for the mutagenicity of malondialdehyde and related acroleins. *Cancer Res., 44,* 2848–2854

Basu, A.K., Marnett, L.J. & Romano, L.J. (1984) Dissociation of malondialdehyde mutagenicity in *Salmonella typhimurium* from its ability to induce interstrand DNA cross-links. *Mutat. Res., 129,* 39–46

Bernheim, F., Bernheim, M.L.C. & Wilbur, K.M. (1984) The reaction between thiobarbituric acid and the oxidation products of certain lipids. *J. biol. Chem., 174,* 257–264

Diczfalusy, U., Falardeau, P. & Hammarstrom, S. (1977) Conversion of prostaglandin endoperoxides to C_{17}-hydroxy acids catalyzed by human platelet thromboxane synthase. *FEBS Lett., 84,* 271–274

Fischer, S.M., Ogle, S., Marnett, L.J., Nesnow, S. & Slaga, T.J. (1983) The lack of initiating and/or promoting activity of sodium malondialdehyde on Sencar mouse skin. *Cancer Lett., 19,* 61–66

George, W.O. & Mansell, V.G. (1968) Nuclear magnetic resonance spectra of acetylacetaldehyde and malondialdehyde. *J. chem. Soc. Br.,* 132–134

Kasai, H., Kumeno, K., Yamaigumi, Z., Nishimura, S., Nagao, M., Fujita, Y., Sugimura, T., Nukaya, H. & Kosuge, T. (1982) Mutagenicity of methylglyoxal in coffee. *Gann, 73,* 681–683

Kondo, S., Ichikawa, H., Iwo, K. & Kato, T. (1970) Base-change mutagenesis and prophase induction in strains of *Escherichia coli* with different DNA repair capacities. *Genetics, 66,* 187–217

Levin, D.E., Hollstein, M., Christman, M.F., Schwiers, E.A. & Ames, B.N. (1982) A new *Salmonella* tester strain (TA102) with A : T base pairs at the site of mutation detects oxidative mutagens. *Proc. natl Acad. Sci. USA, 79,* 7445–7449

Lutz, D., Eder, E., Neudecker, T. & Henschler, D. (1982) Structure-mutagenicity relationship in α,β-unsaturated carbonylic compounds and their corresponding allylic alcohols. *Mutat. Res., 93,* 305–315

Marnett, L.J. & Tuttle, M.A. (1980) Comparison of the mutagenicities of malondialdehyde and the side products formed during its chemical synthesis. *Cancer Res., 40,* 276–282

Marnett, L.J., Hurd, H.K., Hollstein, M.C., Levin, D.E., Esterbauer, H. & Ames, B.N. (1984) Naturally occurring carbonyl compounds are mutagens in *Salmonella* tester strain TA104. *Mutat. Res., 148,* 25–34

Moschel, R.C. & Leonard, N.J. (1976) Fluorescent modification of guanine. Reaction with substituted malondialdehydes. *J. org. Chem., 41,* 294–300

Mukai, F.H. & Goldstein, B.D. (1976) Mutagenicity of malondialdehyde, a decomposition product of peroxidized polyunsaturated fatty acids. *Science (Wash. D.C.), 191,* 868–869

Murayama, I. & Otsuji, N. (1973) Mutations by mitomycins in the ultraviolet light-sensitive mutant of *Escherichia coli. Mutat. Res., 18,* 117–119

Nair, V., Turner, G.A. & Offerman, R.J. (1984) Novel adducts from the modification of nucleic acid bases by malondialdehyde. *J. Am. chem. Soc., 106,* 3370–3371

Orgel, A. & Brenner, S. (1961) Mutagenesis of bacteriophage T4 by acridines. *J. mol. Biol., 3,* 762–768

Osman, M.M. (1972) The acidity of malondialdehyde and the stability of its complexes with nickel (II) and copper (II). *Helv. chim. Acta, 55,* 239–244

Perry, K.L. & Walker, G.C. (1982) Identification of plasmid (pKM101)-coded proteins involved in mutagenesis and UV resistance. *Nature, 300,* 278–280

Seto, H., Akiyama, K., Okuda, T., Hashimoto, T., Takesue, T. & Ikemura, T. (1981) Structure of a new modified nucleoside formed by guanosine-malonaldehyde reaction. *Chem. Lett., 1981,* 707–708

Yau, T.M. (1979) Mutagenicity and cytotoxicity of malonaldehyde in mammalian cells. *Mech. Aging Dev., 11,* 137–144

REACTION OF 2-NITROIMIDAZOLE METABOLITES WITH GUANINE AND POSSIBLE BIOLOGICAL CONSEQUENCES

G.F. WHITMORE, A.J. VARGHESE & S. GULYAS

Physics Division, Ontario Cancer Institute, Toronto, Ontario, Canada

SUMMARY

Nitroimidazoles, under hypoxic conditions, undergo reduction reactions producing a variety of reactive species; at the same time, they exert a variety of biological effects. The purpose of the present study was to investigate the reduction chemistry of 2-nitroimidazoles, the reaction of the reduced metabolites with nucleic acid constituents and the possible biological significance of these reactions. Earlier studies had demonstrated that, following reduction, 2-nitroimidazoles react with guanine derivatives to produce a product identical to that seen on reaction of glyoxal with the same guanine derivatives. We report here the identification of this product in the nucleic acid of cells exposed to 2-nitroimidazoles under hypoxic conditions. Using glyoxal as a model compound, we also demonstrate that cellular formation of such a product could account for a number of the biological properties of 2-nitroimidazoles under hypoxic conditions.

INTRODUCTION

A variety of nitroheterocyclic compounds are currently undergoing evaluation as adjuncts to radiation and chemotherapy in the treatment of malignant disease. The primary aim is to improve the response of hypoxic cells which are intrinsically resistant to radiation and which are probably resistant to many conventional chemotherapeutic agents as a result of either reduced metabolism or poor drug delivery, or both.

The nitroheterocyclic compounds were chosen as possible radiation sensitizers because of their high electron affinity (Adams *et al.*, 1980). What was not originally foreseen was that these compounds would exhibit preferential toxicity to hypoxic cells in the absence of radiation. However, this finding might have been expected because many of these agents had been developed for the treatment of anaerobic infections and because their high electron affinity makes them subject to reduction processes leading to the formation of potentially toxic metabolites and to profound effects on cellular constituents. A major limitation in the clinical use of these agents is the induction of neurotoxicity (Wasserman et al., 1980; Dische et al., 1981) which may also arise as a result of the formation of toxic metabolites.

It has previously been shown that at least one class of nitroheterocyclics, the 2-nitroimidazoles, may undergo reduction reactions typical of those described for nitroaromatic compounds (Varghese & Whitmore, 1981). In particular, such reduction leads to the formation of the hydroxylamine derivative, a highly reactive species capable of undergoing a wide variety of subsequent reactions.

For some time now, we have been investigating the reactivity of the hydroxylamine derivatives of 2-nitroimidazoles in chemical systems, in in-vitro cell systems and in patients undergoing treatment with misonidazole. In each of these instances, we have found evidence for the formation of a reactive species capable of reacting with guanine derivatives to yield a two-carbon addition product identical to the product formed by reaction of glyoxal with the same guanine derivatives (Varghese & Whitmore, 1984a,b; Shapiro et al., these proceedings[1]). Based on this and other evidence, we propose the reaction scheme shown in Figure 1, which illustrates that 2-nitroimidazoles, following reduction to the highly reactive hydroxylamine, may undergo an internal rearrangement plus water addition to yield other reactive intermediates, at least two of which appear capable of reacting with guanine derivatives to yield the two-carbon addition product[2]. One of the reactive species is apparently free glyoxal but this is present in low concentration relative to the other reactive species.

In this paper we demonstrate that when mammalian cells are exposed to misonidazole under hypoxic conditions, cellular macromolecules contain the cyclic derivative of guanine. We further demonstrate that glyoxal in aerobic cells is capable of mimicking several of the biological properties associated with exposure to 2-nitroimidazoles under hypoxic conditions and conclude that the formation of cyclic guanine adducts may account for many of these properties.

MATERIALS AND METHODS

Misonidazole and desmethylmisonidazole were obtained from Roche Products Ltd., Welwyn Garden City, Hertfordshire, England. Azomycin (2-nitroimidazole) was purchased from Aldrich Chemical Co., Montreal, Quebec, Canada. SR-2508 was obtained from the Drug Synthesis and Chemistry Branch, NIH, Bethesda, MD. ^{14}C-guanine (55.5 mCi/mmol) was obtained from New England Nuclear, Boston, MA, and glyoxal (40% solution) from Fisher Scientific Company, Fairlawn, NJ.

---

[1] See p. 165.
[2] The existence of a 4,5-dihydroxy-4,5-dihydroimidazole derivative was first postulated and demonstrated by R. McClelland and R. Panicucci (personal communication).

Fig. 1. Reaction scheme for the formation of the guanine adduct from the hydroxylamine derivative (II) of 2-nitroimidazoles (I). The nitroimidazoles used were azomycin (R = H), misonidazole (R = $CH_2CHOHCH_2OCH_3$), desmethylmisonidazole (R = $CH_2CHOHCH_2OH$) and SR-2508 (R = $CH_2CONHCH_2CH_2OH$)

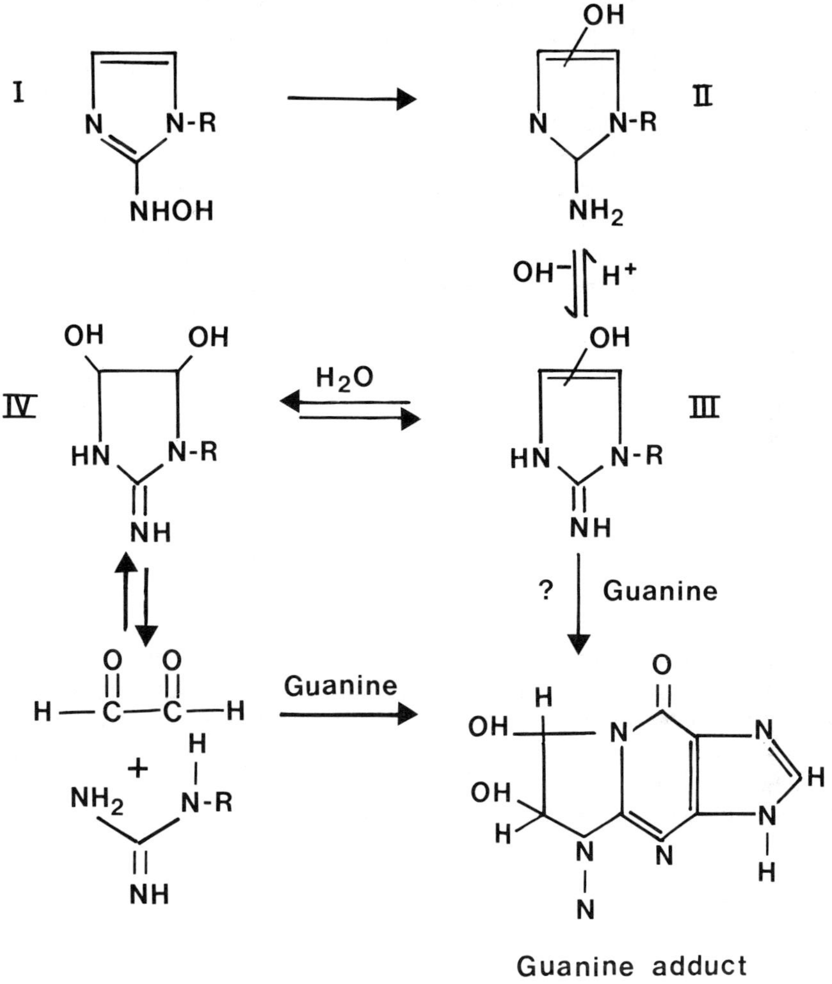

High-performance liquid chromatography (HPLC) was performed on a Waters liquid chromatograph equipped with a Model 440 Absorbance Detector (254 nm) and using a C_{18}-μ Bondapak column (30 cm × 9 mm) obtained from Waters Associates (Milford, MA). The solvent was water : methanol : acetic acid (98 : 1 : 1).

Cell lines used were Chinese hamster ovary (CHO) cells obtained from Dr. W.C. Dewey, plus another wild-type CHO line (AA8-4) and an excision repair-deficient mutant (UV-20), both obtained from Dr L.H. Thompson. All cell lines were grown in α-MEM medium (Stanners et al., 1971) supplemented with 10% fetal calf serum

(FCS) obtained from Flow Laboratories, Rockville, MD. Synchronization was achieved using the ^{3}H-thymidine suicide technique of Whitmore and Gulyas (1966). Exposure to glyoxal and misonidazole was in α-MEM plus FCS plus glyoxal or misonidazole at the concentrations shown. Hypoxia was produced by placing cell suspensions in a glass vial equipped with a magnetic stirrer and passing humidified nitrogen plus 5% CO_2 containing less than four ppm oxygen at a flow rate of approximately one 1 l/min.

To label cells with ^{14}C-guanine, CHO cells were grown for 48 hours (approximately four generations) in α-MEM lacking all nucleosides and containing 10% extensively dialysed FCS plus ^{14}C-guanine (0.02 µCi/ml), washed once and grown for 16 hours in α-MEM plus 10% FCS. Following guanine incorporation, the cells were washed twice in α-MEM, resuspended at 5×10^7 cells/ml in α-MEM and incubated with misonidazole for four hours at 37 °C under hypoxic conditions. Following incubation, the cells were washed once with α-MEM and twice with cold 5% TCA, hydrolysed with 0.1 N HCL at 60 °C for two hours, and subjected to HPLC. Successive aliquots were collected and their radioactivity determined.

RESULTS

We have previously demonstrated that homogenates of cells exposed to misonidazole under hypoxic conditions contain two metabolites of misonidazole capable of either rapid or slow reaction with guanosine (Varghese & Whitmore, 1984a). The rapidly reacting species has been identified as glyoxal. Figure 2 demonstrates that, when cells are exposed to misonidazole under hypoxic conditions, cellular macromolecules contain the cyclic derivative of guanine. In this experiment, cells were incubated with ^{14}C-guanine for 48 hours, exposed to misonidazole (10 mM) under hypoxic conditions, washed to remove nucleic acid precursors, acid hydrolysed and run on HPLC. The guanine derivative is not seen in cells not exposed to misonidazole. The insert in Figure 2 indicates that, on rechromatography, the product cochromatographs with the authentic guanine adduct synthesized by reaction of glyoxal with guanine. The data in Figure 2 indicate that, under these somewhat extreme conditions of exposure to misonidazole, and using hydrolytic conditions which convert approximately 60% of the guanosine product to the guanine product, something in excess of 0.3% of the guanine residues in cellular macromolecules are converted to the cyclic guanine derivative.

In an attempt to investigate the possible biological role of the cyclic guanine adduct resulting from misonidazole exposure, we have used glyoxal to reproduce several previously identified effects of misonidazole, including (1) toxicity in aerobic and hypoxic cells, (2) toxicity as a function of cell age, (3) dose-additive radiosensitization, (4) the enhanced response of CHO mutant cells with increased sensitivity to ultraviolet light and misonidazole.

The data shown in Figure 3 indicate the toxicity observed when CHO cells were exposed to various concentrations of glyoxal under aerobic or hypoxic conditions. Toxicity was somewhat greater for exposure under hypoxic conditions. Under

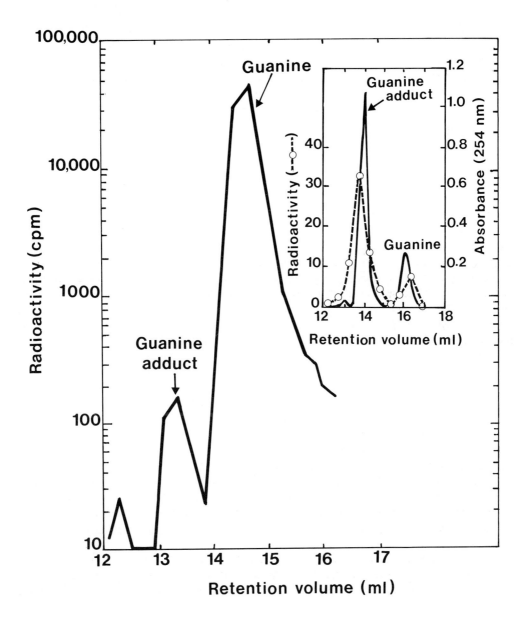

Fig. 2. High-performance liquid chromatographic (HPLC) analysis of acid hydrolysates of Chinese hamster ovary cells incubated with ^{14}C-guanine and exposed to misonidazole (see Materials and Methods). To obtain the data in the insert, material having an initial retention volume of 13 to 13.5 ml was rechromatographed by HPLC together with an authentic marker containing guanine and the guanine adduct produced by reaction of guanine with glyoxal. The dashed line represents radioactivity and the solid line, the absorbance of the authentic sample of guanine plus guanine adduct.

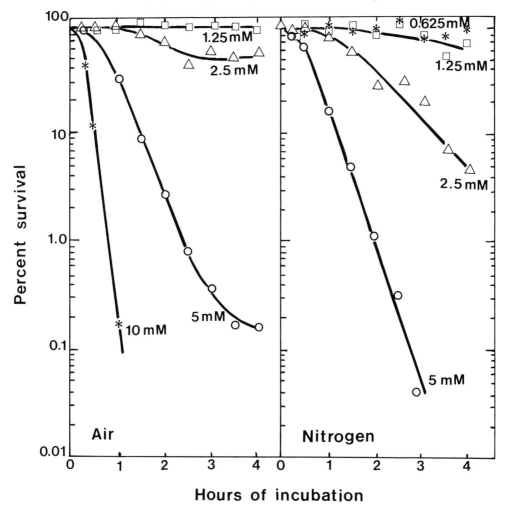

Fig. 3. Survival as a function of time of exposure to glyoxal at the concentrations shown. Left panel, aerobic exposure; right panel, hypoxic exposure; initial cell density, 5×10^5 cells/ml

conditions of aerobic but not hypoxic exposure, the survival curves exhibited an upward concavity.

Synchronized CHO cells exposed to misonidazole under hypoxic conditions show a pronounced variation in toxicity as a function of age in the cell cycle (Whitmore & Gulyas, 1981). To determine the age response for exposure to glyoxal, aliquots of a synchronized cell suspension were removed at various times and exposed to glyoxal for a period of 1.25 hours under aerobic conditions. After exposure, the cells were appropriately diluted and plated to determine colony survival. Figure 4 shows a comparison between survival data obtained in such an experiment and similar data obtained for exposure to misonidazole (Whitmore & Gulyas, 1981). Exposure to

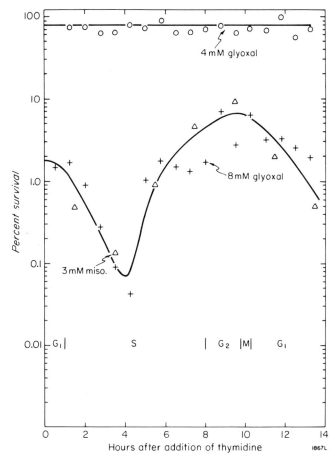

Fig. 4. Cell survival in synchronized populations of Chinese hamster ovary cells exposed to 4 or 8 mM glyoxal for 1.25 hours. Survival is plotted at the mid-point of the exposure interval, and the approximate phases of the cell cycle (G_1, S, G_2, M) are shown. Also shown are survival points for cells exposed to 3 mM misonidazole for three-hour intervals under hypoxic conditions. The misonidazole data are taken from Whitmore and Gulyas (1981) and have been normalized (see text).

8 mM glyoxal produced a marked reduction in cell survival which was most pronounced in early to mid S phase and least pronounced in late G_2 phase. In order to facilitate comparison with glyoxal, misonidazole survival levels previously reported (Whitmore & Gulyas, 1981) have been reduced by a factor of two. Examination of the data in Figure 4 suggests that both agents produce very similar age-dependent variations in cell survival.

It has been reported by Wong et al. (1978) that, when CHO cells are exposed to misonidazole under hypoxic conditions and subsequently exposed to radiation under either aerobic or hypoxic conditions, there is a pronounced reduction in the shoulder

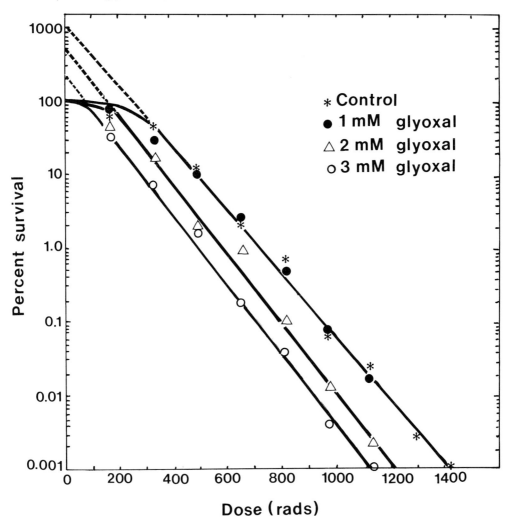

Fig. 5. Radiation survival curves for Chinese hamster ovary cells following a three-hour aerobic exposure to glyoxal at 0, 1, 2 and 3 mM

of the radiation survival curve (dose-additive radiosensitization). Given sufficient exposure to misonidazole, the shoulder on the radiation survival curve is essentially completely removed. This complete removal of the shoulder occurs under approximately the same exposure conditions (time and concentration) as the onset of overt toxicity as measured by loss of survival in the absence of radiation. To determine whether glyoxal could mimic the effect of misonidazole, aerobic cells were exposed to glyoxal (1–3 mM) for three hours at 37 °C and then irradiated with graded doses of radiation to generate complete radiation survival curves. The data in Figure 5 show the resulting survival curves and it is apparent that the principal effect of the

Fig. 6. Survival of AA8-4 (wild-type) and UV-20 (excision repair-deficient) cells following plating in glyoxal at the concentrations shown. Different symbols refer to different sets of experiments.

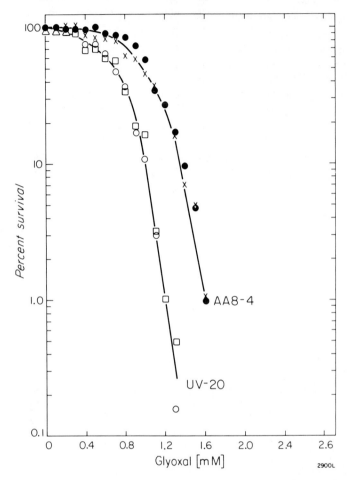

glyoxal exposure is a marked reduction in shoulder width without a concomitant change in slope.

If glyoxal, under aerobic conditions, is capable of mimicking the toxic properties of misonidazole under hypoxic conditions, then DNA repair-deficient mutant cells, which exhibit increased sensitivity to misonidazole, should also exhibit enhanced sensitivity to glyoxal. The CHO mutant, UV-20, is deficient in excision repair and exhibits approximately a seven-fold increased sensitivity to ultraviolet light, a greater than 100-fold increased sensitivity to mitomycin-C and a 1.3- to 1.5-fold increased sensitivity to misonidazole under hypoxic conditions (Thompson et al., 1980). When wild-type (AA8-4) and UV-20 cells were plated in medium containing various concentrations of glyoxal, the survival data shown in Figure 6 were obtained. Under the exposure conditions used, UV-20 exhibits a 1.3- to 1.5-fold increase in sensitivity, very similar to the increased sensitivity seen with misonidazole.

DISCUSSION

A variety of data now strongly suggest that many of the biological phenomena observed following hypoxic exposure to misonidazole and other 2-nitroimidazoles arise as a result of the reduction of these compounds to several potentially reactive species. Much of the evidence suggests that reduction to the hydroxylamine occurs readily (Varghese & Whitmore, 1980; Josephy at al., 1981; Varghese & Whitmore, 1981; Whillans & Whitmore, 1981) and formation of the amine derivative has also been detected (Varghese & Whitmore, 1981). The hydroxylamine derivatives of the 2-nitroimidazoles are a highly reactive species and also undergo transitions to form other potentially reactive species capable of reaction with guanine (Varghese & Whitmore, 1983), glutathione (Varghese, 1983) and other nucleophiles. We have previously reported that reaction with glutathione (Varghese & Whitmore, 1984c) and guanine (Varghese & Whitmore, 1984a) occurs in cellular systems and, in this paper, we have provided evidence for the formation of a two-carbon addition product of guanine in the nucleic acids of cells exposed to misonidazole under hypoxic conditions. Formation of a similar guanine adduct has also been observed with other 2-nitroimidazoles in chemical systems (see Figure 1 and Varghese & Whitmore, 1983) and, while it has not yet been demonstrated, must be presumed to occur in cellular systems exposed to these agents.

Studies with chemical systems and cellular homogenates have indicated that the reduction of 2-nitroimidazoles produces at least two species of products capable of reacting to form the same two-carbon addition product with guanine (Varghese & Whitmore, 1983). While one of these species is apparently free glyoxal, the other product(s), which is present in much larger amounts, appears to be capable of direct reaction with guanine since pretreatment of reduced nitroimidazoles with either acid or base does not appear to markedly affect the kinetics of the subsequent reaction with guanine. The existence of two or more reactive species may be of importance because they may differ markedly in their lipophilicity and pharmacokinetic properties, factors known to affect the human toxicity of the 2-nitroimidazoles (White et al., 1982).

Although the formation of an adduct with guanine is suggestive of subsequent biological effects, it does not prove that formation of such an adduct could account for any or all of the biological phenomena observed with misonidazole under hypoxic conditions. The data presented here suggest that a reduction product(s) with the ability to produce glyoxal-like reactions in mammalian cells could mimic several of the biological processes seen with 2-nitroimidazoles under hypoxic conditions. In particular, we have shown that glyoxal is toxic to mammalian cells, that it shows the same age-dependent toxicity as misonidazole, and that it is also capable of producing the dose-additive or shoulder-reduction type of radiosensitization apparent with prolonged hypoxic exposure to 2-nitroimidazoles.

Since only a fraction of any 2-nitroimidazole undergoes reduction under most cellular exposure conditions, it might be expected that if production of glyoxal or glyoxal-like derivatives was responsible for the biological effects, then smaller concentrations of glyoxal would be required to duplicate the effects seen with 2-nitroimidazoles. Several of our observations would tend to support this contention but comparisons of the concentrations of 2-nitroimidazoles and any putative toxic derivatives required to produce equivalent toxic effects are complicated by the fact that in one instance the reactive species is produced intracellularly whereas with glyoxal the test substance is added exogenously. For this and a variety of other reasons, *a priori* predictions of relative potency are difficult.

Because both glyoxal and the reduction products of the 2-nitroimidazoles are capable of reaction with a wide variety of intracellular chemical species, it is difficult to pinpoint which reaction accounts for biological effects as nonspecific as cell toxicity and possibly dose-additive radiosensitization. However, the finding that reduction products are capable of reacting with nucleic acids supports the contention that such reactions could be responsible for the observed biological effects. This contention is further strengthened by the observation that the UV-20 cell line known to be deficient in excision repair of DNA demonstrates similar increases in sensitivity to both agents.

In summary, our results indicate that, following reduction, 2-nitroimidazoles are capable of producing cyclic addition products with guanine in nucleic acids and that such reactions could account for many of the biological effects seen with 2-nitroimidazoles under reducing conditions. However, it must be stressed that a demonstration that 2-nitroimidazoles produce glyoxal-like adducts with guanine in cellular nucleic acids under hypoxic conditions and that glyoxal can mimic many of the biological effects of the 2-nitroimidazoles, while suggestive, does not constitute proof of the biological importance of the guanine adduct. Furthermore, it must be pointed out that the formation of a cyclic addition product with guanine may not account for the mutagenic properties of these agents since azomycin (2-nitroimidazole), which does produce the guanine adduct, appears to be much less mutagenic than other 2-nitroimidazoles (Chin *et al.*, 1978).

REFERENCES

Adams, G.E., Ahmed, I., Fielden, E.M., O'Neill, P.O. & Stratford, I.J. (1980) The development of some nitroimidazoles as hypoxic cell sensitizers. *Cancer clin. Trials*, *3*, 37–42

Chin, J.B., Sheinin, D.M.K. & Rauth, A.M. (1978) Screening for the mutagenicity of nitro-group containing hypoxic cell radiosensitizers using *Salmonella typhimurium* strains TA-100 and TA-98. *Mutat. Res.*, *58*, 1–10

Dische, S., Saunders, M.I. & Stratford, M.R.L. (1981) Neurotoxicity with desmethylmisonidazole. *Br. J. Radiol.*, *54*, 156–157

Josephy, P.D., Palcic, B. & Skarsgard, L.D. (1981) Reduction of misonidazole and its derivatives by xanthine oxidase. *Biochem. Pharmacol.*, *30*, 849–853

Stanners, C.P., Eliceiri, G.L. & Green, H. (1971) Two types of ribosomes in mouse-hamster cells. *Nature New Biol.*, *230*, 52–54

Thompson, L.H., Rubin, J.S., Cleaver, J.E., Whitmore, G.F. & Brookman, K. (1980) A screening method for isolating DNA repair-deficient mutants of CHO cells. *Somatic Cell Genet.*, *6*, 391–405

Varghese, A.J. (1983) Glutathione conjugates of misonidazole. *Biochem. biophys. Res. Commun.*, *112*, 1013–1020

Varghese, A.J. & Whitmore, G.F. (1980) Binding to cellular macromolecules: a possible mechanism for the cytotoxicity of misonidazole. *Cancer Res., 40,* 2165–2169

Varghese, A.J. & Whitmore, G.F. (1981) Cellular and chemical reduction products of misonidazole. *Chem.-biol. Interactions, 36,* 141–151

Varghese, A.J. & Whitmore, G.F. (1983) Modification of guanine derivatives by reduced 2-nitromidazoles. *Cancer Res., 43,* 78–82

Varghese, A.J. & Whitmore, G.F. (1984a) Detection of a reactive metabolite of misonidazole in hypoxic mammalian cells. *Radiat. Res., 97,* 262–271

Varghese, A.J. & Whitmore, G.F. (1984b) Detection of a reactive metabolite of misonidazole in human urine. *Int. J. Radiat. Oncol. Biol. Phys., 10,* 1361–1363

Varghese, A.J. & Whitmore, G.F. (1984c) Misonidazole-glutathione conjugates in CHO cells. *Int. J. Radiat. Oncol. Biol. Phys., 10,* 1341–1345

Wasserman, T.H., Stetz, J. & Phillips, T.L. (1980) *Clinical trials of misonidazole in the United States.* In: Brady, L.W., ed., *Radiation Sensitizers, Their Use in the Clinical Management of Cancer,* New York, Masson Publishing USA, Inc., pp. 387–396

Willans, D.W. & Whitmore, G.F. (1981) The radiation reduction of misonidazole. *Radiat. Res., 86,* 311–324

White, R., Workman, P. & Owen, L. (1982) The pharmacokinetics in mice and dogs of nitroimidazole radiosensitizers and chemosensitizers more lipophilic than misonidazole. *Int. J. Radiat. Oncol. Biol. Phys., 8,* 473–476

Whitmore, G.F. & Gulyas, S. (1966) Synchronization of mammalian cells with tritiated thymidine. *Science, 151,* 691–694

Whitmore, G.F. & Gulyas, S. (1981) *Lethal and sublethal effects of misonidazole under hypoxic conditions.* In: Brady, L.W., ed., *Radiation Sensitizers, Their Use in the Clinical Management of Cancer,* New York, Masson Publishing USA, Inc., pp. 99–106

Wong, T.W., Whitmore, G.F. & Gulyas, S. (1978) Studies on the toxicity and radiosensitizing ability of misonidazole under conditions of prolonged incubation. *Radiat. Res., 75,* 541–555

STRUCTURE-ACTIVITY RELATIONSHIPS OF
α,β-UNSATURATED CARBONYLIC COMPOUNDS

D. HENSCHLER & E. EDER

*Institute of Toxicology, University of Würzburg,
Würzburg, Federal Republic of Germany*

SUMMARY

The mutagenicity and probably the carcinogenicity of α,β-unsaturated carbonylic compounds such as acrolein are based on direct genotoxic interaction with nucleic acid bases *via* Michael addition or Schiff's base formation. Alkyl and aryl substitution at the α and β carbon atoms reduces or abolishes the mutagenic potential whereas halogen substitution in either position increases mutagenicity. These structure-activity relationships can be predicted from theoretical considerations of well-known electron shift mechanisms. The formation of α,β-unsaturated carbonyls from allylic halides (or similar types of compounds with appropriate leaving groups) has been experimentally demonstrated with appropriate metabolites of allyl bromide in rats.

INTRODUCTION

Acrolein is the prototype of a large series of α,β-unsaturated carbonyls of considerable scientific and practical interest. Some compounds are used as intermediates in organic synthesis, some are formed as combustion products from food and other organic materials, and many have a widespread use as perfumes and flavouring agents.

Acrolein has been shown to be mutagenic (Bignami *et al.*, 1977) in the *Salmonella typhimurium* testing system. We found acrolein to be one of the most potent mutagens in a modified mutagenicity testing procedure (Lutz *et al.*, 1982). This finding prompted us to initiate systematic studies on the structure-activity relationships of α,β-unsaturated carbonyl compounds, in particular on the influence of aliphatic, aromatic and halogen substituents in various positions in the molecular framework (see Fig. 1). Studies on this type of compound are worthwhile, not only because the carbonyls as such are interesting, but also because of the existence of a wide variety of precursor compounds, synthetic as well as naturally occuring, from which the carbonyls can be formed *via* the following metabolic processes (Fig. 2).

Fig. 1. Positions of possible substituents (R_1, R_2, R_3) in the molecular framework of α,β-unsaturated carbonyl compounds

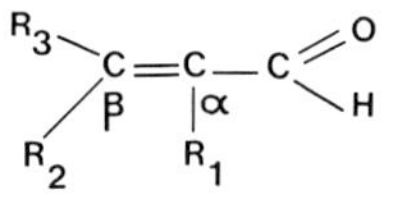

Fig. 2. Metabolic formation of α,β-unsaturated carbonyls from allylic alcohols and the formation of the latter by hydrolysis from allylic compounds having a suitable leaving group (X)

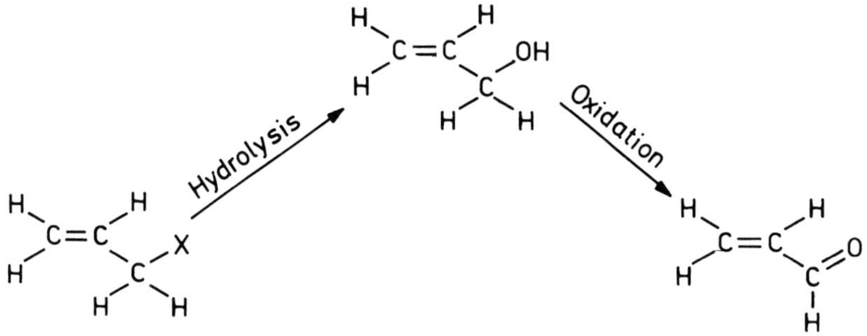

(a) Allylic alcohols can be oxidized by alcohol dehydrogenase (ADH) to the respective aldehydes or ketones; in fact, we could demonstrate mutagenic properties of allyl alcohol and its homologous congeners using *Salmonella typhimurium* strain TA100 which contains ADH (Lutz *et al.*, 1982);

(b) Allyl compounds having an appropriate leaving group in the allylic position can be hydrolysed spontaneously as well as enzymatically to render the respective alcohols which again may be oxidized to the genotoxic carbonyl derivative (Eder *et al.*, 1982).

MATERIALS AND METHODS

Chemicals

The origin, purification and purity of the tested compounds have been described in detail by Lutz *et al.*, (1982).

The mutagenic potential was determined in the *Salmonella typhimurium* strain TA100 as described by Ames *et al.*, (1975) with a modification similar to that proposed by Rannug *et al.* (1976); details are given in a previous paper (Lutz *et al.*, 1982).

RESULTS AND DISCUSSION

Mutagenic properties of acrolein and alkyl and aryl derivatives

In a series of seven structurally related α,β-unsaturated carbonyls, acrolein is the most potent mutagen (see Table I). Its mutagenicity is completely inhibited by the addition of S9 mix (the 9000 *g* supernatant of rat liver homogenate), most probably

Table 1. Mutagenicity of α,β-unsaturated compounds in *Salmonella typhimurium* strain TA100 (according to Lutz et al., 1982)

Compound	Structure	revertants/μmol − S9 mix	revertants/μmol + S9 mix
acrolein	$CH_2=CH-CHO$	2400	0
crotonaldehyde	$CH_3-CH=CH-CHO$	1280	340
2-methylacrolein	$CH_2=C(CH_3)-CHO$	460	220
methyl vinyl ketone	$CH_2=CH-CO-CH_3$	450	250
2-cyclohexen-1-one	(cyclohexenone)	3	1
cinnamaldehyde	$C_6H_5-CH=CH-CHO$	0	0
citral	$CH_3-C(CH_2CH_2CH=C(CH_3)_2)=CH-CHO$	0	0

by reaction with thiol groups in low and high molecular weight compounds (for review, see Izard & Liebermann, 1978).

Simple methyl substitution in the β-position (crotonaldehyde) reduces the direct mutagenic activity by a factor of two, in the α-position (2-methylacrolein) by a factor of five. Methyl substitution at the carbonyl carbon atom (methyl vinyl ketone) also decreases mutagenicity considerably; the activity of this compound clearly demonstrates that α,β-unsaturated ketones may also exert a direct genotoxic effect. However, methyl vinyl ketone, as well as the other methyl-substituted molecules, retain much of their activity in the presence of S9 mix; this could be due to mechanisms other than those involved in the case of simple acrolein. A very low activity is seen with 2-cyclohexen-1-one, a cyclic derivative of acrolein. No activity was observed with the simplest aromatic substitution product, cinnamaldehyde, nor with the flavouring agent, citral.

Fig. 3. Mechanism for the formation of Michael addition-like compounds from acrolein (equation a), and the effect of various alkyl substitutions on the positive partial charge of the β-carbon atom (equations b, c and d)

(a) $\left[CH_2=CH-C\overset{O}{\underset{H}{\diagdown}} \longleftrightarrow {}^{\oplus}CH_2-CH=C\overset{|\overline{O}|^{\ominus}}{\underset{H}{\diagdown}} \right]$

Acrolein

(b) $\left[\underset{\beta}{CH_3}\overset{+I \rightarrow}{-}\underset{\alpha}{CH}=CH-C\overset{O}{\underset{H}{\diagdown}} \longleftrightarrow \underset{\beta}{CH_3}\overset{+I \rightarrow \oplus}{-}\underset{\alpha}{CH}-CH=C\overset{|\overline{O}|^{\ominus}}{\underset{H}{\diagdown}} \right]$

Crotonaldehyde

(c) $\left[\underset{\beta}{CH_2}=\overset{\overset{+I}{\underset{|}{CH_3}}}{\underset{\alpha}{C}}-C\overset{O}{\underset{H}{\diagdown}} \longleftrightarrow {}^{\oplus}CH_2-\overset{\overset{+I}{\underset{|}{CH_3}}}{C}=C\overset{|\overline{O}|^{\ominus}}{\underset{H}{\diagdown}} \right]$

2-Methylacrolein

(d) $\left[\underset{\beta}{CH_2}=\underset{\alpha}{CH}-\overset{O}{\underset{||}{C}}-\overset{+I}{CH_3} \longleftrightarrow {}^{\oplus}CH_2-CH=\overset{|\overline{O}|^{\ominus}}{\underset{|}{C}}-\overset{+I}{CH_3} \right]$

Methyl vinyl ketone

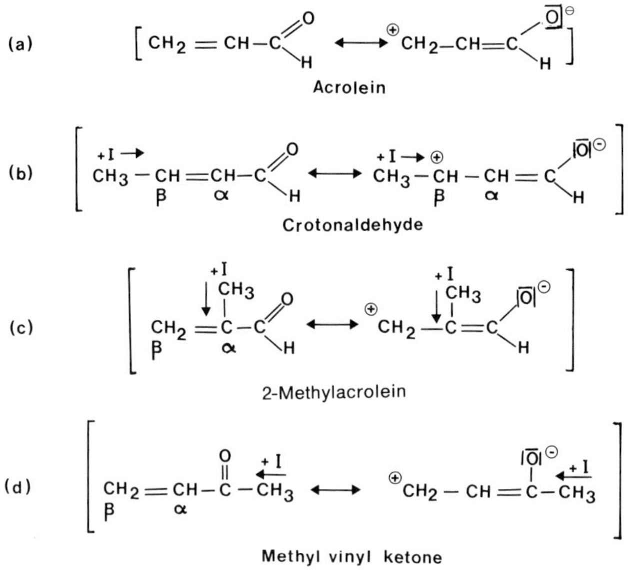

The reduction of the reactivity of acrolein by alkyl and aryl substitutions can easily be explained by their positive inductive (+I) effect. If the genotoxic activity of acrolein is due to the formation of Michael addition-like compounds, the mechanism of which is explained by equation (a) in Figure 3, alkyl substitutions should be expected to reduce the positive partial charge of the β-carbon atom (equations (b) (c) and (d), Fig. 3).

From these few examples of substitutions, simple rules of structure-activity relationships for other derivatives may be readily developed.

Influence of halogen substitutions

Halogen substitutions at both the α- and β-carbon atom of acrolein (and its alkyl and aryl derivatives) should be expected to interfere with the charge distribution in an α,β-unsaturated carbonyl, which is a prerequisite of the Michael addition reaction, in the sense that the positive partial charge at the β-carbon atom will be increased by the electron-withdrawing effect of the halogen (Fig. 4). Also, the reactivity of the aldehyde group for Schiff's base formation will be increased by the electron-withdrawing effect of a halogen substitution.

This hypothesis is confirmed by experimental findings (see Table 2). Rosen *et al.* (1980) found 2-chloroacrolein, 2,3-dichloracrolein, 2,3,3-trichloroacrolein and 2-bromoacrolein to be highly mutagenic in *Salmonella typhimurium* TA100; the increase in activity was in the expected order. Another example has been provided by Neudecker *et al.* (1983). Cinnamaldehyde substituted in the α-C position with

Fig. 4. Effect of halogen substitutions on the positive partial charge of the β-carbon atom

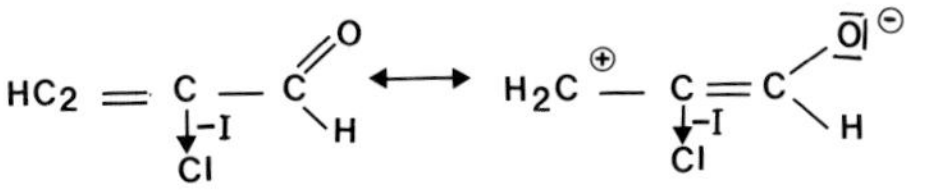

$$HC_2 = \underset{\underset{Cl}{\downarrow -I}}{C} - C\overset{O}{\underset{H}{\nwarrow}} \longleftrightarrow H_2\overset{\oplus}{C} - \underset{\underset{Cl}{\downarrow -I}}{C} = C\overset{\overset{\ominus}{\overline{O}|}}{\underset{H}{\nwarrow}}$$

Table 2. Increase in mutagenicity of acrolein and cinnamaldehyde due to halogen substitutions

Compound	+S9	−S9	Reference
acrolein	−	+	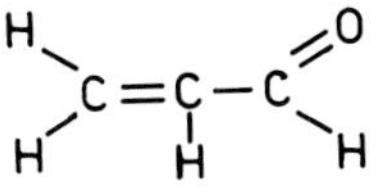 Lutz et al. (1982)
2-chloroacrolein	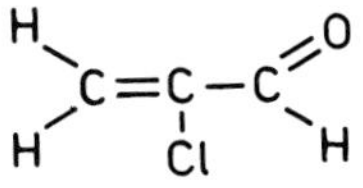 n.d.[a]	+++	Rosen et al. (1980)
2-bromoacrolein	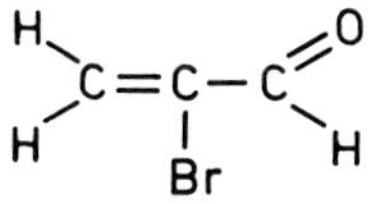 n.d.	+++	Rosen et al. (1980)
2,3-dichloroacrolein	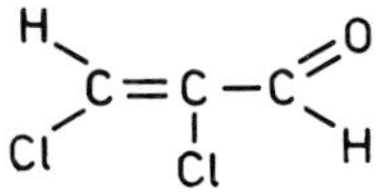 n.d.	+++	Rosen et al. (1980)
2,3,3-trichloroacrolein	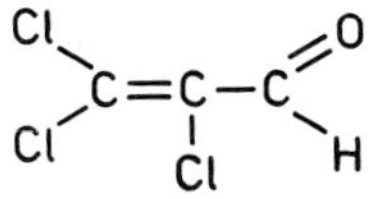 n.d.	+++	Rosen et al. (1980)
cinnamaldehyde	−	−	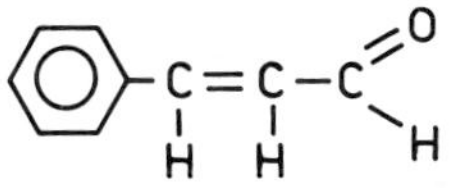 Neudecker et al. (1983)
2-chlorocinnamaldehyde	+	+++	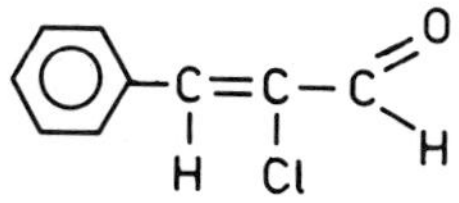 Neudecker et al. (1983)
2-bromocinnamaldehyde	+	+++	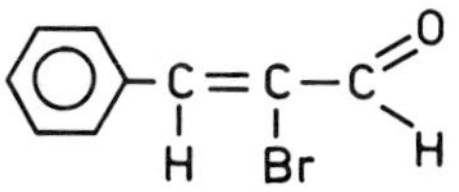 Neudecker et al. (1983)

[a] Not determined

Fig. 5. Pathway for the formation of S-(carboxyethyl) mercapturic acid (shown as the methyl ester) from allyl bromide in rats, with acrolein as an intermediate. M, methylation of urine sample with diazomethane; GSH, glutathione; ADH, alcohol dehydrogenase

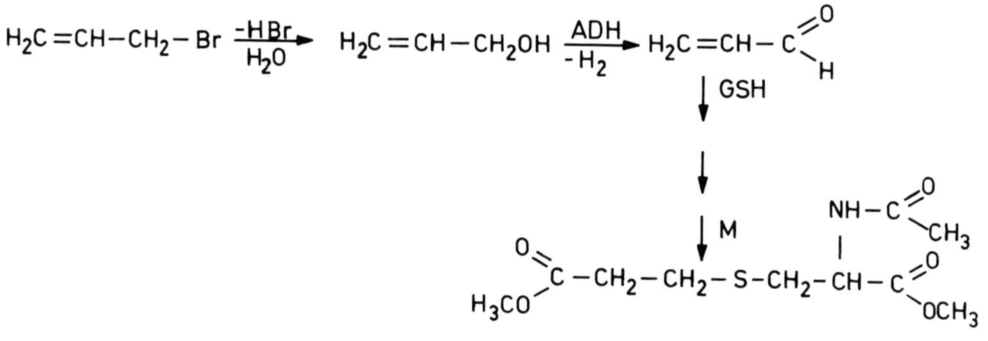

$$H_2C=CH-CH_2-Br \xrightarrow[H_2O]{-HBr} H_2C=CH-CH_2OH \xrightarrow[-H_2]{ADH} H_2C=CH-CHO \xrightarrow{GSH} \xrightarrow{M} H_3CO-CO-CH_2-CH_2-S-CH_2-CH(NH-COCH_3)-COOCH_3$$

chlorine or bromine was mutagenic, whereas α-methylcinnamaldehyde proved to be ineffective, according to expectation.

Formation of α,β-*unsaturated carbonyls from allyl halides*

As indicated in Figure 2, allyl halides or other allyl compounds with appropriate leaving groups may hydrolyse to allylic alcohols, which may subsequently be oxidized by ADH to acrolein or to the corresponding derivatives. Therefore, the genotoxic potential of allyl halides may, at least in part, be due to the formation of α,β-unsaturated carbonylic intermediates. Since, however, other mechanisms of genotoxic interactions of allyl halides may also exist (as described below) it is difficult to determine the importance of the pathway involving α,β-unsaturated carbonyl formation.

We have, however, recently obtained proof for the formation of acrolein from allyl bromide in rats by an analysis of metabolites formed (Eder & Freitag, 1983). S-(Carboxyethyl)mercapturic acid has been identified as a urinary metabolite of allyl bromide, and this can only be explained, as shown in Figure 5, by an intermediate hydrolysis and conjugation with glutathione, as previously reported by Draminski *et al.* (1983).

Whether, and to which extent, allyl halides exert their genotoxicity through α,β-unsaturated carbonyls depends on a variety of influential factors such as chemical structure and reactivity, pharmacokinetics, and suitability as substrates for enzymatic transformations. However, the careful determination of the chemical structure of DNA adducts will permit the quantitation of the major pathways because of characteristic differences in the chemistry of the various possible adducts (see Fig. 7).

Some cyclic nucleic acid adducts from acrolein and crotonaldehyde have been described by Chung *et al.*, these proceedings[1] and by Shapiro *et al.*, these proceedings.[2]

---

[1] See p. 207.
[2] See p. 165.

Fig. 6. Three different pathways for genotoxic interactions of allyl compounds with nucleic acid bases. X, leaving group; Nu, nucleic acid base; ADH, alcohol dehydrogenase; MFO, mixed function oxidase

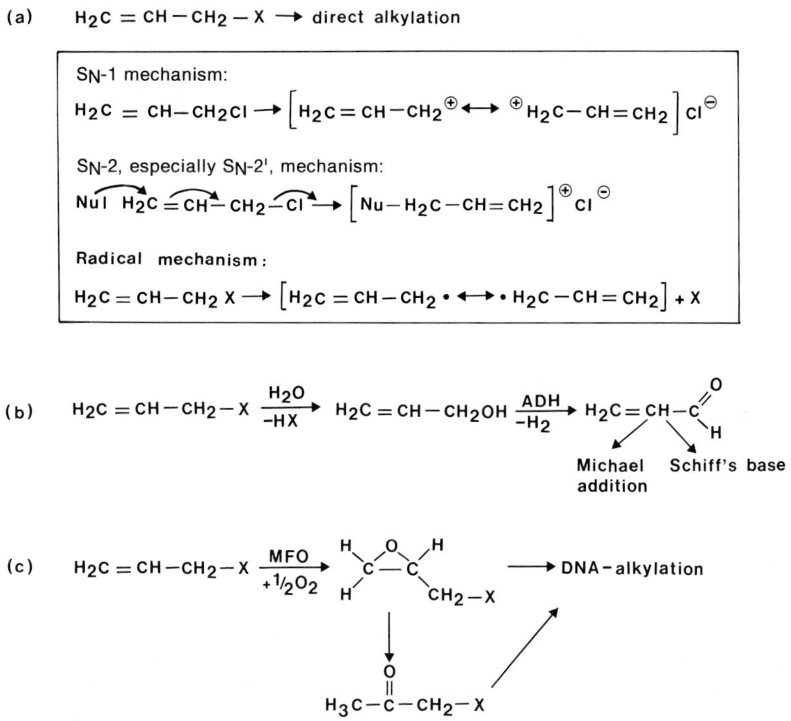

(a) $H_2C=CH-CH_2-X \rightarrow$ direct alkylation

S_N-1 mechanism:
$H_2C=CH-CH_2Cl \rightarrow [H_2C=CH-CH_2^{\oplus} \leftrightarrow {}^{\oplus}H_2C-CH=CH_2] Cl^{\ominus}$

S_N-2, especially S_N-2', mechanism:
$Nu \frown H_2C = CH \frown CH_2 \frown Cl \rightarrow [Nu-H_2C-CH=CH_2]^{\oplus} Cl^{\ominus}$

Radical mechanism:
$H_2C=CH-CH_2 X \rightarrow [H_2C=CH-CH_2 \cdot \leftrightarrow \cdot H_2C-CH=CH_2] + X$

(b) $H_2C=CH-CH_2-X \xrightarrow[-HX]{H_2O} H_2C=CH-CH_2OH \xrightarrow[-H_2]{ADH} H_2C=CH-C\overset{O}{\underset{H}{\diagdown}}$
   Michael   Schiff's base
   addition

(c) $H_2C=CH-CH_2-X \xrightarrow[+\frac{1}{2}O_2]{MFO} \underset{H}{\overset{H}{\diagdown}}C \overset{O}{-} C \underset{CH_2-X}{\overset{H}{\diagup}} \longrightarrow$ DNA-alkylation

$H_3C-\overset{O}{\underset{\|}{C}}-CH_2-X$

Fig. 7. Chemical structures of nucleic acid base adducts formed from allylic compounds by various mechanisms

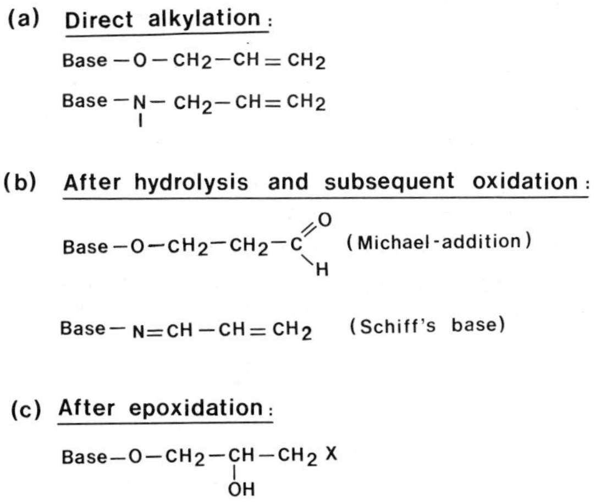

(a) Direct alkylation:

Base $-O-CH_2-CH=CH_2$

Base $-\underset{|}{N}-CH_2-CH=CH_2$

(b) After hydrolysis and subsequent oxidation:

Base$-O-CH_2-CH_2-C\overset{\diagup O}{\underset{H}{\diagdown}}$ (Michael-addition)

Base$-N=CH-CH=CH_2$ (Schiff's base)

(c) After epoxidation:

Base$-O-CH_2-\underset{\underset{OH}{|}}{CH}-CH_2 X$

Fig. 8. Additional pathways for the activation of 2,3-dichloro-1-propene. ADH, alcohol dehydrogenase; GSH, glutathione

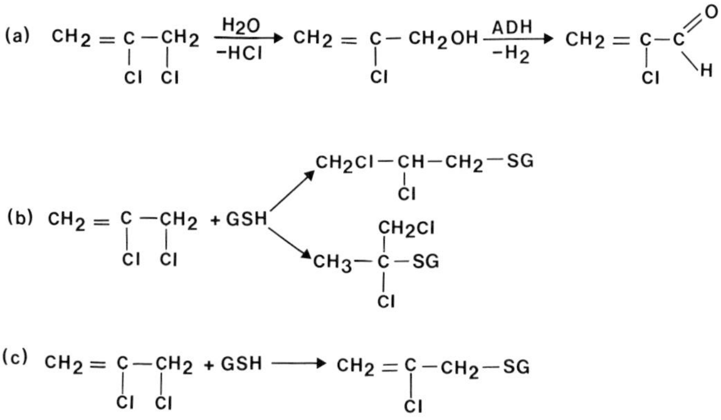

(a) $CH_2 = C-CH_2 \xrightarrow[-HCl]{H_2O} CH_2 = C-CH_2OH \xrightarrow[-H_2]{ADH} CH_2 = C-C\underset{H}{\overset{O}{\diagup}}$
       | | | |
       Cl Cl   Cl   Cl

(b) $CH_2 = C-CH_2 + GSH$ ⟶ $CH_2Cl-CH-CH_2-SG$
       | | |
       Cl Cl   Cl

                                     CH_2Cl
                                     |
                                     CH_3-C-SG
                                     |
                                     Cl

(c) $CH_2 = C-CH_2 + GSH$ ⟶ $CH_2 = C-CH_2-SG$
       | | |
       Cl Cl   Cl

Competing mechanisms of genotoxicity in allyl compounds

Allyl compounds may exert genotoxicity *via* a number of different pathways, as outlined in Figure 6:

(a) direct alkylation if an appropriate leaving group facilitates one of the mechanisms shown in the box in Figure 6,

(b) hydrolysis to allylic alcohols with ensuing oxidation to α,β-unsaturated carbonyls,

(c) epoxidation of the olefinic double bond and alkylation of DNA through opening of the thus formed oxirane ring.

The reaction products with DNA bases will be different for the three main pathways, as shown in Figure 7. In the case of direct alkylation and Schiff's base formation, the allyl structure will remain unchanged, whereas alkylation by the epoxide route will result in hydroxylated adducts. After a Michael addition reaction, the carbonyl function will remain intact.

With some chlorinated allylic compounds, even more possibilities for genotoxic interactions must be taken into consideration. One example has been investigated in more detail: 2,3-dichloro-1-propene. In the Ames test, this compound behaves uniquely in that addition of S9 mix increases considerably the mutagenic activity (Neudecker *et al.*, 1980), despite the fact that the molecule also exerts a direct alkylating effect. This indicates additional activating pathways. Hydrolysis could lead to 2-chloroallyl alcohol, which can be converted to the highly mutagenic (Rosen *et al.*, 1980) 2-chloroacrolein, as shown in Figure 8a. Epoxidation would produce a highly unstable oxirane which, in addition to effecting direct alkylation, would rearrange to 1,3-dichloroacetone, an extremely reactive and mutagenic molecule (Eder, 1982). Enzymatic conjugation with glutathione at the double bond would result in two chemically different alkylating adducts, depending on which of the vinylic carbon atoms was bonded (Fig. 8b). A third alkylating species would be formed if a direct substitution reaction occured at the allylic chlorine residue (Fig. 8c).

The mercapturic acid, $CH_2 = C(Cl)-CH_2-SNAcCyst$, which was synthesized in our laboratory is neither alkylating (in the 4-(*p*-nitrobenzyl)pyridine test) nor mutagenic (in the Ames test).

REFERENCES

Ames, B.N., McCann, J. & Yamasaki, E. (1975) Methods for detecting carcinogens and mutagens with the *Salmonella*/mammalian mutagenicity test. *Mutat. Res., 31,* 347–364

Bignami, M., Cardamone, G., Comba, P., Ortali, V.A., Morpurgo, G. & Carere, A. (1977) Relationship between chemical structure and mutagenic activity in some pesticides: the use of *Salmonella typhimurium* and *Aspergillus nidulans. Mutat. Res., 46,* 243–244

Draminski, W., Eder, E. & Henschler, D. (1983) A new pathway of acrolein metabolism in rats. *Arch. Toxicol., 52,* 243–247

Eder, E., Henschler, D. & Neudecker, T. (1982) Mutagenic properties of allylic and α,β-unsaturated compounds: consideration of alkylating mechanisms. *Xenobiotica, 12,* 831–848

Eder, E. (1982) Structure-activity relationships of mutagenic and carcinogenic allyl compounds: a contribution to the inprovement of toxicological testing strategies (German). Habilitation dissertation, Med. Faculty of the University of Würzburg, Würzburg, Federal Republic of Germany

Eder, E. & Freitag, G. (1983) Investigation of allyl bromide metabolism in the rat: significance for in-vivo genotoxicity. *Naunyn Schmiedeberger's Arch. Pharmacol., 322,* (Suppl.), A490

Izard, C. & Libermann, C. (1978) Acrolein. *Mutat. Res., 47,* 115–138

Lutz, D., Eder, E., Neudecker, T. & Henschler, D. (1982) Structure-mutagenicity relationship in α,β-unsaturated carbonylic compounds and their corresponding allylic alcohols. *Mutat. Res., 93,* 305–315

Neudecker, T., Lutz, D., Eder, E. & Henschler, D. (1980) Structure-activity relationship in halogen and alkyl substituted allyl and allylic compounds: correlation of alkylating and mutagenic properties. *Biochem. Pharmacol., 29,* 2611–2617

Neudecker, T., Öhrlein, K., Eder, E. & Henschler, D. (1983) Effect of methyl and halogen substitutions in the αC position on the mutagenicity of cinnamaldehyde. *Mutat. Res., 110,* 1–8

Rannug, U., Göthe, R. & Wachtmeister, C.A. (1976) The mutagenicity of chloroethylene oxide, chloroacetaldehyde, 2-chloroethanol and chloroacetic acid, conceivable metabolites of vinyl chloride. *Chem.-biol. Interactions, 12,* 251–263

Rosen, J.D., Segall, Y. & Casida, J.E. (1980) Mutagenic potency of haloacroleins and related compounds. *Mutat. Res., 78,* 113–119

FORMATION OF CYCLIC NUCLEIC ACID ADDUCTS FROM SOME SIMPLE α,β-UNSATURATED CARBONYL COMPOUNDS AND CYCLIC NITROSAMINES

F.-L. CHUNG & S.S. HECHT

Division of Chemical Carcinogenesis, Naylor Dana Institute for Disease Prevention, American Health Foundation, Valhalla, NY, USA

G. PALLADINO

Department of Chemistry, The United States Military Academy, West Point, NY, USA

SUMMARY

To determine the structures of the DNA adducts of two cyclic nitrosamines, N-nitrosopyrrolidine and N-nitrosomorpholine, the model compounds, α-acetoxy-N-nitrosopyrrolidine, 4-(carbethoxynitrosamino)butanal and 2-(carbethoxynitrosamino)ethoxyacetaldehyde, were allowed to react with deoxyguanosine in the presence of porcine liver esterase or base. These model compounds are stable precursors of intermediates formed upon metabolic α-hydroxylation of N-nitrosopyrrolidine and N-nitrosomorpholine. The major adducts formed in these reactions were isolated and characterized, on the basis of ultraviolet absorption, mass spectrometry, proton nuclear magnetic resonance and chromatographic properties, as structurally unique $1,N^2$-cyclic deoxyguanosine adducts. Reaction of crotonaldehyde and glyoxal with deoxyguanosine at 37°C, pH 7, also led to the formation of $1,N^2$-cyclic deoxyguanosine adducts identical to those formed from the model compounds for α-hydroxylation of N-nitrosopyrrolidine and N-nitrosomorpholine, respectively. The $1,N^2$-cyclic deoxyguanosine adducts were also formed in DNA, upon incubation with N-nitrosopyrrolidine and rat liver microsomes.

Acrolein, the simplest α,β-unsaturated carbonyl compound, reacted readily with deoxyguanosine to form three major adducts. These adducts were characterized as cyclic $1,N^2$-propanodeoxyguanosine adducts resulting from Michael addition of acrolein to the 1- and N^2-positions of deoxyguanosine followed by ring closure. One of these adducts, as well as the corresponding crotonaldehyde adduct, was formed

Fig. 1. Intermediates and products involved in the α-hydroxylation of N-nitrosopyrrolidine (NPYR), and their reactions with deoxyguanosine (dGuo). *1*, NPYR; *2*, α-acetoxy NPYR; *4*, 4-(carbethoxy-nitrosamino)butanal; *7*, crotonaldehyde; *5, 6, 8* and *9*, deoxyguanosine adducts

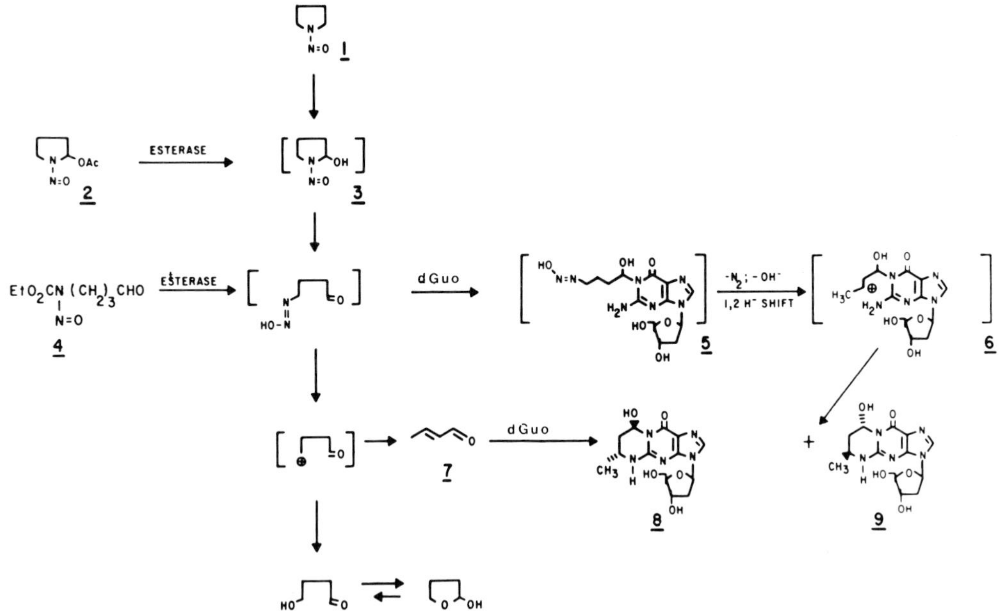

in calf thymus DNA upon reaction with acrolein or crotonaldehyde under physiological conditions. The level of modification of DNA by acrolein was considerably higher than that by either crotonaldehyde or N-nitrosopyrrolidine.

INTRODUCTION

N-Nitrosopyrrolidine (NPYR, Fig. 1, compound 1) and N-nitrosomorpholine (NMOR, Fig. 2, compound 10) are representative cyclic nitrosamines. Unlike acyclic nitrosamines, only limited structural data are available on their binding with nucleic acids. For example, more than a decade ago, Krüger (1972) observed two radioactive peaks in acid hydrolysates of RNA isolated from the livers of rats treated with radiolabelled NPYR. Recently, Hunt and Shank (1982) reported the formation of a thermally-labile fluorescent DNA adduct in rat liver upon treatment with NPYR. Stewart *et al.* (1974) detected several adducts in the hepatic DNA of rats treated with NMOR. However, the structures of the adducts were not elucidated in these studies. In the present study, we have characterized the structures of the adducts from the reaction of α-acetoxyNPYR or 4-(carbethoxynitrosamino)butanal (Fig. 1, compounds 2 and 4) and (2-carbethoxynitrosamino)ethoxyacetaldehyde (Fig. 2, compound 12) with deoxyguanosine. These model compounds are precursors of the

Fig. 2. Intermediates and products involved in the α-hydroxylation of N-nitrosomorpholine (NMOR), and their reactions with deoxyguanosine (dGUO). *10*, NMOR; *12*, (2-carbethoxynitrosamino)ethoxyacetaldehyde; *15*, glyoxal; *13, 14, 16* and *17*, deoxyguanosine adducts

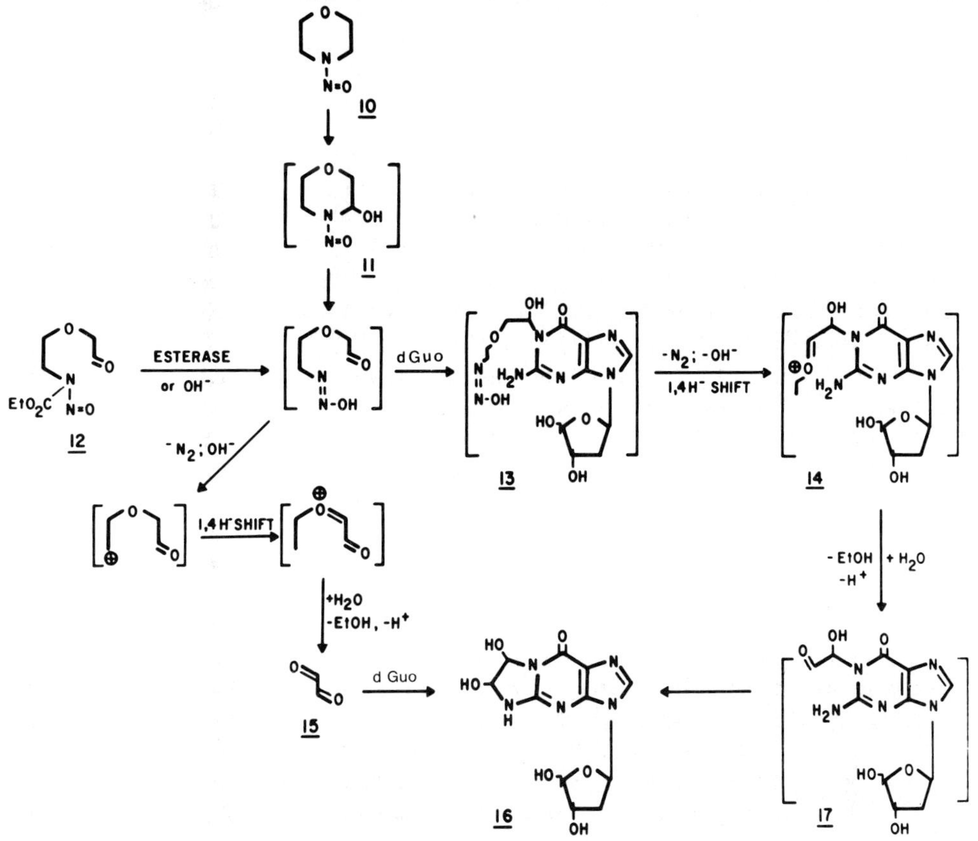

unstable electrophilic intermediates formed in the metabolic α-hydroxylation of NPYR and NMOR, respectively (Hecht et al., 1978; Hecht & Young, 1981). The deoxyguanosine adducts are unique tricyclic derivatives, since reaction of crotonaldehyde (Fig. 1, compound 7) with deoxyguanosine resulted in the formation of adducts identical to those obtained upon reaction of α-acetoxyNPYR or 4-(carbethoxynitrosamino)butanal with deoxyguanosine. We also studied the generality of this $1,N^2$-cyclization reaction by reacting acrolein ($CH_2 = CHCHO$), the simplest α,β-unsaturated carbonyl compound, with deoxyguanosine. Both acrolein and crotonaldehyde are mutagenic in *Salmonella typhimurium* in the absence of an activating system (Neudecker et al., 1981; Eder et al., 1982; Lutz et al., 1982). To investigate the formation of these adducts in DNA, we incubated NPYR with calf thymus DNA in the presence of rat liver microsomes, and also reacted acrolein or crotonaldehyde directly with DNA under physiological conditions.

These studies are essential for understanding the biological significance of these $1,N^2$-cyclic deoxyguanosine adducts in DNA.

MATERIALS AND METHODS

HPLC analysis

HPLC (high-performance liquid chromatography) was performed with a Waters Associates Model ALC/GPC204 high-speed liquid chromatograph equipped with a Model 6000A solvent delivery system, a Model 660 solvent programmer, a Model U6K septumless injector, a Model 440 UV/visible detector operated at 254 nm, or a Perkin-Elmer Model 650-10S fluorescence detector. The following solvent elution systems were used for purification and detection of adducts in DNA as previously reported (Chung *et al.*, 1984).

System 1: A 50-cm × 9.4-mm (inside diameter) Partisil ODS-3 Magnum 9 column (Whatman, Inc., Clifton, NJ) programmed from 0 to 20% methanol in H_2O in 50 min, using Curve 5 or 6, and a flow rate of 5 ml/min.

System 2: Two 25-cm × 4.6-mm Partisil PXS 10/25 ODS columns (Whatman) programmed from 20 to 50% methanol in H_2O in 60 min, using Curve 8 and a flow rate of 1.5 ml/min.

System 3: Two 25-cm × 4.6-mm Partisil PXS 10/25 ODS columns eluted isocratically with 10% methanol in 0.085 M ammonium phosphate pH 3.0, at a flow rate of 1.5 ml/min.

System 4: A 25-cm × 4.6-mm Partisil-10 SCX strong cation-exchange column (Whatman) eluted isocratically with 2% methanol in 0.085 M ammonium phosphate pH 2.0, at a flow rate of 1 ml/min.

Reaction of α-acetoxyNPYR with deoxyguanosine

α-AcetoxyNPYR (2.0 g; 12.7 mmol) was incubated with deoxyguanosine (1.5 g; 5.6 mmol; Sigma Chemical Co., St. Louis, MO.) at 37 °C in 300 ml of phosphate buffer pH 7, in the presence of porcine liver esterase (1350 units; Sigma). The isolation of the products was carried out as previously described (Chung & Hecht, 1983).

Reaction of 4-(carbethoxynitrosamino)butanal with deoxyguanosine

4-(Carbethoxynitrosamino)butanal (2.0 g; 10.6 mmol) was incubated with deoxyguanosine (1.5 g; 5.6 mmol) at 37 °C in 300 ml of phosphate buffer pH 7, in the presence of porcine liver esterase (2 × 1350 units). After three days, the incubation mixture was analysed as previously described (Chung & Hecht, 1983).

Reaction of (2-carbethoxynitrosamino)ethoxyacetaldehyde with deoxyguanosine

(2-Carbethoxynitrosamino)ethoxyacetaldehyde (0.05 g; 0.24 mmol) was incubated with deoxyguanosine (0.13 g; 0.46 mmol) at 37 °C in 25 ml of pH 7 phosphate buffer in the presence of porcine liver esterase (1000 units) or at 37 °C in 25 ml of pH 8 phosphate buffer. After 24 h, the incubation mixture was extracted with ether (2 × 30 ml) or neutralized with 1 N HCl and extracted with ether (2 × 30 ml). The aqueous layer was evaporated *in vacuo* to dryness. The residue was extracted thoroughly with methanol : ethanol (1 : 1, v/v). The extracts were combined and evaporated to a solid which was redissolved in 1.5 ml of H_2O. The sample was analysed by HPLC with a gradient as follows: 0% CH_3OH in H_2O to 30% CH_3OH in H_2O in 45 min at 1 ml/min using Curve 8. The sample showed a major adduct peak at 16.5 ml; this was collected and analysed by ultraviolet (UV) absorption and by proton NMR (nuclear magnetic resonance).

Hydrolysis of deoxyguanosine adducts to guanine adducts

The nucleoside adducts were hydrolysed to the base adducts with 0.1 N HCl (one to two ml) at 37 °C for 16 h or at 90 °C for 45 min. After neutralization, the mixtures were analysed by HPLC. Each product was collected and the UV, proton NMR, mass or CD (circular dichroism) spectra were determined.

Reaction of crotonaldehyde with deoxyguanosine

Crotonaldehyde (70.0 mg; 1.0 mmol; Aldrich Chemical Co., Milwaukee, WIS.) was added to 20 ml of phosphate buffer pH 7, containing deoxyguanosine (53.0 mg; 0.2 mmol). This solution was heated at 37 °C for 48 h and evaporated *in vacuo* to give a solid which was analysed as previously described (Chung *et al.*, 1984).

To obtain larger quantities of the adducts, we followed the published procedures (Chung *et al.*, 1984).

Reaction of acrolein and deoxyguanosine

Small-scale reactions were carried out by adding acrolein (112 mg, 2 mmol; Aldrich Chemical Co., Milwaukee, WI) to 20 ml of phosphate buffer pH 7, containing deoxyguanosine (53 mg, 0.2 mmol), as previously described (Chung *et al.*, 1984).

For larger-scale preparation of acrolein-deoxyguanosine adducts, deoxyguanosine (2 g, 7.6 mmol) was dissolved in 200 ml of phosphate buffer pH 7, by heating at 50 °C. The solution was then cooled to 37 °C and acrolein (8.4 g, 0.15 mol) was added in one batch. The reaction mixture was incubated for two hours at 37 °C with shaking. Another 4.2 g of acrolein were added, and the reaction was continued at 37 °C for one hour longer. The reaction mixture was worked up according to the previously published methods (Chung *et al.*, 1984).

Reaction of acrolein or crotonaldehyde with DNA

Calf thymus DNA (20 mg; Sigma Chemical Co., St. Louis, MO) was dissolved in two ml of phosphate buffer pH 7, at 37 °C. Acrolein (116 mg, 2.1 mmol) or crotonaldehyde (43 mg, 0.61 mmol) was added and the mixture was incubated with shaking for three hours (acrolein), or 16 hours (crotonaldehyde), at 37 °C. Control incubations were carried out without the aldehydes. The DNA was precipitated by addition of 0.1 volume of 3 M sodium acetate and two volumes of cold ethanol. The isolated DNA was reprecipitated twice more using this procedure, and was then dissolved in a solution of 5 mM Tris-HCl (pH 7.1) and 5 mM $MgCl_2$ to give a final DNA concentration of 0.5 to 1 mg/ml. This DNA was hydrolysed enzymatically to nucleosides (Chung et al., 1984).

The final incubation mixture was concentrated to dryness by rotary evaporation at room temperature, and the residue was thoroughly extracted with three 10-ml portions of methanol : ethanol (1 : 1). The extracts were combined, concentrated to 0.3 to 0.5 ml, and the volume was brought to 1.0 ml with H_2O. Samples were analysed by HPLC, using Systems 1–4 sequentially as previously described (Chung et al., 1984).

Incubation of NPYR with rat liver microsomes and calf thymus DNA

NPYR (16 mg, 0.16 mmol) was incubated with calf thymus DNA (5 mg) in an NADPH-generating system (3 ml) for 90 min at 37 °C in the presence of liver microsomes prepared from Aroclor-pretreated F344 rats. The incubation mixture was quenched with ethanol (4 ml) to precipitate DNA. The precipitate was redissolved in sodium p-aminosalicylate containing 1% NaCl, and DNA was isolated by phenol extraction. The purified DNA was hydrolysed enzymatically as described above. The final incubation mixture was evaporated to dryness. The residue was thoroughly extracted with methanol : ethanol (1 : 1, v/v). The combined extracts were concentrated and analysed by HPLC using System 1 and UV detection. The band with the correct retention volume for the adducts was collected, concentrated to dryness, redissolved in H_2O : methanol (4 : 1) and analysed using System 4 with fluorescence detection. The peak that coeluted with the adduct was collected and heated at 90 °C for 45 min. The resulting solution was neutralized, concentrated and analysed for the corresponding guanine adduct using System 4.

RESULTS

A typical HPLC profile of the mixture obtained upon incubation of α-acetoxyNPYR or 4-(carbethoxynitrosamino)butanal with deoxyguanosine in the presence of esterase is shown in Figure 3. Products 4 and 5 were the major deoxyguanosine adducts. None of the products showed UV spectra characteristic of 7- or O^6-substituted guanosine (Singer, 1979). The combined yield of products 4 and 5 was about 1.0 to 1.5% in each reaction. These two products had essentially identical UV absorption, mass and proton NMR spectra. These results indicated that products 4 and 5 were a pair of diastereomers. The UV absorption spectra of products 4 and 5 (with λ_{max} at 262, 258 and 260 nm at pH 1, 7 and 13, respectively) closely resembled that of 1,N^2-dimethylguanosine.

High-resolution mass spectrometry (MS) of the trimethylsilyl derivatives of products 4 and 5 gave molecular ions at m/e 625.2966 and 625.2979, respectively. After correction for the four trimethylsilyl groups, these values support the elemental composition $C_{14}H_{19}O_5N_5$ for both adducts. The MS data were consistent with the addition of one oxobutyl residue to deoxyguanosine.

ADDUCTS OF α,β-UNSATURATED CARBONYLS AND CYCLIC NITROSAMINES 213

Fig. 3. Chromatogram obtained upon high-performance liquid chromatographic analysis of the reaction of 4-(carbethoxynitrosamino)butanal (Fig. 1, compound 4) with deoxyguanosine (dGuo). Elution conditions: 20–50% methanol in H_2O, 50 min at 5 ml/min, using Curve 8 on a Partisil M9 10/50 ODS-3 column (Whatman)

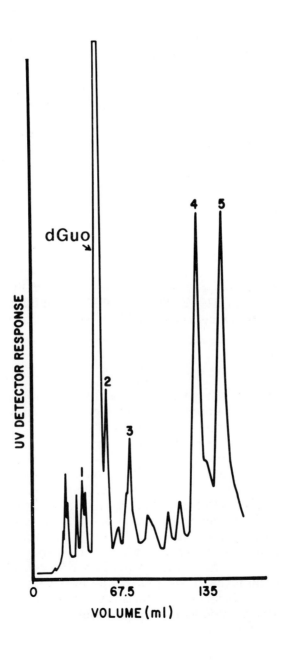

Fig. 4. 300-MHz proton nuclear magnetic resonance spectrum and tentative structural assignment of product 4 (Fig. 3). Spectrum of product 5 was essentially identical to the one shown.

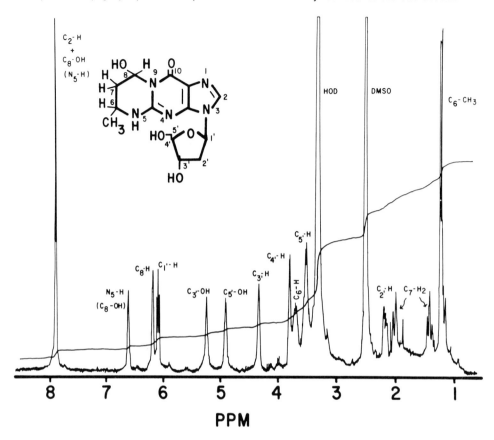

The 300-MHz proton NMR spectra of the adducts (Fig. 4) showed a methyl group at δ 1.21 as a sharp doublet and 2 nonequivalent methylene protons at δ 1.41 and δ 2.00, which appeared as a doublet of doublets and a doublet, respectively. Irradiation of the multiplet at δ 3.70 resulted in the collapse of the methyl doublet into a sharp singlet and the collapse of one of the methylene protons at δ 1.41 into a doublet. These data indicated the presence of a methine proton adjacent to a methyl and methylene group (i.e., CH_3CHCH_2). Two exchangeable protons other than those of the sugar moiety were observed at δ 6.60 and δ 7.90. The latter resonance coincided with the C-2 proton of the deoxyguanosine adduct as shown by integration and by D_2O exchange experiments. The remaining peaks in the spectra were in agreement with the chemical shifts for the deoxyribose protons of deoxyguanosine.

On the basis of these UV absorption, MS, and proton NMR data, the structures of the two adducts (products 4 and 5 in Fig. 3) were tentatively assigned as two diastereomeric 1,N^2-cyclized deoxyguanosine derivatives possessing a new 6-membered ring as shown in Figure 4. Two chiral carbons are introduced at C-6 and

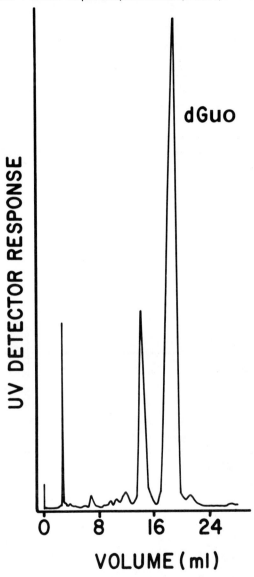

Fig. 5. Chromatogram obtained upon high-performance liquid chromatographic analysis of the reaction of (2-carbethoxynitrosamino)ethoxyacetaldehyde (Fig. 2, compound *12*) with deoxyguanosine (dGuo). Elution conditions: 0–30% methanol in H_2O, 45 min at 1 ml/min, using Curve 8 on a 3.9 mm × 30-cm G8-μBondapak column (Waters)

C-8 upon ring closure at positions 1 and N^2 of deoxyguanosine. Examination of the coupling constants in the proton NMR spectra enabled us to determine the stereochemical properties of these asymmetric carbons. To conform with these coupling constants, the C-6 methyl and the C-8 hydroxy are required to be in the *trans*-configuration as shown in Figure 1, compounds 8 and 9. Consequently, the absolute configurations of C-6 and C-8 in the diastereomeric compounds 8 and 9 were assigned as SS and/or RR, respectively.

The structural assignments were further confirmed by converting these adducts to the corresponding guanine adduct by acid hydrolysis. The $1,N^2$-cyclic guanine adducts were identical in all their properties except that they had CD spectra of opposite polarity. This finding is consistent with the stereochemical assignments for these adducts. None of the above data indicated, however, in which direction ring closure had occurred. To establish this, the $1,N^2$-cyclic guanine adduct was treated with sodium hydroxide and sodium borohydride. A single product was formed. It was characterized as N^2-(3-hydroxy-1-methylpropyl)guanine by UV absorption, MS, and proton NMR analysis. This product could only result from ring opening of the $1,N^2$-cyclic guanine adduct with the methyl group and hydroxy group at C-6 and C-8, respectively. The structures of the adducts were further confirmed by a chemical synthesis in which crotonaldehyde was reacted with deoxyguanosine in phosphate

Fig. 6. Proton nuclear magnetic resonance spectra and structural assignment of the major product from the reaction of (2-carbethoxynitrosamino)ethoxyacetaldehyde with deoxyguanosine (see Fig. 5). Top, deoxyguanosine adduct; bottom, corresponding guanine adduct formed by mild acid hydrolysis

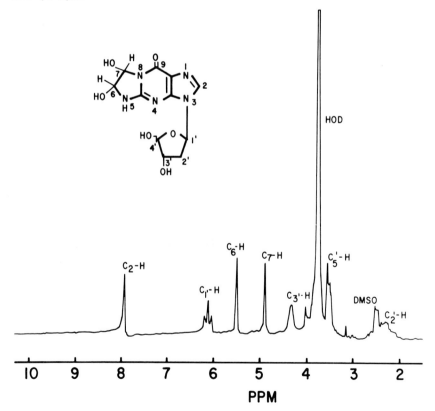

Fig. 6 (contd)

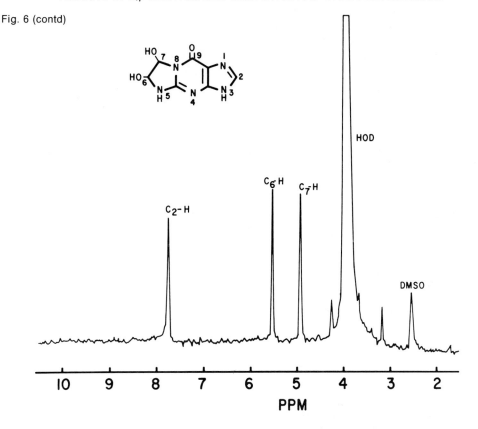

buffer, pH 7.0. Two major adducts were isolated. These products were shown by their spectral properties to be identical to the adducts from the reaction of deoxyguanosine with α-acetoxyNPYR or 4-(carbethoxynitrosamino)butanal.

In an analogous experiment, 2-(carbethoxynitrosamino)ethoxyacetaldehyde, the model compound for α-hydroxylated NMOR, was incubated with deoxyguanosine in the presence of esterase or base. Analysis of this reaction mixture by HPLC showed the formation of a single major adduct which eluted before the deoxyguanosine peak under our conditions (Fig. 5). This product was formed in approximately 10% yield. The UV absorption spectra of this adduct showed λ_{max} similar to those reported for the glyoxal-guanosine adduct (λ_{max} at 254, 246 and 255 nm at pH 1, 7 and 13, respectively) and was thus indicative of a $1,N^2$-cyclized glyoxal-deoxyguanosine adduct (Fig. 2, compound 16). In the proton NMR spectra, two broad singlets were observed at δ 4.9 and δ 5.5. These were assigned as the two methine protons (Fig. 6). Mild acid hydrolysis of this adduct gave the corresponding guanine adduct which was identical, in terms of its chromatographic, UV absorption and proton NMR properties, to the glyoxal-guanine adduct synthesized by the procedures reported by Shapiro and Hachmann (1966).

Fig. 7. Chromatogram obtained upon high-performance liquid chromatographic analysis of the reaction of acrolein with deoxyguanosine (dGuo). Elution conditions were as described in Materials and Methods, System 1.

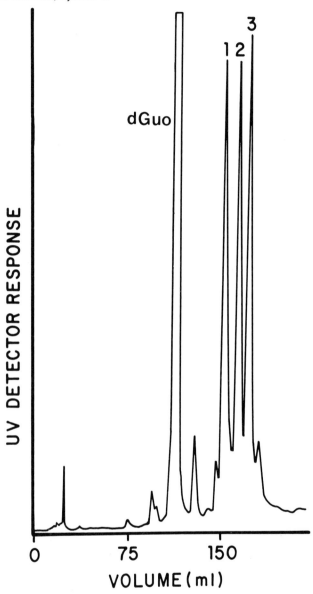

Figure 7 shows a chromatogram obtained by HPLC analysis of the reaction of acrolein and deoxyguanosine; three major products were observed. When either product 1 or product 2 was collected and immediately reanalysed under the same HPLC conditions, a mixture of products 1 and 2 was obtained, indicating that these two products equilibrated rapidly.

Fig. 8. 300-MHz proton nuclear magnetic resonance spectra and structural assignments of the compounds obtained upon acid hydrolysis of product 3 (top) and products 1 and 2 (bottom) from the reaction of acrolein with deoxyguanosine (see Fig. 7)

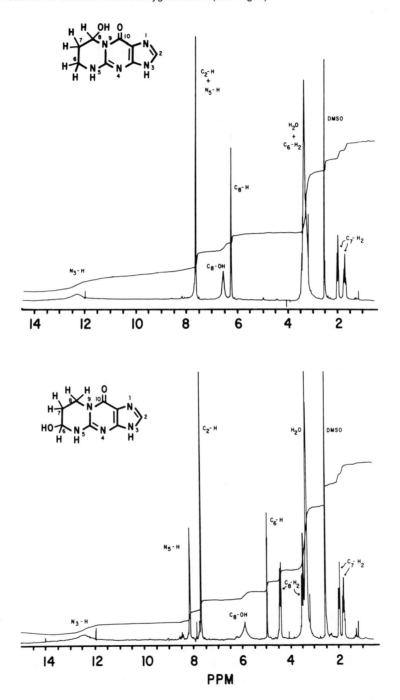

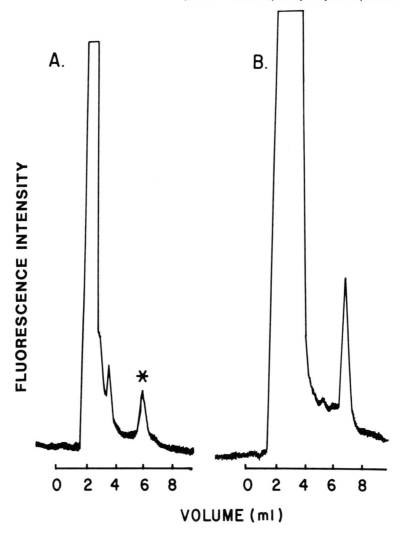

Fig. 9. High-performance liquid chromatogram (System 4, with fluorescence detection) of the purified fraction obtained upon analysis of DNA treated with N-nitrosopyrrolidine (NPYR) in the presence of rat liver microsomes (see Materials and Methods). A. Modified nucleosides from NPYR-treated DNA. Peak with * corresponded in retention volume to products 4 and 5 from the reaction of 4-(carbethoxynitrosamino)butanal or crotonaldehyde with deoxyguanosine (see Fig. 3). B. Profile obtained by treating the peak with * in A with acid. The retention volume of the peak fraction was identical to that of the compound obtained upon hydrolysis of product 4.

The UV absorption spectra of product 3 resembled those of the $1,N^2$-propanodeoxyguanosine adducts formed upon reaction of crotonaldehyde with deoxyguanosine. The UV absorption spectra of products 1 and 2 were identical and were similar to that of product 3. The chemical desorption ionization MS of product 3 showed an M + 1 ion at m/e 324, and a base peak of m/e 208, corresponding to loss

Fig. 10. High-performance liquid chromatogram (System 4, with fluorescence detection) of the purified fraction obtained upon analysis of DNA treated with acrolein (see Materials and Methods for details). A. Modified nucleosides from acrolein-treated DNA. Peak with * corresponded in retention volume to product 3 (see Fig. 7 and top structure in Fig. 8). The peak eluting immediately after this peak corresponded to the compound obtained upon hydrolysis of product 3; this compound was partially formed under the acidic conditions of the high-performance liquid chromatographic system used. B. Profile obtained by acid hydrolysis of the peak with * in A. The retention volume of the peak fraction was identical to that of the compound obtained upon hydrolysis of product 3.

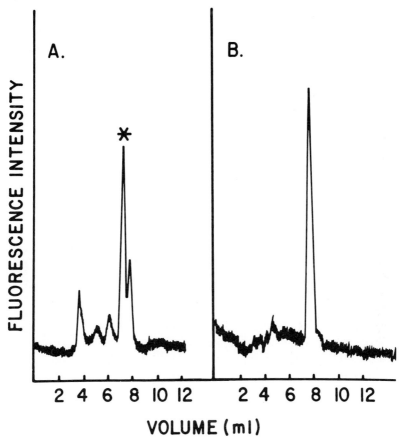

of deoxyribose. These data are consistent with the addition of one molecule of acrolein to deoxyguanosine. The MS of the guanine derivatives from acid hydrolysis of products 1 and 2 did not give the expected molecular ion at m/e 207, but instead gave a peak at m/e 189 (M^+-H_2O). Products 1, 2, and 3 were hydrolysed to the corresponding guanine derivatives. The 300 MHz proton NMR spectra of the 1,N^2-cyclic guanine derivatives are illustrated in Figure 8. The ring-opening reaction with sodium borohydride and sodium hydroxide described for the deoxyguanosine adduct of crotonaldehyde was also carried out to determine the direction of ring-closure by acrolein. Upon this treatment, as expected, the guanine derivative of product 3 was converted in high yield to N^2-(3-hydroxypropyl)guanine. This N^2-substituted guanine derivative was characterized by UV absorption, MS and proton NMR. However, treatment of the guanine derivative of products 1 and 2 with sodium borohydride and sodium hydroxide did not result in a ring-opened product. Instead, a single compound was formed. On the basis of UV absorption, MS and proton NMR data, the structure of this compound was determined as being 5,6,7,8-tetrahydropyrimido[1,2a]purine. All these data are entirely consistent with the structures assigned for products 1–3, as shown in Figure 8.

Calf thymus DNA from incubations with NPYR and rat liver microsomal fraction, or from direct reaction with acrolein or crotonaldehyde, was enzymatically hydrolysed to deoxyribonucleosides. The resulting mixtures were purified by sequential HPLC as described in Materials and Methods. The fractions containing the appropriate 1,N^2-propanodeoxyguanosine adducts were then analysed by HPLC, using a strong cation-exchange column and fluorescence detection. The resulting chromatograms are shown in Figures 9 and 10 for the NPYR-DNA incubation and acrolein-DNA reactions, respectively. The peaks marked with an asterisk had the same retention time, established by coinjection, as the appropriate cyclic 1,N^2-deoxyguanosine adducts. The adduct formed in the NPYR-DNA incubation was chromatographically indistinguishable from the standards prepared from the reaction of deoxyguanosine with α-acetoxyNPYR, 4-(carbethoxynitrosamino)butanal or crotonaldehyde. The adduct formed in acrolein-treated DNA was identical chromatographically to product 3 prepared from acrolein and deoxyguanosine. The identities of these peaks were verified by acid hydrolysis of the fractions containing these nucleoside adducts to their corresponding guanine derivatives. Entirely analogous results were obtained for the analysis of the crotonaldehyde-treated DNA. As in the NPYR-DNA reaction, the diastereometric adducts formed (products 4 and 5 in Figure 3) were detected as one peak in the final HPLC analysis.

The minimal level of modification was 0.2 mmol/mol DNA phosphate for DNA treated with 2.1 mmol of acrolein, and 0.03 mmol/mol DNA phosphate for DNA treated with 0.6 mmole of crotonaldehyde. In the presence of rat liver microsomes, 0.16 mmol of NPYR gave a minimal level of DNA modification of 0.003 mmol/mol DNA phosphate.

DISCUSSION

The nitrosamines, α-acetoxyNPYR and 4-(carbethoxynitrosamino)butanal, are model compounds for α-hydroxyNPYR. We have previously shown that both compounds yield 2-hydroxytetrahydrofuran as a major product upon esterase or base-catalysed hydrolysis (Hecht *et al.*, 1978). The formation of this product is consistent with the intermediacy of α-hydroxyNPYR in the metabolism of NPYR by rat liver microsomes. By analogy, 2-(carbethoxynitrosamino)ethoxyacetaldehyde should be a model compound for α-hydroxyNMOR. The reaction of these compounds with deoxyguanosine in the presence of esterase or base thus provides a practical approach to the synthesis of deoxyguanosine adducts of NPYR and NMOR. Using this approach, we have demonstrated that unique 1,N^2-cyclic deoxyguanosine adducts are formed from the incubation of these model compounds with deoxyguanosine in the presence of esterase or base, as illustrated in Figures 1 and 2. Alternatively, this type of cyclic deoxyguanosine adduct can be prepared from the reaction of deoxyguanosine with α,β-unsaturated carbonyl compounds, such as acrolein and crotonaldehyde, or with dicarbonyl compounds, such as glyoxal. The reactions of crotonaldehyde and glyoxal with deoxyguanosine to give cyclic $1,N^2$-adducts (Figure 1, compounds 8 and 9; Figure 2, compound 16) suggest that crotonaldehyde and glyoxal could be involved in carcinogenesis by NPYR and NMOR, respectively.

In this study, we have also shown that under physiological conditions acrolein, crotonaldehyde and NPYR can modify DNA by forming these cyclic $1,N^2$-propanodeoxyguanosine adducts either directly or with metabolic activation. Since the 1- and N^2-positions of guanine are involved in base pairing, it seems possible that the cyclic adducts described in this study may be involved in the mutagenicity of acrolein and crotonaldehyde and the mutagenicity and carcinogenicity of NPYR. Studies on miscoding properties and persistence in target tissue are needed in order to better understand the role of the $1,N^2$-cyclic deoxyguanosine adducts in mutagenicity and carcinogenicity. It should be noted, however, that adducts with other DNA bases, or cross-linked adducts, might also form upon reaction of DNA with these compounds. These possibilities were not investigated in the present study.

The results of this study also suggest that, as a class, α,β-unsaturated carbonyl compounds may undergo Michael addition to form $1,N^2$-cyclized deoxyguanosine adducts in DNA. Our results show that the reaction of deoxyguanosine with acrolein is more complicated than the reaction with crotonaldehyde. Crotonaldehyde gave cyclic products resulting exclusively from addition of the amino group at position 2 of deoxyguanosine to the C-3 of crotonaldehyde followed by ring closure between the N-1 of deoxyguanosine and the C-1 of crotonaldehyde. However, addition of acrolein to deoxyguanosine occurred in both directions resulting in the formation of the adducts shown in Figure 8. Examination of models suggests that the selectivity in the crotonaldehyde reaction is due to steric crowding caused by the methyl group. Our preliminary evidence also supports the formation of cyclic $1,N^2$-propanodeoxyguanosine adducts upon reaction with other α,β-unsaturated carbonyl compounds. These compounds include methyl vinyl ketone and 2-cyclohexen-1-one.

The number of genotoxic compounds that are known to form cyclic adducts with deoxyribonucleosides or DNA is increasing rapidly. Compounds, other than those mentioned in this report, which form cyclic adducts directly or upon metabolism include 1,3-bis(2-chloroethyl)-1-nitrosourea, ethyl carbamate, chloroacetaldehyde, vinyl chloride, β-propiolactone, glycidaldehyde, triose reductone, misonidazole and substituted malondialdehydes, (Shapiro & Hachmann, 1966; Ludlum et al., 1975; Moschel & Leonard, 1976; Laib & Bolt, 1977; Sattsangi et al., 1977; Goldschmidt et al., 1978; Lee et al., 1979; Gombar et al., 1980; Chen et al., 1981; Laib et al., 1981; Ribovich et al., 1982; Hemminki, 1983; Seto et al., 1983; Varghese & Whitmore, 1983; Marnett et al., these proceedings[1]; Golding et al., these proceedings[2]). This class of nucleoside adducts may have general importance in mutagenesis and carcinogenesis.

---

[1] See p. 175.
[2] See p. 227.

ACKNOWLEDGEMENTS

This study was supported by National Cancer Institute Grant No. 23901.

REFERENCES

Chen, R., Mieyal, J.J. & Goldthwait, D.A. (1981) The reaction of β-propiolactone with derivatives of adenine and with DNA. *Carcinogenesis, 2,* 13–80

Chung, F.-L. & Hecht, S.S. (1983) Formation of cyclic 1,N^2-adducts by reaction of deoxyguanosine with α-acetoxy-*N*-nitrosopyrrolidine, 4-(carbethoxynitrosamino)butanal or crotonaldehyde. *Cancer Res., 43,* 1230–1235

Chung, F-L., Young, R. & Hecht, S.S. (1984) Formation of cyclic 1,N^2-propanodeoxyguanosine adducts in DNA upon reaction with acrolein or crotonaldehyde. *Cancer Res., 44,* 990–995

Eder, E., Henschler, D. & Neudecker, T. (1982) Mutagenic properties of allylic and α,β-unsaturated compounds: consideration of alkylating mechanisms. *Xenobiotica, 12,* 831–848

Goldschmidt, B.M., Biazej, T.P. & Van Duuren, B.L. (1978) The reaction of guanosine and deoxyguanosine with glycidaldehyde. *Tetrahedron Lett., 13,* 1583–1586

Gombar, C.T., Tong, W.P. & Ludlum, D.B. (1980) Reactions of bis-chloroethylnitrosourea and chloroethyl cyclohexyl nitrosourea with deoxyribonucleic acid. *Biochem. Pharmacol., 29,* 2639–2643

Hecht, S.S., Chen, C.B. & Hoffmann, D. (1978) Evidence for metabolic α-hydroxylation of *N*-nitrosopyrrolidine. *Cancer Res., 38,* 215–218

Hecht, S.S. & Young, R. (1981) Metabolic α-hydroxylation of *N*-nitrosomorpholine and 3,3,5,5-tetradeutero-*N*-nitrosomorpholine in the F344 rat. *Cancer Res., 41,* 5039–5043

Hemminki, K. (1983) Nucleic acid adducts of chemical carcinogens and mutagens. *Arch. Toxicol., 52,* 249–285

Hunt, E.J. & Shank, R.C. (1982) Evidence for DNA adducts in rat liver after administration of *N*-nitrosopyrrolidine. *Biochem. biophys. Res. Commun., 104,* 1343–1348

Krüger, F.W. (1972) *New aspects in metabolism of carcinogenic nitrosamines.* In: Nakahara, W., Takayama, S., Sugimura, T. & Odashima, S., eds, *Topics in Chemical Carcinogenesis,* Tokyo, University of Tokyo Press, pp. 213–235

Laib, R.J. & Bolt, H.M. (1977) Alkylation of RNA by vinyl chloride metabolites *in vitro* and *in vivo;* formation of 1,N^6-ethenoadenosine. *Toxicology, 8,* 185–195

Laib, R.J., Gwinner, L.M. & Bolt, H.M. (1981) DNA alkylation by vinyl chloride metabolites: etheno derivatives or 7-alkylation of guanine? *Chem.-biol. Interactions, 37,* 219–231

Lee, J-H., Shinohara, K., Murakami, H. & Omura, H. (1979) Intermediates in the browning reaction of triose reductone with guanine, guanosine or guanylic acid. *Agric. biol. Chem., 43,* 279–286

Ludlum, D.B., Kramer, B.S., Wang, J. & Fenselau, C. (1975) Reaction of 1,3-bis-(2-chloroethyl)-1-nitrosourea with synthetic polynucleotides. *Biochemistry, 14,* 5480–5485

Lutz, D., Eder, E., Neudecker, T. & Henschler, D. (1982) Structure-mutagenicity relationship in α,β-unsaturated carbonylic compounds and their corresponding allylic alcohols. *Mutat. Res., 93,* 305–315

Moschel, R.C. & Leonard, N.J. (1976) Fluorescent modification of guanine reaction with substituted malondialdehydes. *J. org. Chem., 41,* 294–300

Neudecker, T., Lutz, D., Eder, E. & Henschler, D. (1981) Crotonaldehyde is mutagenic in a modified *Salmonella typhimurium* mutagenicity testing system. *Mutat. Res., 91,* 27–31

Ribovich, M.C., Miller, J.A., Miller, E.C. & Timmins, L.G. (1982) Labeled $1,N^6$-ethenocytidine in hepatic RNA of mice given (ethyl-1,2-^{3}H or ethyl-1-^{14}C)ethyl carbamate (urethane). *Carcinogenesis, 3,* 539–546

Sattsangi, P.D., Leonard, N.J. & Frihart, C.R. (1977) $1,N^2$-Ethenoguanine and $N^2,3$-ethenoguanine. Synthesis and comparison of the electronic spectral properties of these linear and angular triheterocycles related to the Y bases. *J. org. Chem., 42,* 3292–3296

Seto, H., Okuda, T., Takesue, T. & Ikemura, T. (1983) Reaction of malondialdehyde with nucleic acid. 1. Formation of fluorescent pyrimido[1,2a]purin-10(3H)-one nucleoside. *Bull. chem. Soc. Jpn., 56,* 1799–1802

Shapiro, R. & Hachmann, J. (1966) The reaction of guanine derivatives with 1,2-dicarbonyl compounds. *Biochemistry, 5,* 2799–2807

Singer, B. (1979) Reaction of guanosine with ethylating agents. *Biochemistry, 11,* 3939–3947

Stewart, B.N., Swann, P.F., Holsman, J.W. & Magee, P.N. (1974) Cellular injury and carcinogenesis. Evidence for the alkylation of rat liver nucleic acids *in vivo* by *N*-nitrosomorpholine. *Z. Krebsforsch., 82,* 1–12

Varghese, A.J. & Whitmore, G.F. (1983) Modification of guanine derivatives by reduced 2-nitroimidazoles. *Cancer Res., 43,* 78–82

REACTION OF GUANOSINE WITH GLYCIDALDEHYDE

B.T. GOLDING[1] & P.K. SLAICH

Department of Organic Chemistry, The University, Newcastle upon Tyne, UK

W.P. WATSON

Shell Research, Sittingbourne, Kent, UK

SUMMARY

The structure and the mechanism of formation of the 1 : 1 adduct from the reaction of guanosine with glycidaldehyde have been investigated. From this reaction, a 2 : 2 adduct has also been isolated, and can be rationalized as a product of self-condensation of the 1 : 1 adduct, with extrusion of formaldehyde.

INTRODUCTION

Van Duuren and his coworkers (Goldschmidt *et al.*, 1968) reported that the reaction between glycidaldehyde and deoxyguanosine at pH 10 yields a 1 : 1 adduct, the structure of which was assigned as either (1a) or (2a), as shown in Figure 1. Glycidaldehyde and guanosine also gave a 1 : 1 adduct, designated as (1b) or (2b) in Figure 1. An analogous modification by glycidaldehyde of guanine residues in calf thymus DNA was subsequently described (Van Duuren & Loewengart, 1977). Recently, Nair and Turner (1984) published spectroscopic evidence for the 1 : 1 adduct from guanosine and glycidaldehyde which, together with mechanistic arguments, led to the assignment of structure (2b) rather than (1b). They also described the conversion of the 9-ethyl analogue of (2b) into 3-ethyl-6-methyl-1,N^2-ethenoguanine, but the yield of this product and details of its characterization were not given.

In an independent investigation of the reaction of glycidaldehyde and guanosine, we have isolated two adducts and have observed putative intermediates *en route* to these. One adduct is identical to that described by Goldschmidt *et al.* (1968) and to that of Nair and Turner (1984). The other is a 2 : 2 adduct (i.e., 2 molecules glycidaldehyde + 2 molecules guanosine → 1 molecule adduct) and its formation releases one molecule of formaldehyde per two molecules of each substrate.

---

[1] To whom correspondence should be addressed.

Fig. 1. Formation of 1 : 1 adducts from the reaction of glycidaldehyde with deoxyguanosine or with guanosine. For (1a) and (2a), R is a 2-deoxyribosyl group; for (1b) and (2b), R is a ribosyl group.

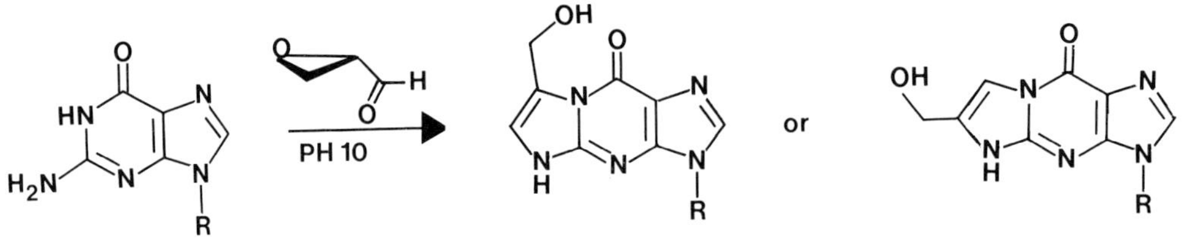

(1a) or (1b)   (2a) or (2b)

RESULTS AND DISCUSSION

For the preparation of the adducts, we react guanosine (0.02 M in aqueous NaOH at pH 9.9 or 11.0) with a slight excess of racemic glycidaldehyde for 60 h at room temperature. The 2 : 2 adduct is filtered off. The material obtained from a reaction carried out at pH 11 is pure (analysis by proton nuclear magnetic resonance spectroscopy, but adduct obtained at pH 9.9 is contaminated with approximately 5% guanosine. The filtrate is neutralized to cause crystallization of the 1 : 1 adduct. The adducts were characterized principally by high field proton and ^{13}C NMR spectroscopy and by fast atom bombardment (FAB) mass spectrometry, which showed an (M-H)-ion for each adduct. The proton NMR data for the 1 : 1 adduct was in excellent agreement with that reported by Goldschmidt *et al.* (1968) and by Nair and Turner (1984). A satisfactory combustion analysis was obtained for the 2 : 2 adduct. The retention times of the adducts relative to that of guanosine in a reverse-phase high-performance liquid chromatographic (HPLC) system (25 cm × 4.6 mm Ultrasphere column, gradient elution starting with water and ending with 60% methanol in water) were 1.21 (1 : 1 adduct), 1.46 (2 : 2 adduct) and 1.00 (guanosine). There is only a subtle difference between the 300 MHz proton NMR spectra of the two adducts taken in $[^2H_6]$-methylsulphoxide + D_2O. The 2 : 2 adduct exhibits a singlet at δ 5.09, whereas the 1 : 1 adduct shows a singlet at δ 4.85 which is twice as intense compared to ribose signals as that of the 2 : 2 adduct. The main difference between the ^{13}C NMR spectra of the adducts is that, whereas the 1 : 1 adduct exhibits a resonance at δ 55.1 (t, CH_2 attached to etheno ring), the 2 : 2 adduct shows a triplet at δ 23.8. The negative-ion FAB mass spectrum of the 2 : 2 adduct shows a peak at m/z 625, assigned to $[M-H]^-$. *Assuming* that the structure of the 1 : 1 adduct is that of (2b) in Figure 1, then the 2 : 2 adduct can be assigned structure (3), as shown in Figure 2, in accord with the spectroscopic data. Structure (4) follows if (1b) is correct. The stoichiometry of the conversion of (1b) into (4), or (2b) into (3), is as follows:

$$2 \cdot (1b) \rightarrow (4) + CH_2O$$
$$\text{or } 2 \cdot (2b) \rightarrow (3) + CH_2O$$

Fig. 2. Possible structures of the 2 : 2 adduct from the reaction between glycidaldehyde and guanosine. R, ribosyl group

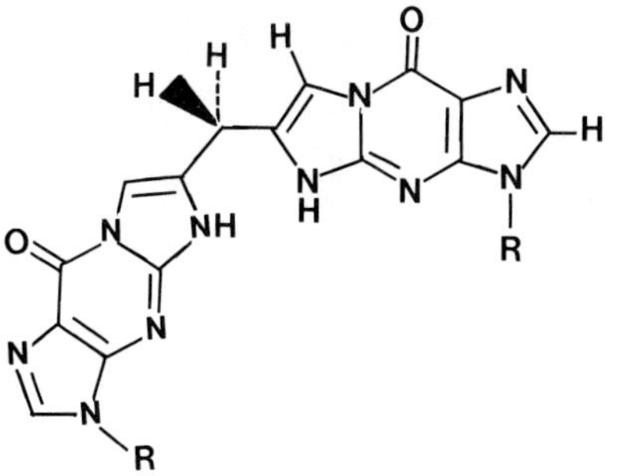

(3) 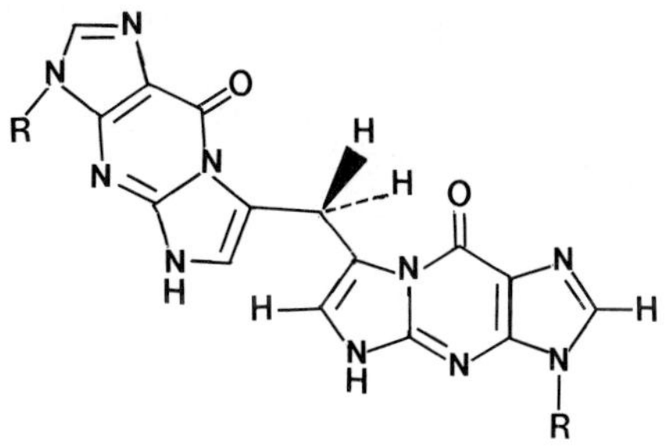 (4)

For a reaction in which the 1 : 1 adduct was converted into 2 : 2 adduct by heating in aqueous solution at pH 11.0 for 4 h at 85 °C, the production of formaldehyde was confirmed by formation of its N-methylbenzothiazolone hydrazone [blue colour test (cf. Paz et al., 1965) and comparison by thin-layer chromatography with an authentic standard].

To establish the mechanisms of formation of the adducts, we have monitored the reaction of guanosine with glycidaldehyde by high field proton NMR spectroscopy and by HPLC (system as given above). During the formation of the 1 : 1 adduct, intermediate species (I_1 and I_2) appear and disappear. Species I_1 is characterized by a singlet at δ 7.98 and a doublet (J 1.5 Hz) at δ 5.50 in the proton NMR spectrum (D_2O). When the reaction was carried out in $NaOD/D_2O$ at pH 11.0, the concentration of I_1 reached a maximum approximately 10 min after mixing the reactants and then decreased over two hours. At pH 10.0, the concentration of I_1 reached a maximum at five hours. Species I_2 is characterized by an AB system at δ 7.25 and 7.61 (J 2.5 Hz), and is 1,N^2-ethenoguanosine (cf. Sattsangi et al., 1977). Its rise follows that of I_1 and it was formed in higher concentration at higher pH. As the concentrations of I_1 and I_2 declined, so the concentration of the 1 : 1 adduct rose and after four days at pH 10.0, it was the major component. The rate of formation of the 1 : 1 adduct was greater at pH 11 than at pH 10, and still lower at pH 7. In water or methylsulphoxide (Me_2SO), the 1 : 1 adduct was converted into the 2 : 2 adduct. Higher concentration of the 1 : 1 adduct (in Me_2SO) or higher pH (in water) favoured the production of the 2 : 2 adduct. Tentative structural assignments for I_1 and some possible reaction pathways are shown in Figures 3 and 4. According to analysis by HPLC, I_1 is a mixture of two diastereoisomers (the one shown in Figures 3 and 4 from (R)-glycidaldehyde, and the isomer from (S)-glycidaldehyde).

A crystal structure analysis will be undertaken to confirm the structures of the 1 : 1 and 2 : 2 adducts. To aid in the elucidation of mechanisms, guanosine and

Fig. 3. Consequences of attack by the N-1 anion of guanosine at the aldehyde of glycidaldehyde

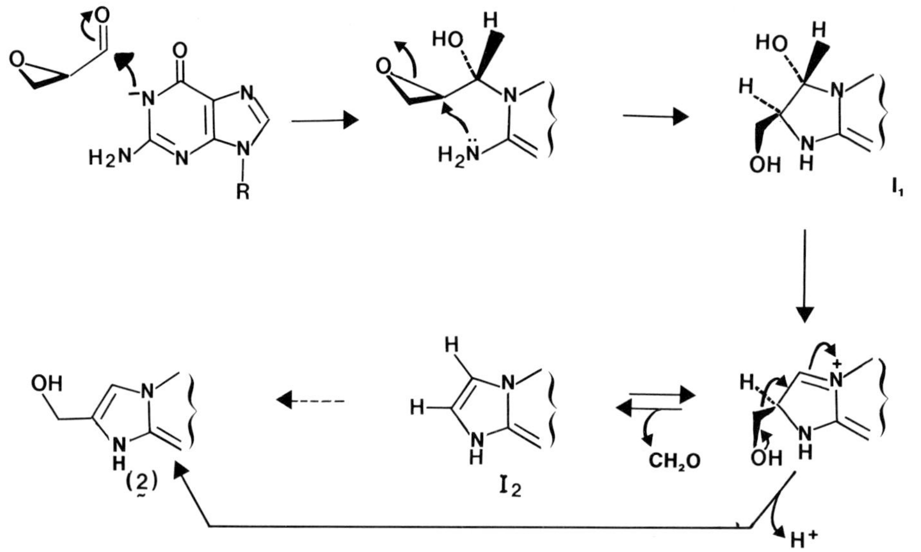

Fig. 4. Consequences of attack by the N² of guanosine at the aldehyde of glycidaldehyde

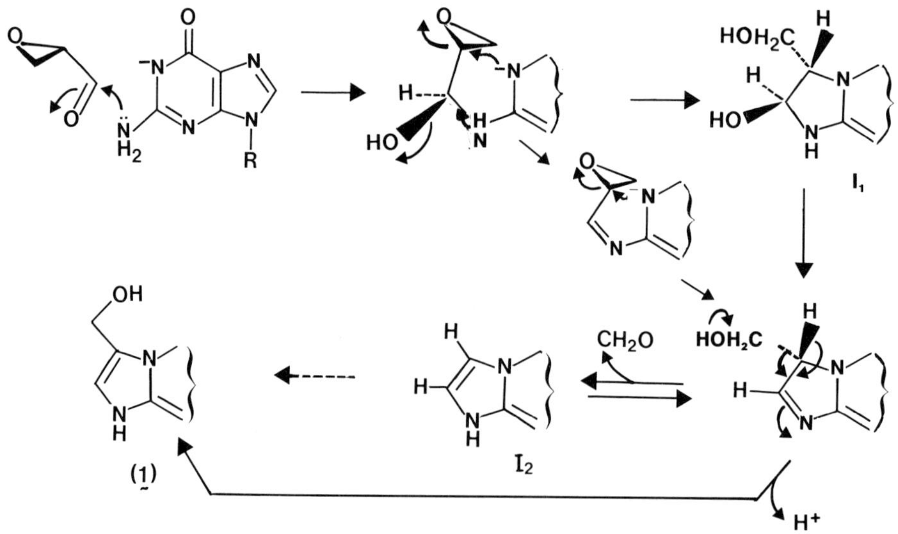

Fig. 5. Laboratory synthesis of glycidaldehyde from but-3-en-1,2-diol. Reaction i: *m*-chloroperbenzoic acid (> 1 mol equiv.) in CH_2Cl_2, 5.5 h at 20 °C; product extracted into water. Reaction ii: sodium metaperiodate (> 1 mol equiv., added to aqueous solution from i), 30 min at 20 °C; glycidaldehyde extracted into dichloromethane and purified by Kugelrohr distillation (bp 60 °C at 120 mm Hg)

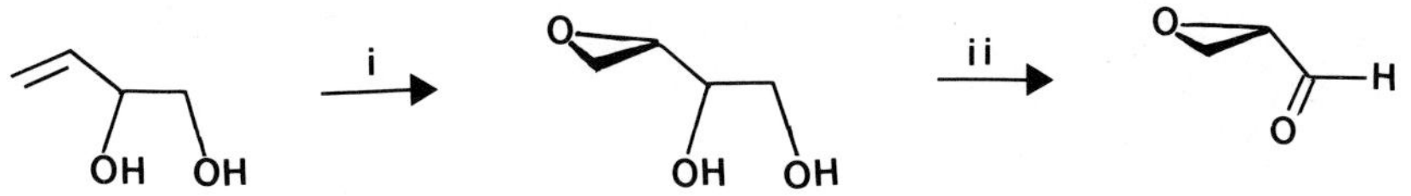

Fig. 6. Reactions of glycidaldehyde with amines. R represents $MeCH_2CH_2$, C_6H_5 or $C_6H_5CH_2$

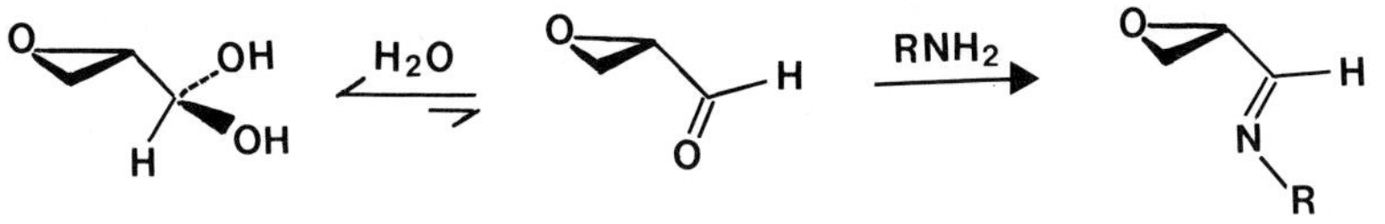

glycidaldehyde specifically labelled with stable isotopes are being synthesized for use in further NMR investigations. A convenient method for the laboratory synthesis of glycidaldehyde has been devised (see Fig. 5).

Concerning the preferred site of reaction of glycidaldehyde with nucleophiles, which is relevant to its reaction with guanosine, we have found that primary amines react smoothly with the aldehyde in either dichloromethane or water to give high yields of imine (see Fig. 6). This supports the postulated initial nucleophilic attack at the aldehyde group of glycidaldehyde (as shown in Figures 3 and 4), rather than alternative mechanisms involving initial nucleophilic attack at an epoxide carbon.

REFERENCES

Goldschmidt, B.M., Blazej, T.P. & Van Duuren, B.L. (1968) The reaction of guanosine and deoxyguanosine with glycidaldehyde. *Tetrahedron Lett.*, 1583–1586

Nair, V. & Turner, G.A. (1984) Determination of the structure of the adduct from guanosine and glycidaldehyde. *Tetrahedron Lett.*, 247–250

Paz, M.A., Blumenfeld, O.O., Rojkind, M., Henson, E., Furfine, C. & Gallop, P.M. (1965) Determination of carbonyl compounds with *N*-methylbenzothiazolone hydrazone. *Arch. Biochem. Biophys.*, **109**, 548–559

Sattsangi, P.D., Leonard, N.J. & Frihart, C.R. (1977) $1,N^2$-Ethenoguanine and $N^2,3$-ethenoguanine. Synthesis and comparison of the electronic spectral properties of these linear and angular triheterocycles related to the Y bases. *J. org. Chem.*, **42**, 3292–3296

Van Duuren, B.L. & Loewengart, G. (1977) Reaction of DNA with glycidaldehyde. *J. biol. Chem.*, **252**, 5370–5371

II. CHEMISTRY AND FORMATION OF CYCLIC AND OTHER ADDUCTS

D. Aromatic Compounds

ELECTROPHILIC ATTACK AT CARBON-5 OF GUANINE NUCLEOSIDES: STRUCTURE AND PROPERTIES OF THE RESULTING GUANIDINOIMIDAZOLE PRODUCTS

R.C. MOSCHEL, W.R. HUDGINS & A. DIPPLE

LBI-Basic Research Program, Chemical Carcinogenesis Program, NCI-Frederick Cancer Research Facility, Frederick, MD, USA

Chemical mutagens and carcinogens react at a variety of sites on the heterocyclic bases of nucleic acids. Simple alkylating agents, such as methyl methanesulfonate, primarily modify ring nitrogen sites (for example, the 7-position of guanine residues) (Lawley, 1966) while more carcinogenic alkylating agents, such as the *N*-alkyl-*N*-nitroso compounds, modify exocyclic oxygen (for example, the O^6 of guanine residues) in addition to ring nitrogen sites (Singer, 1975). Aralkylating carcinogens, such as the reactive derivatives of the polycyclic aromatic hydrocarbons, modify the exocyclic amino groups of the bases (for example, the N^2 of guanine residues) (Dipple et al., 1971). This latter site, as well as the 8-position of guanine residues, is modified by the carcinogenic arylamines (Kriek & Westra, 1979). In an effort to further our understanding of the mechanistic basis for this variation in site selectivity, we have carried out extensive model studies involving reactions between guanosine and several directly reactive benzylating agents (i.e., *p*-Y-Bzl-X where Y = O_2N, Cl, H, CH_3 or CH_3O and X = Br, Cl or $N(NO)CONH_2$) under a variety of solvent conditions. By exploiting changes in benzylating agent reactivity brought about by changes in medium, leaving group and *para*-substituent, we have been able to mimic the reactivity towards nucleic acid components of such different types of carcinogens as simple alkylating agents, the *N*-alkyl-*N*-nitroso compounds and the polycyclic aromatic hydrocarbons. Not only have these studies provided a rationale for the differing site selectivity exhibited by different classes of carcinogens (Moschel et al., 1979; Moschel et al., 1980; Dipple et al., 1982), they have also led to our discovery of a new type of guanosine aralkylation product, i.e., 4-(*p*-Y-benzyl)-5-guanidino-1-β-D-ribofuranosylimidazole products (Moschel et al., 1981). These products result from reaction at carbon-5 of the pyrimidine-imidazole ring junction of guanine residues, indicating that carbon-5 is a site for electrophilic attack and may be a potential site for reaction of a wider variety of ultimate carcinogens with guanine residues in nucleic acids. In this paper, we summarize some of our observations on the formation and properties of these novel guanidinoimidazole-type products.

Fig. 1. Structure of 4-*p*-Y-(benzyl)-5-guanidino-1-β-D-ribofuranosylimidazole

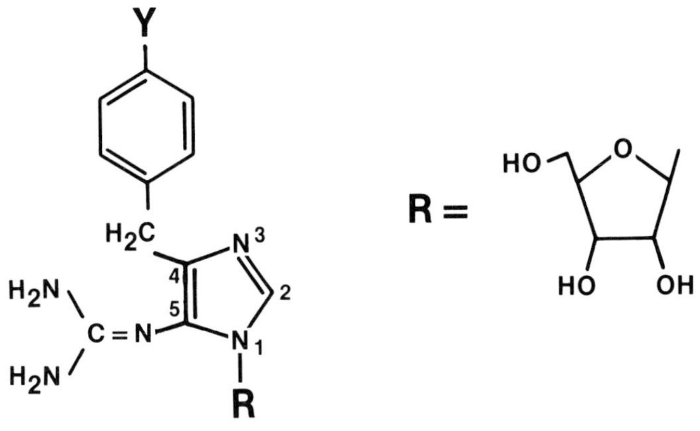

Y = O_2N, Cl, H, CH_3 or CH_3O

The first representative of this class of products was obtained by the reaction of *p*-methylbenzyl chloride with guanosine in neutral aqueous solution (Moschel *et al.*, 1981). Major structural features for the 4-(*p*-methylbenzyl)-5-guanidinoimidazole derivative were defined from a combination of ultraviolet absorption, ^1H and ^{13}C nuclear magnetic resonance and mass spectroscopic studies (Moschel *et al.*, 1981) although the site of attachment of the *p*-methylbenzyl residue could not be assigned unambiguously and the tautomeric form of the guanidino residue present in the structure could not be determined by such investigations. These structural details were later established by X-ray crystallography (Carrell *et al.*, 1982). It is of interest that, in the crystal structure of the *p*-methylbenzyl derivative, the planes of the benzene ring and the guanidino group are approximately perpendicular to the plane of the imidazole ring. Furthermore, the guanidino nitrogen attached to the 5-position of the imidazole ring exists in the imino rather than the amino tautomeric form. Thus, the correct conventional chemical representation of these analogues is as shown in Figure 1.

All of the derivatives indicated in Figure 1 have been isolated from reactions between guanosine and an appropriately substituted benzyl chloride or benzyl bromide in alkaline aqueous solution. Yields are higher under alkaline reaction conditions than in neutral solution (Moschel *et al.*, 1981; Carrell *et al.*, 1982; Moschel *et al.*, 1984) although substantial amounts of these products are produced in benzylations of guanosine at neutral pH (see below). All the guanidinoimidazoles indicated in Figure 1 were purified by Sephadex LH-20 column chromatography using ammoniacal methanol-water mixtures as eluents. Alkaline eluents (pH > 10) are required to recover these products because they fail to elute from Sephadex LH-20 columns when introduced in solutions of pH < 8 and when column elution is carried out with neutral solvent mixtures. This is presumably related to their basicity. Potentiometric titration of the benzyl-, *p*-methylbenzyl- and *p*-methoxybenzyl-

Fig. 2. Possible mechanisms of formation of guanidinoimidazole products from 5-substituted guanosines

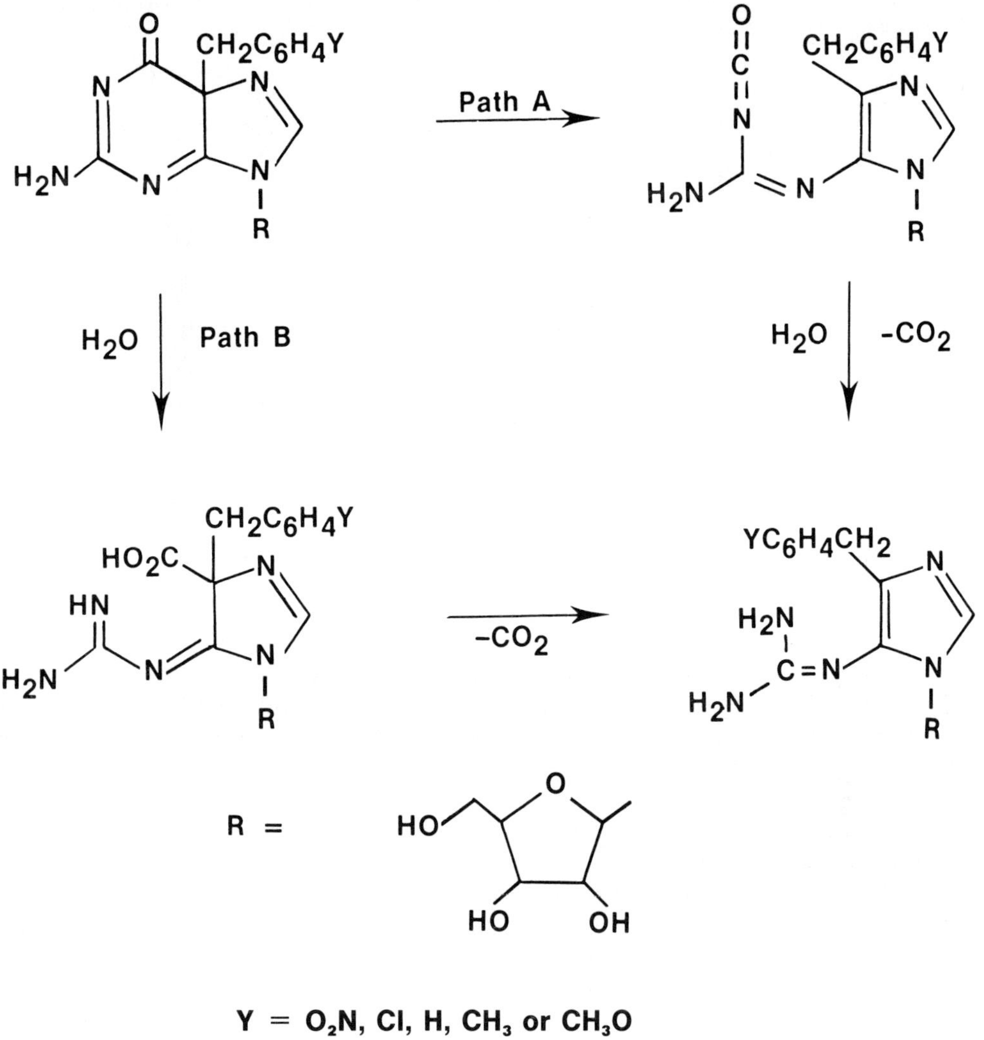

Y = O_2N, Cl, H, CH_3 or CH_3O

derivatives indicate that each has two pK_a values (9.3 and 3.7), irrespective of the nature of the substituent on the respective benzyl group. Under physiological conditions these products exist as singly protonated cationic species. They are doubly protonated at pH < 3 (Moschel et al., 1984).

The formation of these products is, no doubt, a result of pyrimidine ring cleavage in a transiently formed 5-substituted guanosine. Two plausible pathways are shown in Figure 2. Path B illustrates simple hydrolytic cleavage of the pyrimidine ring to yield a substituted imidazole carboxylic acid which may decarboxylate to form the

Fig. 3. Plot of the % yield of 7-(p-Y-benzyl)guanosine (N-7), N^2-(p-Y-benzylguanosine (N^2), O^6-(p-Y-benzyl)guanosine (O^6), and 4-(p-Y-benzyl)-5-guanidino-1-β-D-ribofuranosylimidazole (C-5) as a function of the para-substituent (Y) on benzyl chloride (BzlCl). Y = O_2N, Cl, H, CH_3 or CH_3O

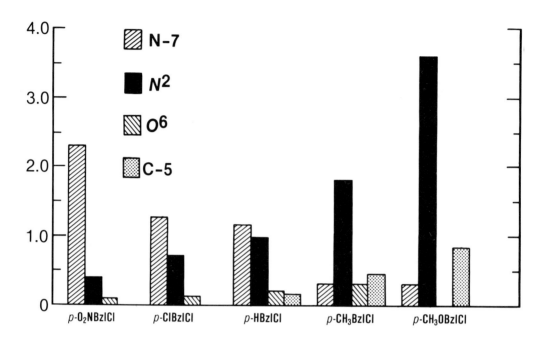

guanidinoimidazole product. Path A illustrates an electrocyclic rearrangement in the pyrimidine ring to form an open-chain isocyanate derivative which may hydrolyse and decarboxylate to form the final product. This latter path involves intermediates similar to those proposed by Leonard and co-workers to account for rearrangements leading to aminoimidazo[1,5-a]-1,3,5-triazinones (Holtwick & Leonard, 1981).

Figure 3 illustrates the changes in product distributions obtained in guanosine reactions with p-substituted benzyl chlorides in neutral aqueous solution. It is apparent that changes from electron-withdrawing to electron-donating p-substituents (i.e., $O_2N \rightarrow Cl \rightarrow H \rightarrow CH_3 \rightarrow CH_3O$) lead to increasing formation of the N^2- and O^6-guanosine derivatives and of the guanidinoimidazole product, whereas reaction at the 7-position of guanosine decreases. Thus, we may surmise that those aspects of chemical reactivity which direct an agent to modify exocyclic sites on guanine residues (particularly the exocyclic amino group) also bring about reaction with carbon-5. This suggests that reaction at carbon-5 of guanine residues might occur with other ultimate carcinogens that are already known to react at the amino group of guanine nucleosides, for example, the halomethylbenz[a]anthracenes (Dipple et al., 1971), polycyclic aromatic hydrocarbon diol-epoxides (Jeffrey et al., 1976; Koreeda et al., 1976; Osborne et al., 1976), reactive metabolites of safrole (Phillips et al., 1981) and reactive derivatives of the arylamines (Kriek & Westra, 1979).

ACKNOWLEDGEMENT

Research sponsored by the National Cancer Institute, DHHS, under contract No. NO1-CO-23909 with Litton Bionetics, Inc. The contents of this publication do not necessarily reflect the views or policies of the Department of Health and Human Services, nor does mention of trade names, commercial products, or organizations imply endorsement by the U.S. Government.

REFERENCES

Carrell, H.L., Zacharias, D.E., Glusker, J.P., Moschel, R.C., Hudgins, W.R. & Dipple, A. (1982) X-ray crystallographic proof of electrophilic attack at the pyrimidine/imidazole ring junction in guanosine. *Carcinogenesis, 3*, 641–645

Dipple, A., Brookes, P., Mackintosh, D.S. & Rayman, M.P. (1971) Reaction of 7-bromomethylbenz[*a*]anthracene with nucleic acids, polynucleotides and nucleosides. *Biochemistry, 10*, 4323–4330

Dipple, A., Moschel, R.C. & Hudgins, W.R. (1982) Selectivity of alkylation and aralkylation of nucleic acid components. *Drug Metab. Rev., 13*, 249–268

Holtwick, J.B. & Leonard, N.J. (1981) Guanine analogues. Allyl-substituted aminoimidazo[1,5-*a*]-1,3,5-triazinones formed by cyclization-rearrangement. *J. org. Chem., 46*, 3681–3685

Jeffrey, A.M., Jennette, K.W., Blobstein, S.H., Weinstein, I.B., Beland, F.A., Harvey, R.G., Kasai, H., Miura, I. & Nakanishi, K. (1976) Benzo[*a*]pyrene-nucleic acid derivative found *in vivo:* structure of a benzo[*a*]pyrene tetrahydrodiol epoxide-guanosine adduct. *J. Am. chem. Soc., 98*, 5714–5715

Koreeda, M., Moore, P.D., Yagi, H., Yeh, H.J.C. & Jerina, D.M. (1976) Alkylation of polyguanylic acid at the 2-amino group and phosphate by the potent mutagen ($\pm$)-7β,8α-dihydroxy-9β,10β-epoxy-7,8,9,10-tetrahydrobenzo[*a*]pyrene. *J. Am. chem. Soc., 98*, 6720–6722

Kriek, E. & Westra, J.C. (1979) *Metabolic activation of aromatic amines and amides and interactions with nucleic acids*. In: Grover, P.L., ed., *Chemical Carcinogens and DNA*, Vol. 2, Boca Raton, FL, CRC Press, pp. 1–28

Lawley, P.D. (1966) Effects of some chemical mutagens and carcinogens on nucleic acids. *Prog. Nucleic Acids Res. mol. Biol., 5*, 89–131

Moschel, R.C., Hudgins, W.R. & Dipple, A. (1979) Selectivity in nucleoside alkylation and aralkylation in relationship to chemical carcinogenesis. *J. org. Chem., 44*, 3324–3328

Moschel, R.C., Hudgins, W.R. & Dipple, A. (1980) Aralkylation of guanosine by the carcinogen *N*-nitroso-*N*-benzylurea. *J. org. Chem., 45,* 533–535

Moschel, R.C., Hudgins, W.R. & Dipple, A. (1981) A novel product from the reaction of *p*-methylbenzyl chloride with guanosine in neutral aqueous solution. *Tetrahedron Lett., 22,* 2427–2430

Moschel, R.C., Hudgins, W.R. & Dipple, A. (1984) Substituent-induced effects on the stability of benzylated guanosines: model systems for the factors influencing the stability of carcinogen-modified nucleic acids. *J. org. Chem., 49,* 363–372

Osborne, M.R., Beland, F.A., Harvey, R.G. & Brookes, P. (1976) The reaction of (±)-7α,8β-dihydroxy-9β,10β-epoxy-7,8,9,10-tetrahydrobenzo[*a*]pyrene with DNA. *Int. J. Cancer, 18,* 362–368

Phillips, D.H., Miller, J.A., Miller, E.C. & Adams, B. (1981) N^2 atom of guanine and N^6 atom of adenine residues as sites for covalent binding of metabolically activated 1′-hydroxysafrole to mouse liver DNA *in vivo*. *Cancer Res., 41,* 2664–2671

Singer, B. (1975) The chemical effects of nucleic acid alkylation and their relation to mutagenesis and carcinogenesis. *Prog. Nucleic Acids Res. mol. Biol., 15,* 219–332

FORMATION OF A CYCLIC ADENINE ADDUCT WITH A 4-NITROQUINOLINE *N*-OXIDE METABOLITE MODEL

N. TOHME, M. DEMEUNYNCK, M.F. LHOMME & J. LHOMME

*Laboratoire de Chimie Organique Biologique, Université de Lille I,
Villeneuve d'Ascq, France*

SUMMARY

Pentacyclic adducts are obtained in the reaction of adenine derivatives with the diacetyl ester of 4-hydroxyamino quinoline oxide, the postulated metabolite of the potent carcinogen, 4-nitroquinoline *N*-oxide (4-NQO).

INTRODUCTION

The synthetic compound, 4-nitroquinoline *N*-oxide (4-NQO, compound *1*, Fig. 1), is a potent carcinogen (Sugimura, 1981), the activity of which was first established by Nakahara *et al.* (1957). The assumed mode of action involves reduction of the nitro group to a hydroxyamino group. Due to the bifunctional character of the 4-NQO derivative, there is a tautomeric equilibrium and the resulting compound *2* (Fig. 2) exists largely in the hydroxyimino form.

A second metabolic activation step is required for binding to DNA. Mono- and diacetyl compounds *3* and *4* (Fig. 1) have been considered as models for the ultimate carcinogenic form of 4-NQO (Kawazoe & Araki, 1967; Enomoto *et al.*, 1968). So far, two adducts with DNA bases have been isolated and characterized (Fig. 2): the quinoline-guanine adduct *5* (Bailleul *et al.*, 1981; Galiegue-Zouitina *et al.*, 1985) and the quinoline-adenine compound *6* (Kawazoe *et al.*, 1975).

In the course of a programme devoted to the preparation and study of the chemical reactivity of the presumed metabolites of 4-NQO, we observed the formation of a cyclic adduct between the quinoline diester *4* and various 9-substituted adenine derivatives (R = *n*-propyl or ribose); the adducts have a highly condensed pentacyclic structure (*7A* or *7B*, Fig. 3).

Fig. 1. Structure of 4-nitroquinoline N-oxide (4-NQO) and three of its derivatives. *1*, 4-NQO; *2*, 4-hydroxyamino quinoline oxide; *3*, monoacetyl ester of 4-hydroxyamino quinoline oxide; *4*, diacetyl ester of 4-hydroxyamino quinoline oxide (quinoline diester)

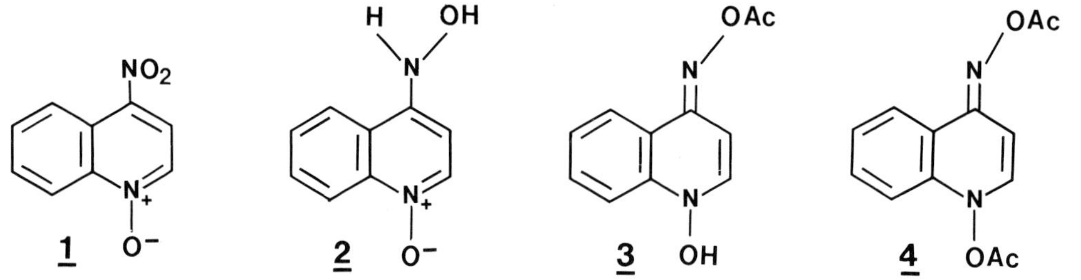

Fig. 2. Adducts of 4-nitroquinoline N-oxide derivatives with DNA bases. *5*, quinoline-guanine adduct; *6*, quinoline-adenine adduct

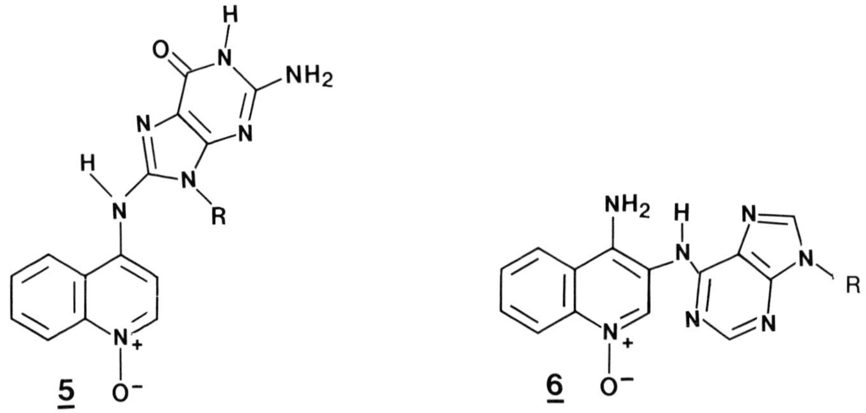

Fig. 3. Reaction of the quinoline diester (*4*) with adenine derivatives to form a cyclic adenine adduct. R, n-propyl or ribose; TFE, trifluoroethanol; *7A* and *7B*, isomeric structures of the cyclic adenine adduct

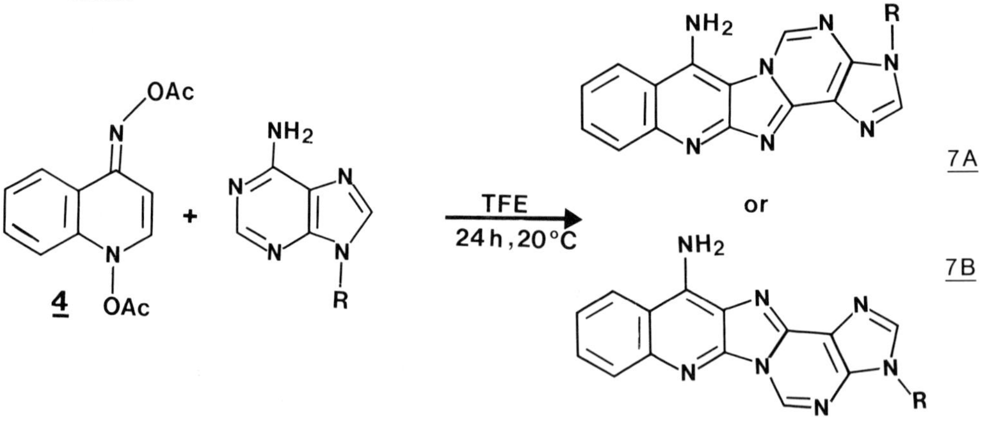

MATERIALS AND METHODS

Materials

1-Acetoxy-4-(acetoxyimino)-1,4-dihydroquinoline (compound 4) was synthesized from (hydroxyamino)-quinoline 1-oxide (compound 2, Enomoto et al., 1968) as previously described by Kawazoe & Araki (1967). 9-Propyladenine was prepared as described by Bolte et al. (1982).

12-Amino-3-propyl-(2',3':5,4)-quinolinyl-(1,2-i)-imidazo purine was prepared as follows. Diacetyl derivative 4 (0.2 mmol) and 9-propyladenine (0.2 mmol) were dissolved in trifluoroethanol (20 ml). The solution was kept at room temperature and in the dark. After 24 h, the solvent was removed and the product was isolated on a silica gel column and purified by crystallizations from methanol (yield: 20%). The purity was checked by high-performance liquid chromatography (HPLC) and thin-layer chromatography.

The riboside analogue, 12-amino-3-ribofuranosyl-(2',3':5,4)-quinolinyl-(1,2-i)-imidazo purine was prepared by analogous treatment of adenosine with diacetyl derivative 4. The structures of these analogues (shown in Fig. 3) were determined by detailed nuclear magnetic resonance (NMR), mass spectral and ultraviolet (UV) absorption analyses.

Methods

Electron mass spectra were recorded on a Riber Mag 10-10 and Varian MAT 311 with a gas-liquid chromatographic (GLC) coupling. NMR spectra were recorded on a Brüker WH 270 with Fourier transform. HPLC separations were carried out on a Waters system consisting of a model U6K injector, two model 6000A pumps, a model 660 solvent programmer and a model 440 dual wavelength detector (254 nm, 365 nm).

RESULTS AND DISCUSSION

In preliminary experiments, diester 4 was reacted with the simple adenine derivative, 9-propyladenine (R = n-propyl). Reaction of the two substances in a non-nucleophilic solvent like trifluoroethanol led to a complex mixture from which compound 7 was isolated by chromatography on silica gel. Mass spectral analysis clearly showed the loss of the two acetyl groups; the molecular peak ($M^+ = 317$) corresponded to the addition of propyladenine ($M^+ = 177$) onto aminoquinoline ($M^+ = 144$) with the formation of two covalent bonds. The NMR spectrum measured at 270 MHz in dimethylsulfoxide and the exchange experiments in the presence of added D_2O indicated that the H-2 and H-3 of quinoline had disappeared. The H-2 and H-8 of adenine appeared as two singlets at 9.4 and 8.3 ppm. The latter signals were unambiguously assigned by exchange experiments in which 8-deuterio-9-propyladenine was reacted with diester 4. The compound isolated after reaction gave the expected molecular ion at $M^+ = 318$ in the mass spectrum and had NMR characteristics identical to those of the previous compound, except for the

Fig. 4. Proposed mode of formation of the cyclic adenine adduct (7)

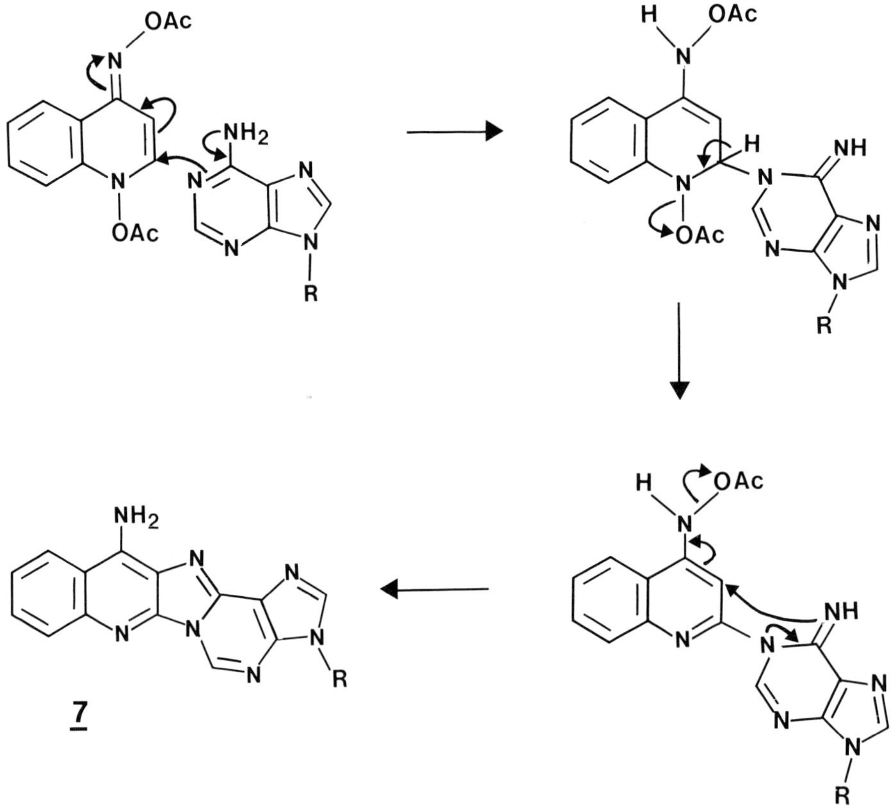

disappearance of the 8.3 ppm signal, which was thus assigned to the H-8. However, the two structures 7A and 7B in Figure 3 are both compatible with these data. More sophisticated NMR experiments at 400 MHz, as well as NOE determinations, could not distinguish between the two isomeric pentacyclic structures.

Similar experimental conditions were used to study the reaction of diester 4 with adenosine. After repeated chromatography and fractional crystallizations, a compound was obtained which was poorly soluble in both water and organic solvents. Again, mass spectral analysis revealed that the compound ($M^+ = 407$) corresponds to an adduct of adenosine ($M^+ = 267$) with aminoquinoline ($M^+ = 144$), involving the formation of two single bonds. The NMR was superposable on that obtained previously in the case of propyladenine, except for the presence of the ribose signals and a slight downfield shift of the H-8 of adenine located in the vicinity of the ribose. The adduct has the structure shown in Figure 3, but again the data do not allow differentiation between the two isomers, 7A and 7B.

It thus appears that the diacetate 4 reacts with the adenine nucleus to form the pentacyclic hetero-aromatic compound, 7. We propose a mode of formation (Fig. 4) in which the two nucleophilic nitrogen atoms of adenine (N-1 and N^6) are involved in condensation steps with the C-2 and C-3 of the quinoline ring. Depending on the sequence of these steps, i.e., whether the N-1 or the N^6 reacts first, one of the two isomers, *7A* or *7B,* will be formed. Further radiocrystallographic study is required to identify the actual isomeric structure.

In the described condensation, the bifunctional character of the 4-NQO derivative is responsible for the "bis-reaction" leading to a cyclic adduct on the adenine nucleus. Additional work is in progress to study the generality of the reaction and the behaviour of the diester with DNA and synthetic polynucleotides.

REFERENCES

Bailleul, B., Galiegue, S. & Loucheux-Lefebvre, M.H. (1981) Adducts from the reaction of *O,O'*-diacetyl or *O*-acetyl derivatives of the carcinogen, 4-hydroxyaminoquinoline-1-oxide, with purine nucleosides. *Cancer Res.,* **41,** 4559–4565

Bolte, J., Demuynck, C., Lhomme, M.F., Lhomme, J., Barbet, J. & Roques, B.P. (1982) Synthetic models related to DNA intercalating molecules. comparison between quinacrine and chloroquine in their ring-ring interaction with adenine and thymine. *J. Am. chem. Soc.,* **104,** 760–765

Enomoto, M., Sato, K., Miller, E.C. & Miller, J.A. (1968) Reactivity of the diacetyl derivative of the carcinogen, 4-hydroxyaminoquinoline-1-oxide, with DNA, RNA and other nucleophiles. *Life Sci.,* **7,** 1025–1032

Galiègue-Zouitina, S., Bailleul, B. & Loucheux-Lefebvre, M.-H. (1985) Adducts from in-vivo action of the carcinogen 4-hydroxyaminoquinoline 1-oxide in rats and from in-vitro reaction of 4-acetoxyaminoquinoline 1-oxide with DNA and polynucleotides. *Cancer Res.,* **45,** 520–525

Kawazoe, Y. & Araki, M. (1967) Studies on chemical carcinogens. V. *O,O'*-diacetyl-4-hydroxyaminoquinoline-1-oxide. *Gann,* **58,** 485–487

Kawazoe, Y., Araki, M., Huang, G.F., Okamoto, T., Tada, M. & Tada, M. (1975) Chemical structure of QA_{II}, one of the covalently bound adducts of carcinogenic 4-nitroquinoline 1-oxide with nucleic acid bases of cellular nucleic acids. *Chem. pharm. Bull.* (Tokyo), **23,** 3041–3043

Nakahara, W., Fukuoka, F. & Sugimura, T. (1957) Carcinogenic action of 4-nitroquinoline *N*-oxide. *Gann,* **48,** 129–137

Sugimura, T. (1981) *Carcinogenesis: A Comprehensive Survey,* Vol. 6, *The Nitroquinolines,* New York, Raven Press

ISOLATION AND CHARACTERIZATION OF PSORALEN PHOTOADDUCTS TO DNA AND RELATED MODEL COMPOUNDS

J. CADET & L. VOITURIEZ

Laboratoires de Chimie, Département de Recherche Fondamentale, Centre d'Etudes Nucléaires de Grenoble, Grenoble, France

F. GABORIAU & P. VIGNY

Laboratoire de Physique et Chimie Biomoléculaire, Institut Curie, Paris, France

SUMMARY

The main products of the photoreaction of 3-carbethoxypsoralen and 8-methoxypsoralen with 2'-deoxyribonucleosides have been isolated and characterized by various spectroscopic measurements involving proton nuclear magnetic resonance and mass spectrometry (fast atom bombardment and ^{252}Cf plasma desorption techniques). Near ultraviolet photolysis of frozen aqueous solutions of thymidine containing 3-carbethoxypsoralen gives rise to two furan-side photocycloadducts having *cis-syn* stereochemistry. The corresponding thymine < > 3-carbethoxypsoralen monoadduct has been shown to be the major photoproduct in DNA. The main *cis-syn* diastereoisomeric [2+2] photocycloadducts which arise from the photoreaction of 8-methoxypsoralen and thymidine in frozen aqueous solutions were shown to involve either the 4',5' furan ring or the 3,4 pyrone moiety and the 5,6-pyrimidine bond. Photobinding of 8-methoxypsoralen to 2'-deoxyadenosine also occurs, with covalent bond formation between carbon 3 or 4 of the pyrone ring and the sugar moiety of the nucleoside.

INTRODUCTION

Various furocoumarin derivatives including the monofunctional 3-carbethoxypsoralen (3-CPs) and the bifunctional 8-methoxypsoralen (8-MOP) have shown interesting properties in the phototreatment of hyperproliferative skin diseases such as psoriasis (Parrish *et al.*, 1974; Dubertret *et al.*, 1979). A likely cellular target for the photoreactions of these psoralen derivatives is deoxyribonucleic acid (DNA).

Fig. 1. Chemical structure of 3-carbethoxypsoralen and 8-methoxypsoralen

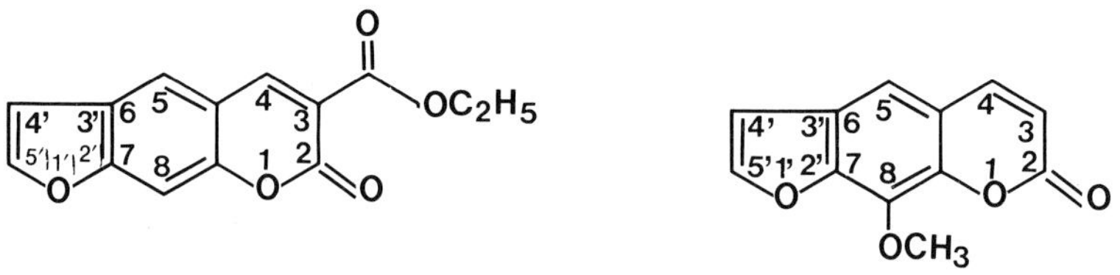

3-CARBETHOXYPSORALEN   8-METHOXYPSORALEN

Intercalation of these compounds within the polynucleotide chains of DNA constitutes the so-called "dark reaction". Psoralen monoadducts to pyrimidine bases involving either the pyrone ring or the 4',5'-ethylenic furan bond are the expected primary photoproducts resulting from exposure to near ultraviolet (UV) light. Further photoreaction of the latter class of photocycloadducts is likely to give rise to DNA cross-links (Song & Tapley, 1979; Hearst, 1981). The bifunctional 8-MOP appears to be more mutagenic and carcinogenic than the monofunctional 3-CPs. However, the exact biological role of each of these different DNA bases lesions remains to be established. In this particular respect, accurate quantitation of these DNA modifications should be provided at the cellular level.

This report describes the isolation and the characterization of monophotoadducts of 8-MOP and 3-CPs (Fig. 1) to DNA and related 2'-deoxyribonucleosides including thymidine (dThd) and 2'-deoxyadenosine (dAdo).

MATERIALS AND METHODS

8-MOP was obtained from Sigma (St. Louis, MO), and 3-CPs was a gift from Dr. Bisagni, Institut Curie, Paris. Photolysis conditions and techniques for the separation of the photoproducts, mainly involving high-performance liquid chromatography (HPLC), have been reported elsewhere (Cadet et al., 1983). Proton NMR (nuclear magnetic resonance) spectra were obtained in the Fourier transform mode, either on a CAMECA TSN 250 spectrometer or a BRUKER WP 250 apparatus. ^{252}Cf plasma desorption mass spectral analyses (PDMS) were carried out on a time-of-flight spectrometer in Orsay, France. Fast atom bombardment (FAB) mass spectra were obtained on a Model MS 50 spectrometer equipped with a commercially available FAB gun.

RESULTS AND DISCUSSION

Characterization of 3-CPs and 8-MOP photocycloadducts to thymidine (dThd)

Near UV photolysis ($\lambda = 365$ nm) of frozen aqueous solutions of 20 mM dThd containing 0.2 mM 3-CPs generated two main photocycloadducts with low yield ($\sim 0.1\%$ with respect to dThd). These two nucleosides have been characterized as the two *cis-syn* diastereoisomers of dThd < > 3-CPs photoadducts involving the furan moiety, by comparison of their FAB and PDMS mass spectra, proton NMR, circular dichroism, UV absorption and phosphorescence properties with those of authentic samples (Cadet et al., 1983).

Photolysis of dThd and 8-MOP under the above conditions gave rise to a more complex mixture of photoproducts. Four major photoproducts isolated by HPLC on an analytical Ultrasphere octadecylsilyl silica-gel column were shown by FAB and PDMS mass spectrometry analyses to be 8-MOP monoaddition products to dThd (Vigny et al., 1983). Two of them which have lost the characteristic UV absorption in the 350 nm region of 3,4-unsaturated benzofuran derivatives were identified as the two *cis-syn* diastereoisomers of dThd < > 8-MOP by considering proton chemical shifts and coupling constants of the cyclobutidyl protons (Cadet et al., 1984).

The two other dThd < > 8-MOP adducts exhibit fluorescence features which are indicative of the saturation of the 4′,5′ furan ethylenic bond. This was confirmed by the observation of a pronounced upfield shift of the proton NMR signal of H-4′ and H-5′. These data, as well as the scalar coupling of H-6 with H-5′, are consistent with a cyclobutidyl structure as the result of a [2 + 2] cycloaddition of the 4′,5′ furan ring to the thymine moiety. The *syn* (head to head) stereochemistry of these two furan-side adducts was deduced from the occurrence of ^3J coupling with vicinal H-4′ and H-5. It has been shown independently that these dThd < > 8-MOP adducts, which were obtained by enzymatic hydrolysis of DNA irradiated in the presence of 8-MOP, have a *cis* stereochemistry on the basis of nuclear Overhauser effect experiments (Kanne et al., 1982). It should also be noted that the conformational properties of the sugar moiety of each of these *cis-syn* diastereoisomers is similar to those of the corresponding *cis-syn* dThd < > 3-CPs photocycloadducts.

Chemical structure of 8-MOP photoadducts to 2′-deoxyadenosine

The photoreaction studies of psoralen derivatives with purine components have received less attention in the past that those concerning the pyrimidines. However, evidence has been reported that adenine and its nucleoside derivatives may undergo photoaddition with 8-MOP (Ou & Song, 1978). In a recent study, occurrence of a photoreaction between 3-CPs and dAdo has been clearly established by FAB and PDMS experiments (Vigny et al., 1983). Near UV photolysis of dAdo and 8-MOP (10 : 1) as a dry film led to the formation of several photoproducts which were separated by reverse-phase HPLC. Four of these compounds show the presence of pseudomolecular ions at m/z 468 (M + H) and m/z 490 (M + Na) in their FAB mass spectra obtained in the positive mode (Fig. 2). This is indicative of a 8-MOP-dAdo monophotoadduct. The loss of the absorption in the near UV region of the electronic spectrum of these adducts strongly suggests the involvement of the pyrone ring in the covalent attachment of 8-MOP to dAdo. This was further supported by the observation of a characteristic AB pattern and its X part in the 250 MHz proton NMR spectra, which are indicative of the presence of a methylene group and a methinic proton at positions 3 and 4 (or 4 and 3), respectively, within the furan ring. Specific decoupling experiments have shown that the latter methinic proton of the main adduct undergoes scalar coupling with the osidic H-5′ proton. It should also

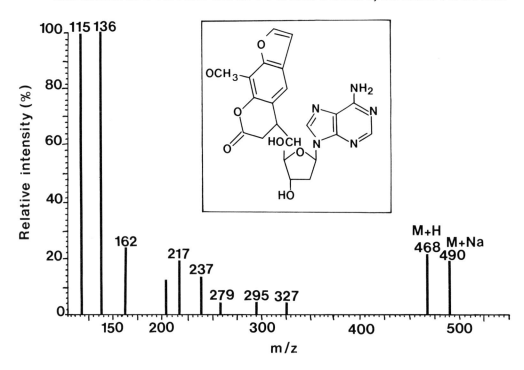

Fig. 2. Fast atom bombardment mass spectrum in the positive mode of the main photoadduct of 8-methoxypsoralen (8-MOP) to 2'-deoxyadenosine (dAdo). Tentative assignment of the covalent bond between the C-4 of 8-MOP and the C-5' of dAdo is shown by the structure in the inset.

be noted that the H-5″ proton of the exocyclic hydroxymethyl group is missing. These observations are consistent with the formation of a covalent bond between the furan carbons, C-3 or C-4, and the osidic carbon, C-5' (see Fig. 2 for the tentative assignment). A reasonable mechanism for the formation of this new type of psoralen <> purine photoadduct would involve radical intermediates such as the osidic 5'-yl radical which would result from initial hydrogen abstraction. Another site for the photobinding reaction of 8-MOP to dAdo is the 1' position.

Photobinding of 3-CPs to native DNA

As an extension of the above model studies dealing with nucleosides, the photoreaction of 3-CPs was investigated in salmon sperm DNA (Sigma, type III). Hydrolysis of DNA was carried out by acidic treatment rather than enzymatic digestion due to quantitative limitations of the latter method (incomplete digestion resulting from the presence of 5,6-dihydropyrimidine derivatives and instability of psoralen<>pyrimidine adducts under slightly alkaline conditions). Acidic hydrolysis (0.4 M hydrogen chloride for four hours at 75 °C) of the DNA irradiated with near UV light in the presence of 3-CPs gave rise to a major thymine <> 3CPs adduct. This compound was characterized as the *cis-syn* isomer of Thy <> 3-CPs by comparison of its HPLC mobility and its fluorescence and PDMS features with those of the authentic sample. Complementary information regarding the internal location and the low mobility of this 3-CPs adduct within the polynucleotide chains was provided by fluorescence quenching experiments and anisotropy measurements. Further studies will involve the search for this Thy <> 3-CPs photoadduct within the DNA of living systems. This should be useful for repair studies and for a better assessment of the biological role of these DNA lesions.

REFERENCES

Cadet, J., Voituriez, L., Gaboriau, F., Vigny, P. & Della Negra, S. (1983) Characterization of photocycloaddition products from reaction between thymidine and the monofunctional 3-carbethoxypsoralen. *Photochem. Photobiol., 37,* 363–371

Cadet, J., Voituriez, L., Ulrich, J., Joshi, P.C. & Wang, S.Y. (1984) Isolation and characterization of the monoheterodimers of 8-methoxypsoralen and thymidine involving the pyrone moiety. *Photobiochem. Photobiophys., 8,* 35–49

Dubertret, L., Averbeck, D., Zajdela, F., Bisagni, E., Moustacchi, E., Touraine, R. & Latarjet, R. (1979) Photochemotherapy (PUVA) of psoriasis using 3-carbethoxypsoralen, a compound noncarcinogenic in mice. *Br. J. Dermatol., 101,* 379–389

Hearst, J.E. (1981) Psoralen photochemistry. *Ann. Rev. Biophys. Bioenerg., 10,* 69–86

Kanne, D., Straub, K., Rapoport, H. & Hearst, J.E. (1982) Psoralen-deoxyribonucleic acid photoreaction. Characterization of the monoaddition products from 8-methoxypsoralen and 4,5′,8-methylpsoralen. *Biochemistry, 21,* 861–871

Ou, C.-N. & Song, P.-S. (1978) Photobinding of 8-methoxypsoralen to transfer RNA and 5-fluouracil-enriched transfer RNA. *Biochemistry, 17,* 1054–1059

Parrish, J.A., Fitzpatrick, T.B., Tanenbaum, L. & Pathak, M.A. (1974) Photochemotherapy of psoriasis with oral methoxsalen and longwave ultraviolet light. *New Engl. J. Med., 291,* 1207–1211

Song, P.-S. & Tapley, L., Jr (1979) Photochemistry and photobiology of psoralens. *Photochem. Photobiol., 29,* 1177–1197

Vigny, P., Spiro, M., Gaboriau, F., Le Beyec, Y., Della Negra, S., Cadet, J. & Voituriez, L. (1983) *Int. J. Mass Spectrom. Ion Phys., 53,* 69–83

III. METABOLISM

METABOLISM AND COVALENT BINDING OF *vic*-DIHALOALKANES, VINYL HALIDES AND ACRYLONITRILE

F.P. GUENGERICH[1], L.L. HOGY, P.B. INSKEEP & D.C. LIEBLER

*Departments of Biochemistry and Pharmacology
and Center in Molecular Toxicology, Vanderbilt University School of Medicine,
Nashville, Tennessee, USA*

SUMMARY

The roles of various metabolic pathways in DNA and protein alkylation are discussed here. Simple vinyl halides are oxidized to 2-haloethylene oxides and 2-haloacetaldehydes, which alkylate DNA and proteins, respectively. Polysubstituted vinyl halides are oxidized with group transfer to yield halocarbonyl compounds which alkylate proteins. Oxidation of *vic*-dihaloalkanes results in protein alkylation while glutathione conjugates alkylate DNA. Acrylonitrile, without previous activation, alkylates proteins and glutathione. Oxidation of acrylonitrile yields a relatively stable epoxide which can react with DNA *in vitro,* but alkylation by this epoxide does not occur readily *in vivo*. Hard-soft acid-base theory is of some use in understanding why some adducts are formed in preference to others.

INTRODUCTION

Chemicals in the classes under consideration are suspect carcinogens and of particular concern because of their large-volume use in industry. The understanding of their mechanisms of action is important in the development of rational risk assessment. Several modes of metabolism and bioactivation occur with these compounds, and much of our effort has been devoted to elucidation of the relative importance of the various pathways. Formation of exocyclic nucleic acid adducts will be discussed in relationship to other pathways.

---

[1] To whom correspondence should be addressed

MONOSUBSTITUTED VINYL HALIDES

Studies on the in-vitro binding of radioactive labels of vinyl halides to proteins in microsomal systems indicated that the 2-haloacetaldehydes are primarily involved, as dehydrogenases added *in situ* could readily block the binding of label to proteins by oxidizing or reducing the 2-haloacetaldehydes (Guengerich *et al.*, 1979, 1981a). However, in similar experiments, epoxide hydrolase was found to block DNA alkylation while the dehydrogenases had no effect, indicating that 2-haloethylene oxides are probably more important than 2-haloacetaldehydes in the formation of DNA adducts. The kinetics of protein alkylation by 2-bromo-[^{14}C]-acetaldehyde are rapid but alkylation of DNA is very slow under physiological conditions. The 2-haloethylene oxides readily convert adenosine to ethenoadenosine while the kinetics using 2-haloacetaldehydes are very slow (Guengerich *et al.*, 1981a). These and other results indicate that 2-haloethylene oxides are more important than 2-haloacetaldehydes in DNA alkylation.

POLYSUBSTITUTED VINYL HALIDES

Our studies support the view that formation of oxidative metabolites of trichloroethylene and vinylidene chloride, as well as olefins in general, proceeds from cytochrome P-450-oxygen-substrate complexes which can collapse *via* several mechanisms, including epoxide ring closure, suicidal heme alkylation and group migration (Miller & Guengerich, 1982; Liebler & Guengerich, 1983). While some covalent binding of metabolites of these compounds to DNA can be found *in vitro*, albeit at low levels, the nature of any adducts formed in this way has not been addressed further.

Metabolism of trichloroethylene does not appear to result in the extensive formation of glutathione (GSH) conjugates *in vitro* or *in vivo*. The major product, chloral, is oxidized to trichloroacetic acid or reduced to trichloroethanol, which is conjugated to form a glucuronide (Miller & Guengerich, 1983). On the other hand, vinylidene chloride oxidation yields *S*-(2,2-dichloro-1-hydroxy)ethyl GSH, *S*-carboxymethyl GSH and *S*-(2-glutathionyl)acetyl GSH, a novel conjugate containing both stable (thioether) and labile (thioester) linkages. This conjugate can bind transiently to added thiols and is eventually hydrolyzed to *S*-carboxymethyl GSH (Liebler *et al.*, 1985).

vic-DIHALOALKANES

We first investigated the roles of oxidative and conjugative reactions in experiments with 1,2-dichloroethane (Fig. 1). NADPH-fortified liver microsomal systems were shown to alkylate proteins but not DNA; cytosolic fractions containing GSH alkylated DNA but not protein. Equivalent results were obtained using bacterial

Fig. 1. Pathways for 1,2-dihaloethane metabolism. X, chloride or bromide ion; RS, cysteine residue in protein; ε-Ad, ethenoadenine; ε-Cy, ethenocytosine; GS, glutathione; Gu, guanine. Dashed lines indicate relatively minor pathways.

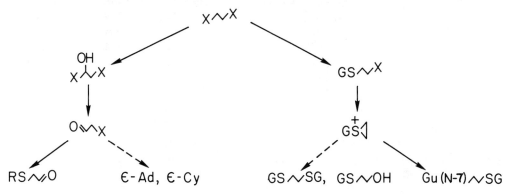

mutagenesis as a means of detecting DNA modification. The protein alkylation could be blocked by GSH or by dehydrogenases, suggesting that 2-chloroacetaldehyde, arising from α-hydroxylation, plays a major role.

Experiments with purified GSH S-transferase, ^{35}S-GSH, and 1,2-dibromo-[^{14}C]-ethane showed that both radiolabels were bound to DNA in equimolar ratios in systems utilizing purified GSH S-transferase or rat hepatocytes (Ozawa & Guengerich, 1983). These results indicate that S-(2-bromoethyl) GSH is the major species which alkylates DNA, and that 2-bromoacetaldehyde, formed by α-hydroxylation of 1,2-dibromoethane, is not involved. Our chemical studies are consistent with S-[2-(7-guanyl)ethyl] GSH as the major adduct formed. Recently, the formation of this adduct *in vivo* has also been demonstrated, using the analogue 1,2-dichloroethane (Svensson & Osterman-Golkar, these proceedings[2]).

The chemical half-life of the adduct in calf thymus DNA is about 150 h at 37 °C in neutral buffer. In rat liver, lung, kidney and stomach, the in-vivo half-life is 70–80 h. The identity of the breakdown products is still under investigation. Experiments with freshly-isolated human hepatocytes indicate that levels of endogenous DNA adduct formation from 1,2-dibromo-[^{14}C]-ethane are roughly half those found in rat hepatocytes prepared under the same conditions; i.e., one adduct is formed per 25,000 bases per hour.

The specificity of this reaction as a mode of DNA adduct formation has been studied further. With 1,2-dibromoethane, rat liver GSH S-transferase B (YaYc) was found to be the most active isozyme. Dihaloalkanes with more than two methylenes separating the halogens are good substrates for GSH conjugation but the conjugates do not lead to DNA alkylation. Two other *vic*-dihaloalkanes, namely, tris-(2,3-dibromopropyl) phosphate and 1,2-dibromo-3-chloropropane, are also activated by this GSH-dependent pathway, although oxidative reactions are important as well (Inskeep & Guengerich, 1984).

---

[2] See p. 269.

Fig. 2. Pathways for acrylonitrile metabolism. GS, glutathione; Prot-S, cysteine residue in protein; ε-Ad, ethenoadenine. Dashed lines indicate relatively minor pathways.

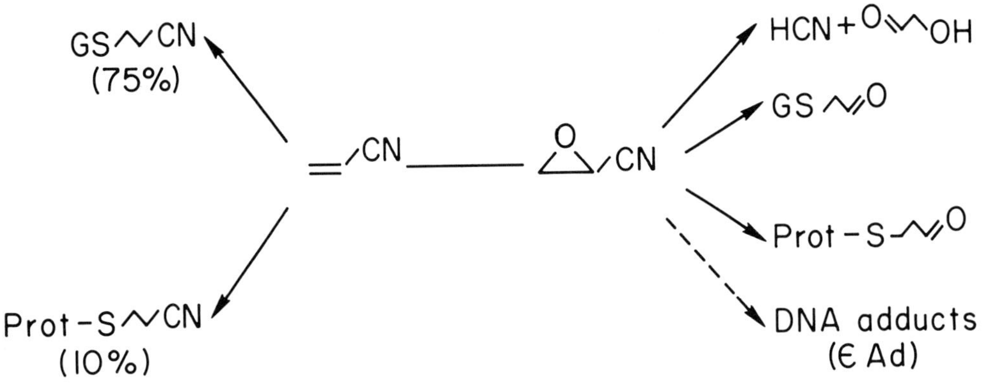

ACRYLONITRILE

In isolated rat hepatocytes, about 75% of the acrylonitrile which disappears is conjugated (enzymatically) with GSH and another 10% is bound to proteins, primarily through cysteine residues (Geiger et al., 1983) (Fig. 2). We demonstrated that cytochrome P-450 oxidizes acrylonitrile to its epoxide, and that hepatocytes accumulate acrylonitrile oxide (to account for 10% of the metabolism) (Guengerich et al., 1981b; Geiger et al., 1983). Acrylonitrile reacts only extremely slowly with DNA but acrylonitrile oxide (which has a half-life of about two hours in neutral buffer) can readily alkylate DNA or react with adenosine to form ethenoadenosine in vitro (Guengerich et al., 1981b). Acrylonitrile oxide also reacts with each of the other nucleosides to form several, yet uncharacterized, products which can be separated by high-performance liquid chromatography (HPLC).

Thus, one might expect that the acrylonitrile oxide formed in vivo might be stable enough to alkylate DNA. Liver microsomes readily form acrylonitrile oxide while the microsomes of brain (a target organ) do not. One possible mechanism which would account for this involves the oxidation of acrylonitrile in the liver and its transport to the brain, where DNA alkylations might not be as readily repaired as in liver. In support of this hypothesis, we found in studies on unscheduled DNA synthesis that, at acrylonitrile levels of 50 mg/kg, liver showed a low but statistically significant repair ratio (Reitz et al., 1980) of 3.07 while brain did not. Acrylonitrile did not cause cytotoxicity under these conditions, as judged by a nonsignificant replicative index.

When 20 µmol of acrylonitrile was recirculated through a perfused rat liver, 0.8 µmol of acrylonitrile oxide was recovered in the perfusate. The kinetics of acrylonitrile accumulation were linear up to about 60 min. Previous studies with acrylonitrile in isolated rat hepatocytes indicated that neither RNA nor DNA was alkylated to the extent of more than one site in 3.5×10^5 residues (Geiger et al., 1983).

Rats were injected intraperitoneally with [2,3-^{14}C]-acrylonitrile oxide, and liver and brain macromolecules were isolated one hour later. Radioactivity was covalently bound to proteins in both liver (8.8 pmol/mg) and brain (7.1 pmol/mg), indicating that the epoxide can indeed migrate throughout the body. However, careful isolation (to minimize protein contamination) of the DNA and RNA from both tissues, and subsequent analysis, indicated that no alkylation occurred to the extent of more than one base in 1.2×10^7.

To circumvent the problems associated with residual protein adducts contaminating nucleic acids and the low levels of radioactivity which can be incorporated into acrylonitrile because of polymerization problems, we tried another approach in which putative adducts were monitored. Since others have shown that guanine can be alkylated at the N-7 position by monosubstituted ethylene oxides arising from compounds such as vinyl chloride (Laib *et al.*, 1981) and vinyl carbamate (Miller & Miller, 1983), we treated rats with unlabelled acrylonitrile or acrylonitrile oxide (50 mg/kg, injected intraperitoneally), isolated the DNA, treated it with tritiated sodium borohydride of high specific radioactivity to reduce any expected 7-(oxoethyl)guanine moieties to labelled alcohols (Miller & Miller, 1983), released the purines by formic acid treatment, and analysed them by HPLC. In this study we again found no radioactivity, which argues against DNA alkylation *via* this mode. We are currently looking for ethenoadenosine adducts *in vivo* using HPLC and fluorescence detection.

The isolated hepatocyte studies and the in-vivo work argue against extensive modification of DNA by acrylonitrile or its epoxide. Apparently both compounds react with proteins and low molecular weight nucleophiles more rapidly than with nucleic acids. Thus, the tumours that result may be due to very low levels of genetic damage or to epigenetic mechanisms.

REFERENCES

Geiger, L.E., Hogy, L.L. & Guengerich, F.P. (1983) Metabolism of acrylonitrile by isolated rat hepatocytes. *Cancer Res., 43,* 3080–3087

Guengerich, F.P., Crawford, W.M., Jr. & Watanabe, P.G. (1979) Activation of vinyl chloride to covalently bound metabolites: roles of 2-chloroethylene oxide and 2-chloroacetaldehyde. *Biochemistry, 18,* 5177–5182

Guengerich, F.P., Geiger, L.E., Hogy, L.L. & Wright, P.C. (1981a) *In vitro* metabolism of acrylonitrile to 2-cyanoethylene oxide, reaction with glutathione, and irreversible binding to proteins and nucleic acids. *Cancer Res., 41,* 4925–4933

Guengerich, F.P., Mason, P.S., Stott, W.T., Fox, T.R. & Watanabe, P.G. (1981b) Roles of 2-haloethylene oxides and 2-haloacetaldehydes derived from vinyl bromide and vinyl chloride in irreversible binding to protein and DNA. *Cancer Res., 41,* 4391–4398

Inskeep, P.B. & Guengerich, F.P. (1984) Glutathione-mediated binding of dibromoalkanes to DNA: specificity of rat glutathione S-transferases and dibromoalkane structure. *Carcinogenesis, 5,* 805–808

Laib, R.J., Gwinner, L.M. & Bolt, H.M. (1981) DNA alkylation by vinyl chloride metabolites: etheno derivatives or 7-alkylation of guanine? *Chem.-biol. Interactions, 37,* 219–231

Liebler, D.C. & Guengerich, F.P. (1983) Olefin oxidation by cytochrome P-450: evidence for group migration in catalytic intermediates formed with vinylidene chloride and *trans*-1-phenyl-1-butene. *Biochemistry, 22,* 5482–5489

Liebler, D.C., Meredith, M.J. & Guengerich, F.P. (1985) Formation of glutathione conjugates by reactive metabolites of vinylidene chloride in microsomes and isolated hepatocytes. *Cancer Res., 54,* 186–193

Miller, R.E. & Guengerich, F.P. (1982) Oxidation of trichloroethylene by liver microsomal cytochrome P-450: evidence for chlorine migration in a transition state not involving trichloroethylene oxide. *Biochemistry, 21,* 1090–1097

Miller, R.E. & Guengerich, F.P. (1983) Metabolism of trichloroethylene in isolated hepatocytes, microsomes, and reconstituted enzyme systems containing cytochrome P-450. *Cancer Res., 43,* 1145–1152

Miller, J.A. & Miller, E.C. (1983) The metabolic activation and nucleic acid adducts of naturally-occurring carcinogens: recent studies with ethyl carbamate and the spice flavors safrole and estragole. *Brit. J. Cancer, 48,* 1–15

Ozawa, N. & Guengerich F.P. (1983) Evidence for formation of an S[2-(N^7-guanyl)ethyl]-glutathione adduct in glutathione-mediated binding of 1,2-dibromoethane to DNA. *Proc. natl Acad. Sci. USA, 80,* 5266–5270

Reitz, R.H., Watanabe, P.G., McKenna, M.J., Quast, J.F. & Gehring, P.J. (1980) Effects of vinylidene chloride on DNA synthesis and DNA repair in the rat and mouse: a comparative study with dimethylnitrosamine. *Toxicol. appl. Pharmacol., 52,* 357–370

METABOLIC ACTIVATION OF VINYL CHLORIDE, FORMATION OF NUCLEIC ACID ADDUCTS AND RELEVANCE TO CARCINOGENESIS

H.M. BOLT

Institut für Arbeitsphysiologie, University of Dortmund, Dortmund 1, Federal Republic of Germany

SUMMARY

Comparative experiments with vinyl chloride and 2,2′-dichlorodiethyl ether confirm that chloroacetaldehyde, which is formed during metabolism of both compounds, binds covalently to proteins, but not to nucleic acids. The putative nucleic-acid-binding agent, chloroethylene oxide, is formed from vinyl chloride, not from 2,2′-dichlorodiethyl ether.

The degree of potential carcinogenicity of epoxides formed from substituted ethylenes is considered to be determined by factors of stability/reactivity and by the rate of breakdown of epoxide and the rate of its formation. Hence, pharmacokinetic data are of considerable importance in assessing the comparative carcinogenic potencies of ethylene derivatives.

INTRODUCTION

Among the halogenated ethylenes, vinyl chloride is the compound with the highest carcinogenic potential. It is metabolized to one or more compound(s) which form adducts with DNA (Laib *et al.*, 1981). Chloroethylene oxide and chloroacetaldehyde are two such intermediary metabolites of proven reactivity towards DNA (Bartsch *et al.*, 1979; Guengerich *et al.*, 1981; Laib *et al.*, 1981). These metabolites are directly mutagenic (although to different degrees) in bacterial test systems (Zajdela *et al.*, 1980), and theoretical calculations of their chemical properties have been published (Politzer & Proctor, 1982). Vinyl bromide, also carcinogenic (Benya *et al.*, 1982), is viewed as an analogue of vinyl chloride (Barbin *et al.*, 1975). Alternative models for the formation of reactive metabolites of vinyl halides are summarized in Figure 1. According to model A, chloroethylene oxide and its rearrangement product,

Fig. 1. Alternative models for the formation of reactive vinyl halide metabolites. P-450, hepatic microsomal cytochrome; X, chloride or bromide ion

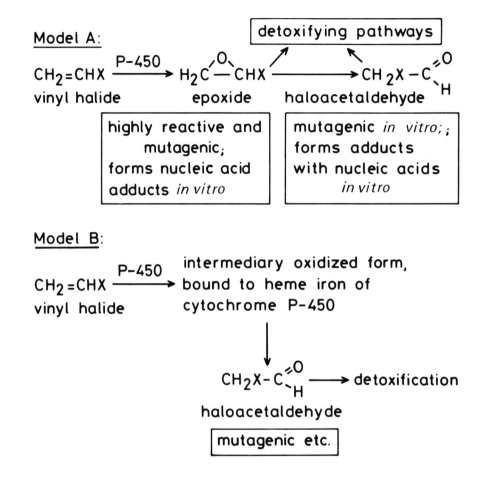

chloroacetaldehyde, are formed in successive reactions, whereas in model B, chloroacetaldehyde is formed directly from vinyl chloride, without proceeding through an epoxide intermediate.

CONSIDERATIONS ON THE DNA-BINDING METABOLITE OF VINYL CHLORIDE

On the basis of various incubation experiments *in vitro*, Guengerich *et al.* (1981) proposed that the bulk of material, which is observed to be covalently bound to proteins after application of radiolabelled vinyl chloride (Bolt & Filser, 1977), is due

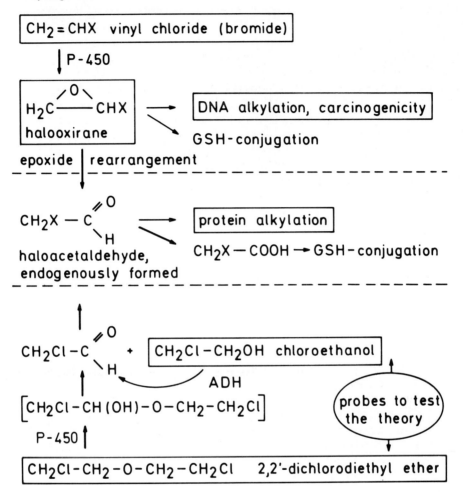

Fig. 2. Metabolic pathways of vinyl halides (top) and 2,2'-dichlorodiethyl ether (bottom). GSH, glutathione; P-450, hepatic microsomal cytochrome; X, chloride or bromide ion; ADH, alcohol dehydrogenase

to chloroacetaldehyde. The much smaller amounts of metabolites covalently bound to DNA were thought to originate directly from the epoxide (chloroethylene oxide).

A direct in-vivo proof for this hypothesis was not possible as the mode of administration of chloracetaldehyde to an animal is critical. To overcome this difficulty, we have used 2,2'-dichlorodiethyl ether and chloroethanol as metabolic precursors which lead to the release of chloroacetaldehyde within the liver cell (Gwinner et al., 1983).

Figure 2 demonstrates the metabolic pathways involved. The critical difference between the metabolism of vinyl halides (vinyl chloride, vinyl bromide) and 2,2'-

dichlorodiethyl ether is that only the former compounds can lead to an intermediary epoxide. The latter compound is oxidatively transformed (by hepatic microsomal cytochrome P-450) to chloroethanol and chloroacetaldehyde; chloroethanol is oxidized in turn to chloroacetaldehyde. Urinary metabolites originating from chloroacetaldehyde are excreted upon exposure to both vinyl chloride and 2,2'-dichlorodiethyl ether (Lingg et al., 1979; Müller & Norpoth, 1979).

After exposure of rats to radiolabelled 2,2'-dichlorodiethyl ether, protein alkylation in the liver was found at a level comparative to that after vinyl chloride (or vinyl bromide) exposure. However, analysis of hydrolysates of liver RNA and DNA gave no indication of the formation of 7-(2-oxoethyl)guanine, $1,N^6$-ethenoadenine or $3,N^4$-ethenocytosine residues within nucleic acids (detection limit: <3 pmol). After application of vinyl chloride, 2,2'-dichlorodiethyl ether or chloroethanol to young rats, only animals exposed to vinyl chloride developed preneoplastic hepatocellular ATPase-deficient foci (Gwinner et al., 1983). Hence, on the basis of these data and the pathways shown in Figure 2, it must be concluded that *chloroethylene oxide is indeed the ultimate carcinogenic metabolite in vinyl-chloride-induced carcinogenicity.*

FACTORS DETERMINING THE EPOXIDE EFFECT

Under the condition (see above) that DNA adduct formation and, hence, carcinogenicity of vinyl halides is due only to the intermediary epoxide, the magnitude of the effect must be a function of chemical stability and reactivity of the epoxide and of its stationary concentration under exposure conditions. The latter is determined by the rate of oxidation of the vinyl halide (which is equal to the rate of epoxide formation) and by the biological half-life of the epoxide (Fig. 3). There is probably some interrelation between chemical reactivity and the biologically relevant half-life, so these two factors are not independent of each other.

Several theoretical studies have focussed on the stability and reactivity of haloethylene epoxides (Politzer et al., 1981; Bolt et al., 1982; Loew et al., 1983; Jones & Mackrodt, 1983) with the idea of predicting their potential carcinogenicities. Some epoxides which are predicted to be mutagenic and carcinogenic due to their optimum 'stability/reactivity' (Jones & Mackrodt, 1983) are shown in Figure 4. These epoxides differ greatly in the rate at which they are biologically formed under relevant exposure conditions.

PHARMACOKINETIC CONSIDERATIONS

The comparative pharmacokinetics of several halogenated ethylenes have been studied in rats (Filser & Bolt, 1979; 1981). It is now clear that the metabolism of vinyl halides is a saturable process and follows first-order kinetics only within a distinct (lower) concentration range. Table 1 shows that the rates of metabolism of unsubstituted and monohalogenated ethylenes differ over a wide range. Vinyl chloride shows the highest V_{max} (maximal metabolic turnover), which is reached at atmospheric concentrations of 250 ppm and higher, whereas the V_{max} of vinyl bromide

Fig. 3. Factors determining the carcinogenicity of an epoxide intermediate

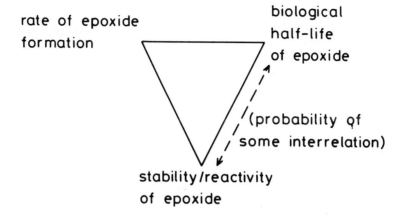

Fig. 4. Some epoxides of optimum stability/reactivity to be mutagenic/carcinogenic (Jones & Mackrodt, 1983)

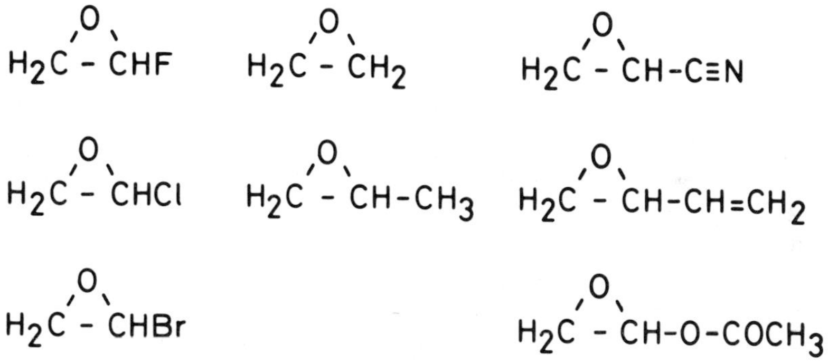

$H_2C - CHF$ (epoxide) $H_2C - CH_2$ (epoxide) $H_2C - CH-C\equiv N$ (epoxide)

$H_2C - CHCl$ (epoxide) $H_2C - CH-CH_3$ (epoxide) $H_2C - CH-CH=CH_2$ (epoxide)

$H_2C - CHBr$ (epoxide) $H_2C - CH-O-COCH_3$ (epoxide)

Table 1. Some kinetic parameters for vinyl halides (Filser & Bolt, 1979) and ethylene (Bolt et al., 1984) in rats

	Atmospheric concentration (in ppm) at which metabolic saturation occurs	Maximal metabolic turnover (V_{max}; $\mu mol \times h^{-1} \times kg^{-1}$)
vinyl fluoride	75 ppm	7
vinyl chloride	250 ppm	110
vinyl bromide	55 ppm	40
ethylene	1,000 ppm	8.5

Fig. 5. Ethylene derivatives which are biotransformed entirely through epoxides (a) or *via* alternative pathways by which epoxide formation is partially (b) or completely (c) prevented. GSH, glutathione

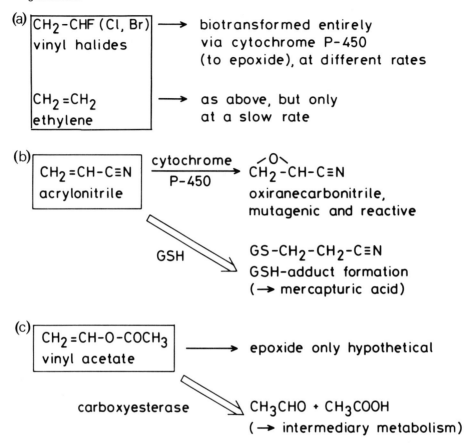

is lower, but is reached at 55 ppm. This means (Bolt & Filser, 1983) that, in fact, vinyl bromide is biotransformed at a higher rate than vinyl chloride if exposure is low (below 55 ppm). At higher exposure levels, vinyl chloride shows the more rapid metabolism. This has consequences for comparisons of the dose-response curves of vinyl chloride and vinyl bromide (Bolt & Filser, 1983).

Vinyl fluoride is metabolized comparatively slowly. This is paralleled by a comparatively low oncogenic potential (Bolt *et al.*, 1982).

Unsubstituted ethylene shows a V_{max} similar to that of vinyl fluoride, but this is reached at very high exposure concentrations only (Filser & Bolt, 1983). The theoretical carcinogenic potential of ethylene remains, therefore, at a magnitude where it cannot be detected by classical animal bioassays (Bolt & Filser, 1984). This is the reason why ethylene was not found to be carcinogenic in an animal bioassay (Hamm *et al.*, 1984) while its established metabolic product, ethylene oxide, is clearly carcinogenic (Filser & Bolt, 1983).

CONSIDERATION OF ALTERNATIVE PATHWAYS

In the case of unsubstituted and halogenated ethylene, the metabolism is entirely oxidative (by cytochrome P-450) with epoxides as likely intermediates (Fig. 5). Hence, the velocity by which cytochrome P-450 is able to biotransform the compound is a main determinant of carcinogenicity. This has already been outlined above.

Other closely related compounds can, however, be metabolized *via* alternative pathways. There is evidence that acrylonitrile is oxidized to its epoxide (which is mutagenic and reactive; Peter *et al.*, 1983), but this is a minor pathway. The main pathway in rats, and probably in man, is direct conjugation with glutathione, which detoxifies the chemical.

Vinyl acetate, *in vivo*, is rapidly split by esterases, especially by those present in hepatic microsomes. Hence, the chance of an epoxide intermediate being formed is very low (H.M. Bolt, unpublished observations).

These examples demonstrate that an individual qualitative and quantitative examination of metabolism is a prerequisite for a comparison of the carcinogenicity of related compounds, such as vinyl halides and their congeners.

ACKNOWLEDGEMENT

The author wishes to thank the "Deutsche Forschungsgemeinschaft" (La 515/1-1) for financial support of part of the studies reported herein.

REFERENCES

Barbin, A., Bresil, H., Croisy, A., Jacquignon, P., Malaveille, C., Montesano, R. & Bartsch, H. (1975) Liver microsome-mediated formation of alkylating agents from vinyl bromide and vinyl chloride. *Biochem. biophys. Res. Commun., 67,* 596–603

Bartsch, H., Malaveille, C., Barbin, A. & Plache, G. (1979) Mutagenic and alkylating metabolites of halo-ethylenes, chlorobutadienes and dichlorobutenes produced by rodent or human liver tissues. *Arch. Toxicol., 41,* 249–277

Benya, T.J., Busey, W.M., Dorato, M.A. & Berteau, P.E. (1982) Inhalation carcinogenicity of vinyl bromide in rats. *Toxicol. appl. Pharmacol., 64,* 367–379

Bolt, H.M. & Filser, J.G. (1977) Irreversible binding of chlorinated ethylenes to macromolecules. *Environ. Health. Perspect., 21,* 107–112

Bolt, H.M. & Filser, J.G. (1983) Quantitative aspects concerning the carcinogenicity of vinylbromide (German). *Verh. Dtsch. Gesund. Arbeitsmed.* (Gentner, Stuttgart), **23,** 433–437

Bolt, H.M. & Filser, J.G. (1984) Olefinic hydrocarbons: a first-risk estimate for ethene. *Toxicol. Pathol., 12,* 101–105

Bolt, H.M., Laib, R.J. & Filser, J.G. (1982) Reactive metabolites and carcinogenicity of halogenated ethylenes. *Biochem. Pharmacol., 31,* 1–4

Bolt, H.M., Filser, J.G. & Störmer, F. (1984) Inhalation pharmacokinetics based on gas uptake studies. V. Comparative pharmacokinetics of ethylene and 1,4-butadiene in rats. *Arch. Toxicol., 55,* 213–218

Filser, J.G. & Bolt, H.M. (1979) Pharmacokinetics of halogenated ethylenes in rats. *Arch. Toxicol., 42,* 123–136

Filser, J.G. & Bolt, H.M. (1981) Inhalation pharmacokinetics based on gas uptake studies. I. Improvement of kinetic models. *Arch. Toxicol., 47,* 279–292

Filser, J.G. & Bolt, H.M. (1983) Exhalation of ethylene oxide by rats on exposure to ethylene. *Mutat. Res., 120,* 57–60

Guengerich, F.P., Mason, P.S., Stott, W.T., Fox, T.R. & Watanabe, P.G. (1981) Roles of 2-haloethylene oxides and 2-haloacetaldehydes derived from vinyl bromide and vinyl chloride in irreversible binding to protein and DNA. *Cancer Res., 40,* 352–356

Gwinner, L.M., Laib, R.J., Filser, J.G. & Bolt, H.M. (1983) Evidence of chloroethylene oxide being the reactive metabolite of vinyl chloride towards DNA: comparative studies with 2,2'-dichloroethyl ether. *Carcinogenesis, 4,* 1483–1486

Hamm, T.E., Guest, D. & Dent, J.G. (1984) Chronic toxicity and oncogenicity bioassay of ethylene in Fischer-344 rats. *Fund. appl. Toxicol., 4,* 473–478

Jones, R.B. & Mackrodt, W.C. (1983) Structure-genotoxicity relationship for aliphatic epoxides. *Biochem. Pharmacol., 32,* 2359–2362

Laib, R.J., Gwinner, L.M. & Bolt, H.M. (1981) DNA alkylation by vinyl chloride metabolites: etheno derivatives or 7-alkylation of guanine? *Chem.-biol. Interactions, 37,* 219–231

Lingg, R.D., Kaylor, W.H., Pyle, S.M. & Tardiff, R.G. (1979) Thiodiglycolic acid: a major metabolite of bis(2-chloroethyl)ether. *Toxicol. appl. Pharmacol., 47,* 23–34

Loew, G.H., Kurkijan, E. & Rebagliati, M. (1983) Metabolism and relative carcinogenic potency of chloroethylenes: a quantum chemical structure-activity study. *Chem.-biol. Interactions, 43,* 33–66

Müller, G. & Norpoth, K. (1979) Identification of S(carboxymethyl)-L-cysteine and thiodiglycolic acid, urinary metabolites of 2,2'-bis(chloroethyl)ether in the rat. *Cancer Lett., 7,* 299–305

Peter, H., Schwarz, M., Mathiasch, B., Appel, K.E. & Bolt, H.M. (1983) A note on synthesis and reactivity towards DNA of glycidonitrile, the epoxide of acrylonitrile. *Carcinogenesis, 4,* 235–237

Politzer, P. & Proctor, T.R. (1982) Calculated properties of some possible vinyl chloride metabolites. *Int. J. quant. Chem., 22,* 1271–1279

Politzer, P., Trefonas, P., Politzer, I.R. & Elfman, B. (1981) Molecular properties of the chlorinated ethylenes and their epoxide metabolites. *Ann. N.Y. Acad. Sci., 367,* 487–492

Zajdela, F., Croisy, A., Malaveille, C., Tomatis, L. & Bartsch, H. (1980) Carcinogenicity of chloroethylene oxide, an ultimate reactive metabolite of vinyl chloride, and bis(chloromethyl)ether after subcutaneous administration and in initiation-promotion experiments in mice. *Cancer Res., 40,* 352–356

COVALENT BINDING OF REACTIVE INTERMEDIATES TO HEMOGLOBIN IN THE MOUSE AS AN APPROACH TO STUDYING THE METABOLIC PATHWAYS OF 1,2-DICHLOROETHANE

K. SVENSSON[1] & S. OSTERMAN-GOLKAR
Department of Radiobiology, Stockholm University, Stockholm, Sweden

SUMMARY

CBA mice were given ^{14}C-labelled 1,2-dichloroethane by intraperitoneal injection. Metabolic pathways leading to the formation of electrophilic intermediates were studied through the determination, by ion-exchange chromatography, of the products of reaction of this labelled substrate with nucleophilic sites in hemoglobin and DNA. The products found in hemoglobin and the pattern of alkylation suggest that chloroacetaldehyde and *S*-(2-chloroethyl)glutathione are important reactive metabolites *in vivo*. The alkyl purines, 7-(2-oxoethyl)guanine and 7-[*S*-(2-cysteine)-ethyl]guanine, were found in DNA hydrolysates, as well as in the urine.

INTRODUCTION

The study of the pattern of alkylated amino acid residues in the hemoglobin of experimental animals has proved to be a useful tool for the identification of reactive metabolites of genotoxic compounds (Osterman-Golkar *et al.*, 1977; Segerbäck, 1983). Furthermore, since the degree of alkylation (or other type of chemical change) of a nucleophilic site is directly proportional to the in-vivo dose of a reactive compound (dose = time integral of concentration; Ehrenberg *et al.*, 1974a), the determination of hemoglobin alkylation provides quantitative information useful in risk estimation (Osterman-Golkar & Ehrenberg, 1983).

The compound, 1,2-dichloroethane (DCE) is of great economic importance. It is used industrially for the production of vinyl chloride. Other applications include use

---

[1] To whom correspondence should be addressed

as a lead scavenger in gasoline, as a solvent, and as a component in fumigants. The widespread human exposure to DCE has been described in several reports (e.g., IARC, 1979). Cancer tests with rodents have given contradictory results. The National Cancer Institute lifetime study (1978), in which animals were given DCE orally (150 mg/kg body weight per day), showed an increased tumour incidence at different sites. No treatment-related tumours were found in the inhalation study of Maltoni et al. (1980) (150 ppm, 7 h per day, 5 days per week).

Although the activation of DCE to genotoxic metabolites has been characterized in in-vitro experiments (see the review by Davidson et al., 1982) little is known about the importance of different metabolic pathways *in vivo*. The purpose of the present study was to identify the reactive metabolites of biological significance by the determination of alkylated residues in hemoglobin. The study also includes determinations of the products formed by reaction with DNA and of the products excreted in the urine after treatment of mice with radiolabelled DCE.

MATERIALS AND METHODS

Chemicals

DCE uniformly labelled with ^{14}C (6.3 mCi/mmol, radiochemical purity 99%, <0.1% vinyl chloride according to gas chromatographic analysis) was obtained from The Radiochemical Centre, Amersham, England. S-(2-Hydroxyethyl)cysteine (HOEtCys) was synthesized according to Zilkha and Rappoport (1963) and the two N^π- and N^τ-(2-hydroxyethyl)histidines (1- and 3-HOEtHis), according to Calleman and Wachtmeister (1979). N^2-(2-Hydroxyethyl)valine (HOEtVal) was synthesized according to Calleman (1984). S,S'-(1,2-Ethylene)bis-cysteine (CysEtCys) was prepared as described by Connors and Ross (1958). N^τ-[S-(2-Cysteine)ethyl]histidine (CysEtHis) was made by reacting N^2-acetylhistidine (1.1 mmol, Sigma, St. Louis, MO) with S-(2-chloroethyl)cysteine (0.2 mmol) in 0.1 M phosphate buffer pH 8 at 100 °C for 30 min. The reaction product was evaporated to dryness and hydrolysed in 6 M HCl for 2 h at 85 °C. The alkylated histidine was isolated by ion-exchange chromatography (see below). 7-(2-Hydroxyethyl)guanine (HOEtGua) was synthesized from guanosine and ethene oxide in acetic acid (Ehrenberg et al., 1974a). 7-[S-(2-Cysteine)ethyl]guanine (CysEtGua) was prepared by reacting S-(2-chloroethyl)cysteine (160 µmol) with guanosine-5'-triphosphate (180 µmol, Sigma) in 1 ml water for 17 h at 40 °C, followed by hydrolysis in 1 M HCl for 1 h at 100 °C. The alkylated guanine was isolated by ion-exchange chromatography (see below). 3,N^4-Ethenocytidine was obtained from Sigma.

Treatment of animals

Radiolabelled DCE, dissolved in corn oil, was given to mice (male CBA, 12 weeks old, average weight 25 g) by intraperitoneal injection (Expt. 1, 0.23 mmol (1.5 mCi)/kg body weight (bw); Expt. 2, 0.37 mmol (2.3 mCi)/kg bw; Expt. 3, 0.16 mmol (1.0 mCi)/

kg bw). The following day (22 h after treatment) the mice were anaesthetized, blood was collected and the mice were killed. Hemoglobin was isolated from the blood according to Osterman-Golkar et al. (1976). DNA was isolated from livers, testes, spleens, kidneys and lungs as described by Segerbäck (1983). Urine was collected and the purines were isolated as described by Wennerberg and Löfroth (1974).

Chromatographic identification of alkylated products in hemoglobin, DNA and urine

Hemoglobin samples (100–200 mg; Expt. 1) were dissolved in 6 M HCl (10 mg/ml) and hydrolysed in evacuated tubes for 15 h at 120 °C. The hydrolysates were evaporated to dryness, redissolved in 5 ml water and incubated for 1 h at 37 °C to hydrolyse 2-chloroethylated products. The hydrolysates were supplied with the carriers HOEtCys, HOEtVal, CysEtCys and CysEtHis and applied to a Dowex 50W-X4 column (77 × 1.0 cm). The column was eluted with 340 ml of 1 M HCl followed by 720 ml of 2 M HCl and 430 ml of 3 M HCl. HOEtCys and HOEtVal eluted after 190 and 215 ml of 1 M HCl, respectively, CysEtCys after 325 ml of 2 M HCl, and CysEtHis after 70 ml of 3 M HCl. The histidine region (155–250 ml of the 2 M HCl eluant), with 1-HOEtHis and 3-HOEtHis added as carriers, was rechromatographed on an Aminex A-5 (54 × 0.9 cm) column connected to a peristaltic pump. 1-HOEtHis and 3-HOEtHis eluted after 50 and 75 ml, respectively, of 0.1 M phosphate buffer pH 7.5. Amino acids in the fractions were identified by ninhydrin using thin-layer chromatography (Osterman-Golkar et al., 1976). Hemoglobin samples of 20–120 mg were dissolved in 10 ml water and treated with 100 mg of sodium borohydride to reduce 2-oxoethyl groups (pH was kept at 6 by automatic titration with 0.1 M HCl). Alkylated products were identified as described above.

DNA from pooled organs (81.0 mg, Expt. 2 and 3) was hydrolysed in the presence of the carriers HOEtGua and $3,N^4$-ethenocytidine in 1 M HCl for 1 h at 100 °C. One mg DNA from the same experiments, with CysEtGua added as a carrier, was incubated in 3 M HCl for 7 h at 100 °C to hydrolyse GSEtGua to CysEtGua. The hydrolysates were applied to a Dowex 50W-X12 column (11 × 1.0 cm) which was eluted with 920 ml of 1 M HCl followed by 500 ml of 3 M HCl. Fractions of 20 ml each were collected. The fractions 1–15 ($3,N^4$-ethenocytosine), 16–46 (HOEtGua and guanine) and 47–71 (CysEtGua and adenine) were evaporated to dryness and rechromatographed separately on an Aminex A-5 column (50 × 0.9 cm). The column was eluted with 0.4 M ammonium formate pH 4.7 at 37 °C. CysEtGua eluted after 128 ml, HOEtGua after 140 ml and $3,N^4$-ethenocytosine after 150 ml of the ammonium formate buffer. The separations were monitored by UV. The carriers were also used as internal standards to account for losses during the work-up procedure. The amount of DNA was calculated from the absorption of guanine ($\varepsilon = 11\,400$ M^{-1} cm^{-1} at $\lambda_{max} = 248$) and adenine ($\varepsilon = 13\,200$ M^{-1} cm^{-1} at $\lambda_{max} = 262.5$) at pH 1. DNA from pooled organs (62.4 mg, Expt. 2 and 3) was reduced with sodium borohydride. Carriers were added and the sample was hydrolysed and chromatographed as described above. Alkylated products in liver DNA (Expt. 1; 1–3 mg samples) and in isolated urinary purines (HOEtGua and CysEtGua; Expt. 2) were identified similarly.

Radioactivity determinations

Radioactivity was measured with an Intertechnique SL 30 liquid scintillation spectrometer with external standardization. Fractions from the ion-exchange columns were counted in equal amounts of Instagel for 250 minutes in the case of liver DNA and hemoglobin samples and for 1200 min in the case of DNA from pooled organs.

Fig. 1a. Ion exchange separation of amino acids in an unreduced hemoglobin sample from mice treated with 1,2-dichloroethane (100 mg; cf. Expt. 1). The following amino acids are indicated: 1, aspartic acid; 2, glycine; 3, alanine; 4, S-(2-hydroxyethyl)cysteine (HOEtCys); 5, N^2-(2-hydroxyethyl)valine (HOEtVal); 6, histidine; 7, S,S'-(1,2-ethylene)bis-cysteine (CysEtCys); 8, N^τ-[S-(2-cysteine)ethyl]histidine (CysEtHis). N^π- and N^τ-(2-hydroxyethyl)histidine (1- and 3-HOEtHis) were eluted in the fractions preceding peak 6.

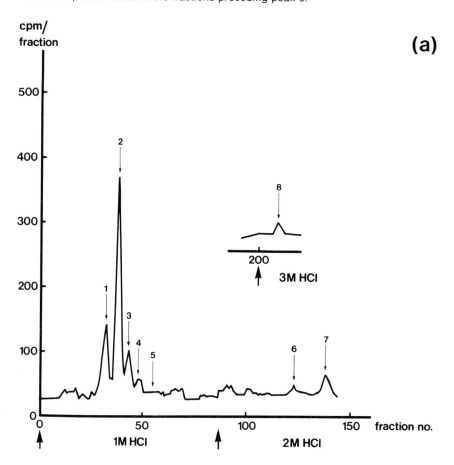

Fig. 1b. The corresponding fractions of a reduced hemoglobin sample (111 mg; cf. Expt. 1)

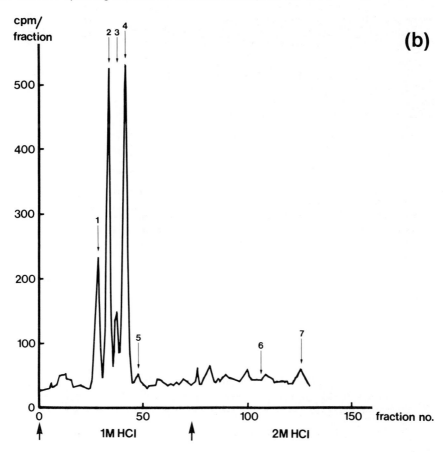

RESULTS

Amino acid separations of hydrolysates – with added reference compounds – of an unreduced hemoglobin sample and of a sodium-borohydride-reduced hemoglobin sample from DCE-treated animals are shown in Figures 1a and 1b, respectively. The radioactivity peaks 4, 5, 7 and 8 correspond to HOEtCys, HOEtVal, CysEtCys and CysEtHis. The three radioactivity peaks eluting before HOEtCys are probably due to metabolic incorporation of ^{14}C fragments into aspartic acid, glycine and alanine. The radioactivity associated with 1-HOEtHis and 3-HOEtHis was determined after rechromatography (cf. Materials and Methods). Products were formed with other reactive groups of the protein as well, but these were not investigated.

The total radioactivity in the hemoglobin, bound through alkylation or metabolic incorporation, corresponded to 50–60 nmol/g Hb per mmol DCE/kg bw. The degree

Table 1a. Degree of alkylation in hemoglobin (nmol product/g Hb) normalized to an injected dose of 0.1 mmol 1,2-dichloroethane (DCE) per kg body weight, assuming a linear dose-response relationship

	Alkylated amino acids[a]					
	HOEtCys	HOEtVal	1-HOEtHis	3-HOEtHis	CysEtCys	CysEtHis
Unreduced Hb	0.05	<0.002	<0.001	<0.002	0.08	0.03
Reduced Hb	0.65	0.03	0.007	0.01	0.06	n.d.[b]

[a] Abbreviations used: HOEtCys, S-(2-hydroxyethyl)cysteine; HOEtVal, N^2-(2-hydroxyethyl)valine; 1-HOEtHis, N^π-(2-hydroxyethyl)histidine; 3-HOEtHis, N^τ-(2-hydroxyethyl)histidine; CysEtCys, S,S'-(1,2-ethylene)bis-cysteine; CysEtHis, N^τ-[S-(2-cysteine)ethyl]histidine
[b] Not determined

Table 1b. Degree of alkylation of DNA (nmol product/g DNA) normalized to an injected dose of 0.1 mmol 1,2-dichloroethane (DCE) per kg body weight, assuming a linear dose-response relationship

	Alkylated bases[a]		
	HOEtGua	CysEtGua	3,N^4-ethenocytosine
Unreduced DNA:			
liver	0.02 ± 0.02[b]	n.d.[c]	<0.02
pooled organs	<0.005	0.6 ± 0.005	
Reduced DNA:			
liver	0.09 ± 0.02	n.d.	<0.02
pooled organs	0.01 ± 0.005		

[a] Abbreviations used: HOEtGua, 7-(2-hydroxyethyl)guanine; CysEtGua, 7-[S-(2-cysteine)ethyl]guanine
[b] 95% confidence limits
[c] not determined

of alkylation of hemoglobin was calculated from the radioactivity of specific alkylation products. The degree of 2-oxoethylation can be determined as the difference between the degree of 2-hydroxyethylation of reduced and unreduced hemoglobin. The results are presented in Table 1a.

The main fraction of the radioactivity in the DNA was due to metabolic incorporation into the bases (e.g., 0.4 nmol ^{14}C in guanine/g DNA per 0.1 mmol DCE/kg bw). The degree of 2-oxoethylation of the N-7 of guanine was low [see Table 1b; 0.07 nmol/g DNA (liver DNA) and 0.005 nmol/g DNA (pooled DNA from different organs) per 0.1 mmol DCE/kg bw] but approximately as expected from the degree of 2-oxoethylation of histidine in hemoglobin (0.008 nmol 3-HOEtHis/g Hb per 0.1 mmol DCE/kg bw) and earlier experience concerning the ratio between DNA alkylation and hemoglobin alkylation for various alkylating agents (for ethene oxide, HOEtGua (liver)/3-HOEtHis = 6/1; cf. Segerbäck, 1983). Also, the amount of CysEtGua was of the magnitude expected (although the value is uncertain).

About 60% of the radioactivity given to the animals was recovered in the urine, in reasonable agreement with earlier observations (Davidson et al., 1982). HOEtGua and CysEtGua were both found in amounts corresponding to about 0.02 nmol per 0.1 mmol DCE/kg bw (fraction of excreted activity: HOEtGua, reduced purines $1.5 \cdot 10^{-6}$, unreduced $0.8 \cdot 10^{-6}$; CysEtGua, reduced $1.1 \cdot 10^{-6}$, unreduced $0.9 \cdot 10^{-6}$), demonstrating that alkylated products of nucleic acids are formed by 2-hydroxyethylating and/or 2-oxoethylating agents as well as by S-(2-chloroethyl)-glutathione.

Efforts were made to detect cyclic derivatives of cytosine in the DNA and the urine but the results were inconclusive.

DISCUSSION

DCE, itself, is a bifunctional alkylating agent with a very low chemical reactivity (the reaction rate constant, k_{H_2O}, equals $5.5 \cdot 10^{-6} \, h^{-1}$; Ehrenberg et al., 1974b). The compound is bioactivated through conjugation with glutathione and through microsomal oxidation. A discussion of possible metabolic pathways leading to the formation of reactive metabolites from DCE may be found in the review by Davidson et al. (1982) and in Guengerich et al. (1980). The suggested pathways include: reductive dehalogenation of DCE to a chloroethyl free radical, dehydrohalogenation to vinyl chloride, microsomal oxidation to chloroacetaldehyde or 1-chloroso-2-chloroethane and conjugation with glutathione to give S-(2-chloroethyl)glutathione. The outline of possible metabolic pathways presented in Figure 2 (adapted from Guengerich et al. 1980) provides a suitable background for the discussion of pathways in vivo. Ethene oxide is included in this scheme since it has been demonstrated that this epoxide is formed in vivo from ethene (Segerbäck, 1983) and since it might be a rearrangement product of chloroethanol at neutral pH (cf. Osterman-Golkar et al., 1970; Jones & Fakhouri, 1979). In addition, the ability of DCE to react directly, i.e., without metabolic or chemical activation, has been indicated.

The chloroethyl radical, 1-chloroso-2-chloroethane, chloroethanol, ethene oxide and DCE itself would all eventually give rise to substitution products of the type Y–CH$_2$CH$_2$OH (Y = nucleophilic site). In the cases of the chloroethyl radical, 1-chloroso-2-chloroethane and DCE, the 2-hydroxyethyl group is formed through hydrolysis subsequent to the introduction of a 2-chloroethyl group (Y–CH$_2$CH$_2$Cl→Y–CH$_2$CH$_2$OH). Chloroethylene oxide (the most important reactive metabolite of vinyl chloride; Osterman-Golkar et al., 1977) and chloroacetaldehyde both introduce 2-oxoethyl groups onto nucleophilic sites of macromolecules. These compounds may react as bifunctional agents and give cyclic products in nucleic acids (Barrio et al., 1972). Finally, the half-mustard, S-(2-chloroethyl)glutathione, gives products of the type RSCH$_2$CH$_2$–Y. (The same type of products may, in theory, be formed via Y–CH$_2$CH$_2$Cl and subsequent substitution of the chlorine for a thiol, e.g., glutathione.)

Electrophilic compounds may be characterized in terms of their selectivity in reactions with nucleophilic centres of varying strengths; in the case of alkylating agents, the s-value (Osterman-Golkar et al., 1970) may be used to describe this

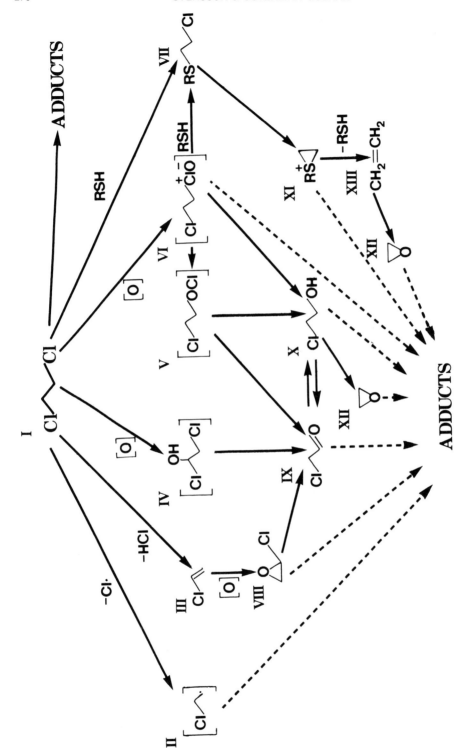

Fig. 2. Proposed reaction pathways for metabolism of 1,2-dichloroethane (I) (adapted from Guengerich *et al.*, 1980). II, a chloroethyl radical; III, vinyl chloride; IV, a *gem*-chlorohydrin; V, a hypochlorite; VI, 1-chloroso-2-chloroethane; VII, a half mustard; VIII, chloroethylene oxide; IX, chloroacetaldehyde; X, chloroethanol; XI, an episulfonium ion; XII, ethene oxide; XIII, ethene

Table 2. Relative degrees of alkylation of cysteine-S, valine-NH_2, histidine-N^π and -N^τ in mouse hemoglobin determined after exposure of the animals to radiolabelled agents. The suggested reactive metabolites and their s-values[a] are given in parentheses.

	Methyl-nitrosourea ($s = 0.4$)	Vinyl chloride (chloro-ethylene oxide; $s = 0.8$)	Methylmethane sulfonate ($s = 0.9$)	Ethene oxide ($s = 0.96$)	1,2-Dichloroethane	
					(chloroethyl-glutathione; $s = 0.9$[b])	(chloroacet-aldehyde; $s = 1.3$)
Cysteine-S	0.55	0.88	5	2.8	3	60
Valine-NH_2	0.08	0.38	–	1.3	–	3
Histidine-N^π	1	1.3	1	1.4	–	1
Histidine-N^τ	1	1	1	1	1	1

[a] The s-value has been used by Osterman-Golkar et al. (1970) to describe the selectivity of electrophilic compounds in reactions with nucleophilic centres of varying strengths.
[b] Assumed to be equal to the s-value of S-(2-chloroethyl)cysteine.

property. Table 2 shows the relative degrees of alkylation of cysteine-S, valine-NH_2 and histidine-N^π and -N^τ in mouse hemoglobin determined after exposure of the animals to radiolabelled alkylating agents. The nucleophilic reactivity increases in the following order: histidine ring nitrogens – valine-NH_2 – cysteine-S. Table 2 shows that shifts of the constant, s, give rise to changes in the alkylation pattern (a high s-value gives a high ratio of cysteine-S (or valine-NH_2) alkylation to histidine alkylation).

The high ratio of cysteine alkylation to histidine alkylation observed for 2-oxoethyl groups in DCE-treated mice (Table 2) is consistent with chloroacetaldehyde ($s = 1.3$; Osterman-Golkar, 1984) as the reactive intermediate (although a contribution from vinyl chloride cannot be excluded). The s-value of S-(2-chloroethyl)glutathione is probably approximately the same as that of S-(2-chloroethyl)cysteine, which has been determined to be 0.9 (unpublished data). In agreement herewith, the ratio of CysEtCys to CysEtHis was found to be 3:1 (see Table 2). HOEtCys constitutes one of the major products in the hemoglobin of DCE-treated mice (see Table 1a). Since, however, the corresponding products of N-terminal valine and histidine-N^π and -N^τ could not be detected, ethene oxide and chloroethanol ($s = 0.9$; unpublished data) can be excluded as important reactive metabolites. Direct reactions of DCE itself are probably unimportant as well ($s = 1$; re-estimated from Ehrenberg et al., 1974b). The origin of HOEtCys remains to be clarified. Reactions of the chloroethyl radical and 1-chloroso-2-chloroethane with cysteine, in-vivo reduction of 2-oxoethylcysteine (hypothetical reaction) and alkylation of carboxylate in hemoglobin by S-(2-chloroethyl)glutathione may account for the formation of this product.

This study indicates that chloroacetaldehyde and S-(2-chloroethyl)glutathione are the predominating reactive metabolites of DCE *in vivo*. These compounds are both of toxicological interest. Chloroacetaldehyde is a bifunctional agent with a mutagenic effectiveness several orders of magnitude higher than expected from its reactivity as an alkylating agent (Hussain & Osterman-Golkar, 1976; studies in *Escherichia coli* Sd. 4). It has been suggested that the cyclic etheno derivatives formed with DNA bases account for this high mutagenic activity. Our studies indicate that 7-(2-oxoethyl)-

guanine is an important product in DNA. The introduction of 2-oxoethyl onto the N-7 of guanine may lead to cyclic products or cross-linking. S-(2-Chloroethyl)-glutathione is a monofunctional alkylating agent and may possibly belong to the class of simple alkylating agents with respect to its mutagenic effectiveness. The impact of the size of the adduct introduced by this compound into DNA is, however, not known.

ACKNOWLEDGEMENTS

We wish to express our thanks to Professor Lars Ehrenberg and Dr. Carl Johan Calleman for valuable discussions. This work was financially supported by the Swedish Natural Science Research Council and the National Swedish Environment Protection Board/Product Control Board.

REFERENCES

Barrio, J.R., Secrist, J.A. & Leonard, N.J. (1972) Fluorescent adenosine and cytidine derivatives. *Biochem. biophys. Res. Commun., 46,* 597–604

Calleman, C.J. (1984) Hemoglobin as a dose monitor and its application to the risk estimation of ethylene oxide. Thesis, University of Stockholm, Sweden

Calleman, C.J. & Wachtmeister, C.A. (1979) Synthesis of N^{π}- and N^{τ}-(2-hydroxyethyl)-L-histidines. *Acta Chem. Scand., B33,* 277–280

Connors, T.A. & Ross, W.C. (1958) Some sulfur-containing amino acids of biological interest. *Biochem. Pharmacol., 1,* 93–100

Davidson, I.W.F., Sumner, D.D. & Parker, J.C. (1982) Ethylene dichloride: a review of its metabolism, mutagenic and carcinogenic potential. *Drug Chem. Toxicol., 5(4),* 319–388

Ehrenberg, L., Hiesche, K.D., Osterman-Golkar, S. & Wennberg, I. (1974a) Evaluation of genetic risks of alkylating agents: tissue doses in the mouse from air contaminated with ethylene oxide. *Mutat. Res., 24,* 83–103

Ehrenberg, L., Osterman-Golkar, S., Singh, D. & Lundquist, U. (1974b). On the reaction kinetics and mutagenic activity of methylating and β-halogenoethylating gasoline additives. *Radiat. Bot., 15,* 185–194

Guengerich, F.P., Crawford, W.M., Jr, Domoradzki, J.Y., Macdonald, T.L. & Watanabe, P.G. (1980) In-vitro activation of 1,2-dichloroethane by microsomal and cytosolic enzymes. *Toxicol. appl. Pharmacol., 55,* 304–317

Hussain, S. & Osterman-Golkar, S. (1976) Comment on the mutagenic effectiveness of vinyl chloride metabolites. *Chem.-biol. Interactions, 12,* 265–267

IARC (1979) *IARC Monographs on the Evaluation of the Carcinogenic Risk of Chemicals to Humans,* Vol. **20,** *Some halogenated hydrocarbons,* Lyon, International Agency for Research on Cancer pp. 429–448

Jones, A.R. & Fakhouri, G. (1979) Epoxides as obligatory intermediates in the metabolism of α-halohydrins. *Xenobiotica, 9,* 595–599

Maltoni, C., Valgimigli, L. & Scarnato, C. (1980) *Long-term carcinogenic bioassays on ethylene dichloride administered by inhalation to rats and mice.* In: Ames, B., Infante, P. & Reitz, R., eds, *Banbury Report 5, Ethylene Dichloride: A Potential Risk?* New York, Cold Spring Harbor Laboratory Press, pp. 3–29

National Cancer Institute (1978) *Bioassay of 1,2-Dichloroethane for Possible Carcinogenicity (Technical Report Series No. 55)*, DHEW Publication No. (NIH) 78-1361, Washington, DC, US Department of Health, Education, & Welfare

Osterman-Golkar, S. (1984) Reaction kinetics in water of chloroethylene oxide, chloroacetaldehyde and chloroacetone. *Hereditas, 101,* 57–68

Osterman-Golkar, S. & Ehrenberg, L. (1983) Dosimetry of electrophilic compounds by means of hemoglobin alkylation. *Ann. Rev. Public Health, 4,* 397–402

Osterman-Golkar, S., Ehrenberg, L. & Wachtmeister C.A. (1970) Reaction kinetics and biological action in barley of mono-functional methanesulfonic esters. *Radiat. Biol., 10,* 303–327

Osterman-Golkar, S., Ehrenberg, L., Segerbäck, D. & Hällström, I. (1976) Evaluation of genetic risks of alkylating agents. II. Haemoglobin as a dose monitor. *Mutat. Res., 34,* 1–10

Osterman-Golkar, S., Hultmark, D., Segerbäck, D., Calleman, C.J., Göthe, R., Ehrenberg, L. & Wachtmeister, C.A. (1977) Alkylation of DNA and proteins in mice exposed to vinyl chloride. *Biochem. biophys. Res. Commun., 76,* 259–266

Segerbäck, D. (1983) Alkylation of DNA and hemoglobin in the mouse after exposure to ethene and ethene oxide. *Chem.-biol. Interactions, 45,* 139–151

Wennerberg, R. & Löfroth, G. (1974) Formation of 7-methylguanine by dichlorvos in bacteria and mice. *Chem.-biol. Interactions, 8,* 339–348

Zilkha, A. & Rappoport, S. (1963) Synthesis of hydroxyalkyl and mercaptoalkyl derivatives of sulfur-containing amino acids. *J. org. Chem., 28,* 1105–1107

IV. BIOLOGICAL EFFECTS

METHYLGLYOXAL IN BEVERAGES AND FOODS: ITS MUTAGENICITY AND CARCINOGENICITY

M. NAGAO[1], Y. FUJITA & T. SUGIMURA

National Cancer Center Research Institute, Tokyo, Japan

T. KOSUGE

Shizuoka College of Pharmacy, Shizuoka-shi, Japan

SUMMARY

Methylglyoxal was found to induce mutations in *Salmonella typhimurium* strains TA100, TA102 and TA104, gene conversion in *Saccharomyces cerevisiae* strain D_7, and diphtheria toxin resistance mutation in Chinese hamster lung cells. Methylglyoxal was detected in coffee, whiskey, a soft drink, toasted bread and soy sauce. The mutagenicity of methylglyoxal was suppressed by sulfite, cysteine, glutathione and dithiothreitol, and enhanced by hydrogen peroxide. Methylglyoxal induced sarcomas in rats at the site of its subcutaneous injection.

INTRODUCTION

Methylglyoxal, which produces a cyclic nucleoside adduct (Shapiro *et al.*, 1969 and these proceedings[2]), has been identified as one of the mutagens in instant coffee and freshly brewed coffee (Kasai *et al.*, 1982). Methylglyoxal is an intermediate in the glycolytic bypass in *Escherichia coli* and in animal tissues. Various foods, including tomatoes (Schormueller & Grosch, 1964), boiled potatoes (Kajita *et al.*, 1972) and roast turkey (Hrdlicka *et al.*, 1965), as well as tobacco smoke (Moree-Testa *et al.*, 1981), were shown to contain methylglyoxal before its mutagenicity was reported, and we recently found methylglyoxal and related dicarbonyls in various beverages and foods.

---

[1] To whom correspondence should be addressed
[2] See p. 165.

This paper reports studies on the mutations induced by methylglyoxal in strains of *Salmonella typhimurium*, *Saccharomyces cerevisiae* D_7 and Chinese hamster lung cells *in vitro*. Modulators of the mutagenicity of methylglyoxal toward *Salmonella typhimurium* and the contribution of methylglyoxal to the total mutagenicity of instant coffee were also studied. In addition, preliminary results on the carcinogenicity of methylglyoxal in rats are reported.

MATERIALS AND METHODS

Chemicals

Methylglyoxal was purchased from Nakarai Chemicals, Kyoto, Japan. Its concentration was determined to be 37% (w/w) by treating it with *o*-phenylenediamine and separating the product, methylquinoxaline, by high-performance liquid chromatography (HPLC) on a C_{18} column of TSK-gel 120A (4 mm × 300 mm, Toyo Soda; eluted with 40% CH_3CN) (Kasai *et al.*, 1982). The compounds used for quantitation, *o*-phenylenediamine and methylquinoxaline, were from Nakarai Chemicals, Osaka, Japan and Aldrich Chemical Co., Milwaukee, WIS, USA, respectively. Methylglyoxal, used for mutation tests, was purified by HPLC on C_{18} columns of TSK-gel 120A (2 columns of 4 mm × 300 mm connected in series; eluted with H_2O).

Mutagenicity test

Salmonella typhimurium TA100, TA98, TA1535, TA1538 (Ames *et al.*, 1975), TA102 and TA104 (Levin *et al.*, 1982) were used. Mutagenicity was tested by the preincubation method (Sugimura & Nagao, 1980) without S9 mix.

Saccharomyces cerevisiae strain D_7 was kindly provided by Dr. H. F. Stich and gene conversion was tested as described by Stich *et al.* (1981).

The diphtheria toxin resistance mutation (DT^r) of Chinese hamster lung cells was tested as reported by Nakasato *et al.* (1984).

Analysis of dicarbonyls in foods

Beverages and soy sauce: Beverages and soy sauce were incubated with 2 mg/ml of *o*-phenylenediamine for 30 min at a pH between 4 and 6, and the reaction mixture was extracted with chloroform. The extract was evaporated and examined for the presence of quinoxaline, methylquinoxaline and ethylquinoxaline, which are derived from glyoxal, methylglyoxal and ethylglyoxal, respectively, by gas chromatography/ mass spectrometry under the following conditions: column, 5% PEG-6000 shimalite W (0.3 × 170 cm); oven temperature, 134 °C; carrier, N_2; detection, flame ionization.

Bread and bean paste: A piece of bread was toasted in the normal way. Its loss of weight on toasting was 14%. Samples of 5 g of toasted and untoasted bread were homogenized in 10 ml of water, and the supernatant obtained by centrifugation at 3000 rpm for 10 min was subjected to the analysis of dicarbonyls. Bean paste was tested directly in the same way.

Carcinogenicity test: Doses of 0.2 ml of a solution of methylglyoxal (not purified) in saline at a concentration of 10 mg/ml were injected subcutaneously twice a week for 10 weeks into 8 male and 10 female Fisher 344 rats. A control group of 21 males and 19 females received only saline in the same way.

RESULTS

Mutagenicities of methylglyoxal and its derivatives

Methylglyoxal was found to induce base-substitution reverse mutations in *Salmonella typhimurium* TA100, TA98, TA104 and TA102 (Table 1). TA104, which has an ochre mutation at the *hisG* gene *(hisG428)*, was more sensitive than TA100, which has a triplet of cytosine residues (-CCC-) at the critical site for base-substitution reversion. TA102, an ultraviolet repair (*uvr*) competent strain, was less sensitive than TA104, although multicopies of the *hisG428* gene are present in its plasmids. This suggests that the DNA damage induced by methylglyoxal and resulting in mutation may be repaired by the *uvr* gene product. TA1535, which is isogenic with TA100 but has no plasmids, did not respond to methylglyoxal.

Glyoxal, ethylglyoxal and propylglyoxal were also mutagenic in TA100, but their mutagenicities were weaker than that of methylglyoxal and they induced 20, 25 and 35 revertants of TA100 per µg, respectively.

Gene conversion at the *trp* locus in *Saccharomyces cerevisiae* was observed at concentrations of 0.5 to 1.5 mg/ml methylglyoxal (Fig. 1).

Methylglyoxal was also mutagenic in Chinese hamster lung cells as shown with DT^r as a marker (Table 2). Its mutagenicity was higher than that of the typical carcinogen, *N*-nitrosodimethylamine (Nakayasu *et al.*, 1983).

Table 1. Mutagenicity of methylglyoxal in *Salmonella typhimurium*

Strain	Genetic background			Mutagenicity (revertants/µg)
	pKM101[a]	*uvr*[b]	LPS[c]	
TA104	+	B	rfa	800
TA102	+	+	rfa	40
TA100	+	B	rfa	100
TA1535	−	B	rfa	0
TA98	+	B	rfa	2
TA1538	−	B	rfa	0

[a] Presence (+) or absence (−) of the plasmid, pKM101
[b] Ultraviolet repair competent (+) or defective (B)
[c] Lipopolysaccharide deficiency, resulting in a rough mutant (rfa)

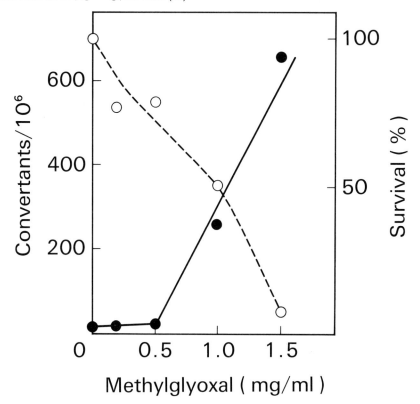

Fig. 1. Gene-converting activity of methylglyoxal in *Saccharomyces cerevisiae* D_7. ●——●, convertants/10^6; ○---○, survival (%)

Table 2. Induction of the diphtheria toxin resistant mutation (DT^r) in Chinese hamster lung cells by methylglyoxal[a]

Methylglyoxal (μg/ml)	DT^r mutants per 2.5×10^5 survivors
24	22.0
36	38.8
48	49.1
60	67.6
66	86.1
0	12.9

[a] From Nakasato et al. (1984)

Table 3. Amounts of methylglyoxal and related dicarbonyls in various beverages and foods

Food or beverage tested	Glyoxal (µg/ml)	Methylglyoxal (µg/ml)	Ethylglyoxal (µg/ml)
Bourbon whiskey	0.39	1.5	0.42
Apple brandy	0.33	0.32	0.43
Wine	0.97	0.57	0.92
Japanese sake	0.29	0.26	0.70
Instant coffee [a]	0.34	1.6	0.70
Brewed coffee [b]	0.87	7.0	1.9
Black tea [c]	0.02	0.05	0.1
Green tea [d]	trace	trace	0.34
Soft drink	–	1.4	–
Bread	0.3 [e]	0.79 [e]	0.4 [e]
Toast	0.5 [e]	2.5 [e]	0.7 [e]
Soy sauce	4.9	8.7	8.4
Soy bean paste	4.2 [e]	5.1 [e]	4.2 [e]

[a] Prepared by dissolving 1.5 g of coffee powder in 100 ml of water
[b] Prepared from 10 g of ground coffee beans and 150 ml of boiling water
[c] Prepared from 4 g of tea leaves and 100 ml boiling water
[d] Prepared from 5 g of tea leaves and 200 ml of hot water
[e] µg/g

Methylglyoxal and related dicarbonyls in various beverages and foods

As we previously reported, instant and freshly brewed coffees contain appreciable amounts of methylglyoxal (Kasai *et al.*, 1982). Coffee brewed with 10 g of roasted coffee beans and 150 ml of boiling water, with a recovery of 100 ml extract, contained 695 µg, 87 µg and 192 µg of methylglyoxal, glyoxal and ethylglyoxal, respectively (Table 3). One kind of soft drink contained an amount of methylglyoxal comparable to that in instant coffee. Black tea and green tea contained less than one-tenth as much methylglyoxal as instant coffee, whereas the amounts of ethylglyoxal in green tea and instant coffee were comparable.

Bourbon whiskey had the highest concentration of methylglyoxal of the alcoholic beverages tested. Since this whiskey is stored in a barrel with a charred surface, the methylglyoxal content of the whiskey may be related to the barrelling process.

The amount of methylglyoxal in bread increased during toasting.

Role of methylglyoxal in the mutagenicity of instant coffee

Figure 2 shows the dose-responses of mutations to methylglyoxal and coffee in *Salmonella tiphimurium* TA100. Methylglyoxal gave a sigmoidal curve, and accounted for only a small part of the total mutagenicity of instant coffee, assuming that its mutagenicity was not affected by other components in instant coffee.

The mutagenicity of coffee was suppressed by catalase (Nagao *et al.*, 1984), but that of methylglyoxal was not. When 0.2 µmol of methylglyoxal was added to 15 mg of

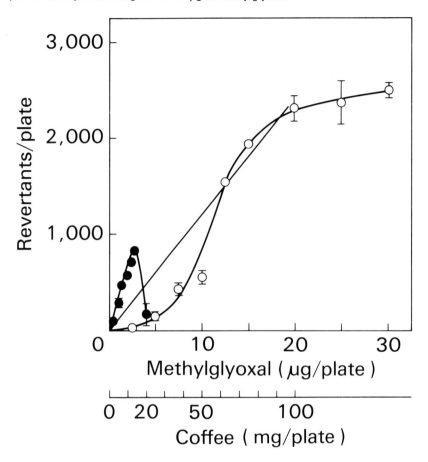

Fig. 2. Dose-dependence of the mutagenicities of methylglyoxal and coffee in *Salmonella typhimurium* TA100. The scales on the abscissa are superimposed to indicate amounts of methylglyoxal in coffee. ○──○, methylglyoxal; ●──●, instant coffee; ──, calculated value based on the specific activity in the range of 12–20 μg of methylglyoxal

instant coffee, only an amount of mutagenicity corresponding to that due to coffee was suppressed, while that due to methylglyoxal was not affected. On the other hand, when cysteine was added to a mixture of methylglyoxal and coffee, it suppressed only the mutagenicity due to the added methylglyoxal, but not that due to coffee (Fujita *et al.*, 1985). The above results support the idea that methylglyoxal is responsible for only a minor part of the total mutagenicity of instant coffee.

Change in the mutagenicity of methylglyoxal by SH-compounds and hydrogen peroxide

The mutagenicity of methylglyoxal in *Salmonella typhimurium* TA100 was suppressed by sulfite. Glutathione is known to conjugate with methylglyoxal in the

Table 4. Modulation of the mutagenicity of methylglyoxal by SH-compounds and hydrogen peroxide

Methylglyoxal (μmole/plate)	Addition	Mutagenicity[a] (revertants/plate)
0.2	None	1,900
	Cysteine, 1 μmol/plate	0
	Glutathione, 6 μmol/plate	100
	Dithiothreitol, 3 μmol/plate	285
	Sulfite, 2 μmol/plate	90
0.02	None	30
	Hydrogen peroxide, 0.3 μmol/plate	970

[a] *Salmonella typhimurium* TA100 without S9 mix was used

presence of glyoxalase I. However, glutathione alone reduced the mutagenicity of methylglyoxal, and dithiothreitol had a similar effect (Table 4). Cysteine was the most effective of the SH-compounds tested for reducing the mutagenicity of methylglyoxal (Table 4). After incubation of methylglyoxal with cysteine at 37 °C for 20 min, methylglyoxal was recovered completely as methylquinoxaline by reaction with *o*-phenylenediamine. There was no evidence for the formation of a stable conjugate by cysteine and methylglyoxal. The mutagenicities of glyoxal and diacetyl were also suppressed by bisulfite (Suwa *et al.*, 1982).

Hydrogen peroxide increased the mutagenicity of methylglyoxal by 30-fold (Table 4); it has also been found to increase the mutagenicity of glyoxal. The mechanism of the enhancing effect of hydrogen peroxide is unknown.

Carcinogenicity of methylglyoxal

After 20 months, four of the methylglyoxal-treated rats (three males and one female) had tumours at the injection site, but no tumours were observed in the control group. The size of the largest tumour was 35 × 35 × 20 mm. Histologically, three were fibrosarcomas and one was a malignant fibrous histiocytoma.

DISCUSSION

Methylglyoxal, which once received much attention because it retards cell growth (Szent-Györgyi, 1968), was recently found to be mutagenic in bacteria, yeast and mammalian cells. Its mutagenicity suggests that it might be carcinogenic. Methylglyoxal was also demonstrated to preferentially induce base-pair substitution mutations, which is in accordance with the results of Marnett *et al.* (these proceedings[3]). We plan to use diagnostic strains of *Salmonella typhimurium*, which

---

[3] See p. 175.

are being developed by B. Ames (personal communication), to obtain more information on the base-specificity of these mutations.

Since formation of adducts of guanosine and methylglyoxal have been reported (Shapiro et al., 1969), it also seems worthwhile to study the relation between the mechanisms of mutation and nucleoside modification by methylglyoxal.

Methylglyoxal is present in various foods, and humans probably ingest an average of about one mg of methylglyoxal per day in coffee. Therefore, it seemed important to determine the carcinogenicity of methylglyoxal. We found that subcutaneous injections of methylglyoxal induced sarcomas in rats. We also found that the mutagenicity of methylglyoxal was decreased by cysteine, dithiothreitol and glutathione and increased by hydrogen peroxide. Since these compounds may influence the effect of methylglyoxal *in vivo*, a long-term test on the effect of oral administration of methylglyoxal to animals seems called for. This experiment is now in progress in our laboratory.

ACKNOWLEDGEMENTS

This work was supported in part by Grants-in-Aid for Cancer Research from the Ministry of Education, Science and Culture, and the Ministry of Health and Welfare of Japan.

REFERENCES

Ames, B.N., McCann, J. &Yamasaki, E. (1975) Methods for detecting carcinogens and mutagens with the *Salmonella*/mammalian-microsome mutagenicity test. *Mutat. Res., 31,* 347–364

Fujita, Y., Wakabayashi, L., Nagao, M. & Sugimura, T. (1985) Characteristics of major mutagenicity of instant coffee. *Mutat. Res., 142,* 145–148

Hrdlicka, J. & Kuca, J. (1965) The changes of carbonyl compounds in the heat-processing of meat. II. Turkey meat. *Poult. Sci., 44,* 27–31

Kajita, T. & Senda, M. (1981) Simultaneous determination of L-ascorbic acid, triose reductone and their related compounds in foods by polarographic method. *Nippon Nogei Kagaku Kaishi, 46,* 137–145

Kasai, H., Kumeno, K., Yamaizumi, Z., Nishimura, S., Nagao, M., Fujita, Y., Sugimura, T., Nukaya, H. & Kosuge T. (1982) Mutagenicity of methylglyoxal in coffee. *Gann, 73,* 681–683

Levin, D.E., Hollstein, M., Christman, M.F., Schwiers, E.A. & Ames, B.N. (1982) A new *Salmonella* tester strain (TA102) with A · T base pairs at the site of mutation detects oxidative mutagens. *Proc. natl Acad. Sci. USA, 79,* 7445–7449

Moree-Testa, P. & Saint-Jalm, Y. (1981) Determination of α-dicarbonyl compounds in cigarette smoke. *J. Chromatogr., 217,* 197–208

Nagao, M., Suwa, Y., Yoshizumi, H. & Sugimura, T. (1984) *Mutagens in coffee*. In: MacMahon, B. & Sugimura, T., eds, *Coffee and Health*. Banbury Report 17, New York, Cold Spring Harbor Laboratory, pp. 69–77

Nakasato, F., Nakayasu, M., Fujita, Y., Nagao, M., Terada, M. & Sugimura, T. (1984) Mutagenicity of instant coffee in cultured Chinese hamster lung cells. *Mutat. Res. Lett., **141**,* 109–112

Nakayasu, M., Nakasato, F., Sakamota, H., Terada, M. & Sugimura, T. (1983) Mutagenic activity of heterocyclic amines in Chinese hamster lung cells with diphtheria toxin resistance as a marker. *Mutat. Res., **118**,* 91–102

Schormueller, J. & Grosch, W. (1964) Aromatics of foods. II. Occurrence of additional carbonyl compounds in the tomato. *Z. Lebensm-Unters.-Forsch., **126**,* 38–49

Shapiro, R., Cohen, B.I., Shiuey, S.-J. & Maurer, H. (1969) On the reaction of guanine with glyoxal, pyruvaldehyde, and kethoxal, and the structure of the acylguanines. A new synthesis of N^2-alkylguanines. *Biochemistry, **8**,* 238–245

Stich, H.F., Rosin, M.P., Wu, C.H. & Powrie, W.D. (1981) A comparative genotoxicity study of chlorogenic acid (3–0-coffeoylquinic acid). *Mutat. Res., **90**,* 201–202

Sugimura, T. & Nagao, M. (1980) *Modification of mutagenic activity.* In: de Serres, F.J. & Hollaender, A., eds, *Chemical Mutagens. Principles and Methods for their Detection*, Vol. 6, New York, Plenum Press, pp. 41–60

Suwa, Y., Nagao, M., Kosugi, A. & Sugimura, T. (1982) Sulfite suppresses the mutagenic property of coffee. *Mutat. Res., **102**,* 383–391

Szent-Györgyi, A. (1968) *Bioelectronics: A Study in Cellular Regulations, Defense and Cancer*, New York & London, Academic Press

EVALUATION OF THE MUTAGENICITY OF 1,N^6-ETHENOADENINE- AND 3,N^4-ETHENOCYTOSINE-NUCLEOSIDES IN *SALMONELLA TYPHIMURIUM*

K. NEGISHI, K. OOHARA, H. URUSHIDANI, Y. OHARA & H. HAYATSU

Faculty of Pharmaceutical Sciences, Okayama University, Tsushima, Okayama, Japan

SUMMARY

To provide supporting evidence for the hypothetical involvement of etheno-derivative formation in DNA in chloroacetaldehyde-mediated mutagenesis, etheno nucleosides were examined for their direct-acting mutagenicity in *Salmonella typhimurium* strain TA100. The results, however, have shown that 1,N^6-etheno-adenosine, 1,N^6-ethenodeoxyadenosine, 3,N^4-ethenocytidine and 3,N^4-etheno-deoxycytidine lack mutagenicity in this test system. A lesson learned in this study is that 3,N^4-ethenocytosine nucleosides prepared synthetically or obtained from commercial sources can give false positive mutagenicity due to mutagenic contaminants.

INTRODUCTION

Mutagenicity of chloroacetaldehyde in *Salmonella typhimurium* has been studied in several laboratories (McCann *et al.*, 1975; Elmore *et al.*, 1976; Kayasuga-Mikado *et al.*, 1980), and has been found to be positive in strain TA100 in the absence of metabolic activation. This suggests that chloroacetaldehyde can induce base-pair substitution mutations by acting directly on cellular DNA. Since chloroacetaldehyde reacts under mild conditions with adenine and cytosine to form etheno derivatives (Thomas & Leonard, 1976), it has been postulated that these chemical modifications in the bases of cellular DNA are the cause of mutation (Elmore *et al.*, 1976).

The most probable mechanism for mutagenicity of a nucleoside analogue is incorporation of the analogue into the cellular DNA, which leads to the generation of mutant cells. For example, N^4-aminocytidine has been shown to be a potent directly-acting mutagen in *Salmonella typhimurium* and in *Escherichia coli* (Negishi *et al.*, 1983), and there is evidence that it is incorporated into the DNA giving rise to base-pair substitution during replication (Negishi et al., 1985). It is therefore worthwhile to examine mutagenic potentials of ethenonucleosides: if they are mutagenic in the Salmonella system, the postulated mechanism for chloroacetaldehyde-mutagenesis would gain strong support. In this report, however, we show that 1,N^6-ethenoadenosine, 1,N^6-ethenodeoxyadenosine, 3,N^4-ethenocytidine and 3,N^4-ethenodeoxycytidine lack mutagenicity in S. typhimurium TA100 in the absence of metabolic activation.

MATERIALS AND METHODS

Chemicals

1,N^6-Ethenoadenosine and 1,N^6-ethenodeoxydenosine were products of Sigma (St. Louis, MO, USA). Samples of 3,N^4-ethenocytidine were obtained from Sigma and Pharmacia PL Biochemicals (Uppsala, Sweden). A sample of 3,N^4-ethenocytidine was prepared by treatment of cytidine (Sigma) with aqueous chloroacetaldehyde (Tokyo Kasei, Japan) according to the method of Bario et al. (1972). In the same way, 3,N^4-ethenodeoxycytidine was synthesized from 2′-deoxycytidine (Sigma); upon recrystallization from aqueous ethanol, the hydrochloride salt was obtained as colourless crystals. The elemental composition of this material was analysed as C, 45.6%; H, 4.9%; N, 14.6% ($C_{11}H_{14}N_3O_4Cl$ requires C, 45.9%; H, 4.9%; N, 14.6%). Ultraviolet absorption analysis showed an absorption maximum in water at 272 nm with characteristic shoulders at about 280 nm and 290 nm. R_F values in cellulose thin-layer chromatography developed by the solvent, isobutyric acid:ammonia:water (75:1:24), were 0.76 for 3,N^4-ethenocytidine, 0.71 for 3,N^4-ethenodeoxycytidine, and 0.59 for deoxycytidine.

High-performance liquid chromatography (HPLC) for purification of the ethenonucleosides

The chromatography was run on a column of μBondapak C_{18} (4 × 300 mm). Aqueous methanol (15%) was used as the elution solvent at a flow rate of 1 ml/min. Fractions of one ml were collected and evaporated to dryness. The residues were dissolved in 0.1 ml water and used in the mutation assay.

Mutation assay

Mutagenicity of compounds was investigated with S. typhimurium TA100 in the absence of metabolic activation (Ames et al., 1975). The preincubation technique (Yahagi et al., 1977) was used. The strain TA100 was a kind gift of Dr. B.N. Ames.

RESULTS

Table 1 shows the results of mutation assays on S. typhimurium TA100 carried out for samples of ethenonucleosides. The samples used were either commercially available materials as such, or those prepared in our own laboratory according to

Table 1. Mutagenicity tests of ethenonucleosides showing false-positive responses in several samples

Compound	Source[a]	Amount (μmol)	His$^+$ revertants formed in Salmonella typhymurium TA100 without metabolic activation (number/plate)[b]	
1,N^6-Ethenoadenosine	S	0.5	67,	87
		0.1	72,	61
		0.05	65,	57
1,N^6-Ethenodeoxyadenosine	S	0.5	89,	80
		0.1	68,	69
		0.05	55,	59
3,N^4-Ethenocytidine	S	0.5	74,	93
		0.1	61,	82
		0.05	73,	61
	P	0.5[c]	170,	205
		0.1[c]	<u>267,</u>	<u>303</u>
		0.05	<u>394,</u>	<u>359</u>
	O	0.5	<u>1383,</u>	<u>1161</u>
		0.1	<u>1207,</u>	<u>1102</u>
		0.05	<u>311,</u>	<u>310</u>
3,N^4-Ethenodeoxycytidine	O	0.5	<u>260,</u>	<u>294</u>
		0.1	<u>176,</u>	<u>174</u>
		0.05	79,	71
Solvent only			47,	65

[a] S, Sigma; P, Pharmacia PL-Biochemicals; O, synthesized in our laboratory
[b] Numbers underlined are positive results
[c] Killing of bacteria took place

known procedures. Both 1,N^6-ethenoadenosine and 1,N^6-ethenodeoxyadenosine gave negative results up to a dose of 0.5 μmol/plate. A commercial sample of 3,N^4-ethenocytidine (Sigma) showed no mutagenicity, whereas another commercial sample (Pharmacia PL Biochemicals), as well as our synthetic sample, produced positive results. The synthetic 3,N^4-ethenodeoxycytidine, which had been crystallized to give correct analytical values, showed a weak but dose-dependent mutagenicity.

The "positive" responses observed were shown to be due to contaminants in these materials by subjecting them to HPLC and assaying the fractions for mutagenicity. As Figure 1 shows, mutagenic components were separated from the ethenonucleosides by this process.

It is concluded that 1,N^6-ethenoadenosine, 1,N^6-ethenodeoxyadenosine, 3,N^4-ethenocytidine and 3,N^4-ethenodeoxycytidine are not directly mutagenic in S. typhimurium TA100.

DISCUSSION

The lack of mutagenicity of a nucleoside analogue in the Ames test system can mean one of several things: (1) the compound cannot penetrate the cell membrane, (2) in the cell, the analogue cannot be converted into the deoxynucleoside 5′-triphosphate,

Fig. 1. High-performance liquid chromatographic fractionation of mutagenic samples. A, 3,N^4-ethenocytidine from Pharmacia PL Biochemicals, 0.6 mg; B, 3,N^4-ethenocytidine prepared in our laboratory by treatment of cytidine with chloroacetaldehyde, 3.1 mg; C, 3,N^4-ethenodeoxycytidine prepared in our laboratory from deoxycytidine, 3.4 mg. The column was μBondapak C_{18}, 4 × 300 mm, and the eluent was 15% methanol. The mutagenicity was assayed in *Salmonella typhimurium* TA100 without metabolic activation. Numbers of spontaneously-formed revertants (55 ± 10) have been subtracted. The absorbance was recorded at 254 nm. The dotted areas represent the ethenonucleoside fractions.

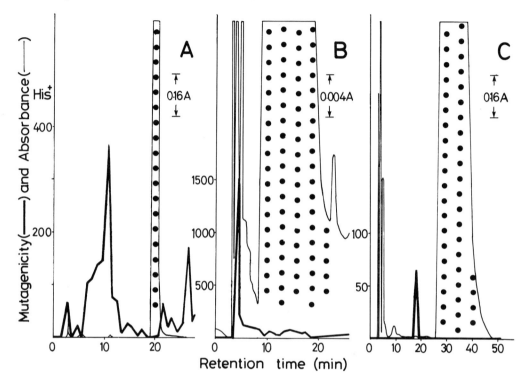

(3) the triphosphate, even if it is formed, is not used by DNA synthesizing enzymes, or (4) although the analogue nucleotide becomes incorporated into newly-formed DNA, it does not produce base-pair substitution. Therefore, the fact that we have obtained negative results for these ethenonucleosides does not disprove the supposed mechanism for chloroacetaldehyde-mutagenesis in which etheno-residue formation in DNA is the key reaction. Proof or disproof of this mechanism is open for future studies. Related to this is the puzzling fact that bromoacetaldehyde, which reacts more readily than chloroacetaldehyde with adenine and cytosine nucleotides to form the ethenoderivatives, shows little, if any, mutagenicity in the Ames test (Kayasuga-Mikado et al., 1980).

The present work also provides a note of caution for mutagenicity testing of compounds: as this study shows, false positive results can easily be obtained due to small amounts of mutagenic components contaminating the test compounds.

ACKNOWLEDGEMENT

This work was supported by a Grant-in-Aid for Scientific Research from the Ministry of Education, Science and Culture (58480446).

REFERENCES

Ames, B.N., McCann, J. & Yamasaki, E. (1975) Methods for detecting carcinogens and mutagens with the *Salmonella*/mammalian-microsome mutagenicity test. *Mutat. Res., 31,* 347–364

Barrio, J.R., Secrist, J.A., III & Leonard, N.J. (1972) Fluorescent adenosine and cytidine derivatives. *Biochem. biophys. Res. Commun., 46,* 597–604

Elmore, J.D., Wong, J.L., Laumbach, A.D. & Streips, U.N. (1976) Vinyl chloride mutagenicity *via* the metabolities chlorooxilane and chloroacetaldehyde monomer hydrate. *Biochim. biophys. Acta, 442,* 405–419

Kayasuga-Mikado, K., Hashimoto, T., Negishi, T., Negishi, K. & Hayatsu, H. (1980) Modification of adenine and cytosine derivatives with bromoacetaldehyde. *Chem. pharm. Bull., 28,* 932–938

McCann, J., Simmon, V., Streitwieser, D. & Ames, B.N. (1975) Mutagenicity of chloroacetaldehyde, a possible metabolic product of 1,2-dichloroethane (ethylene dichloride), chloroethanol (ethylene chlorohydrin), vinyl chloride, and cyclophosphamide. *Proc. natl Acad. Sci. USA, 72,* 3190–3193

Negishi, K., Harada, C., Ohara, Y., Oohara, K., Nitta, N. & Hayatsu, H. (1983) N^4-Aminocytidine, a nucleoside analogue that has an exceptionally high mutagenic activity. *Nucleic Acids Res., 11,* 5223–5233

Negishi, K., Takahashi, M., Yamashita, Y., Nishizawa, M. & Hayatsu, H. (1985) Mutagenesis by N^4-aminocytidine: Induction of AT to GC transition and its molecular mechanism. *Biochemistry* (in press)

Thomas, R.W. & Leonard, N.J. (1976) Examples of the use of fluorescent heterocycles in chemistry and biology. *Heterocycles, 5,* 839–882

Yahagi, T., Nagao, M., Seino, Y., Matsushima, T., Sugimura, T. & Okada, M. (1977) Mutagenicities of *N*-nitrosamines on *Salmonella. Mutat. Res., 48,* 121–130

GENETIC EFFECTS OF DNA MONO- AND DIADDUCTS PHOTOINDUCED BY FUROCOUMARINS IN EUKARYOTIC CELLS

D. AVERBECK & D. PAPADOPOULO

Institut Curie, Section de Biologie, Paris, France

SUMMARY

The genotoxic activity of two bifunctional furocoumarins, 8- and 5-methoxypsoralen (8-MOP and 5-MOP), and the newly developed monofunctional furocoumarin, 7-methyl-pyrido(3,4-c)psoralen (MePyPs), was determined in diploid yeast. In parallel experiments, the DNA photobinding capacity of the compounds was estimated. This capacity was, in decreasing order, MePyPs > 5-MOP > 8-MOP. The lesions induced by 8- and 5-MOP, which consist of a mixture of mono- and diadducts, were found to be more genotoxic than the monoadducts induced by MePyPs. A difference in specificity of MePyPs-induced lesions was observed for the induction of mutations and mitotic gene conversion.

In Chinese hamster V79 cells, 5-MOP showed a higher DNA photobinding activity than 8-MOP. In spite of this, 5-MOP-induced lesions were less genotoxic, on a one-to-one basis, than those induced by 8-MOP. An analysis of the DNA cross-linking capacity of the two furocoumarins in V79 cells by alkaline elution suggests that, at a given dose of 365-nm radiation, 5-MOP produces more photoadducts than 8-MOP but the same number of DNA cross-links, and the ratio of mono- to diadducts induced is higher for samples treated with 5-MOP than for those treated with 8-MOP. This may offer an explanation for differences observed in the mutagenic and carcinogenic activity of these compounds.

INTRODUCTION

Since the discovery that photosensitizing furocoumarins or psoralens are able to photoreact rather specifically with pyrimidine bases in nucleic acids in the presence of long-wave UV radiation (UVA) (Dall'Acqua et al., 1970; Cole, 1970; Musajo & Rodighiero, 1972), these heterocyclic compounds derived from the fusion of a furan ring to a suitably substituted coumarin molecule can be considered as effective tools for studying the relationship between the induction and repair of specific lesions in DNA and the induction of genetic and carcinogenic effects in different organisms (Parsons, 1980; Averbeck, 1982; Averbeck, 1984). Such studies (Bridges & Strauss, 1980; IARC, 1980; Grekin & Epstein, 1981) are of great interest since certain photoreactive furocoumarins such as 8-MOP and 5-MOP are widely employed in the photochemotherapy (PUVA-therapy) of a variety of cutaneous disorders including psoriasis and the T-cell lymphoma mycosis fungoides, as well as in certain sun-tan preparations, despite their known carcinogenic potential in mice and man (see Vainio & Saracci, these proceedings[1]).

The reaction of psoralens with DNA involves the following steps (Dall'Acqua, 1977): 1) noncovalent binding to DNA, i.e., formation of complexes with DNA in the absence of UVA; favourable sites for intercalation of psoralen molecules are alternating sequences of purines and pyrimidines in DNA. 2) Cyclo (C_4) additions to pyrimidines in DNA in the presence of UVA; this reaction involves either the 3,4 or the 4', 5' reactive site of the furocoumarin, and leads to 3, 4 or 4', 5'-monoadducts. 3) Conversion of some of these monoadducts into DNA interstrand cross-links upon further exposure to UVA. The formation of the different types of adducts was found to be dependent on the adenine-thymine content of the DNA and showed a certain sequence specificity (Parsons, 1980). Furthermore, furocoumarins are known to possess a certain specificity depending on their molecular structure (Averbeck, 1982, 1984); the so-called monofunctional furocoumarins such as angelicin, 3-carbethoxypsoralen and certain pyridopsoralens are only able to photoinduce monoadditions in DNA whereas the so-called bifunctional furocoumarins such as psoralen, 8-methoxypsoralen and 5-methoxypsoralen are able to photoinduce monoadditions plus biadditions (interstrand cross-links) in DNA. The type of addition and the isomer formed is highly dependent on the furocoumarin used (Straub *et al.*, 1981; Kanne *et al.*, 1982; Cadet *et al.*, 1983; Vigny *et al.*, 1983; Cadet *et al.*, 1984, and these proceedings[2]).

The aim of the studies reported here was to investigate the relationship between the type of lesion induced by furocoumarins and UVA and the genetic effects, using two eukaryotic cell systems, namely, *Saccharomyces cerevisiae* and Chinese hamster V79 cells.

MATERIALS AND METHODS

Biological material

We used the diploid yeast strain D7 (*Saccharomyces cerevisiae*) (Zimmermann *et al.*, 1975) and Chinese hamster V79 cells in culture (Papadopoulo *et al.*, 1983) for detecting the effects on survival and the induction of mutations. We measured the

---

[1] See p. 15.
[2] See p. 247.

Fig. 1. Chemical structures of furocoumarins

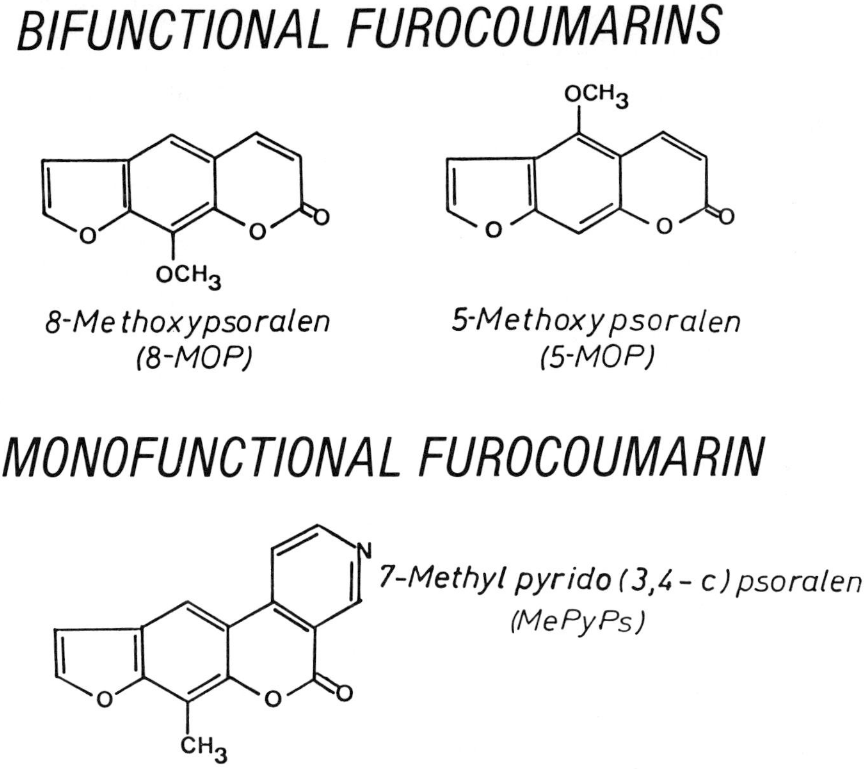

BIFUNCTIONAL FUROCOUMARINS

8-Methoxypsoralen
(8-MOP)

5-Methoxypsoralen
(5-MOP)

MONOFUNCTIONAL FUROCOUMARIN

7-Methyl pyrido (3,4-c) psoralen
(MePyPs)

induction of nuclear reversions (ILV$^+$) and mitotic intragenic recombination (TRP$^+$) in yeast and the induction of 6-thioguanine resistant (6-TG^r) mutants in Chinese hamster cells, according to previously published methods (Papadopoulo et al., 1983).

Chemicals

We used the newly synthesized monofunctional 7-methyl-pyrido(3,4-c)psoralen (MePyPs) (Moron et al., 1983) and the bifunctional furocoumarins, 8-methoxypsoralen (8-MOP) (Chinoin, Milan, Italy) and 5-methoxypsoralen (5-MOP) (Sarsyntex-Interchim, Paris, France). Figure 1 gives the chemical structure of these compounds. The following radioactively labelled psoralens were used for the determination of the in-vivo DNA photobinding capacity: ^{3}H-5-MOP (specific activity, 400 mCi/mmol) and ^{3}H-8-MOP (specific activity, 78 Ci/mmol) from Amersham, England, and ^{3}H-MePyPs (specific activity, 3.8 Ci/mmol) from Centre d'Energie Atomique, Saclay, France, in methanolic solution. All products were chromatographically pure.

Radiation source (UVA)

A HPW 125 Philips lamp with a pyrex-glass/water filter served as the source of 365 nm (UVA) radiation. The incident dose rate was 1.2 kJm^{-2}min^{-1} as determined by the radiometer J-260 (Ultraviolet Products, Inc., San Gabriel, CA, USA).

Media, culture and treatment conditions

The media, culture and treatment conditions were the same as those reported previously (Averbeck *et al.*, 1984a, b). Cells were incubated in the dark in the presence of the furocoumarins (at 5 µM, if not stated otherwise) for 15 to 30 minutes, and aliquots were exposed to increasing doses of 365-nm radiation. After treatment, the cells were plated on complete growth media for the detection of survival and on selective media for the detection of mutants.

Detection of photoadducts in DNA of yeast and V79 cells

Parallel to the genetic experiments, we measured the photoinduced binding of radioactively labelled psoralens to the DNA of yeast and V79 cells following the method of Magaña-Schwencke *et al.*, 1980.

Detection of DNA cross-links in Chinese hamster V79 cells

We performed the alkaline elution technique according to Kohn *et al.*, 1976.

RESULTS

Photobinding of psoralens to the DNA of diploid yeast

Parallel to the genetic experiments in yeast, we measured the DNA photobinding of the monofunctional furocoumarin, MePyPs, and the two bifunctional furocoumarins, 8- and 5-MOP. Figure 2 illustrates the results obtained in the diploid yeast strain D7. The monofunctional furocoumarin, MePyPs, has a photoaffinity towards DNA approximately ten-fold higher than that shown by the bifunctional furocoumarin, 8-MOP. The photoaffinity of 5-MOP is about five times higher than that of 8-MOP. The high photobinding activity of the MePyPs is in accord with results obtained in haploid yeast (Moustacchi *et al.*, 1983) and in mammalian cell cultures (Papadopoulo *et al.*, 1984a, b; Nocentini, 1984).

Induction of nuclear reversions and mitotic intragenic recombination in diploid yeast (D7)

In order to illustrate the relationship between the number of photoadditions induced and the genetic effects observed, we plotted the induction of lethal effects, expressed in terms of survival (A), the induction of TRP$^+$ gene convertants (B) and the induction of ILV$^+$ revertants (C) as functions of photoadditions induced (Fig. 3).

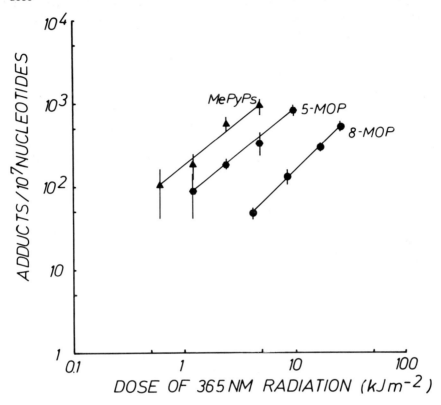

Fig. 2. DNA photobinding of 7-methyl-pyrido(3,4-c)psoralen (MePyPs), 5-methoxypsoralen (5-MOP) and 8-methoxypsoralen (8-MOP) at 5 μM in diploid yeast as a function of long-wave UV-radiation dose

At equal numbers of total photoadditions induced, the lesions induced by the monofunctional compound, MePyPs, (i.e., the 4′,5′-monoadducts) are clearly less lethal, less convertogenic and less mutagenic than the lesions induced by the bifunctional furocoumarins, 5- and 8-MOP.

Interestingly, despite the fact that 5-MOP is known to be more genetically effective per unit dose than 8-MOP (Averbeck, 1984; Averbeck et al., 1984a, b) with regard to the induction of lethal effects, nuclear reversion and mitotic intra- and intergenic recombination (Averbeck et al., 1983a), 5-MOP-induced lesions appear to be less effective than 8-MOP-induced lesions in the production of nuclear genetic events (Fig. 3), the differences in activity being quite comparable for the different genetic endpoints. In contrast, the lesions induced by MePyPs appear to be more convertogenic than mutagenic in relation to the constantly high genetic activity of the lesions induced by 8-MOP or 5-MOP.

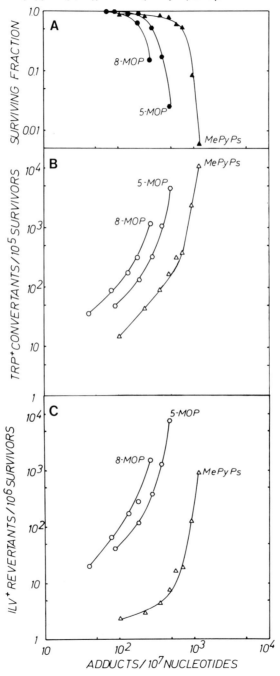

Fig. 3. Survival (A), induction of TRP⁺ gene convertants (B) and induction of ILV⁺ (nuclear) revertants (C) in diploid yeast as functions of the number of photoadditions induced, following treatment of *Saccharomyces cerevisiae* strain D7 with 8-methoxypsoralen (8-MOP), 5-methoxypsoralen (5-MOP) or 7-methyl-pyrido(3,4-c)psoralen (MePyPs) at 5 μM

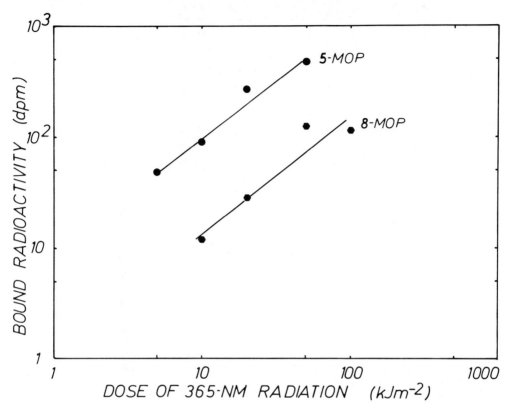

Fig. 4. DNA photobinding of radioactive 5-methoxypsoralen (5-MOP) and 8-methoxypsoralen (8-MOP) (5 μM) in Chinese hamster V79 cells as a function of long-wave UV-radiation dose

Photobinding of 8- and 5-MOP to DNA of Chinese hamster V79 cells

Parallel to experiments on survival and on induction of 6-TG^r mutants, we determined the DNA photobinding activity of the two bifunctional furocoumarins, 8- and 5-MOP, in Chinese hamster V79 cells (Fig. 4). At 10 kJm^{-2} of UVA radiation, the amount of 5-MOP which is photobound to the DNA of V79 cells is about six times greater than the amount of 8-MOP.

Photoinduction of lethal and mutagenic effects in Chinese hamster V79 cells by 8- and 5-MOP

In accord with the results obtained in yeast, 5-MOP was shown to be more effective than 8-MOP in the presence of 365-nm radiation in the induction of lethality and of 6-TG^r mutants in Chinese hamster V79 cells (Averbeck et al., 1983a; Papadopoulo et al., 1984a, b). The relationship between the induction of genotoxic events and the

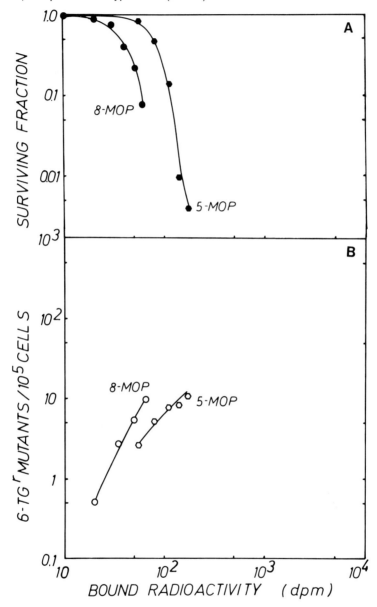

Fig. 5. Survival (A) and induction of 6-thioguanine resistant (6-TG^r) mutants (B) in Chinese hamster V79 cells as functions of DNA photobinding, following treatment with 5 µM 8-methoxypsoralen (8-MOP) or 5 µM 5-methoxypsoralen (5-MOP)

photoaffinity of the compounds towards DNA is illustrated in Figure 5, where the induction of lethal effects, expressed in terms of survival (A), and the induction of mutagenic effects (B) are plotted as functions of the photobinding activity derived from Figure 4.

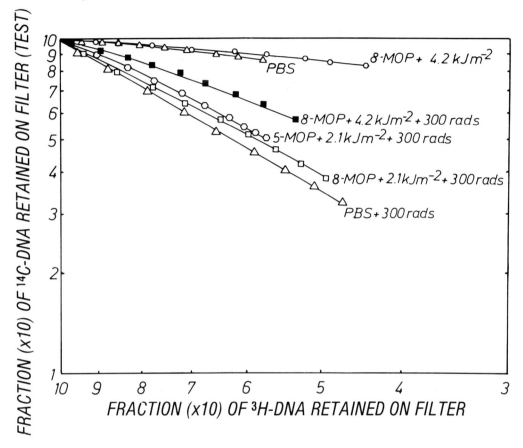

Fig. 6. Alkaline-elution patterns for Chinese hamster V79 cells treated with 5 µM 5-methoxypsoralen (5-MOP) or 8-methoxypsoralen (8-MOP) and UVA radiation, with or without subsequent X-irradiation, and for untreated cells in phosphate-buffered saline (PBS), with and without subsequent X-irradiation

The DNA-bound radioactivity corresponds to DNA-bound radioactive furocoumarin molecules, i.e., to the photoinduced lesions. In spite of the fact that the photoaffinity of 5-MOP was higher than that of 8-MOP (Fig. 4), the lesions induced by 5-MOP are less genotoxic, with regard to both survival and mutation, than those induced by 8-MOP. This may be due either to the induction of different ratios of mono- to diadducts – 8-MOP producing relatively more diadducts than 5-MOP – or to the induction of different isomeric types of photoadditions with different degrees of genotoxicity.

Comparison of the capacities of 8- and 5-MOP to induce DNA interstrand cross-links in Chinese hamster V79 cells

In order to test the above hypothesis, we estimated the relative numbers of cross-links induced by the two bifunctional furocoumarins, 8- and 5-MOP, using the alkaline elution technique. Figure 6 shows the elution profiles obtained for cells labelled with [methyl-^{14}C]-thymidine (0.4 uCi/ml), with tritium-labelled samples included as internal standards (0.125 µCi/ml [methyl-^{3}H]-thymidine) (Kohn et al., 1976).

The elution profile of samples treated with phosphate-buffered saline (PBS) alone appears to be about the same as that of cells treated with 8-MOP (dissolved in PBS) plus 4.2 kJm^{-2} of 365-nm radiation (Fig. 6) and cells treated with 5-MOP plus 4.2 kJm^{-2} (data not shown). This indicates the presence of very few strand breaks in the control samples. A very rapid decrease in the elution profile (see plot labelled as PBS + 300 rads in Figure 6) is observed for cells treated with 300 rads of X-irradiation (250 kV, 2 mm Al and 0.3 mm Cu, 100 rads/min), indicating the presence of a large number of strand breaks. When samples were treated with furocoumarins plus 365-nm radiation followed by exposure to 300 rads of X-rays, the retention on the filter increased. An increased retention of radioactive material (DNA) reflects the presence of double-stranded (cross-linked) DNA (Kohn et al., 1976). Pretreatment with 8-MOP and 4.2 kJm^{-2} of 365-nm radiation resulted in a higher retention than pretreatment with 8-MOP and 2.1 kJm^{-2} of 365-nm radiation, indicating that the induction of DNA interstrand cross-links is dose-dependent. These results are in accord with those of other authors (Cohen et al., 1981).

Interestingly, the retention on filter is very similar for samples treated with equimolar (5 µM) amounts of 5-MOP and 8-MOP (plus 2.1 kJm^{-2} of 365-nm radiation and 300 rads of X-rays in each case). This suggests that at the same dose of 365-nm radiation, approximately equal numbers of DNA cross-links are photoinduced by 5-MOP and 8-MOP. Taken together with the results obtained on the photobinding capacity of these furocoumarins (Fig. 4), this indicates that at a given dose of 365-nm radiation, 5-MOP produces more photoadducts than 8-MOP but the same number of DNA cross-links, and therefore, the ratio of mono- to diadducts induced must be higher for samples treated with 5-MOP than for 8-MOP-treated samples.

DISCUSSION AND CONCLUSION

The results show that the relationship between the genetic effects of photoreactive furocoumarins in eukaryotic cells and the number and type of photoadditions induced in DNA is rather complex. The new monofunctional furocoumarin, MePyPs, demonstrates an extremely high photoaffinity towards DNA, probably related to its very high DNA-complexing capacity (Blais et al., 1984). However, the 4',5'-monoadditions induced by MePyPs are clearly less genetically effective than the mixture of mono- and diadducts induced by the bifunctional furocoumarins, 8- and 5-MOP.

Furthermore, in contrast to the lesions induced by the latter compounds, the lesions induced by MePyPs show a relatively higher capacity for the induction of mitotic intragenic recombination (TRP^+ gene conversion) (Fig. 3B) than for the induction of mutations (ILV^+ reversions) (Fig. 3C). This suggests a certain genetic specificity of the monoadducts induced by MePyPs, which is likely to be related to the fact that these types of monoadducts appear to be difficult to repair in eukaryotic cell systems (Moustacchi et al., 1983; Papadopoulo et al., 1984b; Nocentini, 1984).

The mixture of monoadducts and diadducts induced by 8- or 5-MOP is genetically more active than the monoadducts induced by MePyPs; this suggests that the higher activity of the bifunctional furocoumarins is due to the production of DNA cross-links. Supporting evidence comes from experiments using 8-MOP and different excitation wavelengths. The lesions induced by 8-MOP at wavelengths above 380 nm – probably only 4′,5′-monoadducts – appear to be also less genetically active than the mixture of mono- and diadducts induced at 365-nm (Averbeck & Averbeck, 1984). As shown in re-irradiation experiments with 8-MOP, an increased proportion of DNA cross-links causes increased mutagenic and recombinogenic effects in yeast (Cassier et al., 1984).

The genetic effectiveness of bifunctional furocoumarins in eukaryotic cells appears to be governed not only by the ratio of mono- to diadducts induced but also by the overall photobinding capacity of the compounds. It has been shown that 5-MOP is more effective per unit dose than 8-MOP for the induction of lethal and nuclear genetic events in several eukaryotic cell systems (Averbeck, 1984) as well as for the induction of skin tumours in mice (Grekin & Epstein, 1981; Zajdela & Bisagni, 1981; Young et al., 1983). Our results suggest that the activity of 5-MOP is primarily due to its high capacity to produce photoadditions to DNA, with relatively few cross-links being induced per total number of lesions. This is in accord with results obtained on Chinese hamster ovary cells (Weniger, 1981).

These findings are significant in regard to the recent development of monofunctional, highly photoreactive, angular and linear furocoumarins for photochemotherapeutic use in the treatment of cutaneous skin diseases such as psoriasis and mycosis fungoides, because a lower genotoxic activity can be expected (Averbeck et al., 1984b; Dall'Acqua et al., 1983; Averbeck, 1984). Indeed, two newly synthesized pyridopsoralens, the pyrido(3,4-c)psoralen and MePyPs, have been shown to be highly therapeutically active in the treatment of psoriasis (Dubertret et al., 1984; Averbeck et al., 1983b) and slightly erythematogenic but less tumourigenic in mice (Dubertret et al., 1984; F. Zajdela, personal communication) than the known bifunctional compounds, 8- and 5-MOP, currently in photochemotherapeutic use. Such monofunctional compounds may represent a new generation of photochemotherapeutically active furocoumarins which will reduce the long-term genotoxic side-effects usually connected with photochemotherapy.

In conclusion, it appears that the genotoxic effects of furocoumarins greatly depend on the number of photoadditions induced in DNA, on the type of addition (i.e., monoaddition or DNA interstrand cross-link), as well as on the isomer of the addition formed. Obviously, more research is needed to define those conditions which favour the beneficial effects of furocoumarins but limit the formation of highly genotoxic lesions.

ACKNOWLEDGEMENTS

This work was supported by the research contracts PRC médicament 121001 and 832001 of the Institut National de la Santé et de la Recherche Médicale, the R.C.P. 080572 of the Centre National de la Recherche Scientifique and by the Commissariat à l'Energie Atomique, Saclay, France. We are indebted to Madame S. Averbeck for excellent technical assistance. We thank Drs. E. Moustacchi and R. Latarjet for their interest in our work.

REFERENCES

Averbeck, D. (1982) *Photobiology of furocoumarins.* In: Helene, C., Charlier, M., Montenay-Garestier, Th. & Laustriat, G., eds, *Trends in Photobiology,* New York, Plenum Press, pp. 295–308

Averbeck, D. (1984) Photochemistry and photobiology of psoralens. *Proc. Jpn. Soc. invest. Dermatol., 8,* 52–73

Averbeck, D. & Averbeck, S. (1984) Relationship between DNA damage and genetic effects induced by photoaddition of furocoumarins in diploid yeast (*Saccharomyces cerevisiae*). *Photochem. Photobiol., 39* (Suppl.), 37S

Averbeck, D., Quinto, I. & Papadopoulo, D. (1983a) *On the genotoxic activity of photoreactive psoralens of cosmetological interest.* In. N. Loprieno, ed., *Problemi di Tossicologia dei Prodotti Cosmetici,* Pisa, ETS, pp. 45–61

Averbeck, D., Dubertret, L., Bisagni, E., Moron, J., Papadopoulo, D., Nocentini, S., Blais, J. & F. Zajdela (1983b) Photobiological and phototherapeutic properties of new monofunctional pyridopsoralens. *J. invest. Dermatol., 80,* 306

Averbeck, D., Papadopoulo, D. & Quinto, I. (1984a) Mutagenic effects of psoralens in yeast and V79 Chinese hamster cells. *Natl Cancer Inst. Monogr., 66,* 127–136

Averbeck, D., Averbeck, S., Bisagni, E. & Moron, J. (1984b) Lethal and mutagenic effects photo-induced in haploid yeast (*Saccharomyces cerevisiae*) by two new monofunctional pyridopsoralens compared to 3-carbethoxypsoralen and 8-methoxypsoralen. *Mutat. Res., 148,* 47–57

Blais, J., Vigny, P., Moron, J. & Bisagni, E. (1984) Spectroscopic properties and photoreactivity with DNA of new monofunctional psoralens, the pyridopsoralens. *Photochem. Photobiol., 39,* 145–156

Bridges, B. & Strauss, G. (1980) Possible hazards of photochemotherapy for psoriasis. *Nature, 283,* 523–524

Cadet, J., Voituriez, L., Gaboriau, F., Vigny, P. & Della Negra, S. (1983) Characterization of photocycloaddition products from reaction between thymine and the monofunctional 3-carbethoxypsoralen. *Photochem. Photobiol., 37,* 363–371

Cadet, J., Voituriez, L., Ulrich, J., Joshi, P.C. & Wang, S.Y. (1984) Isolation and characterization of the monoheterodimers of 8-methoxypsoralen and thymidine involving the pyrone moiety. *Photobiochem. Photobiophys., 8,* 35–49

Cassier, C., Chanet, R. & Moustacchi, E. (1984) Mutagenic and recombinogenic effects of DNA cross-links induced in yeast by 8-methoxypsoralen photoaddition. *Photochem. Photobiol., 39,* 799–803

Cohen, L.F., Kraemer, K.H., Waters, H.L., Kohn, K.W. & Glaubiger, D.L. (1981) DNA cross-linking and cell survival in human lymphoid cells treated with 8-methoxypsoralen and long wavelength ultraviolet radiation. *Mutat. Res., 80,* 347–356

Cole, R.S. (1970) Light-induced cross-linking of DNA in the presence of a furocoumarin (psoralen). Studies with phage λ, *Escherichia coli,* and mouse leukemia cells. *Biochim. biophys. Acta, 217,* 30–39

Dall'Acqua, F., Marciani, S. & Rodighiero, G. (1970) Interstrand cross-linkages occurring in the photoreaction between psoralen and DNA. *FEBS Lett., 9,* 121–123

Dall'Acqua, F. (1977) *New chemical aspects of the photoreaction between psoralen and DNA.* In: Castellani, A., ed., *Research in Photobiology,* New York and London, Plenum Press, pp. 245–255

Dall'Acqua, F., Vedaldi, D., Bordin, F., Baccichetti, F., Carlassare, F., Tamaro, M., Rodighiero, P., Pastorini, G., Guiotto, A., Recchia, G. & Cristofolini, M. (1983) 4'-Methylangelicins: new potential agents for the photochemotherapy of psoriasis. *J. med. Chem., 26,* 870–876

Dubertret, L., Averbeck, D., Bisagni, E., Moron, J., Moustacchi, E., Billardon, C., Papadopoulo, D., Nocentini, S., Blais, J. & Zajdela F., (1984) Photobiological and phototherapeutic properties of new monofunctional pyridopsoralens. *Photochem. Photobiol., 39* (Suppl.), 60S

Grekin, D.A. & Epstein, J.H. (1981) Psoralens, UVA (PUVA) and photocarcinogenesis. *Photochem. Photobiol., 33,* 957–960

IARC (1980) *IARC Monographs on the Evaluation of the Carcinogenic Risk of Chemicals to Humans,* Vol. 24, *Some pharmaceutical drugs,* Lyon, International Agency for Research on Cancer, pp. 101–124

Kanne, D., Straub, K., Rapoport, H. & Hearst, J.E. (1982) Psoralen-deoxyribonucleic acid photoreaction. Characterization of the monoaddition products from 8-methoxypsoralen and 4,5',8-trimethylpsoralen. *Biochemistry, 21,* 861–871

Kohn, K.W., Erickson, L.C., Ewig, R.A.G. & Friedman, C.A. (1976) Fractionation of DNA from mammalian cells by alkaline elution. *Biochemistry, 15,* 4629–4637

Magaña-Schwencke, N., Averbeck, D., Pegas-Henriques, J.A. & Moustacchi, E. (1980) Absence of interchain bridges in DNA treated with 3-carbethoxypsoralen and irradiation at 356 nm (Fr.). *C.R. Acad. Sci. (Paris),* ser. D, **291,** 207–210

Moron, J., Nguyen, C.H. & Bisagni, E. (1983) Synthesis of 5-H-furo(3',2':6,7)-(1)-benzopyrano-(3,4-c)pyridin-5-ones and 8-H-pyrano-(3',2':5,6)benzofuro-(3,2-c)-pyridin-8-ones (pyridopsoralens). *J. chem. Soc. Perkin Trans. I,* 225–229

Moustacchi, E., Cassier, C., Chanet, R., Magaña-Schwencke, N., Saeki, T. & Henriques, J.A.P. (1983) *Biological role of photo-induced cross-links and monoadducts in yeast DNA: genetic control and steps involved in their repair.* In: Friedberg, E.C. & Bridges, B.A., eds, *Cellular Responses to DNA Damage,* New York, Alan Liss Inc., pp. 87–106

Musajo, L. & Rodighiero, G. (1972) *Photo-C_4 cycloaddition reactions to nucleic acids.* In: Gallo, U. & Santamaria, L., eds, *Research Progress in Organic, Biological and Medical Chemistry,* Vol. III, Amsterdam, North-Holland Publ. Co., pp. 155–181

Nocentini, S. (1984) DNA photobinding of 7-methyl-(3,4-c)pyridopsoralen and 8-

methoxypsoralen, removal of lesions and biological effects in cultured human and simian cells. *Photochem. Photobiol., **39*** (Suppl.), 55S

Papadopoulo, D., Sagliocco, F. & Averbeck, D. (1983) Mutagenic effects of 3-carbethoxypsoralen and 8-methoxypsoralen plus 365-nm irradiation in mammalian cells. *Mutat. Res., **124***, 287–297

Papadopoulo, D., Desforges, V., Sagliocco, F. & Averbeck, D. (1984a) Genotoxic effects of bifunctional and monofunctional furocoumarins in mammalian cell systems. *Mutat. Res., **130***, 193

Papadopoulo, D., Averbeck, D., Moron, J. & Bisagni, E. (1984b) Genotoxic effects photoinduced in mammalian cells *in vitro* by two new synthetized monofunctional furocoumarins. *Photochem. Photobiol., **39*** (Suppl.), 54S

Parsons, B.J. (1980) Yearly review: psoralen photochemistry. *Photochem. Photobiol., **32***, 813–821

Straub, K., Kanne, D., Hearst, J.E. & Rapoport, H. (1981) Isolation and characterization of pyrimidine-psoralen photoadducts from DNA. *J. Am. chem. Soc., **103***, 2347–2355

Vigny, P., Spiro, M., Gaboriau, F., Le Beyec, V., Della Negra, S., Cadet, J. & Voituriez, L. (1983) ^{252}Cf-Plasma desorption mass spectrometry of covalently bound nucleic acid adducts: psoralen-nucleosides photoadducts. *Int. J. Mass Spectrom. Ion Phys., **53***, 69–83

Weniger, P. (1981) A comparison of the photochemical actions of 5- and 8-methoxypsoralen on CHO cells. *Toxicology, **22***, 53–58

Young, A.R., Magnus, I.A., Davies, A.C. & Smith, N.P. (1983) A comparison of the phototumourigenic potential of 8-MOP and 5-MOP in hairless albino mice exposed to solar-simulated radiation. *Br. J. Dermatol., **108***, 507–518

Zajdela, F. & Bisagni (1981) 5-Methoxypsoralen, the melanogenic additive in sun-tan preparations in tumourigenic in mice exposed to 365 nm UV radiation. *Carcinogenesis, **2***, 121–127

Zimmermann, F.K., Kern, R. & Rasenberger, H. (1975) A yeast strain for simultaneous detection of induced mitotic crossing over, mitotic gene conversion and reverse mutations. *Mutat. Res., **28***, 381–388

INDUCTION OF SISTER CHROMATID EXCHANGE BY ETHYL CARBAMATE AND VINYL CARBAMATE

M.K. CONNER

Departments of Biostatistics and Industrial Environmental Health Sciences, Graduate School of Public Health, University of Pittsburgh, Pittsburgh, PA, USA

SUMMARY

High levels of sister chromatid exchanges (SCEs) can be induced in murine bone marrow, alveolar macrophages and regenerating liver cells by carcinogenic carbamate esters; however, the frequencies observed in the latter two tissues, which are also common tissues for carbamate-induced tumours, are relatively enhanced compared to the frequency in bone marrow. Relative to these tissues, peripheral blood lymphocytes, which do not divide during in-vivo exposure but which can be cultured *in vitro,* exhibit considerably lower levels of ethyl-carbamate-induced SCEs. Lymphocytes, however, are uniquely able to accumulate SCE-inducing lesions produced by multiple treatments with ethyl carbamate. In the present study, significantly elevated levels of SCEs were still apparent in lymphocytes of BDF_1 mice eight weeks after a series of 12 multiple injections (2.2 mmol/kg; 3 times weekly) of ethyl carbamate. Long-term persistence of genetic damage produced by ethyl carbamate in murine lymphocytes is in agreement with our previous findings of highly persistent SCE-inducing damage in murine bone marrow and alveolar macrophage cells. The highly persistent nature of ethyl-carbamate-induced DNA damage and/or its continued ability to induce SCEs is undoubtedly relevant to its tumourigenic activity.

INTRODUCTION

Carbamate ester induction of SCEs

Of all short-term bioassays, in-vivo sister chromatid exchange (SCE) is unique in providing definitive evidence of the genotoxic activity of ethyl carbamate. Many notable parallels between induction of SCE by carbamate ester and its tumourigenic activity have emerged from in-vivo SCE studies. At dose levels previously shown to produce pulmonary tumours in strain A mice (Shimkin *et al.,* 1969), ethyl carbamate

Fig. 1. Dose-response curves for carbamate ester induction of sister chromatid exchange (SCE) in murine alveolar macrophages *in vivo*

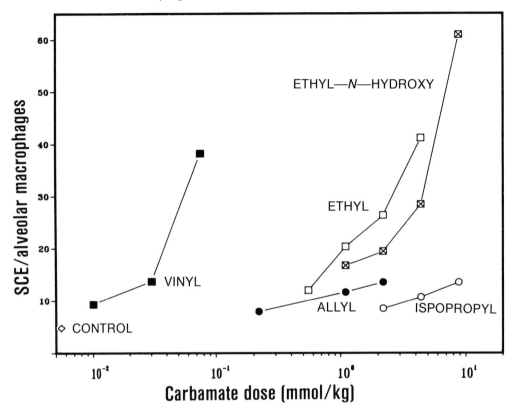

produced highly elevated SCE levels in murine bone marrow, alveolar macrophages (Cheng *et al.*, 1981a) and regenerating liver cells (Roberts & Allen, 1980; Cheng *et al.*, 1981a). Minor structural alterations in the ester moiety result in dramatic and remarkably concordant changes in carbamate-ester-induced lung adenoma incidence in strain A mice (Shimkin *et al.*, 1969; Dahl *et al.*, 1980) and SCE levels in BDF_1 mice (Cheng & Conner, 1982). In both assays, carbamate ester potencies decreased in the order: vinyl > ethyl > ethyl *N*-hydroxy > allyl > isopropyl, and the methyl ester was inactive. The same order of SCE-inducing activities for vinyl carbamate, ethyl carbamate and ethyl *N*-hydroxycarbamate in strain A mice has been confirmed by Allen *et al.* (1982).

The carbamate ester dose-response curves in murine alveolar macrophage cells, illustrated in Figure 1, support the α,β-dehydrogenation biotransformation pathway for ethyl carbamate proposed by Dahl *et al.* (1978). The unsaturated esters, vinyl carbamate and allyl carbamate, effectively induce SCEs at 30- and 10-fold lower doses, respectively, than their corresponding saturated counterparts, ethyl carbamate and isopropyl carbamate. The similarly shaped dose-response curves for vinyl carbamate, ethyl carbamate, and ethyl N-hydroxycarbamate are consistent with the idea that these esters share a common active metabolite. Likewise, parallel dose-response relationships are apparent for the three-carbon esters, allyl carbamate and isopropyl carbamate. However, their considerably lower effectiveness as inducers of SCE relative to the two-carbon esters, may be the consequence of greater steric effects encountered in the genotoxic reactions of their ultimate metabolites.

Tissue specific induction of SCEs by carbamate esters

In our studies of carbamate ester induction of SCEs, we found that at all effective doses ($\geq$ three doses per carbamate) of all active carbamates (five carbamates), each mouse ($\geq$ three hepatectomized and three intact mice/dose) exhibited higher, although not always statistically significant, SCE frequencies in alveolar macrophages and regenerating liver than in bone-marrow cells (Cheng *et al.*, 1981b; Cheng & Conner, 1982). Enhanced SCE responses in alveolar macrophages and regenerating liver parallel previous descriptions of lung and liver as highly susceptible target organs for ethyl-carbamate-induced tumours (IARC, 1974). Exceptional sensitivity of regenerating liver cells has also been observed by Roberts and Allen (1980).

Since they provide an indication of systemic genotoxicity, lymphocyte responses to ethyl carbamate were also examined (Goon & Conner, 1984). The number of induced (total minus baseline) SCEs was 2.5-fold lower in cultured peripheral blood lymphocytes than in bone-marrow and alveolar macrophage cells of mice injected with 3.3 mmol/kg ethyl carbamate. The time-related SCE responses of peripheral blood lymphocytes following acute injection of 3.3 mmol/kg are consistent with the estimated time of 4.3 h for complete clearance of this dose of ethyl carbamate in mice (O'Flaherty & Sichak, 1983). Multiple treatments (10 × 3.3 mmol/kg every other day) produced approximately two-fold higher levels of SCEs in peripheral blood lymphocytes than did a single acute injection (Goon & Conner, 1984). In contrast, bone-marrow and alveolar macrophage cells did not accumulate ethyl-carbamate-induced SCEs upon repeated exposure (Cheng *et al.*, 1981a; Goon & Conner, 1984). The lower acute response of peripheral blood lymphocytes may reflect the fact that they are in the G_o state (i.e., not in the cell-division cycle) during in-vivo exposure and/or that DNA repair occurs prior to their culture *in vitro*. In any case, the higher SCE response following serial treatments indicates that unrepaired genetic damage produced by ethyl carbamate can accumulate in G_o lymphocytes.

Long-term persistence of SCE-inducing damage in peripheral blood lymphocytes has been speculated to be indicative of an increased risk of neoplasia. The aim of the present study was to determine how long ethyl-carbamate-induced damage persists in murine lymphocytes following a treatment protocol known to produce lung adenomas in mice (Shimkin *et al.*, 1969).

MATERIALS AND METHODS

Male BDF_1 mice (3.5-4 months of age) were used in this study. The parental breeders (C57B1/6J and DBA/2J) were purchased from the Jackson Laboratory, Bar Harbor, ME, USA.

Table 1. Sister chromatid exchange (SCE) responses in Concanavalin A (Con A)- and lipopolysaccharide (LPS)-stimulated murine lymphocytes from mice treated with 12 multiple injections (2.2 mmol/kg each, three times weekly) of ethyl carbamate

Time after last injection[a]	Mean SCE response ± SD	
	Con A[b]	LPS[b]
3 h	25.0 ± 0.02 (3)	20.4 ± 0.6 (2)
24 h	21.9 ± 5.9 (2)	[c]
72 h	21.7 ± 2.4 (4)	[c]
1 week	22.6 ± 2.4 (3)	15.3 ± 2.6 (3)
4 week	16.5 ± 0.8 (3)	13.9 ± 0.5 (3)
8 week	18.2 ± 2.0 (3)	12.1 ± 2.8 (3)
Control	8.0 ± 1.8 (9)	8.3 ± 1.1 (5)

[a] At the indicated time, blood was drawn for SCE culture.
[b] The number in parenthesis is the number of mice used.
[c] Inadequate culture growth

BDF$_1$ mice were injected intraperitoneally with 12 multiple injections (2.2 mmol/kg each, three times/week) of ethyl carbamate (Aldrich Chemical Co., Milwaukee, WI, USA) in 0.9% sodium chloride. Groups of three mice each were bled through the orbital sinus at 3, 24, 72 h, 1, 4 and 8 weeks after the last injection. Lymphocytes were isolated and cultured in RPMI 1640 at a bromodeoxyuridine (BUDR) concentration of 4 µM, as described in detail by Goon and Conner (1984). B- and T-lymphocyte cultures were established for each mouse. B-cells were stimulated with lipopolysaccharide (*Escherichia coli* serotype 0111:B4; Sigma Chemical Co.; 60 µM, final concentration) and T-cells with Concanavalin A (type IV; Sigma Chemical Co.; 6 ug/ml, final concentration) in the presence of 25 µM mercaptoethanol. Cells were harvested at 48 h and stained by a modified FPG (fluorescence plus Giemsa) technique, described by Conner *et al.* (1978). SCEs were scored in 20 cells from each culture.

RESULTS

The SCE responses in B- and T-lymphocytes of mice at 3 h through 8 weeks after multiple ethyl carbamate treatments are shown in Table 1. SCE responses decreased slowly with time and remained elevated over control levels for at least 8 weeks. This gradual decline in SCE response suggests a slow rate of repair of genetic damage and turnover of the lymphocyte population. In any case, under conditions which are effective in eliciting lung tumours in mice, ethyl carbamate produces long-lived genetic damage in murine peripheral blood lymphocytes.

DISCUSSION

The ability of ethyl carbamate to produce persistent genetic damage in G_o cells *in vivo* is undoubtedly a major factor in its tumourigenic activity. It is of interest to note that we have also documented highly persistent SCE-inducing damage in bone-marrow and alveolar macrophage cells of mice following acute treatment with 3.3 mmol/kg ethyl carbamate (Conner & Cheng, 1983). In contrast to G_o lymphocytes, bone-marrow and alveolar macrophage cells are labelled with BUDR *in vivo*. Since bone marrow and alveolar macrophages continuously divide *in vivo*, their induced SCE frequencies decrease with time between treatment and BUDR incorporation, as a consequence of DNA repair and dilution of genetic damage with cell division. In the absence of DNA repair, induced (total minus baseline) SCEs are expected to decrease in number by one-half with each cell division. Our observation that the number of ethyl-carbamate-induced SCEs decreases by one-half with each of two cell divisions prior to BUDR infusion indicates lack of repair of the SCE-inducing lesions over at least three cell cycles (Conner & Cheng, 1983). In the same study, persistence of ethyl-carbamate-induced damage was confirmed by data from third-division cells. Following BUDR labelling for three cell cycles, third-division cells can be identified by their distinctive staining pattern. Two types of SCE – nonreciprocal and reciprocal exchanges – can be recognized in these cells. The former are asymmetrical exchanges, visible in only one chromatid, and reflect SCE events that originate during the first and second cycles of BUDR incorportion. Reciprocal SCEs are symmetrical exchanges between sister chromatids and are assumed to occur at the third BUDR cycle as a result of the continued presence of the active agent and/or persistent genetic damage inducing an exchange for the first time at the third cycle. Since clearance of the administered dose of 3.3 mmol/kg ethyl carbamate should be complete within 4.3 h (O'Flaherty & Sichak, 1983), the continued presence of the active agent beyond one cell cycle (approximately 8 h) is highly unlikely. Thus, our observation of highly significant (> four times baseline) elevations in the number of reciprocal SCEs in third-division cells substantiates the persistence of genetic damage in bone marrow and alveolar macrophages for at least three cell cycles.

The longevity of SCE-inducing genetic damage produced by ethyl carbamate in multiple tissues of BDF_1 mice is remarkable in view of the fact that both parental strains exhibit low susceptibility to carbamate-induced pulmonary adenomas. Genotype is a primary determinant in strain susceptibility to lung adenomas. In contrast, strain specificities are not generally apparent for ethyl carbamate induction of SCEs. In spite of their divergent susceptibilities to lung adenomas, similar SCE responses were observed in bone-marrow cells of C57Bl/6J and A/J mice treated with ethyl carbamate or vinyl carbamate (Sharief *et al.*, 1984). Likewise, bone-marrow cells of AKR, C57Bl/6J, C3HJ and BALB/c mice demonstrated comparable SCE responses to ethyl carbamate, whereas DBA mice exhibited a somewhat higher response (Dragani *et al.*, 1983). Regardless of strain susceptibility to lung adenomas, high levels of genetic damage expressed as SCEs are produced by ethyl carbamate.

Characterization of the nature of the persistent SCE-inducing damage produced by ethyl carbamate is of great importance in understanding its mechanism of action. In mice treated with ethyl carbamate, the primary modification of hepatic DNA was found to be 7-(2-oxoethyl)guanine (Scherer et al., 1980 and these proceedings[1]); in RNA, 3,N^4-ethenocytosine and 1,N^6-ethenoadenine have been identified (Ribovich et al., 1982). All three adducts are potentially mutagenic. The presence of etheno adducts in synthetic nucleic acids has been associated with misincorporation of bases during DNA synthesis (Barbin et al., 1981; Hall et al., 1981) and RNA transcription (Spengler & Singer, 1981 and Singer & Spengler, these proceedings[2]) in vitro, whereas the mutagenic effect of 7-(2-oxoethyl)guanine may be the result of depurination. The latter is not likely to be the source of the persistent SCEs observed in our studies, since apurinic sites do not appear to be associated with SCE induction (Jostes, 1981). In view of the relative persistence of ethenocytosine compared with ethenoadenine in rat liver RNA (Laib & Bolt, 1978), the etheno adduct(s) are more likely candidates for the persistent SCE-inducing lesions. However, their presence has not been detected in DNA. Characterization of nucleic acid adducts of ethyl carbamate has been confined to rodent liver, a G_o cell population. Formation of etheno adducts is limited to single-stranded nucleic acids, such as RNA and partially denatured DNA; however, formation of these adducts may be facilitated during DNA replication in dividing cells required for SCE analysis. Such speculation is consistent with our observation of lower SCE levels in peripheral blood lymphocytes compared to dividing cells exposed in vivo. Further studies are required to identify specific SCE-inducing lesions produced by ethyl carbamate.

Regardless of the nature of the genotoxic lesions induced by ethyl carbamate, not only their continued presence but also their ability to induce SCEs may be critical to tumourigenesis. The exact mechanism of SCE formation is not known; however, SCE is a means whereby DNA damage may be by-passed during DNA synthesis. If SCE is simply a common manifestation of general cellular recombinogenic activity, elevated levels of SCEs may be accompanied by enhanced homologous recombination wherein heterozygous recessive genotypes may be converted to homozygous clones in which the recessive phenotype is expressed (Kinsella & Radman, 1978). Alternatively, unequal sister chromatid crossover may lead to gene amplification or altered gene expression. In view of the concordance between the SCE- and tumour-inducing activities presented here, the potential role of the SCE mechanism in neoplastic transformation by carbamate esters cannot be ignored.

ACKNOWLEDGEMENTS

This study was supported in part by National Institute of Environmental Health Sciences Grant 1 R01 CA 33869-01 and March of Dimes Grant 15–34.

---

[1] See p. 109.
[2] See p. 359.

REFERENCES

Allen, J.W., Langenbach, R., Nesnow, S., Sasseville, K., Leavitt, S., Campbell, J., Brock, K. & Sharief, Y. (1982) Comparative genotoxicity studies of ethyl carbamate and related chemicals: further support for vinyl carbamate as a proximate carcinogenic metabolite. *Carcinogenesis, 3,* 1437–1441

Barbin, A., Bartsch, J., Leconte, P. & Radman, M. (1981) Studies on the miscoding properties of 1,N^6-ethenoadenine and 3,N^4-ethenocytosine, DNA reaction products of vinyl chloride metabolites, during in-vitro DNA synthesis. *Nucleic Acids Res., 9,* 375–387

Cheng, M., Conner, M.K. & Alarie, Y. (1981a) Multicellular in-vivo sister chromatid exchanges induced by urethane. *Mutat. Res., 88,* 223–231

Cheng, M., Conner, M.K. & Alarie, Y. (1981b) Potency of some carbamates as multiple tissue sister chromatid exchange inducers and comparison with known carcinogenic activities. *Cancer Res., 41,* 4489–4492

Cheng, M. & Conner, M.K. (1982) Comparison of sister chromatid exchange induction and known carcinogenic activities of vinyl and allyl carbamates. *Cancer Res., 42,* 2165–2167

Conner, M.K., Boggs, S.S. & Turner, J.H. (1978) Comparison of in-vivo BrdU labelling methods and spontaneous sister chromatid exchange frequencies in regenerating liver and bone marrow cells. *Chromosoma (Berlin), 68,* 303–311

Conner, M.K. & Cheng, M. (1983) Persistence of ethyl carbamate-induced DNA damage *in vivo* as indicated by sister chromatid exchange analysis. *Cancer Res., 43,* 965–971

Dragani, T.A., Sozzi, G. & Della Porta, G. (1983) Comparison of urethane-induced sister chromatid exchanges in various murine strains and the effect of enzyme inducers. *Mutat. Res., 121,* 233–239

Dahl, G.A., Miller, J.A. & Miller, E.C. (1978) Vinyl carbamate as a promutagen and a more carcinogenic analog of ethyl carbamate. *Cancer Res., 38,* 3793–3804

Dahl, G.A. Miller, E.C. & Miller, J.A. (1980) Comparative carcinogenicities and mutagenicities of vinyl carbamate, ethyl carbamate, and ethyl N-hydroxycarbamate. *Cancer Res., 40,* 1194–1203

Goon, D. & Conner, M.K. (1984) Simultaneous assessment of ethyl-carbamate-induced SCEs in murine lymphocytes, bone marrow, and alveolar macrophage cells. *Carcinogenesis, 5,* 399–402

Hall, J.A., Saffhill, R., Green, T. & Hathway, D.E. (1981) The induction of errors during in-vitro DNA synthesis following chloroacetaldehyde treatment of poly (dA-dT) and poly (dC-dG) templates. *Carcinogenesis, 2,* 141–146

IARC (1974) *IARC Monographs on the Evaluation of the Carcinogenic Risk of Chemicals to Humans,* Vol. 4, *Some antithyroid and related substances, nitrofurans, and industrial chemicals,* Lyon, International Agency for Research on Cancer, pp. 111–140

Jostes, R.F. (1981) Sister chromatid exchange but not mutation decrease with time in arrested Chinese hamster ovary cells after treatment with ethylnitrosourea. *Mutat. Res., 91,* 371–375

Kinsella, A.R. & Radman, M. (1978) Tumor promoter induces sister chromatid exchanges: relevance to mechanisms of carcinogenesis. *Proc. natl Acad. Sci. USA, 75*, 6149–6153

Laib, R.J. & Bolt, H.M. (1978) Formation of 3,N^4-ethenocytosine moieties in RNA by vinyl chloride metabolites *in vitro* and *in vivo*. *Arch. Toxicol., 39*, 235–240

O'Flaherty, E.J. & Sichak, S.P. (1983) The kinetics of urethane elimination in the mouse. *Toxicol. appl. Pharmacol., 68*, 354–358

Ribovich, M.L., Miller, J.A. & Miller, E.C. (1982) Labeled 1,N^6-ethenoadenosine and 3,N^4-ethenocytidine in hepatic RNA of mice given [ethyl-1,2-^{3}H]- or [ethyl-1-^{14}C]-ethyl carbamate. *Carcinogenesis, 3*, 539–546

Roberts, G.T. & Allen, J.W. (1980) Tissue-specific induction of sister chromatid exchanges by ethyl carbamate in mice. *Environ. Mutagen., 2*, 17–26

Scherer, E., Steward, A.P. & Emmelot, P. (1980) *Formation of precancerous islands in rat liver and modification of DNA by ethyl carbamate: implications for its metabolism.* In: Holmstedt, B., Lauwerys, R., Mercier, M. & Roberfroid, M., eds, *Mechanisms of Toxicity and Hazard Evaluation,* Amsterdam, Elsevier/North Holland Biomedical Press, pp. 249–254

Sharief, Y., Campbell, J., Leavitt, S., Langenbach, R. & Allen, J.W. (1984) Rodent species and strain specificities for sister chromatid exchange induction and gene mutagenesis effects from ethyl carbamate, ethyl *N*-hydroxycarbamate, and vinyl carbamate. *Mutat. Res., 126*, 159–167

Shimkin, M.B., Weider, R., McDonough, M., Fishbein, L. & Swern, D. (1969) Living tumor response in strain A mice as a quantitative bioassay of carcinogenic activity of some carbamates and aziridines. *Cancer Res., 29*, 2184–2190

Spengler, S.J. & Singer, B. (1981) Transcription errors and ambiguity resulting from the presence of 1,N^6-ethenoadenosine or 3,N^4-ethenocytidine in ribonucleotides. *Nucleic Acids Res., 9*, 356–373

ADAPTIVE RESPONSE TO ALKYLATING AGENTS IN MAMMALIAN CELLS

F. LAVAL

Groupe "Radiochimie de l'ADN" (U247 INSERM), Institut Gustave-Roussy, Villejuif, France

INTRODUCTION

Escherichia coli, when exposed to low concentrations of methylating or ethylating agents such as N-methyl-N'-nitro-N-nitrosoguanidine (MNNG) or N-methyl-N-nitrosourea (MNU), becomes resistant to the lethal and mutagenic effects of high doses of these compounds (Samson & Cairns, 1977). Such a response is termed the adaptive response.

Two mechanisms under the control of the *ada* gene are induced as part of the adaptive response in *Escherichia coli*. Decrease in cell killing is due to the induction of a DNA-glycosylase (3-methyladenine glycosylase II), a product of the *alk* gene, which removes 3-methyladenine, 7-methylguanine and 3-methylguanine from methylated DNA. This enzyme is present at about a 20-fold higher level in adapted cells. The adaptive response to mutagenesis involves induction of O^6-alkylguanine-DNA alkyltransferase. Twenty to sixty molecules per cell are present in unadapted *Escherichia coli*, whereas adapted cells or *adc* mutants have about 100 to 200-fold higher levels of this enzyme. These results have been reviewed by Lindahl (1982) and by Walker (1984).

An O^6-alkylguanine-DNA alkyltransferase has been described in mouse (Bogden *et al.*, 1981), rat (Mehta *et al.*, 1981) and human (Pegg *et al.*, 1982) liver cells and in human fetal tissues (Krokan *et al.*, 1983). In addition, 7-alkylguanine, 3-alkylguanine and 3-alkyladenine residues in the DNA of mammalian cells are enzymatically excised by a DNA glycosylase, creating an apurinic site (Cathcart & Goldthwait, 1981). Therefore, mammalian cells contain enzymes that could bring about the response observed in adapted *Escherichia coli* (see also Laval and Boiteux, these proceedings[1]).

---

[1] See p. 381.

ADAPTATION IN ANIMALS

Numerous experiments suggest the existence of an adaptive response in animals. Pretreatment of rats with N-nitrosodimethylamine (NDMA) does not affect the formation or the rate of loss of 3-methyladenine or 7-methylguanine, but does lead to a decrease in the initial amount of O^6-methylguanine in hepatic DNA when the animals are treated with a high dose of NDMA (reviewed by Montesano, 1981). Such pretreatment causes an increase in the O^6-methylguanine-DNA methyltransferase activity of 150–200% over that of control values in rat hepatocytes (Lindamood *et al.*, 1984).

Cross-reactivity between different agents was observed in experiments involving pretreatments with acetylaminofluorene (Buckley *et al.*, 1979) or aflatoxin B1 (Chu *et al.*, 1981); such pretreatment stimulated the repair of O^6-methylguanine in the liver of rats exposed to high doses of NDMA. However, chronic pretreatment of rats with MNU was found to be ineffective (Margison, 1981) and negative results with respect to the adaptive response were obtained with mice (Lindamood *et al.*, 1984) and gerbils (Bamborschke *et al.*, 1983).

ADAPTATION IN MAMMALIAN CELLS *IN VITRO*

An increase in the amount of O^6-alkyltransferase after multiple exposures to MNNG occurs in HeLa cells (Waldstein *et al.*, 1982).

Evidence for an adaptive response to the killing effect of MNNG or MNU has been reported for Chinese hamster ovary (CHO) cells and human skin fibroblasts (Samson & Schwartz, 1980) and for Chinese hamster V79 cells (Durrant *et al.*, 1981; Kaina, 1982). However, varying results have been obtained concerning the adaptive response of mammalian cells to mutagenicity. Pretreatment of CHO cells with MNNG or MNU was not found to modify the mutation frequency in CHO cells (Schwartz & Samson, 1983). It should be noted that CHO cells are unable to remove O^6-methylguanine from their DNA (Goth-Goldstein, 1980). In V79 cells, this absence of adaptation to mutagenicity was observed by Durrant *et al.* (1981), whereas Kaina (1983) found that a nontoxic dose of MNNG of MNU rendered the cells resistant to the mutagenic effect of these agents.

We have studied the adaptive response using a rat hepatoma cell line (H$_4$ cells). Cells were pretreated for 48 h with MNNG (0.27 µM), MNU (4.85 µM) or methyl methanesulfonate (MMS, 25 µM) (Laval & Laval, 1984). The pretreatment induced an adaptive response to cell killing, and cross-reactivity was observed between the various mutagens tested (Table 1). The results given in Table 2 show that a pretreatment with MNNG, but not with MMS, markedly reduced the mutagenic effect of these compounds. Therefore, the cross-reactivity observed for cell survival does not exist for mutagenicity, suggesting that the adaptive responses to killing and mutagenicity are at least two independent processes.

The activities of 7-methylguanine glycosylase and 3-methyladenine glycosylase in H$_4$ cells pretreated with MNNG were similar to their activities in control cells. However, the amount of O^6-methylguanine-DNA methyltransferase was about 4-fold

Table 1. Survival of rat hepatoma (H_4) cells challenged with alkylating agents after various pretreatments[a]

Challenge drug	Adaptive drug	D_0 (µM)[b]
MMS	O	1100
	MMS	2000
	MNNG	1800
MNNG	O	8.5
	MNNG	16.6
	MMS	13.7
MNU	O	78
	MMU	139
	MNNG	125

[a] Cells were pretreated with methyl methanesulfonate (MMS), N-methyl-N'-nitro-N-nitrosoguanidine (MNNG) or N-methyl-N-nitrosourea (MNU) for 48 h then challenged for 1 h with higher concentrations of these compounds.
[b] D_0 values are the drug concentrations which reduce the survival by 1/e in the exponential fraction of the survival curve.

Table 2. Mutation frequency in adapted cells[a]

Challenge drug		Number of mutants / 10^5 survivors		
		No pretreatment	MNNG pretreatment	MMS pretreatment
MNNG	0	1– 2	1–2	1– 2
	5 µM	20–22	2–3	20–22
	10 µM	41–44	4–6	40–45
	15 µM	58–61	6–7	59–61
MMS	1 mM	8–11	3–5	8–11
	2 mM	16–19	5–8	16–19

[a] Cells were pretreated or not for 48 h with either N-methyl-N'-nitro-N-nitrosoguanidine (MNNG, 0.27 µM) or methyl methanesulfonate (MMS, 25 µM), then challenged with these compounds. Cells were subcultured in selective medium containing dialysed fetal calf serum (10%) and 6-thioguanine (2.5 µg/ml).

higher in cells pretreated with MNNG than in control cells. This correlates with the faster rate of O^6-methylguanine removal measured in cells pretreated, and then challenged, with MNNG as compared to the rate measured in control cultures (Laval & Laval, 1984).

In order to explain the different responses observed, with respect to mutagenicity, in cells pretreated with MMS and MNNG, we have measured the amount of O^6-methylguanine induced in the cells by each of the two agents. To induce the same amount of O^6-methylguanine residues in H_4 cells, the concentration of MMS required is 588-fold higher than that of MNNG. This means that if a MNNG dose of 0.27 µM induces the adaptive response, the MMS dose required would be 158 µM. Chronic treatment with such a MMS concentration is toxic for the cells, and therefore pretreatment with the MMS concentration which adapts the cells for survival, but not for mutagenesis, induces much less O^6-methylguanine than does the pretreatment with MNNG.

These differences in the amount of O^6-methylguanine induced during the pretreatment are a probable explanation for the observed variations in the adaptive response to mutagenesis. It should be recalled that the extent of alkylation varies with the nature of the drug (Singer & Grunberger, 1983), the drug concentration (Harris et al., 1981) and the cell metabolism (Lawley & Thatcher, 1970).

ACKNOWLEDGEMENT

This research was supported by grants from Centre National de la Recherche Scientifique et Institut National de la Santé et de la Recherche Médicale, and a Research Contract from "Association pour la Recherche sur le Cancer" (Villejuif).

REFERENCES

Bamborschke, S., O'Connor, P.J., Margison, G.P., Kleihues, P. & Maru, G.B. (1983) DNA methylation by dimethylnitrosamine in the mongolian Gerbil (*Meriones unguiculatus*): indications of a deficient, noninducible hepatic repair system for O^6-methylguanine. *Cancer Res.*, **43**, 1306–1311

Bogden, J.M., Eastman, A. & Bresnick, E. (1981) A system in mouse liver for the repair of O^6-methylguanine lesions in methylated DNA. *Nucleic Acids Res.*, **9**, 3089–3103

Buckley, J.D., O'Connor, P.J. & Craig, A.W. (1979) Pretreatment with acetylaminofluorene enhances the repair of O^6-methylguanine in DNA. *Nature*, **281**, 403–404

Catchcart, R. & Goldthwait, D.A. (1981) Enzymatic excision of 3-methyladenine and 7-methylguanine by a rat liver nuclear fraction. *Biochemistry*, **20**, 273–280

Chu, Y.H., Craig, A.W. & O'Connor, P.J. (1981) Repair of O^6-methylguanine in rat liver DNA is enhanced by pretreatment with single or multiple doses of aflatoxin B1. *Br. J. Cancer*, **43**, 850–855

Durrant, L.G., Margison, G.P. & Boyle, J.M. (1981) Pretreatment of Chinese hamster V79 cells with MNU increases survival without affecting DNA repair or mutagenicity. *Carcinogenesis*, **1**, 55–60

Goth-Goldstein, R. (1980) Inability of Chinese hamster ovary cells to excise O^6-alkylguanine. *Cancer Res.*, **40**, 2623–2624

Harris, G., Lawley, P.D. & Olsen, I. (1981) Mode of action of methylating carcinogens: comparative studies of murine and human cells. *Carcinogenesis*, **2**, 403–411

Kaina, B. (1982) Enhanced survival and reduced mutation and aberration frequencies induced in V79 Chinese hamster cells pre-exposed to low levels of methylating agents. *Mutat. Res., 93,* 195–211

Kaina, B. (1983) Studies on adaptation of V79 Chinese hamster cells to low doses of methylating agents. *Carcinogenesis, 1,* 1437–1443

Krokan, H., Haugen, A., Myrnes, B. & Guddal, P.H. (1983) Repair of premutagenic DNA lesions in human fetal tissues: evidence for low levels of O^6-methylguanine-DNA methyltransferase and uracil-DNA glycosylase activity in some tissues. *Carcinogenesis, 4,* 1553–1564

Laval, F. & Laval, J. (1984) Adaptive response in mammalian cells: crossreactivity of different pretreatments on cytotoxicity as contrasted to mutagenicity. *Proc. natl Acad. Sci. USA, 81,* 1062–1066

Lawley, P.D. & Thatcher, C.J. (1970) Methylation of deoxyribonucleic acid in cultured mammalian cells by N-methyl-N'-nitro-N-nitrosoguanidine. *Biochem. J., 116,* 693–707

Lindahl, T. (1982) DNA repair enzymes. *Ann. Rev. Biochem., 51,* 61–87

Lindamood, C., Bedell, M.A., Billings, K.C., Dyroff, M.C. & Swenberg, J.A. (1984) Dose response for DNA alkylation, (^{3}H)-thymidine uptake into DNA, and O^6-methylguanine-DNA methyltransferase activity in hepatocytes of rats and mice continuously exposed to dimethylnitrosamine. *Cancer Res., 44,* 196–200

Margison, G.P. (1981) Effect of pretreatment of rats with N-methyl-N-nitrosourea on the repair of O^6-methylguanine in liver DNA. *Carcinogenesis, 2,* 431–434

Mehta, J.A., Ludlum, D.B., Renard, A. & Verly, W.G. (1981) Repair of O^6-ethylguanine in DNA by a chromatin fraction from rat liver: transfer of the ethyl group to an acceptor protein. *Proc. natl Acad. Sci. USA., 78,* 6766–6770

Montesano, R. (1981) Alkylation of DNA and tissue specificity in nitrosamine carcinogenesis. *J. Cell. Biochem., 17,* 259–273

Pegg, A.E., Roberfroid, M., von Bahr, C., Foote, S., Mitra, S., Bresil, H., Likhachev, A. & Montesano, R. (1982) Removal of O^6-methylguanine from DNA by human liver fractions. *Proc. natl Acad. Sci. USA., 79,* 5162–5165

Samson, L. & Cairns, J. (1977) A new pathway for DNA repair in *E. coli. Nature, 267,* 281–282

Samson, L. & Schwartz, J.L. (1980) Evidence for an adaptive DNA repair pathway in CHO and human skin fibroblast cell lines. *Nature, 287,* 861–863

Schwartz, J.L. & Samson, L. (1983) Mutation induction in Chinese hamster ovary cells after chronic pretreatment with MNNG. *Mutat. Res., 119,* 393–397

Singer, B. & Grunberger, D. (1983) *Molecular Biology of Mutagens and Carcinogens,* New York, Plenum Press

Waldstein, E.A., Cao, E.H. & Setlow, R.B. (1982) Adaptive resynthesis of O^6-methylguanine-accepting protein can explain the differences between mammalian cells proficient and deficient in methyl excision repair. *Proc. natl Acad. Sci. USA, 79,* 5117–5121

Walker, G.C. (1984) Mutagenesis and inducible responses to DNA damage in *Escherichia coli. Microbiol. Reviews, 48,* 60–93

ASPECTS CONCERNING THE STUDY OF THE QUANTITATIVE CARCINOGENIC EFFECTS OF CHEMOTHERAPEUTIC AGENTS IN MAN

J.M. KALDOR & N.E. DAY

Unit of Biostatistics and Field Studies, Division of Epidemiology and Biostatistics, International Agency for Research on Cancer, Lyon, France

SUMMARY

It is well-known that second primary tumours occur at elevated levels in patients undergoing cancer treatment with cytotoxic drugs. While this is clearly an undesirable outcome, it represents one of the few situations where the relationship between dose and effect of a carcinogen can be studied in humans. Current epidemiological methods for studying this relationship are described, as well as some of the issues involved in the quantification of risk from human data. Finally, the possibility of measuring biochemical markers and other parameters of DNA damage in patients undergoing treatment with antineoplastic agents is discussed.

INTRODUCTION

The public health importance of most of the chemicals known to form cyclic adducts is not yet established, and even those chemicals which have been identified as human carcinogens can only be held responsible for a relatively small fraction of cancer cases. For example, in an extensive investigation, Falk *et al.* (1981) identified a total of 168 cases of angiosarcoma occurring in the US between 1964 and 1974, of which only 12 could be associated with exposure to vinyl chloride.

However, one subgroup of chemicals, the antineoplastic agents, has taken on a special importance in environmental carcinogenesis research, because of the potential for accurately measuring the level to which humans are exposed. For most other human carcinogens, the lack of reliable dose measurements has frustrated attempts to study the quantitative relationship between exposure and disease and to compare the intensity of the carcinogenic activity in humans and animal species.

The IARC Monographs programme has considered a number of the chemotherapeutic agents, and Table 1 summarizes the conclusions of the various

Table 1. Antineoplastic drugs evaluated in IARC Monographs

	Past or present use in treatment of			Evidence for carcinogenicity in		IARC[c] Vol.	Cyclic[d] adducts
	HD[a]	OC[a]	TC[a]	Humans[b]	Animals[b]		
Actinomycin D			+	I	L	10	
Adriamycin	+	+	+	I	S	10	
Azathioprine	+			S	L	26	
N,N-bis (2-chloroethyl)-2-naphthylamine (Chlornaphazine)	+			S	L	4	
BCNU (bis-chloroethylnitrosourea)	+			I	S	26	AF
Bleomycin	+	+	+	I	I	26	
1,4-Butanediol dimethanesulphonate (Busulfan)		+		S	L	4	AF
Chlorambucil	+	+	+	S	S	9,26	AF
CCNU (1-(2-chloroethyl)-3-cyclohexyl-1-nitrosourea)	+	+		I	S	26	AF
Cisplatin	+	+	+	I	L	26	
Cyclophosphamide	+	+	+	S	S	26	
Dacarbazine	+			I	S	26	
5-Fluorouracil		+	+	I	I	26	
Isophosphamide	+	+			L	26	AS
Melphalan		+	+	S	S	9	AS
6-Mercaptopurine			+	I	I	9	AS
Methotrexate	+	+	+	I	I	26	
Mitomycin C	+			I	S	10	
Nitrogen mustard	+	+	+	I	S	9	
Procarbazine HCL	+			I	S	26	
Treosulphan		+		S		26	AS
Trichlorotriethylamine HCl	+				I	9	
Thio-TEPA (Tris(1-aziridinyl)-phosphine sulphide)	+	+		I	S	9	AS
2,4,6-Tris-(1-aziridinyl)-s-triazine	+	+			S	9	AS
Uracil mustard	+	+		I	S	9	AS
Vinblastine sulphate	+		+	I	I	26	
Vincristine sulphate	+		+	I	I	26	

[a] HD, Hodgkin's disease; OC, ovarian cancer; TC, testicular cancer
[b] S, sufficient; L, limited; I, inadequate
[c] See also Supplement 4 to the IARC Monographs (IARC, 1982)
[d] AF, cyclic nuclear acid base adducts formed; AS, cyclic adduct formation suspected

working groups with regard to the carcinogenicity of these agents in humans and animals (IARC, 1974, 1975, 1976, 1981, 1982). The table also indicates those chemicals which have been shown to produce cyclic nucleic acid base adducts, and those which are suspected of doing so on the basis of their chemical structure (H. Bartsch & S.J. Rinkus, personal communication). Further details on the carcinogenicity of some of the compounds are given by Vainio and Saracci (these proceedings[1]).

EPIDEMIOLOGICAL METHODS FOR STUDYING THE CARCINOGENICITY OF CHEMOTHERAPEUTIC AGENTS IN HUMANS

The carcinogenicity of antineoplastic agents has generally been established from a combination of multiple case reports and observations made on patients enrolled in clinical trials of chemotherapeutic treatments. More recently, epidemiological methods of study have been pursued. The United States National Cancer Institute has carried out cohort studies of leukaemia among those recorded with first primary tumours in the cancer registries of the Surveillance, Epidemiology and End Results (SEER) programme, by comparing the number of second primary tumours observed with the number expected on the basis of population age-, sex- and calendar-year-specific rates (Curtis et al., 1984). This methodology is very similar to that used in the first stage of the IARC study of second primary tumours following radiotherapy for cervical cancer (Day & Boice, 1984). The registry-based cohort study offers the possibility of accumulating information on a large number of individuals with first primary tumours, but does suffer from some drawbacks. The main problem is that cancer registries do not generally record detailed information on the treatment of the cancers recorded, so that at best, only a classification into broad groups such as "chemotherapy used" or "radiotherapy and chemotherapy used" can be made. Another criticism often levelled at such studies is that the general population is not a valid comparison group, since individuals who have suffered an initial primary tumour may be at increased risk of being recorded as having a subsequent one, either due to misclassification of metastases or to some true biological phenomenon, such as increased sensitivity to tumour induction.

Case-control studies are also being used to study the long-term effects of chemotherapy. Cases, defined as individuals recorded with two primary tumours separated by a specified time period (usually at least one year), are matched with controls who developed the first tumour but survived as long as the corresponding case without developing the second tumour. The hospital records of the two groups of individuals are then abstracted, and the type and amount of chemotherapy compared between the groups. Using this approach, the comparison group consists of other cancer patients, and it should also be possible to study the effect of specific agents and dose levels on the risk of developing second primary tumours. A potential disadvantage of the case-control study is that only, unless the full cohort size is known, relative, rather than absolute, risk can be measured (Breslow & Day, 1980), so that it is not possible to construct incidence curves or calculate the absolute probability of second tumour onset.

---

[1] See p. 15.

QUANTIFICATION OF THE CARCINOGENIC EFFECT OF CHEMOTHERAPEUTIC AGENTS

Although the potential exists for quantitatively studying the relationship between dose of chemotherapeutic agent administered to patients and the risk of subsequent primary tumours, relatively little work has actually been done in this area. Many of the early studies reported detailed information on the doses administered to those in whom the primary tumours were observed, but not those administered to other comparable patients. Schmähl *et al.* (1982) have compiled a list of all case reports of second tumours believed to be iatrogenically associated with the use of antineoplastic drugs, in which total doses of the drugs were reported. However, in the absence of corresponding information from appropriate control series, no quantitative conclusions can be drawn from this otherwise exhaustive compilation. Greene (1984) has recently reported on ongoing work at the United States National Cancer Institute, in which broad dose groupings are used to investigate radiotherapy/chemotherapy interactions in leukaemogenesis, and this work represents one of the first attempts at quantitative analysis of risk in patients treated with chemotherapy. With a view to quantification of risk, we recently surveyed the literature regarding the development of acute leukaemia in cancer patients treated with melphalan (IARC, 1983). The four most useful studies are summarized in Table 2, and Figure 1 presents the cumulative probability of developing acute leukaemia as a function of time, estimated from one of the studies. Although simple quantitative statements can be made from these data, there is much scope for improvement in their presentation if the results are to be used for more sophisticated statistical analyses. The following points would be of specific interest in such analyses:
(1) the identification of agents or combination therapies associated with increased risk of second primary tumours;
(2) the establishment of dose-response relationships between dose level and risk for these agents;
(3) the elucidation of time relationships, such as the distribution of the latency period and the effect of dose-rate (as opposed to total dose) on cancer risk.

IARC STUDY OF SECOND TUMOURS FOLLOWING CHEMOTHERAPY

The IARC has initiated a study of second tumours following chemotherapy, with the explicit goal of better defining the quantitative relationship between exposure levels of the antineoplastic drugs and the rates of second primary tumours. The study, which follows on from the IARC study of second tumours following radiation treatment for cervical cancer, is a collaboration between over a dozen cancer registries and treatment centres in Europe and North America. In many of the participating centres, cancer registration has been routine since the 1950's, providing long follow-up and relatively large numbers of cases. Moreover, it may be expected that the wide range of treatment methods over time and across countries will enable the study of the long-term effects of single agents – which are still in greater use in Eastern Europe than in the West – as well as of mixed modalities of chemotherapy.

Table 2. Studies of acute leukaemia (AL) following treatment with Melphalan

Reference	First primary tumour	Number of patients reported on	Cases of AL (% crude incidence)	Melphalan dose information
Bergsagel et al. (1979)	Multiple myeloma	98 [a] (3-year survivors)	8 (.08)	— [b]
Buckman et al. (1982)	Multiple myeloma	56 (4-year survivors)	6 (.11)	— [b]
Einhorn et al. (1982)	Ovarian cancer	51 (3-year survivors)	6 (.12)	All patients received at least 300 mg.
Wahlin et al. (1982)	Multiple myeloma	29 (2-year survivors)	4 (.14)	All cases received at least 1100 mg.

[a] 14 cases of AL were reported among all 364 patients enrolled at the start of the study.
[b] Protocol doses available, but not at an individual level

The initial phase of the study, currently in progress, has been to request tabulations of expected and observed numbers of second primary tumours following first primary tumours in the case of three types of cancer for which chemotherapy has been extensively used (see Table 1), namely, Hodgkin's disease and cancers of the testis and ovary. The purpose of this exercise is to determine how many cases of second primary tumours will be available for detailed study. Already three of the four participating Scandinavian countries have reported a total of 144 cases of leukaemia at least one year after diagnosis of the first primary tumour, as compared to about 50 expected on the basis of national rates. Large excesses of second primary cancers of the bladder and lung have also been reported from these countries.

The second phase of this study will proceed using the case-control methods described above. For all individuals diagnosed with one of the second primary tumours of interest (at this stage, leukaemia, bladder and lung cancer), four to five controls will be selected from individuals diagnosed with the same primary tumour, of the same age and who survived the first tumour at least as long as the case, and treatment records of the cases and controls will then be compared.

MEASUREMENT OF ADDITIONAL PARAMETERS OF DNA DAMAGE

The measurement of biochemical markers and of biological endpoints other than cancer itself has received a lot of attention in recent years. Of particular interest are short-term measures of DNA damage, both as a means of monitoring exposure levels in populations potentially exposed to carcinogens, and for the purpose of understanding fundamental carcinogenic mechanisms. Cytogenetic assays, such as those for in-vivo induction of sister chromatid exchange (SCE) and chromosomal aberrations, have been used in the former role to monitor exposure to vinyl chloride, epichlorohydrin, styrene, ethylene oxide, benzene and other known or suspected human carcinogens (see Berlin et al., 1984). SCEs have been observed at an elevated rate in patients undergoing therapy with most alkylating anticancer drugs (Gebhart,

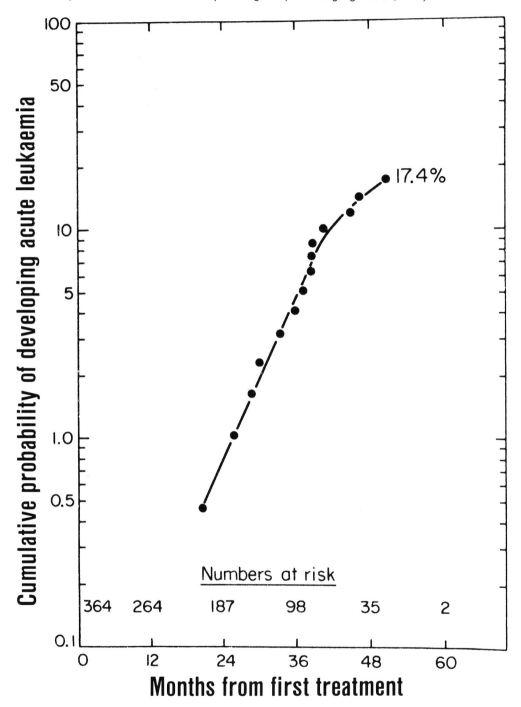

Fig. 1. Probability of developing acute leukaemia in patients with plasma-cell myeloma treated with melphalan and other chemotherapeutic agents (from Bergsagel et al., 1979)

1981; Lambert *et al.*, 1982). The assays for DNA adducts also provide measurements of this kind, but are more specific with regard to the type of damage detected. However, the application of these assays to humans is still very much in the developmental stage, so their practical utility has not yet been established.

Although patients undergoing intensive chemotherapy with alkylating agents are known to be at increased risk of developing tumours, levels of DNA damage are not yet monitored routinely. However, such patients do provide a unique opportunity for studying the relationships between the dose of the chemotherapeutic agents used, the various measures of DNA damage and, ultimately, the incidence of second primary tumours. Such studies may in fact improve the efficacy of the chemotherapeutic regimen. Moreover, as more effective chemotherapies are developed, the decision as to what extent early-stage tumours are to be aggressively treated will have to be weighed against the carcinogenic risk posed by the therapy itself.

In order for such measurements to provide a meaningful addition to our understanding of human carcinogenesis, it must be demonstrated that they are sufficiently sensitive and reproducible to enable differences in groups of individuals to be statistically differentiated. In particular, it is of interest to compare levels of nucleic acid and protein adduct formation, chromosomal aberrations, and other measurements of DNA damage in individuals exposed at various doses, including zero dose. The sensitivity of such comparisons depends on the variability of the measurements being made, both between individuals in the same group and over repeat measurements of the same individual. (In studies of this kind, it may be expected that measurement variation will be smaller within the same individual than between individuals. It is therefore important that measurements be made on patients before treatment begins, so that a baseline is available for each individual.) When the precision of the methodology is such that differences in these groups are detectable, then quantitative relationships may be established, and it will be possible to investigate more subtle questions such as the extent to which DNA damage of various kinds, rather than dose, is predictive of second tumour risk.

ACKNOWLEDGEMENTS

We would like to thank M. Sorsa, H. Vainio and H. Bartsch for helpful comments on the manuscript.

REFERENCES

Bergsagel, D.E., Bailey, A.J., Langley, G.R., MacDonald, R.N., White, D.F. & Miller, A.B. (1979) The chemotherapy of plasma-cell myeloma and the incidence of acute leukemia. *New Eng. J. Med., 301,* 743–748

Berlin, A., Draper, M., Hemminki, K. & Vainio, H., eds (1984) *Monitoring Human Exposure to Carcinogenic and Mutagenic Agents (IARC Scientific Publications No. 59),* Lyon, International Agency for Research on Cancer

Breslow, N.E. & Day, N.E., eds (1980) *Statistical Methods in Cancer Epidemiology, Vol. I, The Analysis of Case-Control Studies (IARC Scientific Publications No. 32)*, Lyon, International Agency for Research on Cancer

Buckman, R., Cuzick, J. & Galton, D. (1982) Long term survivors in myelomatosis. *Br. J. Haematol., 52*, 589–599

Curtis, R.E., Hankey, B.F., Myers, H.H. & Young, J.L., Jr. (1984) Risk of leukemia associated with the first course of cancer treatment: an analysis of the Surveillance, Epidemiology and End Results program experience. *J. natl Cancer Inst., 72*, 531–544

Day, N.E. & Boice, J.D., Jr., eds (1984) *Second Cancer in Relation to Radiation Treatment for Cervical Cancer (IARC Scientific Publications No. 52)*, Lyon, International Agency for Research on Cancer

Einhorn, N., Eklund, G., Franzen, S., Lambert, B., Lindsten, J. & Söderhäll, S. (1982) Late side effects of chemotherapy in ovarian cancer. *Cancer, 49*, 2234–2241

Falk, H. Herbert, J., Crowley, S., Ishak, K.G., Thomas, L.B., Popper, H. & Caldwell, G.G. (1981) Epidemiology of hepatic angiosarcoma in the United States: 1964–1974. *Environ. Health Persp., 31*, 107–113

Gebhart, E. (1981) Sister chromatid exchange (SCE) and structural chromosome aberration in mutagenicity testing. *Hum. Genet., 58*, 235–254

Greene, M.H. (1984) *Interaction between radiotherapy and chemotherapy in human leukemogenesis*. In: Boice, J.D., Jr. & Fraumeni, J.F., eds, *Radiation Carcinogenesis: Epidemiology and Biological Significance*, New York, Raven Press, pp. 199–210

IARC (1974) *IARC Monographs on the Evaluation of the Carcinogenic Risk of Chemicals to Humans*, Vol. 4, *Some aromatic amines, hydrazine and related substances, N-nitroso compounds and miscellaneous alkylating agents*, Lyon, International Agency for Research on Cancer

IARC (1975) *IARC Monographs on the Evaluation of the Carcinogenic Risk of Chemicals to Humans*, Vol. 9, *Some aziridines, N-, S- and O-mustards and selenium*, Lyon, International Agency for Research on Cancer

IARC (1976) *IARC Monographs on the Evaluation of the Carcinogenic Risk of Chemicals to Humans*, Vol. 10, *Some naturally occurring substances*, Lyon, International Agency for Research on Cancer

IARC (1981) *IARC Monographs on the Evaluation of the Carcinogenic Risk of Chemicals to Humans*, Vol. 26, *Some antineoplastic and immunosuppressive agents*, Lyon, International Agency for Research on Cancer

IARC (1982) *IARC Monographs on the Evaluation of the Carcinogenic Risk of Chemicals to Humans*, Suppl. 4, *Chemicals, Industrial Processes and Industries Associated with Cancer in Humans, IARC Monographs, Volumes 1 to 29*, Lyon, International Agency for Research on Cancer

IARC (1983) *Second Report to the EEC: Quantitative Risk Estimation for Human Carcinogens*, Lyon, International Agency for Research on Cancer

Lambert, B., Lindblad, A., Holmberg, K. & Francesconi, D. (1982) *Use of sister chromatid exchange to monitor human populations for exposure to toxicologically harmful agents*. In: Wolf, S., ed., *Sister Chromatid Exchange*, New York, Wiley & Sons, pp. 149–182

Schmähl, D., Habs, M., Lorenz, M. & Wagner, I. (1982) Occurrence of second tumours in man after anticancer drug treatment. *Cancer Treatment Rev., 9,* 167–194

Wahlin, A., Roos, G., Rudolphi, D. & Holm, J. (1982) Melphalan-related leukaemia in multiple myeloma. *Acta med. Scand., 211,* 203–208

V. MECHANISTIC APPROACHES

REACTION OF CHLOROACETALDEHYDE WITH POLY(dA-dT) AND POLY(dC-dG) AND ITS EFFECT UPON THE ACCURACY OF DNA SYNTHESIS

R. SAFFHILL & J.A. HALL

Peterson Laboratories, Christie Hospital and Holt Radium Institute, Manchester, United Kingdom

Chloroethylene oxide (CEO) and chloroacetaldehyde (CAA) react with DNA *in vivo*, and etheno adducts – ethenoadenine (εA) and ethenocytosine (εC) – have been isolated following long term in-vitro exposure to vinyl chloride (VC) (Green & Hathway, 1978). We have investigated (Hall *et al.*, 1981) the reaction of CAA with the DNA-like polynucleotides, poly(dA-dT) and poly(dC-dG), using polynucleotides prelabelled with specific ^{3}H- or ^{14}C-labelled bases; ^{3}H for the determination of cytosine (C) and thymine (T) adducts, and ^{14}C for the determination of adenine (A) and guanine (G) adducts. Unlabelled polynucleotides have also been modified in parallel experiments and used as templates for *Escherichia coli* DNA polymerase I (Pol I) to measure the incorporation of complementary and noncomplementary nucleotides into DNA-like material.

The appropriately labelled polynucleotide was reacted with CAA in acetate buffer (pH 4.5). The release of purines from the polymers during the reaction was determined by chromatography on Aminex A-6, in which system the high molecular weight polymer elutes at the break-through and the guanine, adenine and εA bases elute later. The amount of purine base liberated was calculated from the radioactivity released from the polymer as free base. Only an aliquot from each fraction containing the polynucleotide was counted: the remainder was pooled, lyophilized, digested to deoxynucleosides (Hall *et al.*, 1981) and then separated by Aminex A-6 chromatography. Complementary and noncomplementary nucleotide incorporation was measured as previously described (Hall *et al.*, 1981).

During the reaction there was some release of adenine and guanine bases. This amounted to 0.4% for adenine and 1% for guanine and was independent of the presence of CAA in the reaction mixture. No free εA or any other modified base was detected in any of the reactions. The chromatograms from the analysis of modified adenine and cytosine deoxynucleosides using poly(dA-dT) labelled with ^{14}C-adenine and poly(dC-dG) labelled with ^{3}H-cytosine are shown in Figures 1A and 1B. The only CAA adducts detectable were ethenodeoxyadenosine (εdAdo) and ethenodeoxycytidine (εdCyd), and the amounts present increased with increasing

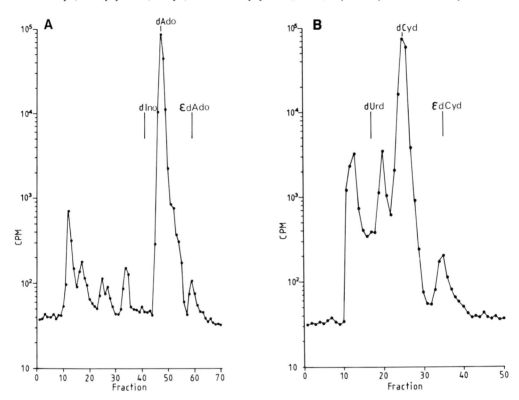

Fig. 1. Analysis of the products of the reaction of chloroacetaldehyde with poly(dA-dT) labelled with ^{14}C-adenine (A), and with poly(dC-dG) labelled with ^{3}H-cytosine (B). Reacted polynucleotides were digested to deoxynucleosides before chromatography on Animex A-6. dAdo, deoxyadenosine; εdAdo, ethenodeoxyadenosine; dIno, expected position of deoxyinosine; dCyd, deoxycytidine; εdCyd, ethenodeoxycytidine; dUrd, expected position of deoxyuridine

concentration of CAA. No deamination products (deoxyinosine or deoxyuridine) were detected in any of the reactions although they were expected. Using poly(dA-dT) labelled with ^{3}H-thymine and poly(dC-dG) labelled with ^{14}C-guanine, no adducts of deoxythymidine or deoxyguanosine were detected in the hydrolysates in initial experiments, although recently a deoxyguanosine adduct (amounting to approximately 10% of the amount of εdCyd) that binds strongly to the column and only elutes at higher ionic strength has been detected. The εdAdo formed has also been detected by Green and Hathway (1978) using fluorescence measurements on the column eluate; the values they obtained are similar to those determined radiochemically.

CAA-modification of the polynucleotides led to a decrease in their ability to act as templates for Pol I. This was accompanied by an increase in the relative incorporation of noncomplementary nucleotides, indicating that reaction with CAA had led to inaccuracies during the in-vitro DNA synthesis. In the case of poly(dA-dT),

Table 1. Incorporation of noncomplementary nucleotides into poly(dA-dT) modified by chloroacetaldehyde (CAA)

CAA (mM)	Ratio of εA to A residues in template[a] ($\times 10^3$)	Relative template activity (%)	Ratio of noncomplementary nucleotides to dTMP[b] ($\times 10^5$)	
			dGMP	dCMP
0	–	100	2	5
4	0.3	79	3	5
8	0.4	68	2	6
17	0.5	49	3	3
26	1.3	35	4	6
34	2.3	24	6	8

[a] εA, ethenoadenine; A, adenine
[b] dTMP, deoxythymidylic acid; dGMP, deoxyguanylic acid; dCMP, deoxycytidylic acid

Table 2. Incorporation of noncomplementary nucleotides into poly(dC-dG) modified by chloroacetaldehyde (CAA)

CAA (mM)	Ratio of εC to C residues in template[a] ($\times 10^3$)	Relative template activity (%)	Ratio of noncomplementary nucleotides to dGMP[b] ($\times 10^5$)	
			dAMP	dTMP
0	–	100	5	4
4	2	71	8	7
8	7	65	30	13
17	12	55	45	19

[a] εC, ethenocytosine; C, cytosine
[b] dGMP, deoxyguanylic acid; dAMP, deoxyadenylic acid; dTMP, deoxythymidylic acid

the increased incorporation was only just detectable at the highest concentrations of CAA used, and amounted to approximately two times background (Table 1). With the poly(dC-dG) templates, the level of noncomplementary nucleotide incorporation increased with increasing concentration of CAA in the reaction, the highest level obtained being nine times the background level (Table 1 & 2). A similar pattern of noncomplementary nucleotide incorporation has been reported by Barbin et al. (1981) using CAA- and CEO-modified homopolynucleotide templates; however, these workers reported no analyses of the templates and worked at high levels of template inactivation. Spengler and Singer (1981) have demonstrated that εC and εA lead to misincorporation during transcription.

The origin of the increased noncomplementary nucleotide incorporation following CAA-modification of poly(dA-dT) and poly(dC-dG) is uncertain. We can discount that it is due to deamination of cytosine and adenine in the polymers since neither deoxyuridine or deoxyinosine were detected in any of the enzyme hydrolysates. Although some apurinic sites were present in the reacted polymers, we can also discount that the observed incorporation of noncomplementary nucleotides arises from these since the apurinic site has only a low mutagenic efficiency (Shearman & Loeb, 1979; Loeb, 1985) and the amounts present in our templates would not lead to significant misincorporation. The results indicate that some of the noncomplementary nucleotide incorporation with CAA-reacted poly(dC-dG) arises from the presence of εC (or a precursor thereof) with one misincorporation of dAMP occuring for approximately each 30 εC residues, and one misincorporation of dTMP, for approximately each 80 εC residues. Although a nearest-neighbour analysis (Hall et al., 1981) of the DNA product has shown that the majority (approximately 90%) of the dAMP incorporation takes place opposite C-related residues in the template strand, there are a small number of errors (approximately 10%) that take place opposite G-related residues. At the same time, the dTMP incorporation appears to take place opposite C- and G-related residues in approximately equal amounts. Even though other adducts have not been detected in the enzyme hydrolysates, we cannot rule out the possibility that they may be present in the polynucleotides (e.g., precursors of εC and εA (Singer et al., 1983)) which may produce the εdCyd and εdAdo observed during the hydrolysis and chromatographic procedures and/or the G adducts described by Sattsangi et al. (1977) which, along with other unknown adducts, may co-chromatograph with other compounds in the system used).

Singer et al. (1983) have shown that the main lesion leading to misincorporation with CAA-reacted poly(dC) templates is εC and not its hydrated precursor and the relatively low mutagenic efficiency of εC in these experiments indicates that the polynucleotides used in our study very likely contain significant amounts of hydrated precursors of the etheno adducts. Our results do confirm the promutagenic potential of CAA and hence, of vinyl chloride itself. If similar situations arise during in-vivo DNA synthesis, they could lead to base sequence changes.

REFERENCES

Barbin, A., Bartsch, H., Leconte, P. & Radman, M. (1981) Studies on the miscoding properties of 1,N^6-ethenoadenine and 3,N^4-ethenocytosine, DNA reaction products of vinyl chloride metabolites, during in-vitro DNA synthesis. *Nucleic Acids Res.,* **9,** 375–387

Green, T. & Hathway, D.E. (1978) Interactions of vinyl chloride with rat liver DNA in vivo. *Chem.-biol. Interactions,* **22,** 211–224

Hall, J.A., Saffhill, R., Green, T. & Hathway, D.E. (1981) The induction of errors during in-vitro DNA synthesis following chloroacetaldehyde-treatment of poly(dA-dT) and poly(dC-dG). *Carcinogenesis,* **2,** 141–146

Loeb, L.A. (1985) Apurinic sites as mutagenic intermediates. *Cell,* **40,** 483–484

Sattsangi, P.D., Leonard, N.J. & Frihart, C.R. (1977) 1,N^2-Ethenoguanine and N^2,3-ethenoguanine. Synthesis and comparison of the electronic spectral properties of these linear and angular triheterocycles related to the Y bases. *J. org. Chem.,* **142,** 3292–3296

Shearman, C.W. & Loeb, L.A. (1979) Effects of depurination on the fidelity of DNA synthesis. *J. mol. Biol.,* **128,** 197–218

Singer, B., Kuśmierek, J.T. & Fraenkel-Conrat, H. (1983) In-vitro discrimination of replicases acting on carcinogen-modified polynucleotide templates. *Proc. natl Acad. Sci. USA, 80*, 969–972

Spengler, S. & Singer, B. (1981) Transcription errors and ambiguity resulting from the presence of 1,N^6-ethanoadenine and 3,N^4-ethenocytosine, DNA reaction products of vinyl chloride, during in-vitro DNA synthesis. *Nucleic Acids Res., 9*, 365–373

MUTAGENIC AND PROMUTAGENIC PROPERTIES OF DNA ADDUCTS FORMED BY VINYL CHLORIDE METABOLITES

A. BARBIN & H. BARTSCH

International Agency for Research on Cancer, Unit of Environmental Carcinogens and Host Factors, Lyon, France

SUMMARY

Published results and work from this laboratory permit the characterization of the possible promutagenic lesions induced by chloroethylene oxide (CEO) and chloroacetaldehyde (CAA), both known as bifunctional alkylating metabolites of vinyl chloride (VC).

1) The mutagenic effectiveness of CEO and CAA in *Escherichia coli,* when compared to their nucleophilic selectivity, suggests that the critical target site in DNA bases is not an oxygen atom, and/or that the reaction mechanism of CEO and CAA is different from a simple alkylation.

2) CEO-mutagenicity in *E. coli* is *recA*-independent, and CEO preferentially induces GC → AT transitions; accordingly, the mutagenicity of CEO in bacteria may result mainly from a miscoding guanosine or cytosine adduct.

3) Two observations argue against the role of $1,N^6$-ethenoadenine (εA) and $3,N^4$-ethenocytosine (εC) in VC-induced mutagenesis/carcinogenesis: i) the lack of detection in double-stranded DNA *in vivo* and *in vitro;* ii) the inconsistency between mutational specificity of CEO and miscoding properties of εA and εC.

4) The lack of miscoding properties of 7-(2-oxoethyl)guanine (oxet-G), the major in-vivo VC-DNA adduct, suggests a minor miscoding base adduct.

5) Several lines of evidence point to N^4-(2-chlorovinyl)cytosine as one possible putative promutagenic lesion produced by VC, but this compound has yet to be identified in DNA.

INTRODUCTION

The carcinogenicity of vinyl chloride (VC) in humans and rodents was recognized in 1974 by Creech and Johnson and by Maltoni *et al.* Subsequently, VC was shown to be mutagenic in prokaryotic and eukaryotic cells, following its activation by microsomal monooxygenases into chloroethylene oxide (CEO) and chloroacetaldehyde (CAA) (Göthe et al., 1974; Barbin et al., 1975; Huberman et al., 1975; Malaveille et al., 1975; Hussain & Osterman-Golkar, 1976; Rannug et al., 1976). CEO and CAA are bifunctional alkylating agents which can bind to nucleic acid bases, eventually yielding cyclic adducts (Kochetkov et al., 1971; Barbin et al., 1975; Sattsangi et al., 1977; Biernat et al., 1978; Guengerich et al., 1979; Laib et al., 1981; Scherer et al., 1981; Oesch & Doerjer, 1982). On the basis of our animal studies (Zajdela et al., 1980), CEO is now considered to be the ultimate carcinogenic metabolite of VC.

In this report, we discuss the role of known VC-DNA adducts as potential promutagenic lesions, in the light of data from carcinogenicity and mutagenicity studies and from misincorporation assays. Some preliminary data are also presented on the nature of a new putative DNA adduct which could play a role in VC-induced mutagenesis and carcinogenesis.

RESULTS AND DISCUSSION

Carcinogenic and mutagenic potencies of reactive VC-metabolites

An insight into the mechanism of action of VC can be obtained by analysing the relationship between alkylating properties of the reactive metabolites CEO and CAA, on the one hand, and their mutagenic and carcinogenic potencies, on the other. As shown in Figure 1 (open circles), Bartsch et al. (1983) established a positive relationship between the carcinogenic potency (TD_{50}) in rodents and the nucleophilic selectivity (initial ratio of 7-alkylguanine/O^6-alkylguanine in double-stranded DNA) for a series of nine monofunctional alkylating agents. Thus, alkylation of oxygen atoms of DNA bases appeared to determine the overall carcinogenicity of these alkylating agents in rodents. However, this general relationship did not hold true for bifunctional alkylating agents (Fig. 1, filled circles). This discrepancy suggests that the critical target site in DNA for alkylation by these agents is not the O^6 of guanine (or any other oxygen atom of DNA bases) but another position of higher nucleophilicity. Because in one mouse bioassay (Zajdela et al., 1980), the carcinogenic potency of CEO was of the same order as that of bis(chloromethyl)ether, one of the most potent carcinogenic bifunctional alkylating agents (Gold et al., 1984), the above mechanism may also apply to CEO (Swain-Scott constant, $s = 0.71$; Barbin et al., 1985b) (Fig. 1, open square).

A similar conclusion, i.e., that CEO does not act exclusively via simple alkylation, was reached by Hussain and Osterman-Golkar (1976) based on their mutagenicity studies. Hussain (1984) demonstrated a proportionality between efficiency (A) of inducing forward mutations to streptomycin independence in Escherichia coli sd-4 ('A' being expressed as the number of mutants per 10^8 survivors per mmol per h) and the reactivity (B) of an agent towards nucleophilic groups in DNA of low nucleophilicity in the Swain-Scott scale ('B' is the second order rate constant at $n = 2$, expressed in $M^{-1} \times h^{-1}$, at 37°C.) For simple alkylating agents, including epoxides, the A/B ratio is about 100–200, but for CEO and CAA, it is several orders

Fig. 1. Carcinogenic potency and nucleophilic selectivity of alkylating agents. Data were collated from IARC Monographs (1974, 1975, 1981), Zajdela et al. (1980), Spears (1981), Bartsch et al. (1983), Gold et al. (1984), Barbin et al. (1985b.). ENU, N-ethyl-N'-nitrosourea; MNU, N-methyl-N'-nitrosourea; MNNG, N-methyl-N'-nitro-N-nitrosoguanidine; MNUT, N-methyl-N-nitrosourethane; PS, 1,3-propane sultone; PL, β-propiolactone; GA, glycidaldehyde; MMS, methyl methane-sulfonate; ECH, epichlorohydrin; CEO, chloroethylene oxide; BCNU, 1,3-bis-(2-chloroethyl)-1-nitrosourea; HN2, methyl-bis-(2-chloroethyl)amine hydrochloride. The line represents the linear regression of TD_{50} (median) values for 9 monofunctional alkylating agents (o) vs their nucleophilic selectivity:

$$\ln (TD_{50}) = 1.394 \ln \left[\frac{\text{7-alkyl-G}}{\text{O}^6\text{-alkyl-G}} \right] + 0.134$$

($r^2 = 0.86$)

The TD_{50}-value for CEO was derived from a simple mouse bioassay with no dose-response curve (Zajdela et al., 1980). (TD_{50} is defined as the total dose of carcinogen in mg/kg bw required to reduce by one-half the probability of the animal being tumour-free (all sites) throughout a standard lifetime.)

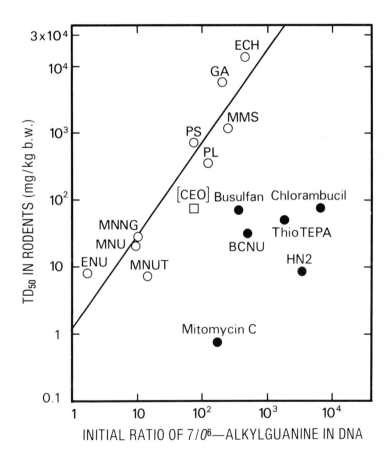

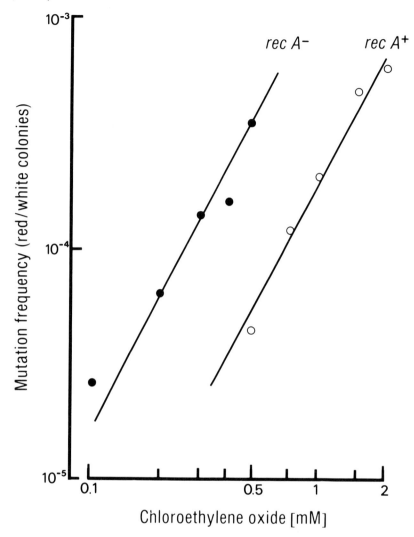

Fig. 2. Genetic effects of chloroethylene oxide (CEO) in *Escherichia coli* strains developed by Toman *et al.* (1980), demonstrating that CEO-mutagenicity is essentially *recA*-independent (Barbin *et al.*, 1985d)

of magnitude higher: 3.7×10^3 for CEO and 1.5×10^6 for CAA (Hussain & Osterman-Golkar, 1976; Hussain, 1984). Hussain & Osterman-Golkar (1976) tentatively explained the high mutagenic effectiveness of CAA by Schiff's base formation on the exocyclic amino groups of adenine or cytosine, yielding $1,N^6$-ethenoadenine (εA) or $3,N^4$-ethenocytosine (εC), respectively (see Fig. 4).

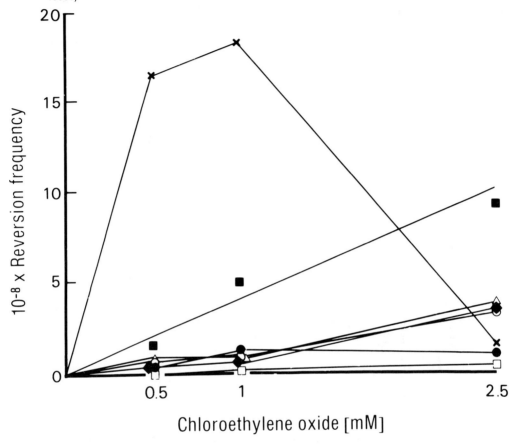

Fig. 3. Analysis of the types of base-pair substitution induced in *Escherichia coli* by chloroethylene oxide. The strains, mutated in the *trpA* gene, were developed by Yanofsky (1971). For strains which have one simple reversion pathway, the corresponding base-pair substitution is indicated in parentheses: x--x, *A446* (GC → AT); ■--■, *A3* (AT → TA), △--△, *A46;* ○--○, *A88* (AT → CG); ●--●, *A23;* ◆--◆, *A58* (AT → GC); □--□, *A11* (CG → GC). (Barbin *et al.*, 1985c)

Mechanism of CEO-mutagenesis in E. coli

In order to elucidate the mechanism involved in CEO-mutagenesis, we investigated whether CEO acts as a "direct" mutagen through a miscoding DNA adduct, or as an "indirect" mutagen, inducing SOS functions. Using *E. coli* strains developed by Toman *et al.* (1980), we showed that CEO is not recombinogenic and that it mutates the $recA^-$ strain more efficiently than the $recA^+$ strain (Fig. 2; Barbin *et al.*, 1984; Barbin *et al.*, 1985d). Therefore, the mutagenicity of CEO in *E. coli* must be attributed mainly to the formation of one or more miscoding DNA adducts.

Furthermore, as CEO is known to induce base-pair substitution mutations (Malaveille *et al.*, 1975; Rannug *et al.*, 1976), we investigated the mutational

Fig. 4. A summary of nucleic acid adducts produced by vinyl chloride metabolites. Dotted lines indicate possible secondary lesions formed by scission of the N-glycosidic bond or by cleavage of the 7-N-alkylimidazole ring. (d)R stands for either deoxyribose or ribose. (Collated from: Kochetkov et al., 1971; Barrio et al., 1972; Barbin et al., 1975; Laib & Bolt, 1977; Osterman-Golkar et al., 1977; Sattsangi et al., 1977; Biernat et al., 1978; Green & Hathway, 1978; Laib & Bolt, 1978; Guengerich et al., 1979; Laib et al., 1981; Scherer et al., 1981; Kuśmierek & Singer, 1982; Oesch & Doerjer, 1982; Barbin et al., 1984; O'Neill et al., these proceedings, p. 57)

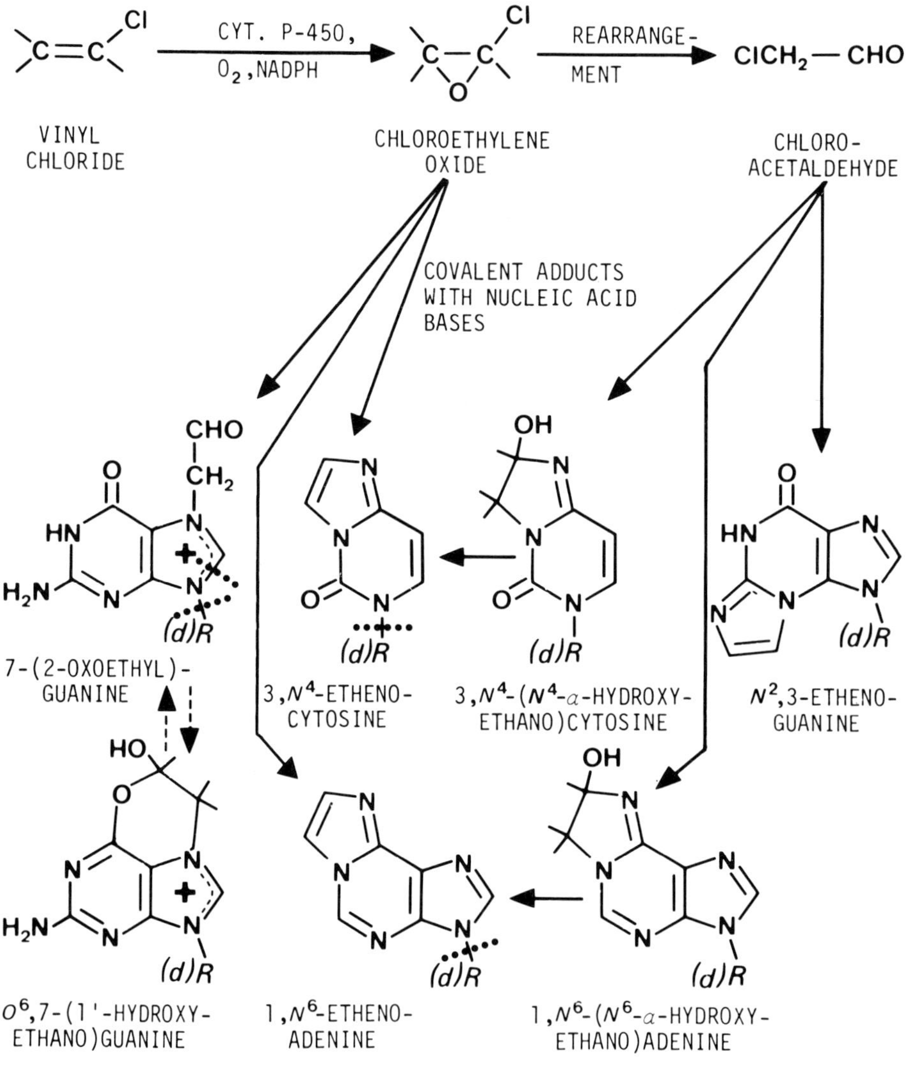

specificity of CEO in *E. coli trpA* mutants, a well-defined reversion system developed by Yanofsky (1971). Among the set of mutant strains, *trpA446* was significantly more sensitive than the others to CEO-induced reversion, followed by strain *trpA3* (Fig. 3). Therefore, CEO preferentially causes GC → AT transitions in the *trpA* gene of *E. coli* and, to a lesser extent, AT → TA transversions. Other types of substitution were also observed in the other strains investigated, but at much lower frequencies (Fig. 3; Barbin et al., 1984; Barbin et al., 1985c).

From these data, we concluded that CEO-mutagenicity results from the formation of a miscoding guanine or cytosine adduct.

Miscoding properties of known VC-DNA adducts

The formation of VC-DNA adducts and their miscoding properties have been recently investigated by several laboratories: CEO and CAA are both capable of forming cyclic adducts with nucleic acid bases (Fig. 4) to yield εA, εC and N^2,3-ethenoguanine (N^2,3-εG) (Kochetkov et al., 1971; Barrio et al., 1972; Barbin et al., 1975; Sattsangi et al., 1977; Guengerich et al., 1979; Oesch & Doerjer, 1982; O'Neill et al., these proceedings[1]). Two cyclic hydrated intermediates that occur in the course of the reaction of CAA with adenine and cytosine derivatives have been characterized (Biernat et al., 1978). The εA and εC adducts, which can be formed *in vitro* either with CEO (Barbin et al., 1975; Guengerich et al., 1979; O'Neill et al., these proceedings[1]) or CAA (Kochetkov et al., 1971), have also been identified in liver RNA (Laib & Bolt, 1977, 1978) but not in liver DNA (Laib et al., 1981) of VC-exposed rats, except in one study (Green & Hathway, 1978). This specific binding pattern could be explained by "inaccessibility" in double-stranded DNA of some nucleophilic sites, like the *N*-3 of cytosine and the *N*-1 of adenine, to alkylating agents (Singer & Grunberger, 1983); it is known that neither CEO (Barbin et al., 1984; Barbin et al., 1985a) nor CAA (Kimura et al., 1977; Kuśmierek & Singer, 1982) can form εA or εC in double-stranded DNA *in vitro*. The N^2,3-εG adduct has been isolated from CAA-treated DNA *in vitro* (Oesch & Doerjer, 1982) and, recently, from liver DNA of VC-exposed rats (Laib, these proceedings[2]). Oesch and Doerjer (1982) have suggested that this adduct could miscode for thymine. The major guanine adduct found in liver DNA of VC-exposed rodents is 7-(2-oxoethyl)guanine (oxet-G) (Osterman-Golkar et al., 1977; Laib et al., 1981) which is specifically derived from CEO (Scherer et al., 1981). Therefore, CEO and CAA may produce distinctly different patterns of DNA adducts (Fig. 4), depending on metabolic factors. Scherer et al. (1981) found that under strong acidic conditions oxet-G can form a cyclic hemiacetal at the O^6-position, which has been proposed to be able to mispair with thymine. In addition to these primary adducts, secondary lesions like apurinic/apyrimidinic (AP) sites and imidazole-ring-opened guanine products could possibly be derived from oxet-G, εA and εC (Green & Hathway, 1978; Laib et al., 1981; Scherer et al., 1981; Barbin et al., 1984; Barbin et al., 1985a).

The base-pairing properties of εA and εC (Barbin et al., 1981; Hall et al., 1981; Spengler & Singer, 1981; Singer & Spengler, these proceedings[3]), as well as of the hydrated intermediate of εC (εC · H_2O) (Singer et al., 1983) and of oxet-G (Barbin et al., 1984; Barbin et al., 1985a) have been investigated in in-vitro misincorporation

---

[1] See p. 57.
[2] See p. 101.
[3] See p. 359.

Table 1. Replicational ambiguity and expected mutational specificity of vinyl chloride-DNA adducts[a]

Base adduct[b]	Mispair observed with	Ratio of mispair to adduct	Expected mutational specificity
εA	G	1:(60–550)	AT → CG
εC	T	1:(20–80)	CG → AT
εC·H$_2$O	None (A, T, C)[c]	–	–
oxet-G	None (A, T)[c]	–	–

[a] The data reported in this table were obtained with chloroethylene oxide- and chloroacetaldehyde-modified polydeoxyribonucleotide templates in the presence of *Escherichia coli* DNA polymerase I (collated from: Barbin *et al.*, 1981; Hall *et al.*, 1981; Singer *et al.*, 1983; Barbin *et al.*, 1984; Barbin *et al.*, 1985a; Singer & Spengler, these proceedings p. 359)
[b] Abbreviations: εA, 1,N^6-ethenoadenine; εC, 3,N^4-ethenocytosine; εC·H$_2$O, 3,N^4-(N^4-α-hydroxyethano)cytosine; oxet-G, 7-(2-oxoethyl)guanine
[c] Brackets indicate bases for which mispairing was tested.

assays using CEO- or CAA-treated polynucleotides and different DNA or RNA polymerases. The data reported in Table 1 were obtained with modified polydeoxyribonucleotides and *E. coli* DNA polymerase I in the presence of the four nucleoside 5′-triphosphates and Mg^{2+}. The εA adduct may direct a limited level of guanine misincorporation (Barbin *et al.*, 1981; Hall *et al.*, 1981; Singer & Spengler, these proceedings[3]), which would lead to AT → CG transversions; εC can miscode for thymine (Barbin *et al.*, 1981; Hall *et al.*, 1981; Singer *et al.*, 1983), which would generate CG → AT transversions; εC·H$_2$O does not miscode for adenine, thymine or cytosine (Singer *et al.*, 1983). In comparison with O^6-methylguanine (Abbott & Saffhill, 1979), εA and εC are not efficient miscoding lesions, as evident from the ratios of mispair to adduct (Table 1).

In order to test the miscoding properties of oxet-G, possibly involving the hemiacetal form (Scherer *et al.*, 1981), we used CEO-modified poly(dG-dC) templates and *E. coli* DNA polymerase I (Barbin *et al.*, 1984; Barbin *et al.*, 1985a). The templates contained up to 27% oxet-G and only trace amounts of εC; no N^2,3-εG could be detected. The dose-response curves that were obtained for dAMP and dTMP misincorporations paralled the inhibition of DNA synthesis. However, nearest-neighbour analyses established that the majority of the errors (approximately 80%) occurred opposite cytosine and were mainly C:A mispairs. (C:A mispairs have also been observed by Hall *et al.* (1981) using CAA-treated poly(dG-dC) templates.) Furthermore, misincorporation of dAMP and dTMP was reduced roughly by one-half after nicking of the templates with *Micrococcus luteus* AP endonuclease B. This experiment demonstrated that at neutral pH oxet-G miscodes for neither adenine nor thymine.

The expected genetic consequences of secondary lesions possibly formed from VC-DNA adducts are listed in Table 2. The imidazole-ring-opened form of oxet-G is probably not a miscoding lesion since Boiteux & Laval (1983) have shown that the imidazole-ring-opened form of 7-methylguanine blocks DNA synthesis *in vitro* and miscodes for neither adenine nor thymine. AP sites could be generated from εA, εC

Table 2. Expected replicational ambiguity and mutational specificity of secondary lesions possibly derived from vinyl chloride-DNA adducts[a]

Secondary lesion[b]	Expected miscoding for:	Expected mutational specificity
Imidazole-ring-opened form of oxet-G	None (tested for A, T)	–
AP sites via loss of:		
εA	A	AT → TA
εC	A	GC → AT
oxet-G	A	CG → AT

[a] Collated from: Boiteux & Laval, 1982; Strauss et al., 1982; Boiteux & Laval, 1983; Schaaper et al., 1983; Barbin et al., 1984; Barbin et al., 1985a
[b] Abbreviations: oxet-G, 7-(2-oxoethyl)guanine; AP, apurinic/apyrimidinic; εA, 1,N^6-ethenoadenine; εC, 3,N^4-ethenocytosine

and oxet-G. AP sites are known to block the replication machinery and can be bypassed by the DNA polymerase in *E. coli* following induction of the SOS functions; in this case, dAMP is preferentially incorporated opposite AP sites (Boiteux & Laval, 1982; Strauss et al., 1982; Schaaper et al., 1983). AP sites cannot be regarded as miscoding lesions but rather as non-informational lesions (Strauss et al., 1982).

The results of the misincorporation assays may be summarized as follows: i) εA and εC have weak miscoding effects which do not explain the observed CEO-induced GC → AT transitions in *E. coli;* ii) oxet-G, the major VC- and CEO-DNA adduct, is not miscoding in our system; iii) in CEO- and CAA-modified poly(dG-dC) templates, most errors (C:A mispairs) occur opposite minor, unidentified, cytosine lesions; iv) C:A mispairs are generated only in part by AP sites. These observations suggest the formation of other minor VC-DNA adducts, possibly cytosine adducts, which have yet to be identified in DNA.

Search for new VC-DNA adducts and identification of a N^4-(alkyl)-3-methylcytidine

When considering the reaction mechanisms of CEO and CAA, it is clear that they can react either as monofunctional or bifunctional reagents towards nucleic acid bases; e.g., CEO can yield oxet-G as well as εC (Table 3). Thus, the nature of the resulting alkylation products is in part controlled by the secondary structure of DNA: CEO and CAA produce εA and εC in RNA and single-stranded DNA, but not in double-stranded DNA (Table 3). These findings imply that because the *N*-3 position of cytosine and the *N*-1 position of adenine are involved in base-pairing in double-stranded DNA, they are "inaccessible" to alkylation, thus preventing εA and εC from being formed. However, in double-stranded DNA, CEO and CAA may still react with exocyclic amino groups of adenine or cytosine to yield noncyclic monoadducts.

In order to identify such noncyclic monoadducts, we reacted 3-methylcytidine with CEO and CAA, in methanol. Alkylation products were silylated and analysed by gas

Table 3. Reaction of chloroethylene oxide and chloroacetaldehyde with nucleic acid bases as affected by their accessibility and conformation[a]

VC-metabolite[b]	Base-adduct formed[c]	Free base or nucleoside	RNA	Single-stranded DNA	Double-stranded DNA
CEO or CAA	εA	+	+	+	−
CEO or CAA	εC	+	+	+	−
CAA	1,N^2-εG	+; [−][d]	ND[e]	−	ND
CAA	N^2,3-εG	−; [+][d]	ND	+	ND
CEO	oxet-G	+	+	+	+

[a]Collated from Kochetkov et al., 1971; Barrio et al., 1972; Barbin et al., 1975; Kimura et al., 1977; Laib & Bolt, 1977; Osterman-Golkar et al., 1977; Sattsangi et al., 1977; Biernat et al., 1978; Green & Hathway, 1978; Laib & Bolt, 1978; Guengerich et al., 1979; Laib et al., 1981; Scherer et al., 1981; Kuśmierek & Singer, 1982; Oesch & Doerjer, 1982; Barbin et al., 1984; Barbin et al., 1985a
[b]VC, vinyl chloride; CEO, chloroethylene oxide; CAA, chloroacetaldehyde
[c]εA, 1,N^6-ethenoadenine; εC, 3,N^4-ethenocytosine; 1,N^2-εG, 1,N^2-ethenoguanine; N^2,3-εG, N^2,3-ethenoguanine; oxet-G, 7-(2-oxoethyl)-guanine
[d]Only when O^6 is substituted
[e]Not determined

Fig. 5. Mispairing of the putative N^4-(2-chlorovinyl)cytosine residue expected to lead to GC→AT transitions (in analogy to N^4-methoxycytosine; Singer et al., 1984). dR, deoxyribose

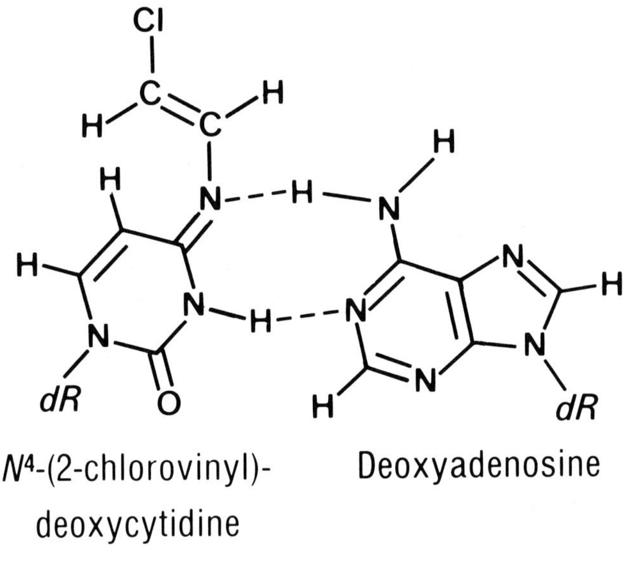

N^4-(2-chlorovinyl)-deoxycytidine   Deoxyadenosine

chromatography/mass spectrometry. In brief, the following results were obtained; details are described by O'Neill et al. (these proceedings[4]). The patterns of reaction products produced by CEO and CAA were similar, consisting of one major base adduct carrying one ClC_2H_2-substituent together with one minor isomer, three pairs

---

[4] See p. 57.

of isomeric ribose adducts with one $ClCH_2$–$CH(OTMS)$-group substituted at any of the hydroxyl groups, and 9 out of 12 possible isomeric diadducts, with one ClC_2H_2-group on the base and one $ClCH_2$–$CH(OTMS)$-group on the ribose. Under similar reaction conditions, when compared with 3-methylcytidine, CEO and CAA were much less reactive towards the base moiety (but equally reactive towards the sugar moiety) of 3-methyluridine. From these preliminary studies, we concluded that the free amino group in 3-methylcytidine is the main target for alkylation by CEO or CAA. The spectral data are consistent with the structure of the major 3-methylcytidine adduct being a N^4-(2-chlorovinyl)-3-methylcytidine, existing as *trans* and *cis*-isomers (see also Fig. 5).

Should CEO and CAA react with cytosine in double-stranded DNA in a similar fashion, the resulting putative adduct, N^4-(2-chlorovinyl)cytosine, could be a potential miscoding lesion. The imino tautomer would, by analogy to N^4-methoxycytosine (Singer *et al.*, 1984), be expected to mispair with adenine during DNA replication (Fig. 5). Thereby, GC → AT transition mutations should be generated; as mentioned before, such mutations actually are induced by CEO. What is now required is a search for this adduct in double-stranded DNA treated with CEO or CAA.

ACKNOWLEDGEMENTS

We wish to thank the following persons for their help or support: Mr J.-C. Béréziat, Drs M. Friesen, I. O'Neill (IARC, Lyon, France), Mlle M.-H. Perrard, Drs F. Besson, G. Michel (Université Claude Bernard, Lyon, France), Dr A. Croisy (Institut Curie, Orsay, France), Dr J. Laval (Institut Gustave Roussy, Villejuif, France), Drs M. Radman, P. Lecomte, Z. Toman, L. Tenenbaum (Université Libre de Bruxelles, Rhodes-St-Genèse, Belgium), Dr R. Laib (Universität Dortmund, Dortmund, FRG), Dr C. Yanofsky (Stanford University, Stanford, USA). We are indebted to Mrs M. Wrisez for typing the manuscript and to Mrs E. Heseltine for editorial assistance.

REFERENCES

Abbott, P.J. & Saffhill, R. (1979) DNA synthesis with methylated poly(dC-dG) templates. Evidence for a competitive nature to miscoding by O^6-methylguanine. *Biochim. biophys. Act.*, **562,** 51–61

Barbin, A., Brésil, H., Croisy, A., Jacquignon, P., Malaveille, C., Montesano, R. & Bartsch, H. (1975) Liver microsome-mediated formation of alkylating agents from vinyl bromide and vinyl chloride. *Biochem. biophys. Res. Commun.*, **67,** 596–603

Barbin, A., Bartsch, H., Lecomte, P. & Radman, M. (1981) Studies on the miscoding properties of 1,N^6-ethenoadenine and 3,N^4-ethenocytosine, DNA reaction products of vinyl chloride metabolites, during in-vitro DNA synthesis. *Nucleic Acids Res.*, **9,** 375–387

Barbin, A., Laib, R. & Bartsch, H. (1984) Studies on the mechanism of chloroethylene oxide (CEO)-induced mutagenesis. *Mutat. Res.*, **130,** 165–166

Barbin, A., Laib, R. & Bartsch, H. (1985a) Lack of miscoding properties of 7-(2-oxoethyl)guanine, the major vinyl chloride-DNA adduct. *Cancer Res.*, **45**, 2440–2444

Barbin, A., Béréziat, J.-C., Croisy, A., O'Neill, I. & Bartsch, H. (1985b) Reaction kinetics and nucleophilic selectivity of chloroethylene oxide assessed by ^{1}H-NMR and by the 4-(*p*-nitrobenzyl)pyridine assay (submitted for publication)

Barbin, A., Besson, F., Perrard, M.-H., Béréziat, J.-C., Kaldor, J., Michel, G. & Bartsch, H. (1985c) Induction of specific base-pair substitutions in *E. coli trpA* mutants by chloroethylene oxide, a carcinogenic vinyl chloride metabolite. *Mutat. Res.*, **152**, 147–156

Barbin, A., Tenenbaum, L., Toman, Z., Radman, M. & Bartsch, H. (1985d) *RecA*-independent mutagenicity induced by chloroethylene oxide in *E. coli. Mutat Res.*, **152**, 157–159

Barrio, J.R., Secrist, J.A., III & Leonard, N.J. (1972) Fluorescent adenosine and cytidine derivatives. *Biochem. Biophys. Res. Commun.*, **46**, 597–604

Bartsch, H., Terracini, B., Malaveille, C., Tomatis, L., Wahrendorf, J., Brun, G. & Dodet, B. (1983) Quantitative comparison of carcinogenicity, mutagenicity and electrophilicity of 10 direct-acting alkylating agents and of the initial O^6:7-alkylguanine ratio in DNA with carcinogenic potency in rodents. *Mutat. Res.*, **110**, 181–219

Biernat, J., Ciesiołka, J., Górnicki, P., Adamiak, R.W., Krzyżosiak, W.J. & Wiewiórowski, M. (1978) New observations concerning the chloroacetaldehyde reaction with some tRNA constituents. Stable intermediates, kinetics and selectivity of the reaction. *Nucleic Acids Res.*, **5**, 789–804

Boiteux, S. & Laval, J. (1982) Coding properties of poly(dC) templates containing uracil or apyrimidinic sites: in-vitro modulation of mutagenesis by DNA repair enzymes. *Biochemistry*, **21**, 6746–6750

Boiteux, S. & Laval, J. (1983) Imidazole open ring 7-methylguanine: an inhibitor of DNA synthesis. *Biochem. biophys. Res. Commun.*, **110**, 552–558

Creech, J.L. & Johnson, M.N. (1974) Angiosarcoma of the liver in the manufacture of polyvinyl chloride. *J. occup. Med.*, **16**, 150–151

Gold, L.S., Sawyer, C.B., Magaw, R., Bachman, G.M., de Veciana, M., Levinson, R., Hooper, N.K., Havender, W.R., Bernstein, L., Peto, R., Pike, M.C. & Ames, B.N. (1984) A carcinogenic potency data-base of the standardized results of animal bioassays. *Environ. Health Perspect.*, **58**, 9–319

Göthe, R., Calleman, C.J., Ehrenberg, L. & Wachtmeister, C.A. (1974) Trapping with 3,4-dichlorobenzenethiol of reactive metabolites formed *in vitro* from the carcinogen, vinyl chloride. *Ambio*, **3**, 234–236

Green, T. & Hathway, D.E. (1978) Interactions of vinyl chloride with rat liver DNA *in vivo. Chem.-biol. Interactions*, **22**, 211–224

Guengerich, F.P., Crawford, W.M., Jr & Watanabe, P.G. (1979) Activation of vinyl chloride to covalently bound metabolites: roles of 2-chloroethylene oxide and 2-chloroacetaldehyde. *Biochemistry*, **18**, 5177–5182

Hall, J.A., Saffhill, R., Green, T. & Hathway, D.E. (1971) The induction of errors during in-vitro DNA synthesis following chloroacetaldehyde-treatment of poly(dA-dT) and poly(dC-dG) templates. *Carcinogenesis*, **2**, 141–146

Huberman, E., Bartsch, H. & Sachs, L. (1975) Mutation induction in Chinese hamster V79 cells by two vinyl chloride metabolites, chloroethylene oxide and 2-chloroacetaldehyde. *Int. J. Cancer*, **16**, 639–644

Hussain, S. (1984) Dose-response relationships for mutations induced in *E. coli* by some model compounds. *Hereditas*, **101**, 57–68

Hussain, S. & Osterman-Golkar, S. (1976) Comment on the mutagenic effectiveness of vinyl chloride metabolites. *Chem.-biol. Interactions*, **12**, 265–267

IARC (1974) *IARC Monographs on the Evaluation of the Carcinogenic Risk of Chemicals to Humans*, Vol. 4, *Some aromatic amines, hydrazine and related substances,* N-*nitroso compounds and miscellaneous alkylating agents*, Lyon, International Agency for Research on Cancer, pp. 247–252

IARC (1975) *IARC Monographs on the Evaluation of the Carcinogenic Risk of Chemicals to Humans*, Vol. 9, *some aziridines,* N-, S- *and* O-*mustards and selenium*, Lyon, International Agency for Research on Cancer, pp. 193–207

IARC (1981) *IARC Monographs on the Evaluation of the Carcinogenic Risk of Chemicals to Humans*, Vol. 26, *Some antineoplastic and immunosuppressive agents*, Lyon, International Agency for Research on Cancer, pp. 79–95 and 115–136

Kimura, K., Nakanishi, M., Yamamoto, T. & Tsuboi, M. (1977) A correlation between the secondary structure of DNA and the reactivity of adenine residues with chloroacetaldehyde. *J. Biochem.*, **81**, 1699–1703

Kochetkov, N.K., Shibaev, V.N. & Kost, A.A. (1971) New reaction of adenine and cytosine derivatives, potentially useful for nucleic acids modification. *Tetrahedron Lett.*, **22**, 1993–1996

Kuśmierek, J.T. & Singer, B. (1982) Chloroacetaldehyde-treated ribo- and deoxyribopolynucleotides. 2. Errors in transcription by different polymerases resulting from ethenocytosine and its hydrated intermediate. *Biochemistry*, **21**, 5273–5278

Laib, R.J. & Bolt, H.M. (1977) Alkylation of RNA by vinyl chloride metabolites *in vitro* and *in vivo:* formation of $1,N^6$-ethenoadenosine. *Toxicology*, **8**, 185–195

Laib, R.J. & Bolt, H.M. (1978) Formation of $3,N^4$-ethenocytidine moieties in RNA by vinyl chloride metabolites *in vitro* and *in vivo*. *Arch. Toxicol.*, **39**, 235–240

Laib, R.J., Gwinner, L.M. & Bolt, H.M. (1981) DNA alkylation by vinyl chloride metabolites: etheno derivatives or 7-alkylation of guanine? *Chem.-biol. Interactions*, **37**, 219–231

Malaveille, C., Bartsch, H., Barbin, A., Camus, A.-M., Montesano, R., Croisy, A. & Jacquignon, P. (1975) Mutagenicity of vinyl chloride, chloroethylene oxide, chloroacetaldehyde and chloroethanol. *Biochem. biophys. Res. Commun.*, **63**, 363–370

Maltoni, C., Lefemine, G., Chieco, P. & Carretti, D. (1974) Vinyl chloride carcinogenesis: current results and perspectives. *Med. Lav.*, **65**, 421–444

Oesch, F. & Doerjer, G. (1982) Detection of N^2,3-ethenoguanine in DNA after treatment with chloroacetaldehyde *in vitro*. *Carcinogenesis*, **3**, 663–665

Osterman-Golkar, S., Hultmark, D., Segerbäck, D., Calleman, C.J., Göthe, R., Ehrenberg, L. & Wachtmeister, C.A. (1977) Alkylation of DNA and proteins in mice exposed to vinyl chloride. *Biochem. biophys. Res. Commun.*, **76**, 259–266

Rannug, U., Göthe, R. & Wachtmeister, C.A. (1976) The mutagenicity of chloroethylene oxide, chloroacetaldehyde, 2-chloroethanol and chloroacetic acid, conceivable metabolites of vinyl chloride. *Chem.-biol. Interactions, 12,* 251–263

Sattsangi, P.D., Leonard, N.J. & Frihart, C.R. (1977) 1,N^2-ethenoguanine and N^2,3-ethenoguanine. Synthesis and comparison of the electronic spectral properties of these linear and angular triheterocycles related to the Y bases. *J. org. Chem., 42,* 3292–3296

Schaaper, R.M., Kunkel, T.A. & Loeb, L.A. (1983) Infidelity of DNA synthesis associated with bypass of apurinic sites. *Proc. natl Acad. Sci., USA, 80,* 487–491

Scherer, E., Van Der Laken, C.J., Gwinner, L.M., Laib, R.J. & Emmelot, P. (1981) Modification of deoxyguanosine by chloroethylene oxide. *Carcinogenesis, 2,* 671–677

Singer, B. & Grunberger, D. (1983) *Reactions of directly acting agents with nucleic acids.* In: Singer, B. & Grunberger, D., eds, *Molecular Biology of Mutagens and Carcinogens,* New York, Plenum Press, pp. 45–96

Singer, B., Kuśmierek, J.T. & Fraenkel-Conrat, H. (1983) In-vitro discrimination of replicases acting on carcinogen-modified polynucleotide templates. *Proc. natl Acad. Sci., USA, 80,* 969–972

Singer, B., Fraenkel-Conrat, H., Abbott, L.G. & Spengler, S.J. (1984) N^4-Methoxydeoxycytidine triphosphate is in the imino tautomeric form and substitutes for deoxythymidine triphosphate in primed poly d[A-T] synthesis with *E. coli* DNA polymerase I. *Nucleic Acids Res., 12,* 4609–4618

Spears, C.P. (1981) Nucleophilic selectivity ratios of model and clinical alkylating agents by 4-(4′-nitrobenzyl)pyridine competition. *Mol. Pharmacol., 19,* 496–504

Spengler, S. & Singer, B. (1981) Transcriptional errors and ambiguity resulting from the presence of 1,N^6-ethenoadenosine or 3,N^4-ethenocytidine in polyribonucleotides. *Nucleic Acids Res., 9,* 365–373

Strauss, B., Rabkin, S., Sagher, D. & Moore, P. (1982) The role of DNA polymerase in base substitution mutagenesis on non-instructional templates. *Biochimie, 64,* 829–838

Toman, Z., Dambly, C. & Radman, M. (1980) *Induction of a stable heritable epigenetic change by mutagenic carcinogens. a new test system.* In: Montesano, R., Bartsch, H. & Tomatis, L., eds, *Molecular and Cellular Aspects of Carcinogen Screening Tests,* Lyon, International Agency for Research on Cancer, pp. 243–255

Yanofsky, C. (1971) *Mutagenesis studies with* Escherichia coli *mutants with known amino acid (and base-pair) changes.* In: Hollaender, A., ed., *Chemical Mutagens. Principles and Methods for their Detection,* Vol. 1, New York, Plenum Press, pp. 238–287

Zajdela, F., Croisy, A., Barbin, A., Malaveille, C., Tomatis, L. & Bartsch, H. (1980) Carcinogenicity of chloroethylene oxide, an ultimate reactive metabolite of vinyl chloride, and bis(chloromethyl)ether after subcutaneous administration and in initiation-promotion experiments in mice. *Cancer Res., 40,* 352–356

REPLICATION AND TRANSCRIPTION OF POLYNUCLEOTIDES CONTAINING ETHENOCYTOSINE, ETHENOADENINE AND THEIR HYDRATED INTERMEDIATES

B. SINGER[1]

Laboratory of Chemical Biodynamics, Lawrence Berkeley Laboratory, University of California, Berkeley, CA, USA

S.J. SPENGLER

Department of Molecular Biology, University of California, Berkeley, CA, USA

SUMMARY

Replication with *Escherichia coli* DNA polymerase I (Pol I) and transcription with DNA-dependent RNA polymerase from *Escherichia coli* or calf thymus, using as templates synthesized ribo- or deoxyribopolynucleotides containing $1,N^6$-ethenoadenine (εA) or $3,N^4$-ethenocytosine (εC), showed that only εC could direct significant misincorporation. The hydrated intermediate of εC caused errors only upon transcription, but not upon replication. εA was a very poor mutagen as assessed by replication with Pol I. Transcription of polynucleotides containing εA under error-prone conditions caused frequent A misincorporation which could not be detected in replication assays. It is concluded that εC may lead to point mutations, specifically directing the misincorporation of thymine. The analogous derivative, εA, is bulky and is likely to be bypassed rather than read. This mechanism could cause frameshift mutation, as generally found for other bulky adducts.

INTRODUCTION

Transcription of RNA → RNA or DNA → RNA is generally accurate with normal bases and faithfully reflects those modified bases with changed pairing, such as tautomers.

---

[1] To whom correspondence should be addressed

When no stable base-pair can be formed due to the loss of hydrogen-bonding sites, many errors are found in the transcript, up to a stoichiometric number. This occurs regardless of whether transcription takes place in the presence of the normal cation, Mg^{2+}, or in the presence of Mn^{2+}, which is more error-prone. Similarly, the differences are small between prokaryotic and eukaryotic transcriptases. However, experimental data are probably not completely valid if the extent of transcription is very low compared to that of the control (unmodified polymer). If only a few bases are transcribed, this may be due to inhibition or termination of elongation by the modified base, or to significant alterations in the structure of the polynucleotide due to a high proportion of modified base.

Replication of DNA or deoxyribopolynucleotides can be achieved with many types of DNA polymerases. These vary in their fidelity *in vitro* (Loeb & Kunkel, 1982). *Escherichia coli* DNA polymerase I (Pol I) and phage T_4 DNA polymerase replicate with high fidelity: only one guanine (G) residue is misincorporated per 10^5 adenine (A) or thymine (T) residues using a poly d[A-T] template. One reason for lower fidelity, as exemplified by avian myeloblastosis virus (AMV) reverse transcriptase, is the lack of either $5' \rightarrow 3'$ or $3' \rightarrow 5'$ exonuclease activity (Batulla & Loeb, 1974). These activities have the ability to correct errors in base selection or excise an incorrect nucleotide (proofreading). Mismatch or post-replicative repair has not yet been achieved *in vitro* with purified proteins. The enzymes which make few errors have correction mechanisms that do not occur in transcription, which could account for the generally low error rate resulting from Pol I-catalysed replication of deoxyribopolynucleotides containing modified bases. In all instances, the type of enzyme appears to play an important role in the selection of bases and their stereoisomers, particularly when the modified derivative is bulky and unable to fit into a helix or to form base-pairs.

In our experiments testing for mutational ability of various modified nucleosides, we have found only one carcinogen-modified derivative, from those examined so far, which has the same changed pairing, both qualitatively and quantitatively, in transcription and replication. This is O^4-methyluracil (m^4U), or O^4-methylthymine (m^4T), which can substitute for either U/T or C (Singer *et al.*, 1978, 1983b, 1984b). Although this derivative can only form weak hydrogen bonds with A or G, it does not distort the helix in poly d[A-T, m^4T] (a deoxyribopolynucleotide having alternating A and T residues, with some of the T residues being replaced by m^4T), nor interfere with normal stacking interactions (Singer *et al.*, 1983b).

In contrast, the etheno-substituted adenine and cytosine derivatives, $3,N^4$-ethenocytosine (εC), $1,N^6$-ethenoadenine (εA) and their hydrated intermediates (Biernat *et al.*, 1978; Kuśmierek & Singer, 1982a; Leonard, 1983) exhibit a variety of responses when transcribed or replicated (Spengler & Singer, 1981a; Kuśmierek & Singer, 1982b; Singer *et al.*, 1984b). Experimental evidence suggests that upon replication with Pol I, εA and its hydrate are not read, but do not terminate replication, and thus may cause frameshifts, while εC causes point mutations and its hydrate does not cause detectable errors. However, DNA or RNA polymerases lacking proofreading or used under error-prone conditions reveal and amplify misincorporations.

Table 1. Fidelity in transcription with *Escherichia coli* DNA-dependent RNA polymerase [a]

Polymer	Frequency of misincorporation (incorrect NTP/modified base)			
	CTP	ATP	UTP	GTP
poly r(C, εC)	1/65	1/10	1/3	–
poly d(C, εC) [b]	1/40	1/9	1/3	–
poly r(C, m³C)	1/5	1/6	1/3	–
poly d(C, m³C)	1/3	1/14	1/3	–
poly d(A, εA)	nd [c]	nd	–	nd
poly d(C, T)	nd	1/1	nd	1/1

[a] Transcription is in the presence of Mg^{2+} and equimolar concentrations of all four nucleoside triphosphates (NTPs: CTP, cytosine triphosphate; ATP, adenosine triphosphate; UTP, uridine triphosphate; GTP, guanosine triphosphate) with one being [$\alpha^{32}P$]-labelled. Misincorporation is determined by nearest-neighbour analysis. The percent of modified (or second minor) base was determined by high-performance liquid chromatographic analysis and ranged from 3–17%. All data are derived from multiple polymers (ribo- or deoxyribopolynucleotides) containing 2–4 different amounts of εC (3,N^4-ethenocytosine), εA (1,N^6-ethenoadenine) or m³C (3-methylcytosine), together with the corresponding unmodified base. In this table and all others, the ratio of misincorporated base to the base in the polymer is calculated from the percent modified base and the number of NTPs misincorporated into the complementary polymer.
[b] The hydrated intermediate yields similar results.
[c] No transfer of ^{32}P above that of the background

RESULTS

Transcription

DNA-dependent RNA polymerase has been used in transcriptional studies of both ribo- and deoxyribopolynucleotides containing etheno derivatives. In Table 1, polynucleotides containing εC are shown to preferentially misincorporate UTP and to a lesser extent, ATP. Little CTP is misincorporated. These apparently specific errors are the same for ribo- and deoxyribopolynucleotides, suggesting that the transcribing enzyme does not differentiate between the sugars.

The most significant misincorporation directed by εC is that of UTP, which indicates that this etheno derivative is perceived as A. Figure 1 shows possible ways in which such misincorporation could occur. It should be noted in (1) that εC and U residues are sterically compatible and do not distort the Watson-Crick dimensions. If the calculations proposed by Ornstein & Fresco (1983) for hydrogen bonding through carbons (open dots) could be extended to polynucleotides, the base-pairs hypothesized in Figure 1 (1), (2), and (3) might be stabilized by an additional bond. In the case of the εC-A pair (2), appropriate bonding can only be constructed if the A is in the *syn* configuration, which occurs with a frequency of 10^{-2}. Thus, the significantly higher incorporation of ATP ($\sim 1/10$, Table 1) may be a reflection of the enzyme's preference for inserting purines, particularly A, (as reported by Rabkin *et al.*, 1983; Sagher & Strauss, 1983; Kunkel, 1984) and not a result of any pairing.

Fig. 1. Possible structures for the approach of the base moiety of each of the four nucleoside triphosphates to the 3,N^4-ethenocytosine residue (εC). The position of the sugar indicates the position of either the sugar-phosphate moiety in the nucleoside triphosphate or the helix chain in the polymer.

(1) possible pairing of εC with either uracil (U) or thymine (T), involving one standard hydrogen bond (...) and one weak carbon-hydrogen bond (ooo) (Ornstein & Fresco, 1983).

(2) εC with adenine (A) in its normal *anti* configuration: no bonds possible; εC with *Asyn* can be stabilized in the same manner as (1). The frequency of a nucleoside triphosphate rotating to the *syn* configuration is approximately 10^{-2}.

(3) εC with cytosine (C), with the same possible bonding as in (1). (Incorporation of U/T, A, or C has been observed in transcription (see Table 1).)

(4) εC does not form bonds with guanine (G) in either its *anti* or its *syn* configuration.

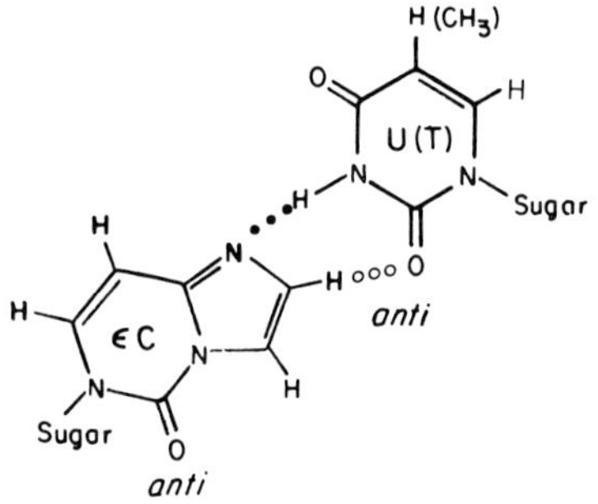

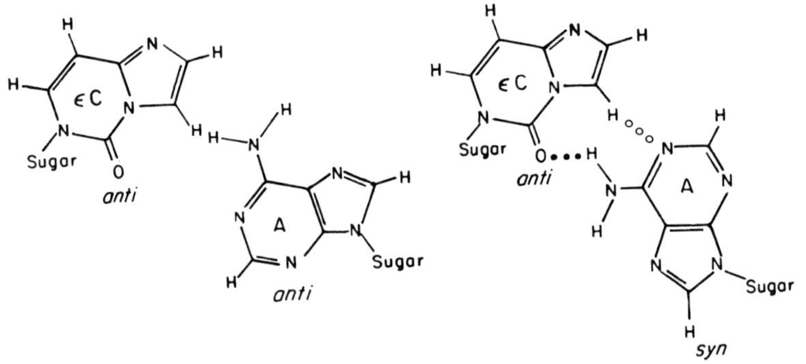

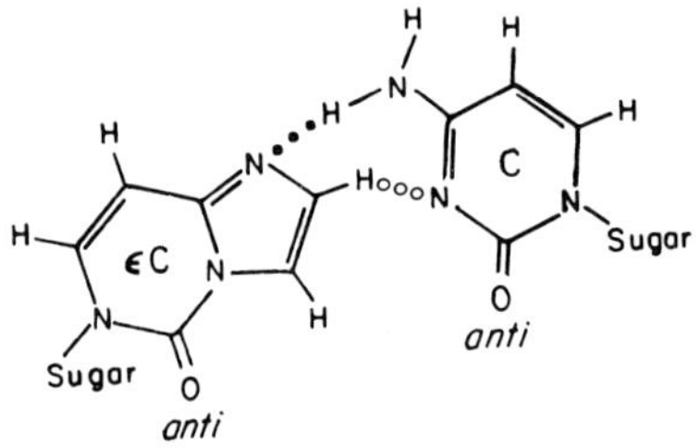

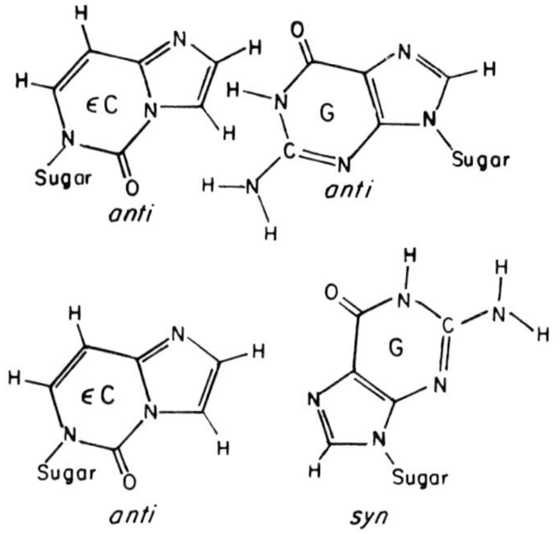

Fig. 2. Possible structure for the approach of the adenine moiety (A) of adenosine triphosphate to the 1,N^6-ethenoadenine residue (εA). The position of the sugar indicates the position of either the sugar-phosphate moiety in the nucleoside triphosphate or the helix chain in the polymer. The A has been rotated *syn* in order to be accommodated in the helix and to present the possibility of bonding through its N^6 (Spengler & Singer, 1981a).

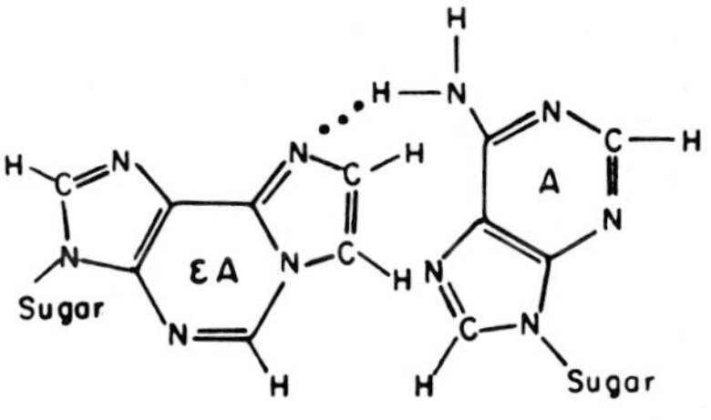

amino, *syn* (10^{-2})

Table 2. Fidelity in transcription of polynucleotides containing promutagenic bases [a]

Modified base in ribopolynucleotide	Frequency of misincorporation (incorrect NTP/modified base)			
	CTP	ATP	UTP	GTP
mo^{4}C [b]	nd [f]	1/1.1	nd	nd
m^{4}U [c]	[g]	1/3	nd	1/3
mo^{6}A [d]	1/30	nd	1/3	—
m^{6}G [e]	1/14	1/9	1/4	—

[a] See Table 1, footnote a. The enzyme used for transcription was *E. coli* DNA-dependent RNA polymerase.
[b] N^4-methoxycytosine; Spengler & Singer (1981b)
[c] O^4-methyluracil; Singer *et al.* (1984b)
[d] N^6-methoxyadenine; Singer & Spengler (1982)
[e] O^6-methylguanine; the data in the table are from the authors' laboratory. Gerchman & Ludlum (1973) also reported misincorporation of ATP and UTP, using Mn^{2+}. The corresponding deoxyribopolynucleotide, also transcribed using Mn^{2+}, was reported to incorporate UTP, but ATP incorporation was not studied (Mehta & Ludlum, 1978).
[f] No transfer of ^{32}P above that of the background
[g] Misincorporation of CTP found in the presence of Mn^{2+} (Singer *et al.*, 1978; Singer, 1982)

Data on 3-methylcytosine (m^{3}C) are included to illustrate that misincorporation by DNA-dependent RNA polymerase from *E. coli* or calf thymus (data not shown) is also a function of the modified base and not only of the polymerase. Both εC and m^{3}C have blocked Watson-Crick pairing sites, yet the incorporation of CTP is high only with m^{3}C.

Incorporation directed by εA must be very different since the εA-containing deoxyribopolynucleotide does not cause detectible misincorporation of any of the nucleoside triphosphates (Table 1). Earlier data from this laboratory indicated that, in the presence of Mn^{2+}, the εA-containing ribopolynucleotide directed incorporation

Table 3. Fidelity in replication with *Escherichia coli* DNA polymerase I[a]

Polymer	Frequency of misincorporation (incorrect dNTP/modified base)			
	dTTP	dGTP	dATP[b]	dCTP
poly d(C, εC)	1/20	–		
poly d(C, εC·H$_2$O)[c]	nd	–		
poly d(C, m³C)	1/80	–		
poly d(A, εA)		1/433		nd
poly d(A, C)[d]		1/16		
poly d(C, U)[d,e]			1/0.7	

[a] See Singer *et al.* (1984a, b) for replication conditions, and Table 1, footnote a, for abbreviations of modified bases and determination of misincorporation. The extent of replication was usually 20–50% that of the control polymer. dNTP, deoxynucleoside triphosphate
[b] The deamination of C, observed with all polymers, allows quantitation only above a high background. None of the modified polymers tested showed dATP incorporation.
[c] εC·H$_2$O, 3,N^4-(N^4-hydroxyethano)cytosine
[d] These copolymers contained 0.8–9% of the second base.
[e] Boiteux & Laval (1982) reported a direct relationship between the presence of uracil residues in heat-deaminated poly (dC) and dAMP incorporation.

of considerable amounts of ATP (Spengler & Singer, 1981a). Neither UTP nor GTP incorporation was detected. As with εC-containing polynucleotides, we attribute the ATP incorporation mostly to the enzyme preference. However, it is possible to form a very weak interaction between εA and A if the A is rotated *syn* (Fig. 2).

In contrast to the data for bases with blocked Watson-Crick sites, promutagenic lesions (Table 2) such as N^4-methoxycytosine (mo^{4}C), N^6-methoxyadenine (mo^{6}A), O^4-methyluracil (m^{4}U) and O^6-methylguanine (m^{6}G) all show specificity of incorporation based on their ability to form stable base pairs. Even the tautomeric shift of the amino-imino exocyclic nitrogen of A in mo^{6}A is observed in these data. Three well-studied products of mutagen reaction are shown in Figure 3, with the probable structures which could occur during the observed change in transcription. We believe, therefore, that transcription does give valid data on the possibility of modified bases causing mutation.

Replication

We have extended our studies of fidelity to systems utilizing DNA polymerases and deoxyribopolynucleotides. *E coli* DNA polymerase I (Pol I) is considered to be one of the most accurate polymerases *in vitro,* yet when replicating εC-containing deoxyribopolynucleotides, an increased error rate is easily detected; approximately one dTTP is misincorporated per 20 εC residues (Table 3). The hydrated intermediate (εC·H$_2$O) does not cause misincorporation at a detectable level. The methylated cytosine, m³C, directs misincorporation of a measurable amount of dTTP but this amount is less than that detected for εC. εA is very poorly mutagenic in this system,

Fig. 3. Possible structures for hydrogen-bonded base pairs of modified promutagenic bases. The position of the sugar indicates the position of either the sugar-phosphate moiety in the nucleoside triphosphate or the helix chain in the polymer. Data in Table 2.

(1) O^6-Methylguanine (m^6G) has been observed to pair with cytosine (C), uracil or thymine (U/T) and adenine (A) (Gerchman & Ludlum, 1973; Mehta & Ludlum, 1978; S.J. Spengler & B. Singer, unpublished results).

(2) O^4-Methylthymine (m^4T) can replace either C or T, to form base pairs with G or A (Singer et al., 1983a, b; 1984b).

(3) N^4-Methoxycytosine (mo^4C), as the keto tautomer, forms normal hydrogen bonds with A (Spengler & Singer, 1981b). Similarly, the substitution of the N^6 of A to yield N^6-methoxyadenine (mo^6A) can involve either the amino or imino tautomer and mo^6A can hydrogen-bond with U or C (Singer & Spengler, 1982).

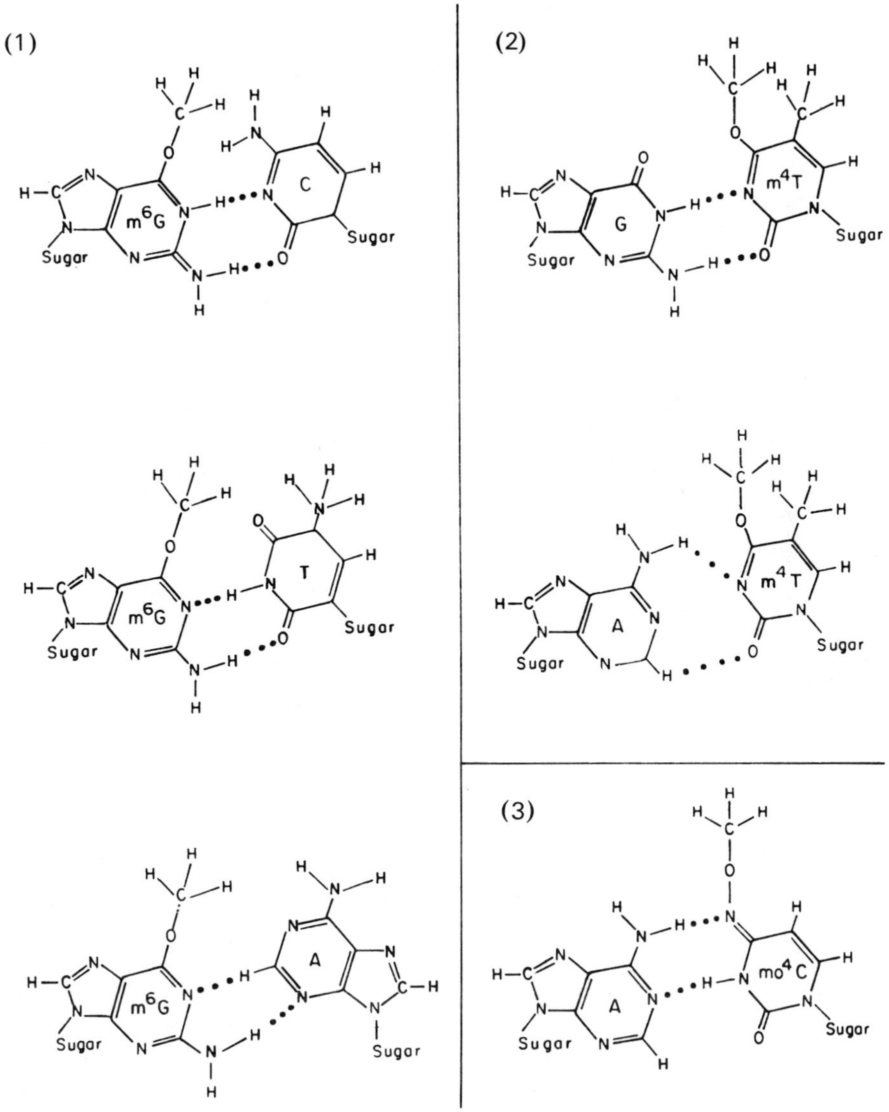

and only about one dGTP is misincorporated per 433 εA residues. In contrast, the presence of small amounts of normal bases in deoxyribopolynucleotides is reflected by specific and essentially stoichiometric incorporation of complementary bases (Table 3 and other data not shown). This shows that under the experimental conditions, Pol I can read through accurately. Since the presence of εA does not depress synthesis greatly, we conclude that the polymerase does not recognize εA as it does unmodified bases, but generally excludes it in order to continue copying the template.

It has been reported that εA is mutagenic, leading to the incorporation of significant dGTP. This conclusion was drawn from replication experiments in which poly(dA) or poly d[A-T], which had been treated extensively with chloroacetaldehyde, was reacted for long time periods (Barbin et al., 1981; Hall et al., 1981). Our data are apparently in conflict with these reports, since we find only about one dGTP misincorporated per 500 εA residues. For this reason, we also tested for the level of errors using polymerases of lower fidelity. Quantitative data are shown in Table 4. It is clear that only an error-prone enzyme leads to εA-directed errors, but the errors are still an order of magnitude lower than those caused by the presence of a significantly mutagenic derivative, m^4T (Singer et al., 1983b).

DISCUSSION

There are several variables to be considered when evaluating the effects of chloroacetaldehyde-modification in mutation. These include the presence and stability of the hydrated intermediates of εA and εC, the size of these compounds, the type of enzyme, and the ability of the template to be both primed and replicated to an appreciable extent. Such variables apply to both natural substrates and repetitive polymers. In addition, there are technical difficulties in searching for each possible misincorporation (Singer et al., 1984b). Thus, various studies of the possible mutagenic effect of vinyl chloride in vivo are not necessarily in agreement.

Previous studies of errors attributed to εC or εA have utilized chloroacetaldehyde-treated poly (dC) (Barbin et al., 1981), poly d[C-G] (Hall et al., 1981), poly (dA) (Barbin et al., 1981), or poly d[A-T] (Hall et al., 1981), which were not analysed for the amount of hydrate or, in the case of poly d[G-C], for modified G. Most of these studies have focused on identifying, but not quantitating, an error in replication of modified templates. When all of this is taken into consideration, we can make some tentative conclusions based primarily on the data in this paper and references therein.

The only hydrogen-bonded structures which can be formed by εC or εC · H_2O are those with U/T or A_{syn} (Fig. 1), the latter of necessity being infrequent, but not excluded by the polymerase (Fersht et al., 1983). What do we actually observe in transcription and replication of polynucleotides containing only one of these compounds? Ribo- and deoxyribopolynucleotides containing εC and its hydrate primarily direct UTP incorporation in transcription (1 UTP/3 εC, Table 1). Replication under conditions enhancing fidelity shows that εC also directs dTTP incorporation (1 dTTP/20 εC, Table 3). But the hydrated intermediate does not lead to infidelity in replication; we attribute this to its structure being flexible enough so

Fig. 4. Possible structures for the association between a 1,N^6-ethenoadenine (εA) residue and a guanine (G) residue. G is shown in each orientation and tautomer. The arrows at two of the sugars indicate the correct position in a helix. The number in parentheses is the calculated frequency of occurrence. There can be no hydrogen bonds formed between εA and G.

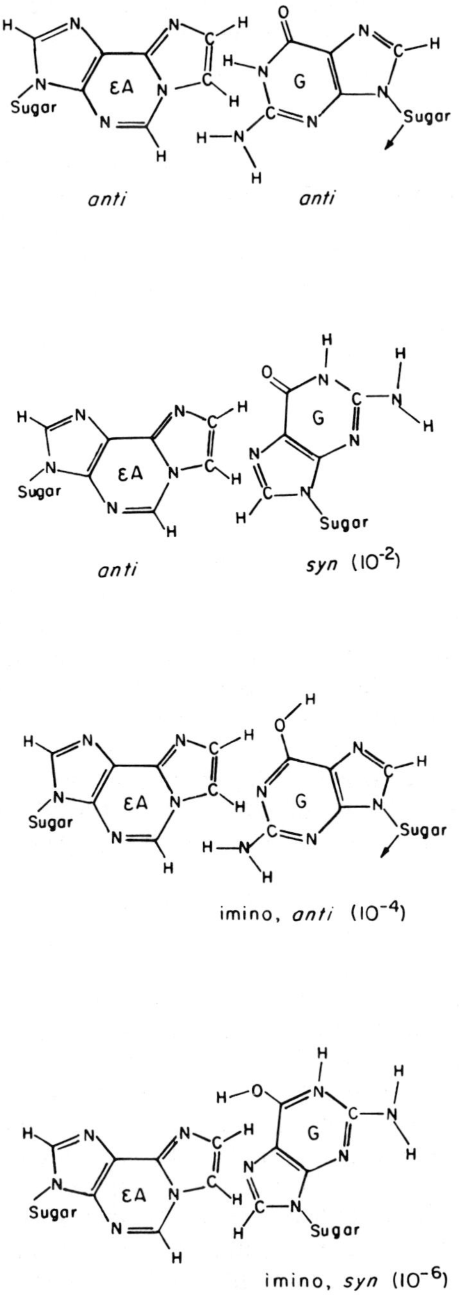

anti   anti

anti   syn (10^{-2})

imino, anti (10^{-4})

imino, syn (10^{-6})

Table 4. Fidelity in replication with various polymerases[a]

Polymer	Frequency of misincorporation[b] (dGTP/modified base)		
	Pol I	Klenow fragment	Reverse transcriptase
poly d(A, εA)[c]	1/433	1/550	1/25
poly d(A, m⁴T)[d]	1/12		1/3
poly d(A, C)	1/1.6		1/1

[a] Data from Singer et al. (1984b)
[b] See Table 1, footnote a
[c] εA, 1,N^6-ethenoadenine
[d] m⁴T, O^4-methylthymine

Fig. 5. Representation of 8-acetylaminofluorene deoxyguanosine (the guanine moiety being rotated out of the helix) with unmodified deoxyguanosine to show the size effect and lack of potential hydrogen bonds. The arrow at the sugar indicates the correct position in a helix.

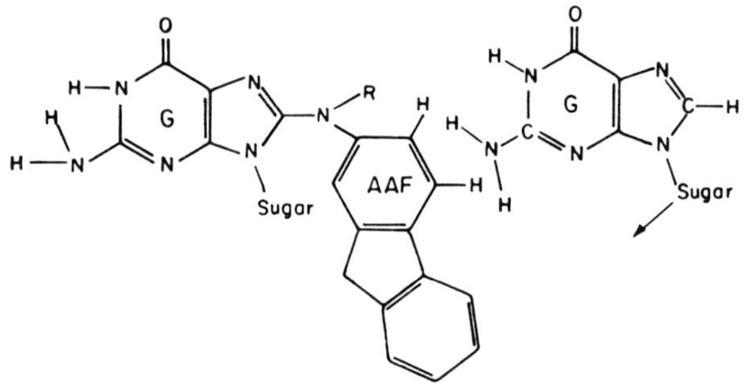

that a type of C · G pair can result. *In vivo*, due to the stability of εC · H_2O (half-life, 13 h at 37 °C) (Kuśmierek & Singer, 1982a), it is likely that this compound predominates during replication and is replicated as C, which could account for the lower error rate (1 dTTP/80 εC) reported by Hall et al. (1981). None of the investigators were able to prove that A misincorporation is not due to the presence of some U, the result of deamination which occurs in all cytosine-containing polynucleotides (Singer & Grunberger, 1983).

While there is little difficulty in postulating εC · U/T pairs, the structure of εA does not permit any type of accommodation with G within a Watson-Crick helix, as would be necessary to account for the misincorporation data. Even allowing the sugar to be distorted or invoking the rare tautomer of G, or rotating the G to the *syn* configuration, cannot lead to a stable structure (Fig. 4). It would seem that the infrequent incorporation of dGTP is only a manifestation of Pol I preferring purines opposite non-coding lesions. We can, however, visualize that εA and A_{syn} might fit within the helix (Fig. 2). Using poly (dA) or poly d[A-T], it was not possible to detect low dATP incorporation due to either reiteration (poly (dA)) or normal incorporation (poly d[A-T]). We did find that transcription of the polyribonucleotide, poly (C, εA), in the presence of Mn^{2+} led to a high misincorporation of ATP (Spengler & Singer, 1981a). Thus, the possibility that A can be the favoured noncomplementary base is not completely ruled out in the present experiments. No dCTP incorporation was detected even though the smaller pyrimidine might fit into the helix.

Perhaps the best way to evaluate how εA is seen by polymerases is to illustrate its size using acetylaminofluorene (AAF) which, when bound to the C-8 of G, rotates to replace G in the helix (Grunberger & Weinstein, 1978) (Fig. 5), and blocks replication by Pol I, leading to bypass (Rabkin et al., 1983). With the less stringent polymerase from AMV or with transcriptases, the frequency of misincorporation is increased. While the chemistry remains the same for all substrates, whether natural or synthetic, the actual level of infidelity which is tolerated *in vivo* cannot be evaluated.

ACKNOWLEDGEMENT

This work was supported by Grant CA 12316 from the National Cancer Institute, National Institutes of Health, Bethesda, MD, USA. SJS is a senior Postdoctoral Fellow of the American Cancer Society, California Division, #S-46-83.

REFERENCES

Barbin, A., Bartsch, H., Leconte, P. & Radman, M. (1981) Studies on the miscoding properties of 1,N^6-ethenoadenine and 3,N^4-ethenocytosine, DNA reaction products of vinyl chloride metabolites, during in-vitro DNA synthesis. *Nucleic Acids Res., 9,* 375–387

Battula, N. & Loeb, L.A. (1974) The infidelity of avian myeloblastosis virus DNA polymerase in polynucleotide replication. *J. biol. Chem., 249,* 4086–4093

Biernat, J., Ciesiołka, J., Górnicki, P., Adamiak, R.W., Krzyżosiak, W.J. & Wiewiórowski, M. (1978) New observations concerning the chloroacetaldehyde reaction with some tRNA constituents. Stable intermediates, kinetics and selectivity of the reaction. *Nucleic Acids Res., 5,* 789–804

Boiteux, S. & Laval, J. (1982) Coding properties of poly(deoxycytidylic acid) templates containing uracil or apyrimidinic sites: in-vitro modulation of mutagenesis by deoxyribonucleic acid repair enzymes. *Biochemistry, 21,* 6746–6751

Fersht, A.R., Shi, J.-P. & Tsui, W.-C. (1983) Kinetics of base misinsertion by DNA polymerase I of *Escherichia coli*. *J. mol. Biol., 165,* 655–667

Gerchman, L.L. & Ludlum, D.B. (1973) The properties of O^6-methylguanine in templates for RNA polymerase. *Biochim. biophys. Acta, 308,* 310–316

Grunberger, D. & Weinstein, I.B. (1978) *Conformational changes in nucleic acids modified by chemical carcinogens.* In: Grover, P.L., ed., *Chemical Carcinogens and DNA, Part 2,* Boca Raton, Fla., CRC Press, pp. 60–93

Hall, J.A., Saffhill, R., Green, T. & Hathway, D.E. (1981) The induction of errors during in-vitro DNA synthesis following chloroacetaldehyde-treatment of poly(dA-dT) and poly(dC-dG) templates. *Carcinogenesis, 2,* 141–146

Kunkel, T.A. (1984) The mutation specificity of depurination. *Proc. natl Acad. Sci. USA*, **81**, 1494–1498

Kuśmierek, J.T. & Singer, B. (1982a) Chloroacetaldehyde-treated ribo- and deoxyribopolynucleotides. 1. Reaction products. *Biochemistry*, **21**, 5717–5722

Kuśmierek, J.T. & Singer, B. (1982b) Chloroacetaldehyde treated ribo- and deoxyribopolynucleotides. 2. Errors in transcription by different polymerases resulting from ethenocytosine and its hydrated intermediate. *Biochemistry*, **21**, 5723–5728

Leonard, N.J. (1984) Etheno-substituted nucleotides and coenzymes: fluorescence and biological activity. *CRC Crit. Rev. Biochem.*, **15**, 125–199

Loeb, L.A. & Kunkel, T.A. (1982) Fidelity of DNA synthesis. *Ann. Rev. Biochem.*, **52**, 429–457

Mehta, J.R. & Ludlum, D.B. (1978) Synthesis and properties of O^6-methyldeoxyguanylic acid and its copolymers with deoxycytidylic acid. *Biochim. biophys. Acta*, **521**, 770–778

Ornstein, R.L. & Fresco, J.R. (1983) Correlation of crystalographically determined and computationally predicted hydrogen-bonded pairing configurations of nucleic acid bases. *Proc. natl Acad. Sci. USA*, **80**, 5171–5175

Rabkin, S.D., Moore, P.D. & Strauss, B.S. (1983) In-vitro bypass of UV-induced lesions by *Escherichia coli* DNA polymerase I: specificity of nucleotide incorporation. *Proc. natl Acad. Sci. USA*, **80**, 1541–1545

Sagher, D. & Strauss B. (1983) Insertion of nucleotides opposite apurinic/apyrimidinic sites in deoxyribonucleic acid during in-vitro synthesis: uniqueness of adenine nucleotides. *Biochemistry*, **22**, 4518–4526

Singer, B. (1982) *Mutagenesis from a chemical perspective: nucleic acid reactions, repair, translation and transcription.* In: Lemontt, J.F. & Generoso, W.M., eds, *Molecular and Cellular Mechanisms of Mutagenesis,* New York, Plenum Press, pp. 1–42

Singer, B. & Spengler, S. (1982) Reaction of O-methylhydroxylamine with adenosine shifts tautomeric equilibrium to cause transitions. *FEBS Lett.*, **139**, 69–71

Singer, B. & Grunberger, D. (1983) *Molecular Biology of Mutagens and Carcinogens,* New York, Plenum Press

Singer, B., Fraenkel-Conrat, H. & Kuśmierek, J.T. (1978) Preparation and template activities of polynucleotides containing O^2- and O^4-alkyluridine. *Proc. natl Acad. Sci. USA*, **75**, 722–726

Singer, B., Kuśmierek, J.T. & Fraenkel-Conrat, H. (1983a) In-vitro discrimination of replicaces acting on carcinogen-modified polynucleotide templates. *Proc. natl Acad. Sci. USA*, **80**, 969–972

Singer, B., Sagi, J. & Kuśmierek, J.T. (1983b) *Escherichia poli* polymerase I can use O^2-methyldeoxythymidine or O^4-methyldeoxythymidine in place of deoxythymidine in primed poly(dA-dT) · poly(dA-dT) synthesis. *Proc. natl Acad. Sci. USA*, **80**, 4884–4888

Singer, B., Fraenkel-Conrat, H., Abbott, L.G. & Spengler, S.J. (1984a) N^4-methoxydeoxycytidine triphosphate is in the imino tautomeric form and substitutes for deoxythymidine triphosphate in primed poly d[A-T] synthesis with *E. coli* DNA polymerase I. *Nucleic Acids Res.*, **12**, 4609–4619

Singer, B., Abbott, L.G. & Spengler, S.J. (1984b) Assessment of mutagenic efficiency of two carcinogen-modified nucleosides, $1,N^6$-ethenodeoxyadenosine and O^4-methyldeoxythymidine, using polymerases of varying fidelity. *Carcinogenesis,* **5,** 1165–1171

Spengler, S.J. & Singer, B. (1981a) Transcription errors and ambiguity resulting from the presence of $1,N^6$-ethenoadenosine or $3,N^4$-ethenocytidine in polyribonucleotides. *Nucleic Acids Res.,* **9,** 356–373

Spengler, S. & Singer, B. (1981b) The effect of tautomeric shift on mutation: N^4-methoxycytidine forms hydrogen bonds with adenosine in polymers. *Biochemistry,* **20,** 7290–7294

REPAIR OF ETHENO DNA ADDUCTS BY N-GLYCOSYLASES

F. OESCH, S. ADLER, R. RETTELBACH & G. DOERJER

Institute for Toxicology, University of Mainz, Mainz, Federal Republic of Germany

SUMMARY

After incubation of chloroacetaldehyde-treated DNA with cell-free homogenates, the excision of N^2,3-ethenoguanine and 1,N^6-ethenoadenine was observed with a rat brain tumour cell line. The repair mechanism was that of an N-glycosylase. The high specifity of all known DNA N-glycosylases and some unique properties of the enzymatic reaction indicate the existence of N-glycosylases specific for the repair of etheno, or similar, adducts.

INTRODUCTION

Covalent products of the reactions of chemical carcinogens with DNA can lead to miscoding during DNA replication and may therefore be the cause of mutations in the genetic information. Positive correlations have been obtained between the formation, persistance and accumulation of promutagenic DNA lesions such as O^6-alkylguanine and O^4-alkylthymine and carcinogenic events (Bedell *et al.*, 1982; Swenberg *et al.*, 1982). For all known etheno adducts which result from treatment of DNA with chloroacetylaldehyde (CAA) (Oesch & Doerjer, 1982), miscoding potential has been demonstrated in various systems (Hall *et al.*, 1981; Barbin *et al.*, 1981; Kuśmierek & Singer, 1982). For N^2,3-ethenoguanine (εG), a miscoding scheme can be formulated which is analogous to that of O^6-methylguanine; however, such analogies are not feasible for 1,N^6-ethenoadenine (εA) or 3,N^4-ethenocytosine (εC), as indicated in Figure 1.

In addition to the miscoding action of DNA adducts, another important factor concerning their biological relevance is the mechanism of repair, especially since it is known that repair processes may also introduce errors into the DNA, as, for example, in bacteria. A recent review of DNA repair enzyme systems (Lindahl, 1982) allows the classification of known DNA repair mechanisms into three categories. First, the incision-excision mechanism of nucleases creates single-strand breaks by cleavage of

Fig. 1. Miscoding potential of modified bases in DNA. A, adenine; T, thymine; G, guanine; C, cytosine; εG, N^2,3-ethenoguanine; MeG, O^6-methylguanine; εA, 1,N^6-ethenoadenine; εC, 3,N^4-ethenocytosine

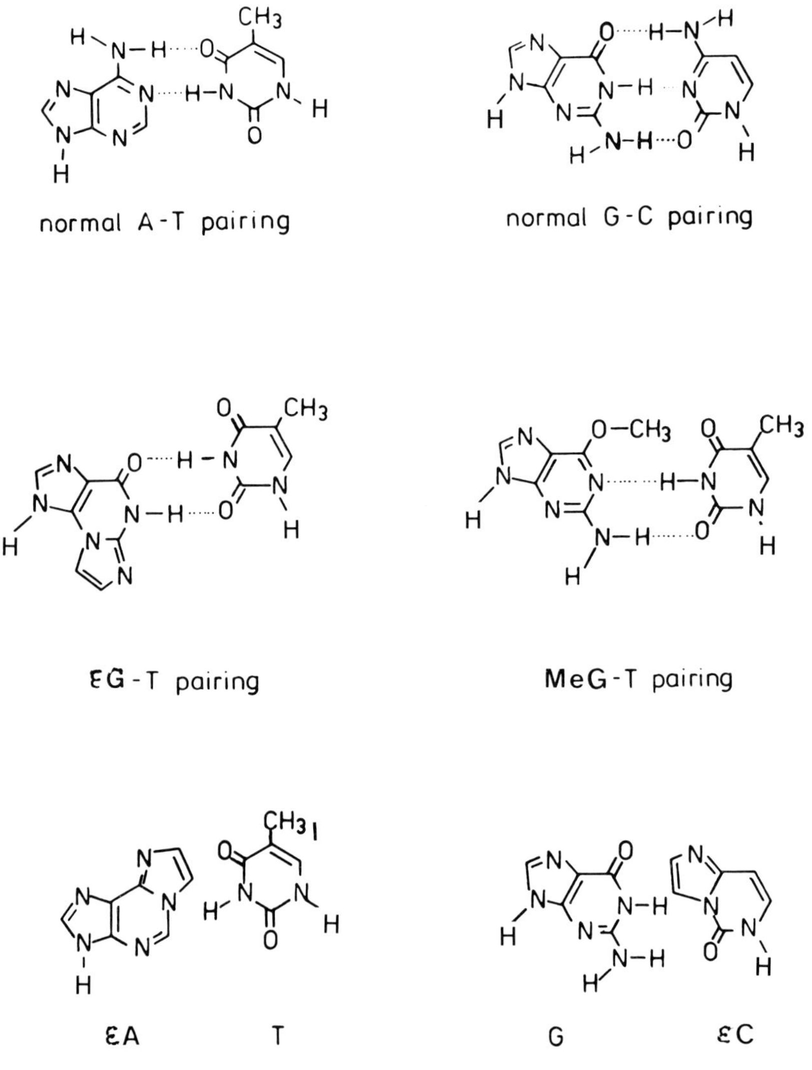

normal A-T pairing

normal G-C pairing

εG-T pairing

MeG-T pairing

εA   T

G   εC

phosphodiester bonds. After removal of an oligonucleotide containing the DNA lesion, the DNA sequence is reconstructed by polymerases. This process seems to be responsible for the repair of many different types of DNA lesions. A second mechanism starts with the action of DNA *N*-glycosylases and results in apurinic/apyrimidinic sites in the DNA which in turn are repaired by the first mechanism or by DNA base insertases. All known DNA *N*-glycosylases show a high substrate specificity either towards endogenously formed DNA lesions (e.g., uridine

N-glycosylase) or towards DNA bases modified by chemical carcinogens, as has been demonstrated for methylated DNA bases. A third repair mechanism involves the removal of the DNA adduct moiety from an altered base, which restores the original base in the DNA. Examples of this third mechanism are the dealkylations of O^6-alkylguanine and O^4-alkylthymine by the transfer of the alkyl groups to specific alkyl acceptor proteins (Lindahl, 1982; Ahmmed & Laval, 1984).

So far, no repair mechanisms for etheno DNA adducts formed from bifunctionally reactive species have been described in the literature. In this study, we report the excision of εG and εA by DNA N-glycosylases.

MATERIALS AND METHODS

Chloroacetaldehyde (CAA)-treated DNA

DNA (salmon sperm) was incubated in 0.12 M sodium phosphate at room temperature for 24 h with an excess of freshly prepared CAA. After incubation, DNA was precipitated with ethanol. CAA-treated DNA was depurinated in 0.1 N HCl or enzymatically digested as described previously (Doerjer et al., 1978) and analysed by high-performance liquid chromatography (HPLC).

Incubations

After trypsinization, BT3/Caf cells were ultrasonicated in buffer (100 mM sodium phosphate pH 7.2, 1 mM dithiothreitol, 0.1 mM EDTA). CAA-treated DNA was added and the homogenate (10 mg protein) was incubated at 37 °C for 2 h. When heat-denatured (100 °C, 10 min) homogenate was used for the control, no excision of modified bases was observed. After incubation, the mixture was heated in boiling water for 30 min and centrifuged at 10 000 × g. The supernatant fraction was diluted with 2 volumes of ethanol and stored overnight at −20 °C. After centrifugation, the ethanol was evaporated by a stream of nitrogen and aliquots of the sample were analysed by HPLC.

HPLC analysis

Samples were analysed on a HPLC system (Milton Roy) equipped with a 3 cm RP 18 (LiChrosorb C 18, Merck, 10 μ) precolumn, a 10 cm RP 8 (LiChrosorb C8, Merck, 10 μ) column and a 10 cm anion-exchange column (SB, Machery-Nagel, 10 μ). 50 mM ammonium formate (pH 8.0) was used as the eluant at flow rates of 0.4 ml/min (0–35 min) and 0.8 ml/min (35–80 min). Compounds were detected by their fluorescence (FS 970, Kratos) at 230 nm excitation with a 389 nm cut-off filter in the emission beam. The identity of εG, εA, 1,N^6-ethenodeoxyadenosine (εdAdo) and N^2,3-ethenodeoxyguanosine (εdGuo) was confirmed by chromatography with authentic standards.

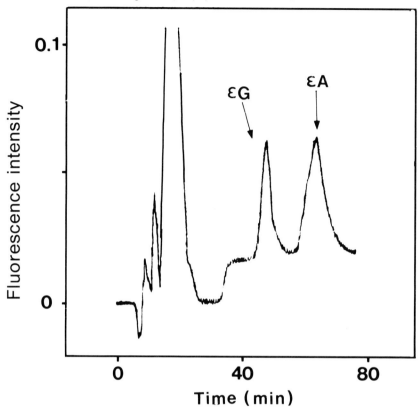

Fig. 2. High-performance liquid chromatographic analysis of liberated etheno adducts after incubation of CAA-treated DNA with homogenates of BT3/Caf cells. Details are given in 'Materials and Methods'. εG, N^2,3-ethenoguanine; εA, 1,N^6-ethenoadenine

RESULTS AND DISCUSSION

For the analysis of CAA-modified DNA bases, a combination of reverse-phase and strong anion-exchange columns was necessary to achieve satisfactory resolution of εG and εA for all purposes of the present study. With reverse-phase columns alone, a complete separation of εG from adenine or a separation of the corresponding deoxyribonucleosides was not obtained. With strong anion-exchange columns alone, the separation of εG and εA was poor at pH values around 8. Under more alkaline conditions, the retention time increased for εG, but decreased for εA with the result that separation of εA from proteins in samples after incubation of CAA-treated DNA with cell homogenates was not possible. Because εG and εA give maximal fluorescence at neutral pH and their fluorescence is almost completely lost at acid pHs, other separation conditions have not been examined. With its fluorescence optimum in acidic solutions and its relatively high polarity, a possible excision of εC

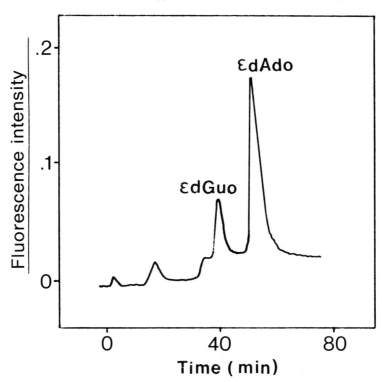

Fig. 3. High-performance liquid chromatographic analysis of CAA-treated DNA after enzymatic digestion by DNase, diesterase and alkaline phosphatase. Details are given in "Materials and Methods". εdGuo, N^2,3-ethenodeoxyguanosine; εdAdo, 1,N^6-ethenodeoxyadenosine

cannot be observed in our system. Due to the alkaline pH, the effectiveness of the reverse-phase columns decreases slightly with time. With new columns, the addition of up to 5% acetonitrile to the eluent is of advantage.

The ratio of εG to εA (Fig. 2) obtained after incubation of CAA-treated DNA with homogenates from BT3/Caf cells (a rat brain tumour cell line) differs from the ratio of adducts present in the modified DNA as determined after enzymatic digestion with DNase, diesterase and alkaline phosphatase (Fig. 3). Acid depurination of the CAA-treated DNA gives a ratio of εG to εA similar to that of the corresponding deoxyribonucleosides after enzymatic hydrolysis of modified DNA. However, quantitation of the peaks and comparisons with available standards indicate a substantial decrease in the fluorescence of both εG and εA during the acid depurination of DNA. This decrease in fluorescence intensities indicates a partial decay or a rearrangement to less fluorescent derivatives during the acid depurination. Since this does not occur during enzymatic hydrolysis, the ratio of the modified deoxyribonucleosides shown in Figure 3 gives a more objective impression about the ratios of adducts in the DNA used as substrate. The ratios of the excised modified bases compared to the ratios of adducts present in the DNA suggest a preferential excision of εG compared to εA. The excision was linear with protein concentration (2–5 mg/ml).

When larger amounts of homogenates were prepared and frozen at $-20\,°C$ for use in more extensive studies on the enzyme characteristics, the N-glycosylase activity towards εG and εA in DNA was completely lost (after storage of some weeks). This lability is exceptionally uncommon for known DNA N-glycosylases.

The existence and specificity of DNA N-glycosylases responsible for the repair of methylated DNA bases suggests that such adducts can be formed by endogenous events, as well, and that there may be a continuous process of endogenous DNA methylation and repair of the resulting adducts. This concept has been supported by recent reports of the formation of methylated DNA bases during treatment with a no-carbon-containing carcinogen, hydrazine (Quintero-Ruiz et al., 1981), and of the non-enzymatic DNA methylation by S-adenosylmethionine (Barrows & Magee, 1982). By analogy, if the DNA N-glycosylases for the removal of etheno adducts observed in this study are specific for these and similar anellated DNA bases, a possible endogenous formation of such DNA adducts could also be suggested. In addition to reactions which lead to the formation of etheno derivatives such as the 'Y'-bases in RNA (Kasai et al., 1976), many bifunctional compounds have to be considered. One example is malondialdehyde, a common product of lipid peroxidation, which leads to the formation of propeno adducts (Seto et al., 1983). Research with such annellation products would supplement studies concerning aliphatically alkylated DNA bases.

Our results show the existence of DNA N-glycosylases which excise etheno purines from chemically modified DNA in homogenates of mammalian cells. The existence of these enzymes suggests that etheno adducts can be of similar importance as methylated DNA bases for our understanding of chemical carcinogenesis.

ACKNOWLEDGEMENT

These investigations were supported by the Bundesministerium für Forschung und Technologie and the Deutsche Forschungsgemeinschaft. BT3/Caf cells were a generous gift of Drs N. Huh and M.F. Rajewsky.

REFERENCES

Ahmmed, Z. & Laval, J. (1984) Enzymatic repair of O-alkylated thymidine residues in DNA: involvement of a O^4-methylthymidine-DNA methyltransferase and a O^2-methylthymine-DNA glycosylase. *Biochem. biophys. Res. Commun., 120,* 1–8

Barbin, A., Bartsch, H., Leconte, P. & Radman, U. (1981) Studies on the miscoding properties of 1,N^6-ethenoadenine and 3,N^4-ethenocytosine, DNA-reaction products of vinyl chloride metabolites, during in-vitro DNA synthesis. *Nucleic Acids Res., 9,* 375–387

Barrows, L.A. & Magee, P.N. (1982) Nonenzymatic methylation of DNA by S-adenosylmethionine *in vitro*. *Carcinogenesis, 3,* 349–351

Bedell, M.A., Lewis, J.G., Billings, K.C. & Swenberg, J.A. (1982) Cell specifity in hepatocarcinogenesis: O^6-methylguanine preferentially accumulates in target cell during continuous exposure of rats to 1,2-dimethylhydrazine. *Cancer Res., 42,* 3079–3083

Doerjer, G., Diessner, H., Buecheler, J. & Kleihues, P. (1978) Reaction of 7,12-dimethylbenz(a)anthracene with DNA of fetal and maternal rat tissues *in vivo*. *Int. J. Cancer, 22,* 288–291

Hall, J.A., Saffhill, R., Green, T. & Hathway, D.E. (1981) The induction of errors during in-vitro synthesis following chloroacetaldehyde treatment of poly(dA-dT) and poly(dG-dC) templates. *Carcinogenesis, 2,* 141–145

Kasai, H., Goto, M., Ikeda, K., Zama, M., Mizuno, Y., Matsuursa, S., Sugimoto, T. & Goto, T. (1976) Structure of wye (Yt base) and wyosine (Yt) from *Torulopsis utilis* phenylalanine transfer ribonucleic acid. *Biochemistry, 15,* 898–904

Kuśmierek, J.T. & Singer, B. (1982) Chloroacetaldehyde-treated ribo- and deoxyribonucleotides. Errors in transcription by different polymerases resulting from ethenocytosine and its hydrated intermediate. *Biochemistry, 21,* 5723–5728

Lindahl, T. (1982) DNA repair enzymes. *Ann. Rev. Biochem., 51,* 61–87

Oesch, F. & Doerjer, G. (1982) Detection of N^2,3-ethenoguanine after treatment of DNA with chloroacetaldehyde *in vitro*. *Carcinogenesis, 3,* 663–665

Quintero-Ruiz, A., Paz-Neri, L.L. & Villa-Trevino, S. (1981) Indirect alkylation of mouse liver DNA and RNA by hydrazine *in vivo:* a possible mechanism of action as a carcinogen. *J. natl Cancer Inst., 67,* 613–618

Seto, H., Okuda, T., Takesue, T. & Ikemura, T. (1983) Reaction of malondialdehyde with nucleic acid. I. Formation of fluorescent pyrimido(1,2-a)purin-10(3H)-one nucleosides. *Bull. chem. Soc. Jpn., 56,* 1799–1802

Swenberg, J.A., Bedell, M.A., Billings, K.C., Umbenhauer, D.R. & Pegg, A.E. (1982) Cell-specific differences in O^6-alkylguanine DNA repair activity during continuous exposure to carcinogen. *Proc. natl Acad. Sci. USA, 79,* 5499–5502

REPAIR OF CYCLIC NUCLEIC ACID ADDUCTS AND ADVERSE EFFECTS OF APURINIC SITES

J. LAVAL & S. BOITEUX

Groupe «Réparation des lésions chimio et radioinduites», Centre National de la Recherche Scientifique (LA 147), Institut National de la Santé et de la Recherche Médicale (U 140), Institut Gustave-Roussy, 94805 Villejuif, France

INTRODUCTION

Since the proposal by Loveless (1969) that a specific base modification in DNA could lead to mutagenicity and carcinogenicity, considerable efforts have been made to correlate a specific DNA lesion to mutation and/or to cancer induction. Base modifications introduced by alkylating agents and other carcinogenic compounds have been extensively studied (for review, see Lawley, 1976; Singer & Grunberger, 1983).

In order to maintain the integrity of the genome, cells have efficient DNA repair systems. The mechanisms of repair differ according to the type of lesion (for review, see Laval & Laval, 1980; Lindahl, 1982).

DNA REPAIR OF LESIONS INTRODUCED BY MONOFUNCTIONAL ALKYLATING AGENTS

Simple alkylating agents such as the dialkylsulfates and alkylalkanesulfonates react mainly with nitrogens of the bases and hardly, if at all, with oxygens. N-Nitroso compounds, either directly or after metabolic activation, react with both the nitrogens and the oxygens of the bases. They are potent carcinogens (Singer & Grunberger, 1983).

Most of these lesions are actively repaired. For some lesions, such as 3-methyladenine, 3-methylguanine, 7-methylguanine, the imidazole-ring-open form of 7-methylguanine and others, the base excision system is involved. This system, first detected in *Micrococcus luteus* (Laval, 1977) and later in human cells (Brent, 1984), involves the sequential action of two enzymes. A DNA glycosylase excises the modified base yielding an apurinic/apyrimidinic site (AP site) which in turn is excised by an endonuclease specific for AP sites. A different system operates for guanine modified at the O^6 position; this lesion is repaired by an alkyl acceptor protein which transfers the alkyl group from the modified guanine to a cysteine of the protein. During this process, the protein is inactivated (Lindahl, 1982). A similar mechanism is involved in the repair of O^4-methylthymidine (Ahmmed & Laval, 1984).

Bulky lesions such as pyrimidine dimers, platinum adducts, and aflatoxin adducts are repaired in *Escherichia coli* by the *uvr* ABC endonuclease (Sancar & Rupp, 1983). Cross-link repair is also dependent on the *uvr* ABC gene product. A similar mechanism for excising bulky lesions has not been described in mammalian cells.

DNA REPAIR OF LESIONS GENERATED BY BIFUNCTIONAL ALKYLATING AGENTS

The nitrosoureas used in chemotherapy, 1,3-bis-(2-chloroethyl)-1-nitrosourea and 1-(2-chloroethyl)-3-cyclohexyl-1-nitrosourea, are bifunctional since they carry a chloride group. Tong *et al.* (1982) have shown that guanine substituted at positions O^6 and N-7 by hydroxyalkylation or haloethylation, as well as diguanyl ethane, are present in DNA modified by nitrosoureas. The molecular structure of the nitrosourea used modulates the relative amount of these various modifications. Chemotherapeutic action of these nitrosoureas is believed to be due to the occurence of various cross-links in DNA: the di(7-guanyl)ethane, which is an intrastrand one, and an ethyl link between the N-3 position of a cytosine and the N-1 of guanine. Robins *et al.* (1984) have shown that *E. coli* O^6-methylguanine-DNA methyltransferase prevents the appearance of cross-links in DNA exposed to bis(chloroethyl)nitrosourea. The methyl acceptor protein removes the chloroethyl residue present at the O^6 position of guanine, thus preventing the internal rearrangement of this residue to the N-1 position, which is a prerequisite for cross-link formation. Similar results were obtained using partially purified protein of human leukaemic lymphoblasts (Brent, 1984). Once the cross-links are formed, they are expected to be repaired in *E. coli* by the *uvr* ABC endonuclease.

β-PROPIOLACTONE

β-Propiolactone is a carcinogen which interacts with DNA yielding 1-(2-carboxyethyl)-2′deoxyadenosine and 3-(β-D-2-deoxyribosyl)-7,8-dihydropyrimido-[2,1-*i*]purine-9-one (Chen *et al.*, 1981). Both lesions appear to be excised by the *E. coli* 3-methyladenine-DNA glycosylase II (Thomas *et al.*, 1982).

VINYL CHLORIDE

Vinyl chloride has been shown to be carcinogenic in man (Creech & Johnson, 1974), as well as mutagenic (Huberman *et al.*, 1975; Barbin *et al.*, 1984), and due to the large amounts that are processed in the world, it is a major industrial concern. It has been postulated that vinyl chloride is activated by the P-450 microsomal system to an epoxide, chloroethylene oxide, which chemically rearranges to form chloroacetaldehyde (Guengerich et al., 1979; Zajdela et al., 1980). This latter compound reacts with DNA in vitro and in vivo yielding 3,N^4-ethenocytidine, 1,N^6-ethenoadenosine and N^2,3-ethenoguanosine (Green & Hathway, 1978; Laib et al., 1981; Oesch & Doerjer, 1982; Sherer et al., 1981). These lesions show miscoding properties during in-vitro synthesis (Barbin et al., 1981; Hall et al., 1981; Kuśmierek et al., 1982). It does not appear that the repair of these lesions have been investigated in micro-organisms. However, Oesch et al. (these proceedings[1]) have reported the excision of ethenoadenine and ethenoguanine residues by N-glycosylases present in a rat brain tumour cell line.

GENETIC EFFECTS OF APURINIC/APYRIMIDINIC SITES

Most of the DNA-damaging agents which modify bases labilize the glycosidic bond (Lawley, 1976; Drinkwater et al., 1980; Singer & Grunberger, 1983), thereby increasing the rate constant of depurination (estimated by Lindahl and Nyberg (1972) as 1.8×10^{-9} per min) by three to four orders of magnitude. In addition, repair of alkylated bases by the base-excision mechanism described above involves the generation of an AP site. It has been postulated that AP sites are potentially mutagenic (Bautz & Freese, 1960). Using an in-vitro system, we have obtained evidence in support of this hypothesis (Boiteux & Laval, 1982). Deoxyuridylic (dUMP) residues were generated by heating poly(deoxycytidylic acid); these residues were then excised by pure uracil-DNA glycosylase in order to yield AP sites. When this modified polymer was replicated by E. coli DNA polymerase I, deoxyadenylic (dAMP) residues were preferentially incorporated, opposite the AP sites. Deoxythymidylic (dTMP) residues were also incorporated at a much reduced rate. The incorporation of dAMP, as well as dTMP, was abolished when the polymer was pretreated with pure AP-endonuclease. These results emphasize the key importance of DNA repair enzymes in preventing mutagenesis. The premutagenic properties of AP sites have been demonstrated in natural DNA (Sagher & Strauss, 1983), and also in experiments performed in vivo using the E. coli phage ΦX 174 (Shaaper et al., 1983) and in SV-40 infected mammalian cells (Gentil et al., 1984).

In conclusion, we would like to propose that some DNA lesions, if not repaired, could be lethal by blocking the progression of DNA polymerase; relevant studies have been carried out with 3-methyladenine (Boiteux et al., 1984) and the ring-open form of 7-methylguanine (Boiteux & Laval, 1983). The search for anticancer drugs is based upon this property. Other lesions, such as O^6-methylguanine (Loveless, 1969) and AP sites, are premutagenic and, unless eliminated from the genome, are believed to lead to cancer.

REFERENCES

Ahmmed, Z. & Laval, J. (1984) Enzymatic repair of O-alkylated thymidine residues in DNA: involvement of a O^4-methylthymine-DNA methyltransferase and a O^2-methylthymine-DNA glycosylase. *Biochem. biophys. Res. Commun.*, **120**, 1–8

---

[1] See p. 373.

Barbin, A., Bartsch, H., Lecomte, P. & Radman, M. (1981) Studies on the miscoding properties of 1,N^6-ethenoadenine and 3,N^4-ethenocytosine, DNA reaction products of vinyl chloride metabolites, during in-vitro DNA synthesis. *Nucleic Acids Res., 9,* 375–387

Barbin, A., Laib, R. & Bartsch, H. (1984) Studies on the mechanism of chloroethylene oxide (CEO)-induced mutagenesis. *Mutat. Res., 130,* 165–166

Bautz, E. & Freese, E. (1960) On the mutagenic effect of alkylating agents. *Proc. natl Acad. Sci. USA, 46,* 1585–1594

Boiteux, S. & Laval, J. (1982) Coding properties of poly(dC) templates containing uracil or apyrimidinic sites: in-vitro modulation of mutagenesis by DNA repair enzymes. *Biochemistry, 21,* 6746–6750

Boiteux, S. & Laval, J. (1983) Imidazole-open-ring 7-methylguanine: an inhibitor of DNA synthesis. *Biochem. biophys. Res. Commun., 110,* 552–559

Boiteux, S., Huisman, O. & Laval, J. (1984) 3-Methyladenine residues in DNA induces the SOS function *sfi* A in *E. coli EMBO J., 3,* 2569–2573

Brent, T.P. (1984) Suppression of cross-link formation in chloroethylnitrosourea-treated DNA by an activity in extracts of human leukemic lymphoblasts. *Cancer Res., 44,* 1887–1892

Chen, R., Mieyal, J. & Goldthwait, D.A. (1981) The reaction of β-propiolactone with derivatives of adenine and with DNA. *Carcinogenesis, 2,* 73–80

Creech, J.L. & Johnson, M.N. (1974) Angiosarcoma of the liver in the manufacture of polyvinyl chloride. *J. occup. Med., 16,* 150–151

Drinkwater, N.R., Miller, E.C. & Miller, J.A. (1980) Estimation of apurinic/apyrimidinic sites and phosphotriesters in deoxyribonucleic acid treated with electrophilic carcinogens and mutagens. *Biochemistry, 19,* 5087–5092

Gentil, A., Margot, A. & Sarasin, A. (1984) Apurinic sites cause mutation in simian virus 40. *Mutat. Res., 129,* 141–147

Green, T. & Hathway, D.E. (1978) Interactions of vinyl chloride with rat liver DNA in vivo. *Chem.-biol. Interactions, 22,* 211–224

Guengerich, F.P., Crawford, M.W., Jr. & Watanabe, P.G. (1979) Activation of vinyl chloride to covalently bound metabolites: roles of 2-chloroethylene oxide and 2-chloroacetaldehyde. *Biochemistry, 18,* 5177–5182

Hall, J.A., Saffhill, R., Green, T. & Hathway, D.E. (1981) The induction of errors during in-vitro DNA synthesis following chloroacetaldehyde-treatment of poly(dA-dT) and poly(dC-dG) templates. *Carcinogenesis, 2,* 141–146

Huberman, E., Bartsch, H. & Sachs, L. (1975) Mutation induction of Chinese hamster V79 cells by two vinyl chloride metabolites, chloroethylene oxide and 2-chloroacetaldehyde. *Int. J. Cancer, 16,* 639–644

Kuśmierek, J.T. & Singer, B. (1982) Chloroacetaldehyde-treated ribo- and deoxyribopolynucleotides. 2. Errors in transcription by different polymerases resulting from ethenocytosine and its hydrated intermediate. *Biochemistry, 21,* 5273–5278

Laib, R.J., Gwinner, L.M. & Bolt, H.M. (1981) DNA alkylation by vinyl chloride metabolites: etheno derivatives or 7-alkylation of guanine? *Chem.-biol. Interactions, 37,* 219–231

Laval, J. (1977) Two enzymes are required for strand incision in repair of alkylated DNA. *Nature, 269,* 829–832

Laval, J. & Laval, F. (1980) *Enzymology of DNA repair.* In: Montesano, R., Bartsch, H. & Tomatis, L., eds, *Molecular and Cellular Aspects in Carcinogen Screening Tests (IARC Scientific Publications No. 27),* Lyon, International Agency for Research on Cancer, pp. 55–73

Lawley, P.D. (1976) *Methylation of DNA by carcinogens: some applications of chemical analytical methods.* In: Montesano, R., Bartsch, H. & Tomatis, L., eds, *Screening Tests in Chemical Carcinogenesis (IARC Scientific Publications No. 12),* Lyon, International Agency for Research on Cancer, pp. 181–210

Lindahl, T. (1982) DNA repair enzymes. *Ann. Rev. Biochem., 51,* 61–87

Lindahl, T. & Nyberg, B. (1972) Rate of apurination of native DNA. *Biochemistry, 11,* 3610–3618

Loveless, A. (1969) Possible relevance of O^6-alkylation of deoxyguanosine to the mutagenicity and carcinogenicity of nitrosamines and nitrosamides. *Nature, 223,* 206–207

Oesch, F. & Doerjer, G. (1982) Detection of N^2,3-ethenoguanine in DNA after treatment with chloroacetaldehyde *in vitro. Carcinogenesis, 3,* 663–665

Robins, P., Harris, A.L., Goldsmith, I. & Lindahl, T. (1984) Cross-linking of DNA induced by chloroethylnitrosourea is prevented by O^6-methylguanine methyltransferase. *Nucleic Acids Res., 11,* 7743–7757

Sagher, D. & Strauss, B. (1983) Insertion of nucleotides opposite apurinic/apyrimidinic sites in deoxyribonucleic acid during in vitro synthesis: uniqueness of adenine nucleotides. *Biochemistry, 22,* 4518–4526

Sancar, A. & Rupp, W.D. (1983) A novel repair enzyme: *uvr* ABC excision nuclease of *Escherichia coli* cuts a DNA strand on both sides of the damaged region. *Cell, 33,* 249–260

Schaaper, R.M., Kunkel, T.A. & Loeb, L.A. (1983) Infidelity of DNA synthesis associated with bypass of apurinic sites. *Proc. natl Acad. Sci. USA, 80,* 487–491

Scherer, E., Van Der Laken, C.J., Gwinner, L.M., Laib, R.J. & Emmelot, P. (1981) Modification of deoxyguanosine by chloroethylene oxide. *Carcinogenesis, 2,* 671–677

Singer, B. & Grunberger, D. (1983) *Molecular Biology of Mutagens and Carcinogens,* New York, Plenum Press

Thomas, L., Yang, C.H. & Goldthwait, D.A. (1982) Two DNA glycosylases in *Escherichia coli* which release primarily 3-methyladenine. *Biochemistry, 21,* 1162–1169

Tong, W.P., Kirk, M.C. & Ludlum, D.B. (1982) Formation of the cross-link 1-[N^3-deoxycytidyl],2-[N^1-deoxyguanosinyl]-ethane in DNA treated with N,N'-bis(2-chloroethyl)-N-nitrosourea. *Cancer Res., 42,* 3102–3105

Zajdela, F., Croizy, A., Barbin, A., Malaveille, C., Tomatis, L. & Bartsch, H. (1980) Carcinogenicity of chloroethylene oxide, an ultimate reactive metabolite of vinyl chloride, and bis(chloromethyl)ether after subcutaneous administration and in initiation-promotion experiments in mice. *Cancer Res., 40,* 352–356

TERMINATION OF SYNTHESIS RESULTING FROM MODIFYING BASES IN DNA

B. STRAUSS, K. LARSON, S. RABKIN, D. SAGHER & J. SAHM

Department of Molecular Genetics and Cell Biology, University of Chicago, Chicago, IL, USA

Termination of DNA synthesis by lesions in the template depends on the lesion itself, the kind and concentration of nucleotides present, the identity of the polymerase and the site at which the lesion occurs. Bypass is particularly affected by the template sequence 5′ to the lesion. Nucleotides can be added opposite non-instructional lesions and it is possible that this type of addition has in-vivo significance.

The factors involved in addition of a base at any site have been discussed by Fersht (1979). Addition of an incorrect nucleotide opposite a particular site (N_1, Fig. 1) depends on: the rate of addition at that site (K_{mis}), the rate of removal of the incorrect nucleotide by an editing function when such exists (K_{ex}), and the rate at which elongation to the next nucleotide occurs (K_{el}) since elongation protects misincorporated nucleotides from the proofreading exonuclease. Most studies on fidelity (Loeb & Kunkel, 1982) have used unmodified templates. Creation of a non-instructional or pseudoinformational lesion (Walker, 1984) at the N_1 site makes addition of a base more difficult, provides more time for editing and hinders elongation. In this report, we summarize some experiments with an in-vitro system which illustrate these points.

The system has been described previously (Strauss *et al.*, 1982) and is based on the Sanger sequencing methodology (Sanger *et al.*, 1977). Instead of synthesis being terminated with dideoxy analogues, it is terminated because of lesions on the template strand. Synthesis with T_4 DNA polymerase terminates immediately before the lesion; isolation of the terminated (first-stage) product and use in a second reaction (stage 2) permits the study of the conditions necessary to add opposite the lesion and to proceed beyond it. A variation in the method involves the use of a nicked double-stranded template as a model for the replication of double-stranded DNA. We see no difference between the termination pattern obtained in single- and double-stranded DNA in our system after either UV-irradiation or treatment with dimethylsulfate using deoxynucleoside triphosphate (dNTP) concentrations of about 100 μM and polymerase concentrations in molar excess (Fig. 2).

Fig. 1. Factors involved in DNA chain elongation (modified from Fersht, 1979). N_1 and N_2, nucleotide residues in DNA; M, incorrect nucleotide; N_2^c, complementary nucleotide; K_{mis}, rate of misincorporation; K_{ex}, rate of removal of M; K_{el}, rate of elongation

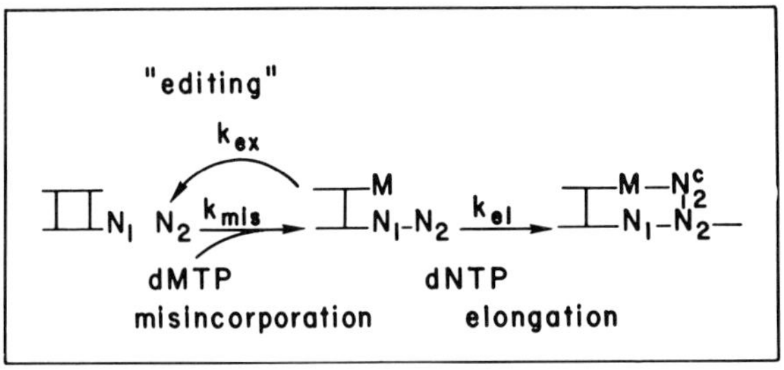

Fig. 2. Termination of synthesis on alkylated double-stranded templates. Double-stranded M13mp2 nicked at the EcoR1 restriction enzyme site was prepared by annealing EcoR1 cleaved replicative-form DNA with excess (+)-strand DNA and purifying the mixture by electrophoresis and electroelution. The isolated double-stranded DNA was reacted with 10mM dimethylsulfate (DMS) for 30 min at 25 °C. The DNA (0.1 µg) was incubated with 0.3 units E. coli DNA polymerase I (holoenzyme) and 0.6 µM 32P-labelled deoxynucleoside triphosphates for 10 min at 30 °C followed by additional incubation for 5, 20 or 80 min with 100 µM deoxynucleoside triphosphates as indicated on the figure. Heating was at 65 °C for 60 min before polymerase treatment. After reaction, the polymerase was inactivated at 65 °C and the DNA was restricted with EcoR1 for 45 min at 37 °C and then electrophoresed on 14% denaturing polyacrylamide gels as described previously (Rabkin & Strauss, 1984). The dideoxy sequence standards (Sanger et al., 1977) were run in the lanes marked "dd". The template sequence, numbered in the centre of the figure, starts from the EcoR1 digestion site as follows:
3'A(0)GCATT(5)AGTAC(10)CAGTA(15)TCGAC(20)AAAGG(25)ACACA(30)CTTTA(35)ACAAT-(40)AGGCG(45)AGTGT(50)TAAGG(55)TGTGT(60)TGTAT(65)GCTCG(70)GCCTT(75)GGTAT(80)-TTCAC(85)ATTTC(90)GG5'.

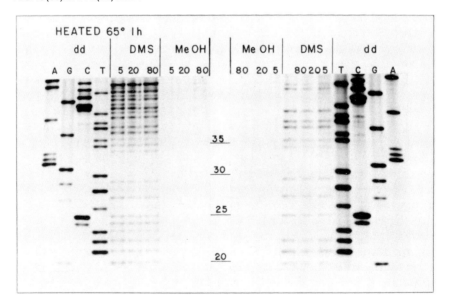

For synthesis on a dimethylsulfate-treated template, termination occurs immediately before adenine (A) sites in the template; only after heating do blocks appear before guanine (G) sites. We interpret this data as indicating that 3-methyladenine and apurinic (AP) sites, but not 7-methylguanine (the major alkylation product) block DNA synthesis.

The incorporation of nucleotides opposite a non-instructional lesion depends on at least four factors: 1) the lesion itself, since special factors may make lesions partly informational; 2) the nucleotides present; 3) the enzyme, since there is a protein preference for particular nucleotides; 4) the site at which incorporation occurs, since template nucleotides 5' to the lesion affect the ability to elongate.

The importance of the lesion itself is demonstrated by the study of acetylaminofluorene-guanine (AAF-G) lesions. In contrast to the other lesions studied, deoxycytidine triphosphate (dCTP) is incorporated most readily except when Mn^{2+} is present, in which case deoxyadenosine triphosphate (dATP) is preferentially incorporated, as observed with other lesions. We assume that incorporation occurs most readily when the adduct is in its *anti* (informational) configuration (Rabkin & Strauss, 1984). The role of enzyme and nucleotide is shown in Figure 3, in which incorporation is studied opposite AP sites formed by the conversion of cytosine to uracil with bisulfite followed by treatment with uracil-DNA glycosylase. In stage 1 reactions, *Escherichia coli* DNA polymerase I (pol I) is more likely to terminate opposite the AP site than is T_4 DNA polymerase, possibly due to the higher activity of the T_4 3'→5' exonuclease. In stage 2 reactions, the role of the pol I exonuclease is seen in the pattern of both elongation and degradation. The eukaryotic polymerases present a clearer picture because of their lack of exonuclease activity. There is a preference for the insertion of purines but with site-specific differences; polymerase beta appears to recognize 2-aminopurine deoxynucleoside triphosphate better than dATP, in contrast to the behaviour of avian myeloblastosis virus (AMV) reverse transcriptase (Fig. 3).

A molecule with a nucleotide added opposite a lesion is not necessarily a good substrate for further elongation. AMV reverse transcriptase adds cytosine residues opposite all AAF-G lesions, but DNA synthesis is completely blocked (Moore *et al.*, 1982). Only at certain sites is rapid bypass seen, for example, at the sequence 3'TGCCCTACTTGT5' where the G residues are reacted with AAF. Addition by pol I/Mn^{2+} of a mismatched G opposite the lesion is rate-limiting but once a base is inserted, synthesis proceeds rapidly presumably because of the three complementary C residues to the 5' of the added G (Fig. 4). Additional factors are likely to be involved since, in second-stage reactions, there must be a distributive association of polymerase at the site of the lesion. This may pose a different enzymological problem than when synthesis is started some bases removed from a termination site, as in first-stage synthesis (see Ollis *et al.*, 1985).

We have been attempting to use bacterial extracts for synthesis in an effort to approach the in-vivo system more closely. Although there is no problem in using published techniques to produce extracts which convert single-stranded *E. coli* phage M13 to the double-stranded replicative form, termination bands are not produced in experiments using reacted templates. Instead there seems to be complete degradation of the product. We suspect this to be analogous to the in-vivo situation. In bacteria,

Fig. 3. Incorporation of deoxynucleoside triphosphates opposite apurinic (AP) sites produced by the removal of cytosine residues from DNA
Left: Formation of first-stage product from bisulfite-treated M13 DNA with (AP DNA) or without (U DNA) uracil-DNA glycosylase treatment (dd, dideoxy sequence standards; I, *E. coli* DNA polymerase I (pol I); T_4, T_4 DNA polymerase).
Right: Elongation of labelled first-stage product synthesized with T_4 DNA polymerase and ^{32}P-labelled deoxynucleoside triphosphates. Isolated first-stage product was incubated with 200 μM non-labelled deoxynucleoside triphosphate and one of the following enzymes: *E. coli* pol I, (Klenow fragment), 0.2 units, 10 min at room temperature (lanes "I"); polymerase alpha (a preparation from the human lymphoid line Daudi, 0.6 μg of which catalyzed incorporation on a control DNA template equal to 20% of that of 0.2 units of pol I), 15 min at 35°C (lanes "α"); polymerase beta (Novikoff hepatoma), 0.03 units, 30 min. at 35°C (lanes "β"); and AMV reverse transcriptase, 8 units, 20 min at 35°C (lanes "rt"). 2 APdTP: 2-aminopurine deoxynucleoside triphosphate (kindly provided by Dr. Myron Goodman)

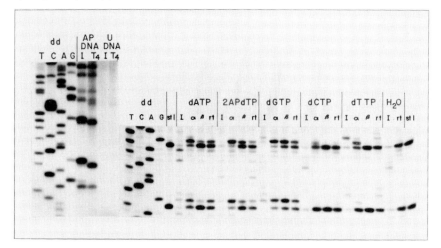

it has long been known that degradation of growing points occurs when synthesis is no longer possible (Reiter & Ramareddy, 1970). If synthesis cannot go forward and it is not possible to "turn over" the terminal nucleotide, then the product will be degraded.

We think these in-vitro results may have some significance for the in-vivo situation. Carcinogens can block DNA synthesis completely *in vitro* and yet organisms survive and mutate with adducts in their DNA. In some way, the replication apparatus can get past the lesion while preserving the continuity of the polynucleotide chain. A variety of mechanisms have been proposed which permit such synthesis, but at some point it is necessary to account for a nucleotide being added opposite a damaged site. Park and Cleaver (1979) suggested that, in a multi-replicon system, synthesis is terminated by the lesion. The critical step is then addition of a base opposite the lesion followed by ligation. The mechanisms described above could accomplish this addition.

Fig. 4. Time course for the incorporation of deoxyguanosine triphosphate (dGTP) opposite an acetylaminofluorene-guanine adduct (G*) by *E. coli* DNA polymerase I (Klenow fragment) with Mn^{2+} as the divalent ion. Stage-I product was synthesized with ^{32}P-labelled deoxynucleoside triphosphates to just before the lesion and the isolated stage-I product was incubated with dGTP (180 μM). The percent of the initial stage-I product as determined by scanning the autoradiograph of the sequencing gel (insert), is plotted for positions 33, 34 and 36 (from Strauss *et al.*, 1984).

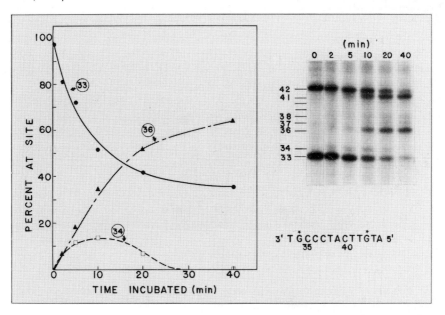

ACKNOWLEDGEMENT

This work was supported in part by grants from the National Institutes of Health (GM 07816, CA32436) and the US Department of Energy (DE-AC02-76).

REFERENCES

Fersht, A. (1979) Fidelity of replication of phage phiX 174 DNA by DNA polymerase III holoenzyme: spontaneous mutation by misincorporation. *Proc. natl Acad. Sci. USA,* **76,** 4946–4950

Loeb, L. & Kunkel, T. (1982) Fidelity of DNA Synthesis. *Ann. Rev. Biochem.,* **52,** 429–457

Moore, P., Rabkin, S., Osborn, A., King, C. & Strauss, B. (1982) Effect of acetylated and deacetylated 2-aminofluorene adducts on in-vitro DNA synthesis. *Proc. natl Acad. Sci. USA,* **79,** 7166–7170

Ollis, D., Brick, P., Hamlin, R., Xuang, N. & Steitz, T. (1985) Structure of a large fragment of *Escherichia coli* DNA polymerase I complexed with dTMP. *Nature*, *313*, 762–766

Park, S. & Cleaver, J. (1979) Postreplication repair: questions of its definition and possible alteration in xeroderma pigmentosum cell strains. *Proc. natl Acad. Sci. USA*, *76*, 3927–3931

Rabkin, S. & Strauss, B. (1984) A role for DNA polymerase in the specificity of nucleotide incorporation opposite *N*-acetyl-2-aminofluorene adducts. *J. mol. Biol.*, *178*, 569–594

Reiter, H. & Ramareddy, G. (1970) Loss of DNA behind the growing point of thymine-starved *Bacillus subtilis* 168. *J. mol. Biol.*, *50*, 533–548

Sanger, F., Nicklen, S. & Coulson, A. (1977) DNA sequencing with chain-terminating inhibitors. *Proc. natl Acad. Sci. USA*, *74*, 5463–5467

Strauss, B., Rabkin, S., Sagher, D. & Moore, P. (1982) The role of DNA polymerase in base substitution mutagenesis on non-instructional templates. *Biochimie*, *64*, 829–838

Strauss, B., Rabkin, S. & Sagher, D. (1984) *Interaction of DNA polymerase and adduct conformation in the specificity of incorporation opposite carcinogenic lesions in DNA*. In: Bishop, J.M., Graves, M. & Rowley, J., eds, *UCLA Symposia on Molecular and Cellular Biology*, New Series, Vol. 17: *Genes and Cancer*, New York, Alan R. Liss, pp. 157–166

Walker, G. (1984) Mutagenesis and inducible responses to deoxyribonucleic acid damage in *Escherichia coli*. *Microbiol. Rev.*, *48*, 60–93

MUTAGENESIS AND REPAIR OF O^6-SUBSTITUTED GUANINES

J.M. ESSIGMANN[1], E.L. LOECHLER[2] & C.L. GREEN[3]

Laboratory of Toxicology, Department of Applied Biological Sciences, Massachusetts Institute of Technology, Cambridge, MA, USA

SUMMARY

The mutagenic activity of O^6-methylguanine has been investigated using a single-stranded M13mp8 phage DNA molecule in which a single O^6-methylguanine residue was positioned in the unique recognition site for the restriction endonuclease, *Pst*I. After introduction of this vector into *Escherichia coli*, progeny phage were produced, of which 0.4% were mutated in their *Pst*I site. To determine the impact of DNA repair on mutagenesis, levels of O^6-methylguanine-DNA methyltransferase (an O^6-methylguanine repair protein) were depleted in host cells by treatment with N-methyl-N'-nitro-N-nitrosoguanidine prior to viral DNA uptake. In these cells, the mutation frequency due to O^6-methylguanine increased with increasing N-methyl-N'-nitro-N-nitrosoguanidine dose (the highest mutation frequency observed was 20%). DNA sequence analysis of mutant genomes revealed that O^6-methylguanine induced G to A transitions, exclusively.

INTRODUCTION

The O^6-substituted guanines have received much attention in recent years because their presence or persistence in DNA can often be correlated with the mutagenic and carcinogenic effects of alkylating agents (Loveless, 1969; Goth & Rajewsky, 1974; Margison & Kleihues, 1974; Schendel & Robins, 1978). In the work described here, we have examined the mutagenic activity of O^6-methylguanine (O^6MeGua) *in vivo*, using a viral genome containing this adduct at a unique site.

---

[1] To whom reprint requests should be addressed
[2] Present address: Biology Department, Boston University, MA 02215, USA
[3] Present address: AMGen, Inc., Thousand Oaks, CA 91320–1789, USA

394   ESSIGMAN ET AL.

Fig. 1. A. Structures of M13mp8 genomes containing a single O^6-methylguanine residue (O^6MeGua) in the unique *Pst*I site. The numbers within the phage genomes indicate genome positions modified by O^6MeGua.
B. Isolation of O^6MeGua-derived mutants. Double-stranded O^6MeGua-M13mp8 was base-denatured to give single-stranded (ss) O^6MeGua-M13mp8 (step 1), which then was used to transform *Escherichia coli* MM294A cells (step 2). A mixture of wild-type and mutant phage was produced (X denotes position of mutation) and used to infect JM103 cells. Replicative form (RF) DNA prepared from these phage (step 3) was treated with *Pst*I (step 4); mutant DNA remained circular, whereas wild-type molecules were linearized and subsequently selectively degraded with exonuclease III (step 5). Steps 4 and 5 were repeated, and the remaining DNA (an essentially pure mutant population) was retransfected into JM103 cells to produce phage (step 6). The ss phage genome was isolated for DNA sequencing (steps 7–9).

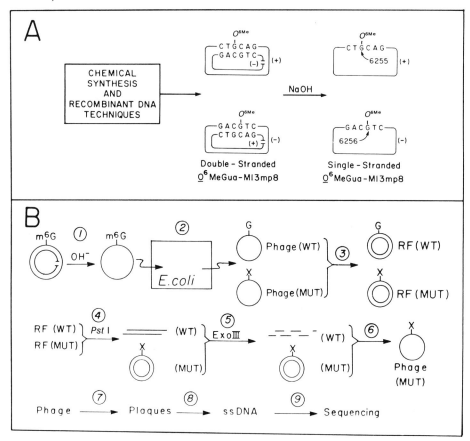

Using a combination of chemical synthesis and recombinant DNA techniques (Fowler *et al.*, 1982; Green *et al.*, 1984), genomes of the *Escherichia coli* phage M13mp8, were modified to contain a single O^6MeGua residue (these DNA molecules are referred to herein as O^6MeGua-M13mp8). Initially, double-stranded (ds) genomes were constructed with O^6MeGua at either position 6255 or 6256 (Fig. 1A);

both sites are located within the unique *Pst* I recognition sequence of M13mp8. This product was subjected to a series of characterization experiments, which in sum demonstrated that O^6MeGua was located within the *Pst* I site, that the adduct was structurally intact, and that the molecules shown in Figure 1A were essentially free of contamination from genomes containing the unmodified base (guanine) at the adduction target. This DNA molecule had a nick situated in the strand opposite that containing the adduct, and thus alkali denaturation yielded single-stranded (ss) monoadducted genomes. It was desirable to have the option of removing the complementary strand, because it has been reported that the major repair protein for O^6MeGua is less active on ss DNA than on ds substrates (Lindahl *et al.*, 1982).

THE MUTATION FREQUENCY OF O^6MeGua *IN VIVO*

To investigate mutagenesis by O^6MeGua, ss O^6MeGua-M13mp8 was introduced into *E. coli* MM294A cells, where the phage genomes were acted upon by the endogenous replication and repair systems (Fig. 1B, and Loechler *et al.*, 1984). A mixture of wild-type and mutant phage was produced; this mixture was used to prepare the ds replicative form (RF) DNA. The method used to differentiate mutant and wild-type phage was based on the fact that the progenitor phage DNA molecule (step 1 of Fig. 1B) contained the adduct in the unique *Pst* I recognition site. Mutations affecting this site rendered the RF DNA insensitive to cleavage by this endonuclease. This property made it possible to isolate a pure mutant phage population for DNA sequencing and for calculation of mutation frequencies.

The mutation frequency of O^6MeGua was expressed as the fraction of phage produced after step 2 of Figure 1B with mutations in the *Pst* I site (Loechler *et al.*, 1984). As indicated by the DNA sequencing results (see below), the mutations induced by O^6MeGua were at either of two guanines in this site, because O^6MeGua-M13mp8 was an equal mixture of genomes with the single adduct located either at position 6255 in the (+) strand or at position 6256 in the (−) strand (Fig. 1A). As shown in Table 1, the mutation frequencies of the adduct in the (+) and (−) strands (MF^+ and MF^-) were determined to be 0.36% and 0.08%, respectively. (In calculating these values, we assumed that an adduct at a given site gave rise to a mutation at that site; i.e., ss O^6MeGua-M13mp8 with the adduct in the (+) strand gave rise to the mutation observed at position 6255.) The sum of the values for the mutation frequencies in the individual strands is defined as the total mutation frequency (MF^t) of O^6MeGua, which was determined to be approximately 0.4%.

ROLE OF DNA REPAIR IN PROTECTING AGAINST O^6MeGua MUTAGENESIS

Other studies concerning the miscoding characteristics of O^6MeGua *in vitro* have typically involved random incorporation of O^6-substituted guanines into DNA or RNA polymers, which then were copied with polymerases (Gerchman & Ludlum, 1973; Mehta & Ludlum, 1978; Abbott & Saffhill, 1979). Subsequently, the replication

Table 1. Mutation frequencies (in %) due to the presence of a single O^6-methylguanine residue (O^6MeGua) at the PstI site of phage M13mp8

MNNG challenge[a] (μg/ml)	MF⁻ [b]	MF⁺ [b]	MFᵗ [b]	MF [c]
0	0.08	0.36	0.4	≤0.03
17	1.3	4.1	5.4	≤0.11
33	3.6	4.7	8.3	≤0.20
50	4.1	13.7	17.8	≤0.16

[a] Some host cells for O^6MeGua-M13mp8 replication were challenged with N-methyl-N'-nitro-N-nitrosoguanidine (MNNG) prior to DNA uptake; details are given in the text.
[b] MF⁺ and MF⁻ are the percentage of progeny phage with O^6MeGua-derived mutations that originated in the (+) and (−) strands, respectively; the sum of these quantities is MFᵗ
[c] The upper limit of the mutation frequency of a control, which was the (+) strand of M13mp8 (This DNA did not contain O^6MeGua at the PstI site.)

products were analysed for the presence of noncomplementary nucleotides. The results of such experiments have demonstrated that DNA and RNA polymerases misreplicate O^6-alkylguanines approximately one-third of the time. Although less work has been done on estimating the mutation frequency *in vivo*, the few data that are available indicate a similar (Lawley & Martin, 1975) or slightly higher (Guttenplan, 1984) mutation frequency, as compared to that observed *in vitro*. However, some uncertainty must exist as to the accuracy of the in-vivo estimates, because these experiments were done with DNAs containing the full range of adducts created by treatment with alkylating agents.

The mutation frequency we observed *in vivo* is several orders of magnitude less than that determined or predicted by the experiments cited above. The most likely reason for this apparent discrepancy is the fact that in our studies a single adduct was built into the genome, and this single lesion probably was removed quickly in *E. coli* by repair proteins associated with the adaptive response (specifically, the O^6-methylguanine-DNA methyltransferase (MT)). This protein acts by transferring the methyl group from O^6MeGua in DNA to itself (Olsson & Lindahl, 1980); the alkylated protein is not thought to turn over, and thus it is irreversibly inactivated in the process of dealkylating the genome.

We took advantage of the suicidal property of the MT to diminish the intracellular capacity to repair O^6MeGua. Two minutes before the O^6MeGua-M13mp8 uptake step (step 2 of Fig. 1B), host cells were treated with N-methyl-N'-nitro-N-nitrosoguanidine (MNNG), which introduced O^6MeGua residues into the host chromosome (Schendel & Robins, 1978). Repair of these lesions depleted endogenous levels of the MT, and thus diminished the ability of cells to repair the single adduct in O^6MeGua-M13mp8. (MNNG is known to induce several DNA repair systems, but we estimated that fixation of O^6MeGua as a mutation would occur before their induction, and thus their effect on mutagenesis by this adduct should be minimal.) Table 1 presents the results of an experiment in which the mutation frequency of

O^6MeGua was examined in a series of cell populations that had been pretreated with a range of increasing doses of MNNG, i.e., treatments that created a range of reduced MT activities within the host cells. As expected, the mutant fraction of O^6MeGua-M13mp8 increased with the level of MNNG treatment. At the highest level of MNNG treatment (50 μg/ml), MF^t had increased to almost 50 times the comparable value in unchallenged cells. The mutation frequency of this sample (approximately 20%) does not necessarily represent the inherent mutation efficiency of O^6MeGua, because the function relating MNNG dose and mutagenesis (Table 1) was still ascending at this level of MNNG challenge; rather, this value represents the lower limit of the true mutation frequency of this lesion. Interestingly, at its present value, this mutation frequency is at the lower end of the range measured for mutagenesis of O^6-alkyl guanines *in vitro* (Gerchman & Ludlum, 1973; Mehta & Ludlum, 1978; Abbott & Saffhill, 1979), and it is within a factor of two or three of the level predicted by indirect in-vivo measurements (Lawley & Martin, 1975; Guttenplan, 1984).

NATURE OF MUTATIONS INDUCED BY O^6MeGua

After the mutant isolation procedure was completed (i.e., after step 6 of Fig. 1B) and individual mutant plaques were obtained, these mutant species were characterized by DNA sequencing. Figure 2 shows an autoradiogram of a DNA sequencing gel and reveals the DNA sequences of the wild type and of the two mutant species in the vicinity of the *Pst*I site of M13mp8. The only sequence change in the leftmost lanes was a guanine to adenine (G to A) transition at position 6256. We assume this was due to misreplication of the ss O^6MeGua-M13mp8 genome in which the adduct was located in the (−) strand (see Fig. 1A). Presumably, O^6MeGua base-paired with thymine (T) during synthesis of the (+) strand *in vivo*, creating a phage population with T rather than C (cytosine), opposite the original position of the lesion. The rightmost lanes show the sequence of the mutant we assume issued from the O^6MeGua-M13mp8 genome in which the adduct was in the (+) strand. The sequencing data indicate that a C to T change occurred at position 6255. Since it is the sequence of the (−) strand which appears in the autoradiogram, this means that in the complementary (+) strand a G to A change occurred. Thus, the sequencing data are consistent with the original prediction, based on model building (Loveless, 1969) and on in-vitro data (cited above), that O^6MeGua would induce G to A transitions. A total of 60 mutants has been sequenced, and all have shown this base change.

Our general approach to investigating the mechanisms of mutagenesis by carcinogen adducts was based on the assumption that the adduct that leads ultimately to mutations was formed by the attack of an electrophilic carcinogen moiety on the DNA molecule. Topol and coworkers have recently proposed a novel alternative mechanism by which alkylating agents such as MNNG and methylnitrosourea may act (Eadie *et al.*, 1984). Their hypothesis suggests that these agents may alkylate the DNA synthesis precursor pool and that abnormal base-pairing of altered precursors (including O^6MeGua residues) during DNA synthesis could result in mutation. By this mechanism, they predict that O^6MeGua would induce mutations at AT base pairs. Our model system, as described in this manuscript, cannot directly address the

Fig. 2. DNA sequencing of O^6-methylguanine-induced mutants, and of wild-type phage. Phage DNA was isolated and sequenced by the method of Sanger et al. (1977). Shown here are autoradiograms of sequencing gels in the region of the M13mp8 genome containing the PstI sites of (from left to right) (−) strand mutant, wild-type and (+) strand mutant phage.

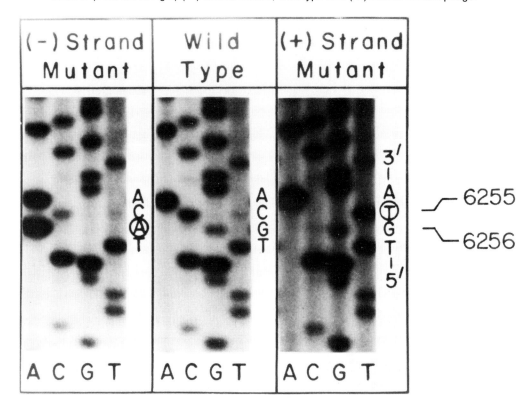

issue of whether the precursor pool serves as a source of premutagenic DNA lesions. However, we can say definitely that the pattern of mutagenesis we observed quantitatively and qualitatively parallels what has been observed in cells treated with alkylating agents (i.e., GC to AT transitions, which were modulated by the endogenous levels of the O^6-methylguanine-DNA methyltransferase).

ACKNOWLEDGEMENTS

This work was supported by NIH Grants 5 PO1 ES00597, T 32 ES07020, CA 33821, and by the Monsanto Fund.

REFERENCES

Abbott, P.J. & Saffhill, R. (1979) DNA synthesis with methylated poly(dC-dG) templates: evidence for a competitive nature to miscoding by O^6-methylguanine. *Biochim. biophys. Acta,* **562,** 51–61

Eadie, J.S., Conrad, M., Toorchen, D. & Topal, M.D. (1984) Mechanism of mutagenesis by O^6-methylguanine. *Nature, 308* 201–203

Fowler, K.W., Buchi, G. & Essigmann, J.M. (1982) Synthesis and characterization of an oligonucleotide containing a carcinogen-modified base: O^6-methylguanine. *J. Am. chem. Soc., 104* 1050–1054

Gerchman, L.L. & Ludlum, D.B. (1973) The properties of O^6-methylguanine in templates for RNA polymerase. *Biochim. biophys. Acta., 308,* 310–316

Goth, R. & Rajewsky, M.F. (1974) Persistence of O^6-ethylguanine in rat brain: correlation with nervous system-specific carcinogenesis by ethylnitrosourea. *Proc. natl Acad. Sci. USA, 71,* 639–643

Green, C.L., Loechler, E.L., Fowler, K.W. & Essigmann, J.M. (1984) Construction and characterization of extrachromosomal probes for mutagenesis by carcinogens: site-specific incorporation of O^6-methylguanine into viral and plasmid genomes. *Proc. natl Acad. Sci. USA, 81,* 13–17

Guttenplan, J. (1984) Mutagenesis and O^6-ethylguanine levels in DNA from *N*-nitroso-*N*-ethylurea-treated *Salmonella typhimurium*: evidence for a high mutational efficiency of O^6-ethylguanine. *Carcinogenesis, 5,* 155–159

Lawley, P. & Martin, C. (1975) Molecular mechanisms in alkylation mutagenesis: induced reversion of bacteriophage T4rII AP72 by ethyl methanesulphonate in relation to extent and mode of ethylation of purines in bacteriophage deoxyribonucleic acid. *Biochem. J., 145,* 85–91

Lindahl, T., Demple, B. & Robins, P. (1982) Suicide inactivation of the *E. coli* O^6-methylguanine-DNA methyltransferase. *EMBO J., 1,* 1359–1363

Loechler, E.L., Green, C.L. & Essigmann, J.M. (1984) In-vivo mutagenesis by O^6-methylguanine built into a unique site in a viral genome. *Proc. natl Acad. Sci. USA, 81,* 6271–6275

Loveless, A. (1969) Possible relevance of O^6 alkylation of deoxyguanosine to the mutagenicity and carcinogenicity of nitrosamines and nitrosamides. *Nature, 223,* 205–207

Margison, G.P. & Kleihues, P. (1974) Carcinogenicity of *N*-methyl-*N*-nitrosourea: possible role of excission repair of O^6-methylguanine from DNA. *J. natl Cancer Inst., 53,* 1839–1841

Mehta, J.R. & Ludlum, D.B. (1978) Synthesis and properties of O^6-methyl-deoxyguanylic acid and its copolymers with deoxycytidilic acid. *Biochim. biophys. Acta, 512,* 770–778

Olsson, M. & Lindahl, T. (1980) Repair of alkylated DNA in *Escherichia coli*. Methyl group transfer from O^6-methylguanine to a protein cysteine residue. *J. biol. Chem., 255,* 10569–10571

Sanger, F., Nicklen, S. & Coulson, A.R. (1977) DNA sequencing with chain-terminating inhibitors. *Proc. natl Acad. Sci. USA, 74,* 5463–5467

Schendel, P.F. & Robins, P.E. (1978) Repair of O^6-methylguanine in adapted *Escherichia coli*. *Proc. natl Acad. Sci. USA, 75,* 6017–6020

VI. SENSITIVE METHODS FOR DETECTION OF NUCLEIC ACID ADDUCTS

MONOCLONAL ANTIBODY-BASED IMMUNOANALYTICAL METHODS FOR DETECTION OF CARCINOGEN-MODIFIED DNA COMPONENTS

J. ADAMKIEWICZ, O. AHRENS, G. EBERLE, P. NEHLS &
M.F. RAJEWSKY

*Institute for Cell Biology (Tumour Research), Essen University, Essen,
Federal Republic of Germany*

SUMMARY

Hybridoma cell lines secreting monoclonal antibodies (Mab) directed against the products formed by reaction of alkylating *N*-nitroso carcinogens with DNA have been established by fusion of rat or mouse myeloma cells, respectively, with spleen cells of rats or mice immunized either with conjugates of various alkyl-ribonucleosides with suitable carrier proteins, or with alkylated DNA electrostatically complexed to carrier proteins. Due to their high affinity and specificity, some of these Mab detect very low amounts of the respective alkyl-deoxynucleosides (e.g., O^6-methyl-2'-deoxyguanosine, O^6-ethyl-2'-deoxyguanosine, O^6-*n*-butyl-2'-deoxyguanosine, O^6-isopropyl-2'-deoxyguanosine, O^4-methyl-2'-deoxythymidine, O^4-ethyl-2'-deoxythymidine) and can be used in various types of immunoassays. With a competitive radioimmunoassay (RIA), specific DNA alkylation products can be quantitated in hydrolysates of cellular DNA, in body fluids, or in urine. The RIA is routinely applicable, reproducible, and sufficiently sensitive to permit the quantitation of femtomole amounts of modified nucleosides in small samples of DNA. When the alkyl-deoxynucleosides in question are separated from bulk DNA by high-performance liquid chromatography prior to analysis by RIA, very low levels of modification in DNA can be detected. The immuno-slot-blot (ISB), a noncompetitive solid-phase immunoassay, is more sensitive than the RIA. For analysis by ISB, alkylated DNA is heat-denatured and immobilized on nitrocellulose filters prior to exposure to the respective Mab and subsequent binding of a second (^{125}I-labelled or biotinylated) antibody. In immunocytological analysis (ICA), the binding of Mab to alkyl-deoxynucleosides is visualized in individual cells by immunostaining of denatured nuclear DNA *in situ* (direct immunofluorescence; peroxidase-staining). With the aid of combined image intensification and microprocessor-based image analysis, fluorescence signals can be evaluated quantitatively. Finally, immuno-electron-microscopy (IEM) permits the visualization – via Mab binding sites – of specific alkyl-deoxynucleosides in individual DNA molecules.

INTRODUCTION

Based on the earlier work of Erlanger and Beiser (1964), Plescia et al. (1964) and Stollar (1973) on antibodies directed against naturally-occuring DNA constituents, immunological methods have recently been introduced for the detection of specific carcinogen-modified DNA components. It has been shown that even deoxynucleosides which are structurally altered only by the covalent attachment of single small alkyl groups (e.g., methyl, ethyl or butyl residues) can be recognized by the respective antibodies in the presence of a large excess of their normal unmodified counterparts (for review, see Müller & Rajewsky, 1981; Müller et al., 1982). Antisera or monoclonal antibodies (Mab) have been raised against various products from the reactions of cellular DNA with alkylating N-nitroso compounds, N-acetoxy-N-2-acetyl-aminofluorene, benzo[a]pyrene and aflatoxin B_1 (Müller & Rajewsky, 1981; Müller et al., 1982; Poirier, 1981). Selected high-affinity antibodies, in particular Mab, can now be used for the sensitive detection and quantitation of DNA components structurally modified by chemical agents that need not be radioactively labelled (as previously required for analysis by radiochromatography). Depending on the characteristics of the antigen (i.e., the modified DNA component recognized by the antibody), and on the aims of the respective analysis, various types of immunological assays can be applied. With the use of the appropriate assays, alkyl-deoxynucleosides can be quantitated in small samples of DNA or in DNA hydrolysates, in the DNA of individual cells, or in single DNA molecules (Adamkiewicz et al., 1982, 1984, 1985; Nehls et al. 1984b). In this report, we describe the properties of Mab specific for the following alkyl-deoxynucleosides: O^6-methyl-2'-deoxyguanosine (O^6-MedGuo), O^6-ethyl-2'-deoxyguanosine (O^6-EtdGuo), O^6-butyl-2'-deoxyguanosine (O^6-BudGuo), O^6-isopropyl-2'-deoxyguanosine (O^6-iProdGuo), O^4-methyl-2'-deoxythymidine (O^4-MedThd) and O^4-ethyl-2'-deoxythymidine (O^4EtdThd). The principles, as well as the advantages and disadvantages, of different types of immunoassays (RIA, ISB, ICA, IEM) are discussed, and their respective detection limits and sensitivities compared.

MATERIALS AND METHODS

Production of monoclonal antibodies (Mab)

Alkyl-ribonucleosides were coupled to keyhole limpet hemocyanin (KLH; Calbiochem, Marburg, FRG) as a carrier protein following the procedure of Erlanger and Beiser (1964), as described by Müller and Rajewsky (1980) and Adamkiewicz et al. (1982). Alternatively, N-ethyl-N-nitrosourea (EtNU)-treated calf thymus DNA with an O^6-EtdGuo/dGuo molar ratio in DNA of $\sim 1.1 \times 10^{-2}$, was complexed with methylated horseshoe crab hemocyanin (HCH). Adult female Balb/c mice or BDIX rats were immunized by intracutaneous injections of emulsions containing the immunogen, aluminium hydroxide (Alugel S; Serva, Heidelberg, FRG), and Freund's adjuvant (Behring-Werke, Marburg, FRG). Spleen cells were isolated from the immunized mice or rats and fused with cells of the murine myeloma cell line P3-X63-Ag8.653 (Kearney et al., 1979) or the rat myeloma line X3-Ag1.2.3 (Galfré et al., 1979), respectively, using polyethylene glycol (PEG 4000; Roth, Karlsruhe, FRG) as a fusion reagent (Rajewsky et al., 1980; Adamkiewicz et al., 1982). Hybridoma cell lines secreting Mab with the expected specificity were selected by an enzyme immunoassay (Adamkiewicz et al., 1982) or by RIA (see below).

Competitive radioimmunoassay (RIA)

The competitive RIA was carried out as described by Müller and Rajewsky (1978, 1980). In a total volume of 100 μl of Tris-buffered saline (containing 1% bovine serum albumin [w/v] and 0.1% bovine immunoglobulin G [w/v]), each sample contained $\sim 2.5 \times 10^3$ dpm of ^{3}H-labelled tracer, an antibody solution diluted to give 50% binding of tracer in the absence of inhibitor, plus varying amounts either of alkylated DNA hydrolysed to mono-deoxynucleosides (see below), or of other natural or modified DNA constituents to be analysed for cross-reactivity (inhibitor). After incubation for two hours at room temperature (equilibrium), 100 μl of a saturated ammonium sulphate solution (pH 7.0) was added. After 10 min, the samples were centrifuged for 3 min at 10 000 × g. The ^{3}H-activity in 150 μl of supernatant was measured by liquid scintillation spectrometry. The degree of inhibition of tracer-antibody binding (ITAB) was calculated as described by Müller and Rajewsky (1980). For the quantitation of unknown amounts of modified nucleosides in DNA samples, DNA isolated from tissues or cultured cells was hydrolysed enzymatically to mono-deoxynucleosides with DNase I (EC 3.1.4.5; Boehringer Mannheim, Mannheim, FRG), snake venom phosphodiesterase (EC 3.1.4.1; Boehringer) and alkaline phosphatase (EC 3.1.3.1; Boehringer), as described by Müller and Rajewsky (1980). The concentrations of dGuo and dThd in the DNA hydrolysates were determined by peak integration after separation by high-performance liquid chromatography (HPLC). HPLC was also used for the separation of different alkylation products from the same DNA sample, prior to analysis by RIA (Adamkiewicz et al., 1982). Antigen concentrations in the respective DNA samples were determined by comparing the ITAB-values of DNA hydrolysates with those of standard curves for the particular modified nucleosides to be analysed.

Immuno-slot-blot (ISB)

Samples of alkylated DNA (3 μg of DNA in 100 μl) were heat-denatured for 10 min, immediately chilled on ice, and mixed with equal volumes of 2 M ammonium acetate. The single-stranded DNA was then immobilized on nitrocellulose (NC) filters (BA 52; Schleicher & Schüll, Dassel, FRG) using a 72-slot Minifold II vacuum filter device (Schleicher & Schüll, Dassel, FRG). Prior to use, the NC filters were presoaked in 1 M ammonium acetate. After application of DNA, the slots were rinsed with 1 M ammonium acetate (200 μl/slot). The NC filters were then soaked in 5 × SSC (0.75 M NaCl, 0.075 M trisodium citrate) for 5 min, dried, and baked in a vacuum oven for two hours at 80 °C. Prior to incubation with an anti-(alkyl-deoxynucleoside) Mab (first antibody), the NC filters were treated for two hours with phosphate-buffered saline (PBS) containing 0.1%–0.5% casein (Sigma, St. Louis, MO, USA) and 0.1% deoxycholate to prevent nonspecific binding of Mab ("blocking"). The NC filters were then incubated for one hour in the same solution as above in a heat-sealed plastic bag with a Mab solution containing 15 µg of Mab per ml and per 10–15 cm^2 of NC filter area (first antibody). Mab concentrations were determined by RIA titration as described by Müller (1980). After extensive washing in PBS supplemented with 0.16 M NaCl and 0.1% Triton X-100 (with several buffer changes), the NC filters carrying specifically-bound first antibody molecules were again sealed in plastic bags containing PBS supplemented with 0.1%–0.5% casein, 0.1% deoxycholate and an ^{125}I-labelled second antibody specific for the immunoglobulin of the Mab. After one hour, the NC filters were washed as above, dried, and exposed to Kodak X-Omat AR films for different lengths of time. Alternatively, the degree of binding of the anti-alkyl-deoxynucleoside Mab to the DNA on the NC filters was determined by a three-step procedure, which ultimately results in a coloured precipitate. Following incubation with the first Mab, excess antibody was removed by extensive washing in PBS supplemented with 0.16 M NaCl (PBS-NaCl). The NC filters were then incubated for 30 min at room temperature with a biotinylated second antibody specific for the first Mab (Vectastain; biotinylated rabbit anti-(rat IgG) Ig; Camon, Wiesbaden, FRG; diluted 1:200 in PBS-NaCl). After three washes of 10 min each in PBS-NaCl, and a further wash (5 min) with Tris-NaCl buffer (0.1 M Tris-HCl pH 7.5, 1.0 M NaCl, 2 mM MgCl$_2$, 0.05% Triton X–100), the NC filters were incubated for 30 min at room temperature with a complex of avidin and biotinylated, polymerized alkaline phosphatase (Leary et al., 1983). Following a final washing step with Tris-NaCl buffer (four times 5 min at room temperature), the NC-filters were incubated in a solution of nitro blue tetrazolium (0.33 mg/ml) and 5-bromo-4-chloro-3-indolyl phosphate (0.17 mg/ml) in 0.1 M Tris-HCl, 0.1 M NaCl, 5 mM MgCl$_2$, pH 9.5 (Leary et al., 1983). Colour development was stopped by washing in PBS containing 10 mM ethylenediaminetetraacetic acid (EDTA). The NC filters were stored wet in heat-sealed plastic bags prior to densitometric evaluation (Nehls et al., 1984a).

Immunocytological analysis (ICA)

Cytocentrifuged cell samples, cell smears, frozen or paraffin-embedded sections, or squash preparations of small tissue samples were fixed for 10 min in Carnoy's ethanol:chloroform:acetic acid (6:3:1), washed in ethanol, and rehydrated in 2 × SSC (0.3 M NaCl, 0.03 M sodium citrate). Following treatment with RNase A (EC 3.1.4.22; Sigma, München, FRG; 200 µg/ml of 2 × SSC) and T1 RNase (EC 3.1.4.8; Boehringer; 50 units/ml of 2 × SSC) for one hour at 37 °C, cells were washed in 0.15 M NaCl, and nuclear DNA was denatured by dipping the slides into 0.07 M NaOH for exactly 4 min at room temperature. Immediately thereafter, the slides were washed for 10 min in cold TEA buffer (10 mM triethanolamine, 150 mM NaCl, 100 mM MgCl$_2$, 10 mM EDTA, 0.02% NaN$_3$, pH 7.2) and the antibody solution was layered onto the preparations. For *in situ* detection of O^6-EtdGuo in cellular DNA by direct immunofluorescence, cell preparations were incubated overnight at 4 °C with a tetramethylrhodamine isothiocyanate (TRITC)-labelled Mab (ER-14; J. Adamkiewicz & M.F. Rajewsky, in preparation) at a concentration of 20 µg/ml of TEA buffer supplemented with 4% polyethylene glycol 4000 (Roth, Karlsruhe, FRG) and 0.25 mg of the Fc-fragment of rat IgG/ml. Finally, the cell preparations were washed for two times 10 min in TEA buffer and embedded in 10% (w/v) Elvanol (DuPont, Niagara Falls, NY), 30% glycerol, dissolved in PBS pH 8, containing p-phenylenediamine (1 mg/ml). For quantitation of nuclear fluorescence, fluorescence images were amplified by an image intensifier and fed into an image analysis system *via* a high sensitivity television camera (Adamkiewicz *et al.*, 1983; J. Adamkiewicz, O. Ahrens & M.F. Rajewsky, in preparation).

Immuno-electron-microscopy (IEM)

To visualize O^6-EtdGuo in EtNU-treated, double-stranded DNA by immuno-electron-microscopy, 80 µg/ml of the anti-(O^6-EtdGuo) Mab ER-6 (Rajewsky *et al.*, 1980) was incubated with DNA (50 µg/ml) in TMS buffer (10 mM triethanolamine, 100 mM Mg-acetate, 150 mM NaCl, 10 mM EDTA, 0.02% NaN_3, pH 7.2) for 30 min at 37 °C, or for varying periods (2.5 to 30 h) at 4 °C. In most cases, glutardialdehyde was subsequently added (final concentration, 0.01%). Control DNA, or DNA carrying bound Mab, was separated from unreacted Mab molecules by gel filtration on Sephacryl S-1000 (Pharmacia, Uppsala, Sweden) in TMS buffer. Aliquots of DNA-containing fractions were diluted with TBS buffer (10 mM Tris-HCl, 150 mM NaCl, 1 mM EDTA, pH 7.2) to give a concentration of 5–10 µg of DNA/ml, and immediately mounted onto freshly cleaved mica as described by Nehls *et al.* (1984b). The mica supports, with the adherent material, were then washed in water for one min, stained for 30 sec in a 1% aqueous solution of uranyl acetate, and washed again with water. Thereafter, the DNA preparations were shadowed with Pt/C at 7 °C, and the replica were enforced with carbon conditioned by NaCl. Transmission electron microscopy was carried out with a Philips EM 400 instrument.

RESULTS AND DISCUSSION

Some of the characteristics of a series of Mab with binding specificities for the alkyl-deoxynucleosides O^6-MedGuo, O^6-EtdGuo, O^6-BudGuo, O^6-iProdGuo, O^4-MedThd, and O^4-EtdThd are listed in Table 1. An important precondition for the sensitive immunological detection of carcinogen-modified DNA constituents is a high degree of specificity of the antibodies for the respective antigens; i.e., their ability to discriminate between the modified deoxynucleosides and their unmodified, naturally-occuring counterparts. The antibodies must recognize the modified deoxynucleoside even when (i) the modification only represents a small (alkyl) residue covalently bound to the purine or pyrimidine base and (ii) only small amounts of modified nucleosides are present in a very large excess of unmodified nucleosides. The binding characteristics of the Mab shown in Table 1 were determined by measuring their cross-reactivities with other modified or natural DNA constituents by competitive

Table 1. Properties of selected groups of anti-(alkyl-deoxynucleoside) monoclonal antibodies

Alkyldeoxynucleoside	Immunogen	No. of antibodies	Range of antibody affinity constants (l/mol)	Detection limit (fmol)	
				RIA[a] (at 50% ITAB)[b]	ISB[c]
O^6-Methyl-2′-deoxyguanosine	O^6-EtGuo-KLH[d]	6	$3 \times 10^7 - 1 \times 10^9$	250	~50
O^6-Ethyl-2′-deoxyguanosine	O^6-EtGuo-KLH	39	$1 \times 10^8 - 3 \times 10^{10}$	40	0.3
O^6-Ethyl-2′-deoxyguanosine	O^6-EtdGuo-DNA	17	$9 \times 10^7 - 2 \times 10^9$	210	n.d.[f]
O^6-n-Butyl-2′-deoxyguanosine	O^6-EtGuo-KLH	6	$5 \times 10^8 - 2 \times 10^{10}$	60	n.d.
O^6-Isopropyl-2′-deoxyguanosine	O^6-iProGuo-KLH	3	$3 \times 10^9 - 2 \times 10^{10}$	50	n.d.
O^4-Methyl-2′-deoxythymidine	O^4-MerThd-KLH	3	$8 \times 10^6 - 4 \times 10^7$	7050	n.d.
O^4-Methyl-2′-deoxythymidine	O^4-EtrThd-KLH	9	$3 \times 10^6 - 2 \times 10^7$	8000[e]	n.d.
O^4-Ethyl-2′-deoxythymidine	O^4-EtrThd-KLH	9	$1 \times 10^8 - 1 \times 10^9$	240	0.1

[a] Competitive radioimmunoassay
[b] Inhibition of tracer-antibody-binding
[c] Immuno-slot-blot (3 μg of DNA/slot)
[d] Keyhole limpet hemocyanin
[e] ^{3}H-Labelled O^4-EtdThd used as a tracer in the RIA
[f] Not yet determined

RIA. Some of the Mab exhibit extremely low degrees of cross-reactivity. For example, with the best of the anti-(O^6-EtdGuo) Mab, dGuo at concentrations exceeding the concentration of O^6-EtdGuo by more than 10^7-fold does not interfere with the quantitation of O^6-EtdGuo by RIA. This demonstrates that Mab indeed represent very specific reagents for the detection of carcinogen-modified DNA components. However, not all Mab exhibit this extreme degree of specificity. It is, therefore, recommended that a sufficient number of Mab directed against a given structurally-modified DNA component be produced and characterized so that Mab exhibiting a high specificity can be selected from a larger Mab collection.

For the sensitive detection of alkyl-deoxynucleosides in cellular DNA, the type of immunoassay (e.g., competitive versus noncompetitive assay) and the affinity of the antibody play an important role. With the competitive RIA, modified nucleosides are quantitated in DNA enzymatically hydrolysed to monodeoxynucleosides (Rajewsky et al., 1980; Adamkiewicz et al., 1982). The sensitivity of this assay, which is carried out under equilibrium conditions, depends primarily on the affinity constants of the antibodies applied. Higher antibody affinity constants will improve the detection limit; i.e., a lower amount of antigen will be required for inhibition of tracer-antibody binding (ITAB) in the RIA.

Table 1 shows the range of affinity constants of a selected group of anti-alkyl-deoxynucleoside Mab, and the detection limits of the Mab with the highest affinity constant in each group. The best of these Mab, i.e., the anti-(O^6-EtdGuo) Mab ER-6 (Rajewsky et al., 1980), detects, at 50% ITAB, 40 fmoles of O^6-EtdGuo in a 100 μl RIA-sample. If the sigmoidal ITAB-curves are transformed into straight lines by plotting the ITAB values in a probability grid (Müller & Rajewsky, 1980), reading at 20% ITAB becomes possible, which lowers the detection limit, in this case, to ~10 fmol/100 μl sample. Thus, Mab ER-6 is able to select O^6-EtdGuo at an O^6-EtdGuo/dGuo molar ratio of ~3×10^{-7} in a hydrolysate of 100 μg of DNA, corresponding to ~700 O^6-EtdGuo residues per diploid genome. The sensitivity of the RIA can be further improved by separating O^6-EtdGuo from the complete DNA hydrolysate by HPLC prior to analysis by RIA. In this case, the sensitivity of the assay is limited only by the amount of DNA available for analysis (and by possible "background" contamination due to the separation and concentration procedure). HPLC separation prior to RIA analysis also offers the possibility of simultaneously quantitating more than one alkyl-deoxynucleoside in a single DNA sample.

As shown in Table 1, antibody affinity constants and detection limits comparable to those of Mab ER-6 (see above) are not always reached by the Mab of the other groups. However, depending on their particular design, immunoassays other than the RIA permit highly sensitive detection by Mab of lower affinity as well. A typical example is the recently established ISB, a noncompetitive solid-phase immunoassay (Nehls et al., 1984a). In this assay, the measured signals (i.e., colour development on an NC filter or silver grains on a sensitive X-ray film) can be amplified by increasing the time of incubation or exposure, respectively. This results in a considerably improved detection limit (see Table 1). For example, the detection limit for O^6-EtdGuo obtained with Mab ER-6 in the RIA (40 fmol of O^6-EtdGuo at 50% ITAB) is reduced to $\geqslant 0.3$ fmol in the ISB when 3 µg of DNA, or less, are analysed per slot. Note that in comparison with the RIA, much less DNA is required for ISB analysis (3 µg DNA instead of ~ 100 µg for a DNA sample containing O^6-EtdGuo at an O^6-EtdGuo/dGuo molar ratio of $\sim 3 \times 10^{-7}$). As the (noncompetitive) ISB does not operate under equilibrium conditions, the influence of the antibody affinity constant on the detection limit is less pronounced than in the RIA. For example, the best of the anti-(O^4-EtdThd) Mab (ER-01; Adamkiewicz et al., 1982) is characterized by an antibody affinity constant of 1×10^9 l/mol and a detection limit of 240 fmol of O^4-EtdThd in the RIA (Table 1). In contrast, the detection limit of Mab ER-01 in the ISB ($\geqslant 0.1$ fmol in a sample of 3 µg DNA, or less) is even lower than the ISB-detection limit of Mab ER-6 (antibody affinity constant, 2×10^{10} l/mol). The ISB was designed particularly for the sensitive quantitation of modified deoxynucleosides in very small samples of DNA, e.g., in DNA from selected fractions of chromatin or in DNA restriction fragments. It can be applied only to DNA containing modified structures that are heat- or alkali-stable during the DNA denaturation step.

In comparison to the RIA and the ISB, immunocytological analysis (ICA) requires more sophisticated equipment and experience (Adamkiewicz et al., 1983). The detection limit of ICA (which also represents a noncompetitive, solid-phase type of immunoanalysis) is defined by the number of modified nucleosides detectable in the genomic DNA of single cells, and is strongly dependent on the nonspecific background binding produced by the individual Mab applied for analysis. The important advantage of ICA lies in the fact that individual cells and squash preparations or cultures of small cell populations can be analysed, even – in the case of frozen tissue sections – in their proper histological environment. Only a few of the Mab listed in Table 1 can be used in ICA, since the majority of Mab produce either no staining or nonspecific staining, probably due to stickiness of the Mab molecules or to their binding to cellular epitopes not identical with the antigenic determinant on the particular DNA alkylation product to be detected. At present, $\sim 10^3$-O^6-EtdGuo residues per diploid genome can be detected using a TRITC-labelled Mab (ER-14; Adamkiewicz et al., 1983).

With the use of immuno-electron-microscopy (IEM) in conjunction with a protein-free DNA spreading technique (Nehls et al., 1984b), single alkyl-deoxynucleosides can be visualized in double-stranded DNA molecules *via* antibody-binding sites. Using this technique, nonrandom formation of O^6-EtdGuo has recently been demonstrated in rat brain chromosomal DNA after in-vivo administration of the *N*-nitroso carcinogen EtNU (Nehls et al., 1984b). IEM can thus be applied to localize possible "hot spots" of specific structural modifications caused by carcinogens and mutagens within DNA strands (genes) of known nucleotide sequence.

ACKNOWLEDGEMENTS

This research was supported by the Deutsche Forschungsgemeinschaft (SFB 102/A9), the Commission of the European Communities (ENV-544-D[B]), and by the Wilhelm and Maria Meyenburg Stiftung. We are grateful to Ms U. Schauer, I. Schmidt, I. Spratte and G. Jost for reliable and precise technical assistance.

REFERENCES

Adamkiewicz, J., Drosdziok, W., Eberhardt, W., Langenberg, U. & Rajewsky, M.F. (1982) *High-affinity monoclonal antibodies specific for DNA components structurally modified by alkylating agents.* In: Bridges, B.A., Butterworth, B.E. & Weinstein, I.B., eds, *Indicators of Genotoxic Exposure, Banbury Report 13,* Cold Spring Harbor, NY, Cold Spring Harbor Laboratory, pp. 265–276

Adamkiewicz, J., Ahrens, O., Huh, N. & Rajewsky, M.F. (1983) Quantitation of alkyldeoxynucleosides in the DNA of individual cells by high-affinity monoclonal antibodies and electronically intensified, direct immunofluorescence. *J. Cancer Res. clin. Oncol., **105**,* A15

Adamkiewicz, J., Ahrens, O., Huh, N., Nehls, P., Spiess, E. & Rajewsky, M.F. (1984) *High-affinity monoclonal antibodies for the specific recognition and quantitation of deoxynucleosides structurally modified by N-nitroso compounds.* In: O'Neill, I.K., Von Borstel, R.C., Miller, C.T., Long, J. & Bartsch, H., eds, N-*Nitroso Compounds: Occurrence and Biological Effects (IARC Scientific Publications No. 57),* Lyon, International Agency for Research on Cancer, pp. 581–587

Adamkiewicz, J., Nehls, P. & Rajewsky, M.F. (1985) *Immunological methods for detection of carcinogen-deoxyribonucleic acid adducts.* In: Berlin, A., Draper, M., Hemminki, K. & Vainio, H., eds, *Monitoring Human Exposure to Carcinogenic and Mutagenic Agents (IARC Scientific Publ. No. 59),* Lyon, International Agency for Research on Cancer, (in press)

Erlanger, B.F. & Beiser, S.M. (1964) Antibodies specific for ribonucleosides and ribonucleotides and their reaction with DNA. *Proc. natl Acad. Sci. USA,* 52, 68–74

Galfré, G., Milstein, C. & Wright, B. (1979) Rat × rat hybrid myelomas and a monoclonal anti-Fd portion of mouse IgG. *Nature (London),* **277,** 131–133

Kearney, J.F., Radbruch, A., Liesegang, B. & Rajewsky, K. (1979) A new mouse myeloma cell line that has lost immunoglobulin expression but permits the construction of antibody-secreting hybrid cell lines. *J. Immunol.*, **123**, 1548–1550

Leary, J.J., Brigati, D.J. & Ward, D.C. (1983) Rapid and sensitive colorimetric method for visualizing biotin-labelled DNA probes hybridized to DNA or RNA immobilized on nitrocellulose: Bio-blots. *Proc. natl Acad. Sci. USA*, **80**, 4045–4049

Müller, R. (1980) Calculation of average antibody affinity in anti-hapten sera from data obtained by competitive radioimmunoassay. *J. immunol. Methods*, **34**, 345–352

Müller, R. & Rajewsky, M.F. (1978) Sensitive radioimmunoassay for detection of O^6-ethyldeoxyguanosine in DNA exposed to the carcinogen ethylnitrosourea *in vivo* or *in vitro*. *Z. Naturforsch.*, **33c**, 897–901

Müller, R. & Rajewsky, M.F. (1980) Immunological quantification by high-affinity antibodies of O^6-ethyldeoxyguanosine in DNA exposed to N-ethyl-N-nitrosourea. *Cancer Res.*, **40**, 887–896

Müller, R. & Rajewsky, M.F. (1981) Antibodies specific for DNA components structurally modified by chemical carcinogens. *J. Cancer Res. clin. Oncol.*, **102**, 99–113

Müller, R., Adamkiewicz, J., Rajewsky, M.F. (1982) *Immunological detection and quantification of carcinogen-modified DNA components*. In: Armstrong, B. & Bartsch, H., eds, *Host Factors in Human Carcinogenesis (IARC Scientific Publications No. 39)*, Lyon, International Agency for Research on Cancer, pp. 463–479

Nehls, P., Adamkiewicz, J. & Rajewsky, M.F. (1984a) Immuno-slot-blot: a highly sensitive immunoassay for the quantitation of carcinogen-modified nucleosides in DNA. *J. Cancer Res. clin. Oncol.*, **108**, 23–29

Nehls, P., Rajewsky, M.F., Spiess, E. & Werner, D. (1984b) Highly sensitive sites for guanine-O^6-ethylation in rat brain DNA exposed to N-ethyl-N-nitrosourea *in vivo*. *EMBO J.*, **3**, 327–332

Plescia, O.J., Braun, W. & Palczuk, N.C. (1964) Production of antibodies to denatured deoxyribonucleic acid (DNA). *Proc. natl Acad. Sci. USA*, **52**, 279–285

Poirier, M.C. (1981) Antibodies to carcinogen-DNA adducts. *J. natl Cancer Inst.*, **67**, 515–519

Rajewsky, M.F., Müller, R., Adamkiewicz, J. & Drosdziok, W. (1980) *Immunological detection and quantification of DNA components structurally modified by alkylating carcinogens (ethylnitrosourea)*. In: Pullman, B., Ts'o, P.O.P. & Gelboin, H., eds, *Carcinogenesis: Fundamental Mechanisms and Environmental Effects*, Dordrecht, Reidel, pp. 207–218

Stollar, B.D. (1973) *Nucleic acid antigens*. In: Sela, M., ed., *The Antigens, Vol I*, New York, Academic Press, pp. 1–28

DETECTION AND IDENTIFICATION OF MUTAGENS BY THE ADDUCTS FORMED UPON REACTION WITH GUANOSINE DERIVATIVES

H. KASAI, Z. YAMAIZUMI & S. NISHIMURA

Biology Division, National Cancer Center Research Institute, Chuo-ku, Tokyo 104, Japan

SUMMARY

Mutagens present in crude samples such as heated glucose can be detected or identified by means of the adducts formed upon reaction with a fluorescent guanosine derivative (FG) or isopropylideneguanosine (IPG). After the reaction of IPG with heated glucose, two adducts were isolated by high-performance liquid chromatography. One of the adducts was identified as the cyclic adduct formed between IPG and glyoxal. Mesoxaldialdehyde, which is structurally related to glyoxal, also produced a cyclic IPG-adduct and showed mutagenic activity in *Salmonella typhimurium* strain TA100. The other adduct isolated from the reaction mixture of IPG and heated glucose was 8-hydroxy-IPG. Various reagents which generate oxygen radicals were effective in the hydroxylation of guanosine derivatives at the C-8 position. These reagents also cause hydroxylation of guanine residues in DNA.

INTRODUCTION

We have isolated three potent mutagens from broiled fish and fried beef based on a bacterial mutagenicity assay and have characterized their chemical structures as 2-amino-3-methylimidazo[4,5-*f*]quinoline (IQ), 2-amino-3,4-dimethylimidazo[4,5-*f*]-quinoline (MeIQ) and 2-amino-3,8-dimethylimidazo[4,5-*f*]quinoxaline (MeIQx) (Kasai *et al.*, 1980, 1981a, 1981b). Carcinogenicity of IQ was demonstrated in mice and rats when given in the diet (Ohgaki *et al.*, 1984; Takayama *et al.*, 1984). It is important to characterize many other unknown mutagens which exist in cooked foods. Particularly, direct-acting mutagens in foods seem to be important in relation to cancers of the digestive tract, such as stomach cancer.

Fig. 1. Structure of fluorescent 2'-deoxy-2'-(2'',3''-dihydro-2'',4''-diphenyl-2''-hydroxy-3''-oxo-1''-pyrrolyl)guanosine (FG)

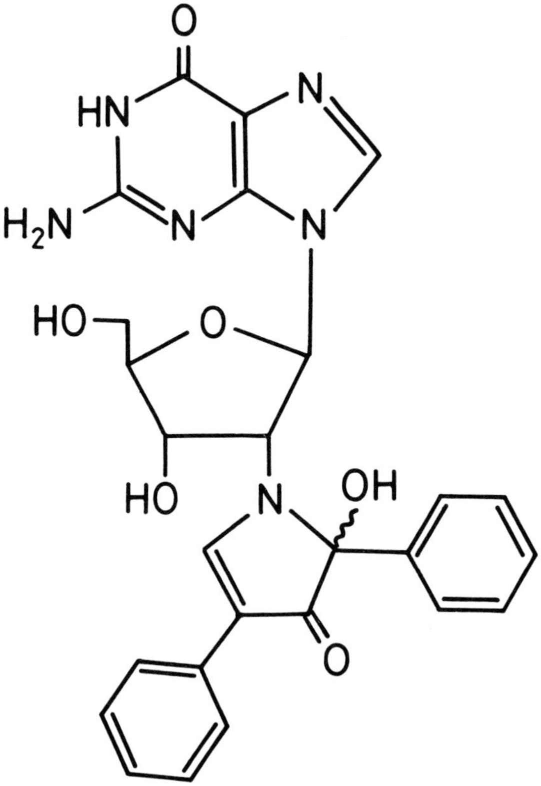

FG

For this reason, we have recently developed methods to detect and identify mutagens and carcinogens by means of the adducts formed upon reaction with guanosine derivatives, based on the principle that many mutagens and carcinogens react with guanine residues in DNA (Kasai et al., 1984). The method was found to be particularly useful for characterization of unstable, direct-acting mutagens which decompose during successive purification procedures in a conventional method. The derivatives used were fluorescent 2'-deoxy-2'-(2'',3''-dihydro-2'',4''-diphenyl-2''-hydroxy-3''-oxo-1''-pyrrolyl)guanosine (FG), the structure of which is shown in Figure 1, and 2',3'-O-isopropylideneguanosine (IPG). In this communication, we show that direct-acting mutagens in heated glucose are easily detected as adducts with FG and easily identified as adducts with IPG.

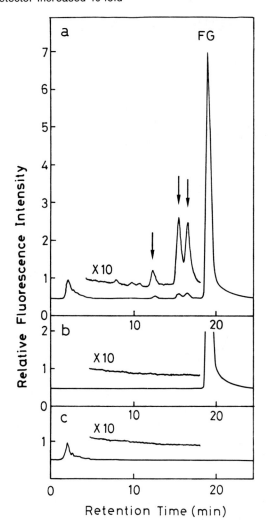

Fig. 2. Detection, by high-performance liquid chromatography, of mutagens in heated glucose as adducts of the fluorescent guanosine derivative (FG). a) Reaction mixture of FG and heated glucose, b) FG control, c) heated glucose control. Peaks corresponding to adducts are indicated by arrows. Excitation wavelength, 365 nm; emission wavelength, 480 nm; "X10", sensitivity of fluorescence detector increased 10-fold

MATERIALS AND METHODS

Reaction of heated glucose with FG

An extract (10 µl) of heated glucose (300 °C, 30 min; 1 mg/10 µl dimethylsulfoxide) was mixed with 10 µl of a solution of FG (5 µg/10 µl H_2O) and 50 µl of phosphate buffer (100 mM, pH 7.4) and incubated at 37 °C for 20 h. The reaction mixture

(50 μl) was injected into a high-performance liquid chromatography (HPLC) apparatus equipped with a fluorescence detector. The conditions for HPLC were as previously described (Kasai et al., 1984).

RESULTS AND DISCUSSION

Detection of mutagens in heated glucose by reaction with FG

We have previously shown that when FG is reacted with various mutagens in the presence or absence of S9 mix, formation of adducts can be detected by the HPLC-fluorescence detector system (Kasai et al., 1984). These mutagens include glyoxal, methylglyoxal, 4-nitroquinoline-N-oxide and 2-(2-furyl)-3-(5-nitrofuryl)acrylamide. This method has now been used to detect the mutagens in heated glucose, which has previously been shown to be mutagenic in *Salmonella typhimurium* TA100 in the absence of S9 mix (Nagao et al., 1977). When the reaction mixture of FG and heated glucose was analysed by HPLC, at least three peaks were detected in the chromatogram (Fig. 2a) which were not apparent in the FG or heated glucose controls (Fig. 2b and c). These peaks seem to be adducts produced by the reaction between FG and heated glucose. The sensitivity of this method in detecting mutagens is much higher than the bacterial mutagenicity test, because formation of the FG-adduct was clearly detected using 1 mg of heated glucose as shown in Figure 2, while in the mutation assay using TA100, only 200 revertants over a background of 100 revertants were induced by 10 mg of heated glucose.

Identification of possible mutagens in heated glucose by reaction with IPG

The identity of the mutagens in crude samples can be determined by characterizing the structure of adducts produced by the reaction of the samples with IPG. IPG-adducts can be easily extracted from the reaction mixture and purified by HPLC. For example, from the reaction mixture of IPG and heated glucose, two adducts were isolated by HPLC (Kasai et al., 1984). One of the adducts was identified as a cyclic glyoxal-IPG-adduct (Fig. 3a), based on UV- and mass spectral data and on a comparison of its retention time in HPLC with that of an authentic sample. From this result, it is evident that one of the mutagens in heated glucose is glyoxal, which was previously shown to be mutagenic in TA100. Mesoxaldialdehyde, which is structurally related to glyoxal, also produced a cyclic IPG-adduct (Fig. 3b), and showed mutagenic activity in TA100 (specific activity: 1400 revertants/mg, $-$ S9 mix).

The other adduct isolated from the reaction mixture of IPG and heated glucose was identified as 8-hydroxy-IPG (Fig. 3a), on the basis of spectral measurements (Kasai et al., 1984). It is interesting to study the mechanism of this hydroxylation reaction and its relation to mutagenesis. Various reagents were tested for their capacity to hydroxylate deoxyguanosine or guanine residues in DNA at the C-8 position (Fig. 4). Reducing agents or metal ions were effective for this hydroxylation reaction in the presence of oxygen (Kasai & Nishimura, 1984a). Polyphenol-Fe-H_2O_2 (Kasai & Nishimura, 1984b) and asbestos-H_2O_2 systems were also effective. The hydroxylation reaction by these reagents may proceed *via* generation of an oxygen radical.

Fig. 3. Structure of adducts produced by a) the reaction of 2′,3′-O-isopropylideneguanosine (IPG) with heated glucose and b) the reaction of IPG with mesoxaldialdehyde

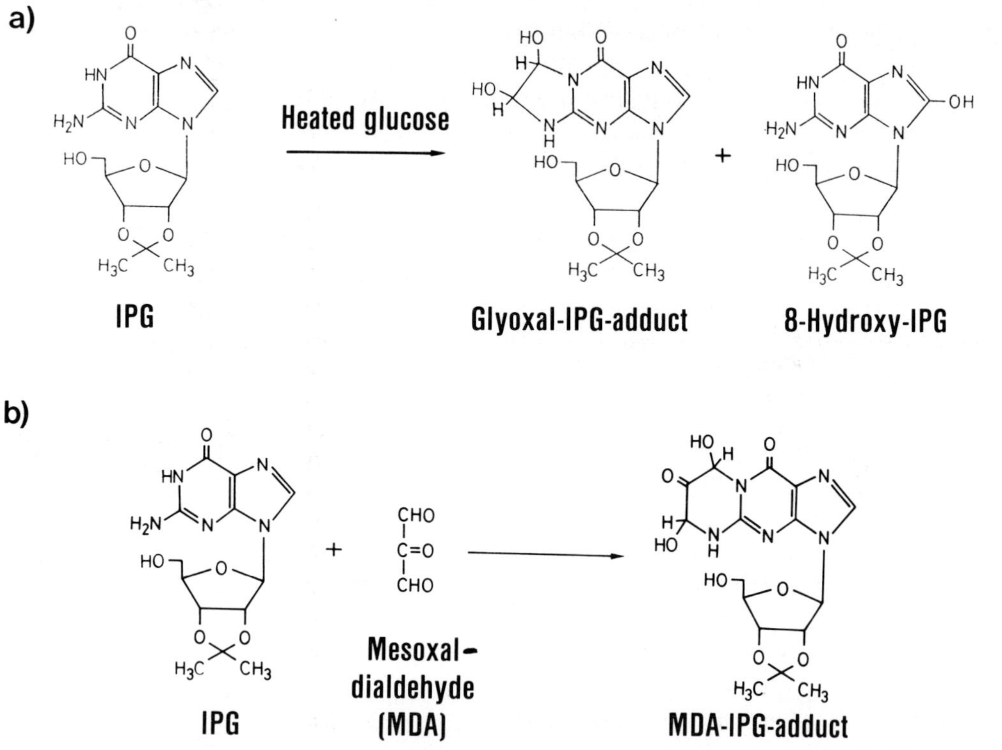

Fig. 4. Formation of 8-hydroxydeoxyguanosine by various reagents which produce an oxygen radical

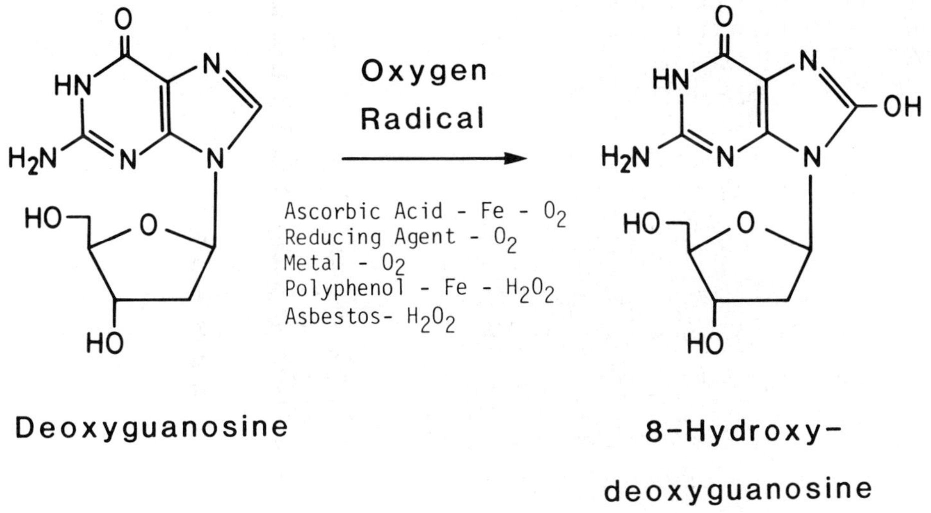

Formation of 8-hydroxyguanine would therefore be a new type of DNA damage induced by an oxygen radical, in addition to DNA strand scissions and formation of thymine glycol and 5-hydroxymethyluracil. The 8-hydroxyguanine residue in DNA may possibly induce mutation during the replication or repair processes because 8-hydroxydeoxyguanosine favours *syn*-conformation and may disturb the tertiary structure of DNA as do other carcinogen-DNA adducts. Since 8-hydroxydeoxyguanosine can be easily analysed by HPLC, its detection should also be useful for monitoring the extent of damage to DNA by oxygen radicals.

REFERENCES

Kasai, H. & Nishimura, S. (1984a) Hydroxylation of deoxyguanosine at the C-8 position by ascorbic acid and other reducing agents. *Nucleic Acids Res.*, **12**, 2137–2145

Kasai, H. & Nishimura, S. (1984b) Hydroxylation of deoxyguanosine at the C-8 position by polyphenols and aminophenols in the presence of hydrogen peroxide and ferric ion. *Gann*, **75**, 565–566

Kasai, H., Yamaizumi, Z., Wakabayashi, K., Nagao, M., Sugimura, T., Yokoyama, S., Miyazawa, T. & Nishimura, S. (1980) Structure and chemical synthesis of Me-IQ, a potent mutagen isolated from broiled fish. *Chem. Lett.*, 1391–1394

Kasai, H., Yamaizumi, Z., Nishimura, S., Wakabayashi, K., Nagao, M., Sugimura, T., Spingarn, N.E., Weisburger, J.H. Yokoyama, S. & Miyazawa, T. (1981a) A potent mutagen in broiled fish. Part 1. 2-Amino-3-methyl-3H-imidazo[4,5-*f*]-quinoline. *J. chem. Soc. Perkins Trans.*, *I*, 2290–2293

Kasai, H., Yamaizumi, Z., Shiomi, T., Yokoyama, S., Miyazawa, T., Wakabayashi, K., Nagao, M., Sugimura, T. & Nishimura, S. (1981b) Structure of a potent mutagen isolated from fried beef. *Chem. Lett.*, 485–488

Kasai, H., Hayami, H., Yamaizumi, Z., Saitô, H. & Nishimura, S. (1984) Detection and identification of mutagens and carcinogens as their adducts with guanosine derivatives. Nucleic Acids Res., **12**, 2127–2136

Nagao, M., Yahagi, T., Kawachi, T., Seino, Y., Honda, M., Matsukura, N., Sugimura, T., Wakabayashi, K., Tsuji, K. & Kosuge, T. (1977) *Mutagens in foods, and especially pyrolysis products of protein*. In: Scott, D., Bridges, B.A. & Sobels, F.H., eds, *Progress in Genetic Toxicology*, Amsterdam, Elsevier Press, pp. 259–264

Ohgaki, H., Kusama, K., Matsukura, N., Morino, K., Hasegawa, H., Sato, S., Sugimura, T. & Takayama, S. (1984) Carcinogenicity in mice of a mutagenic compound, 2-amino-3-methylimidazo[4,5-*f*]quinoline, from broiled sardine, cooked beef and beef extract. *Carcinogenesis,* **5**, 921–924

Takayama, S., Nakatsuru, Y., Masuda, M., Ohgaki, H., Sato, S. & Sugimura, T. (1984) Demonstration of carcinogenicity in F344 rats of 2-amino-3-methylimidazo[4,5-*f*]quinoline from broiled sardine, fried bef and beef extract. *Gann,* **75**, 467–470

THREE-DIMENSIONAL FLUOROMETRY FOR THE DETECTION OF DNA ADDUCTS

G. DOERJER, E. NIES, J. MERTES & F. OESCH

Institute for Toxicology, University of Mainz, Mainz, Federal Republic of Germany

SUMMARY

We describe the interfacing of a fluorometer to a desk-top computer by means of a commercially available interface box, for the purpose of generating three-dimensional fluorescence spectra. The important features of a self-designed program in BASIC are discussed in detail.

INTRODUCTION

Sensitive and specific methods are required to detect and quantitate DNA adducts of chemical carcinogens. These methods include the use of radioactively labelled compounds with a high specific activity, immunoassays and postlabelling of isolated modified nucleotides. We now report the usefulness of three-dimensional (3-D) fluorometry for qualitative and quantitative determination of DNA adducts of chemical carcinogens.

In fluorescence spectroscopy, two spectra are usually determined for a chemical compound. In one spectrum, the fluorescence intensity is monitored while the excitation wavelength is scanned at a constant emission wavelength. In the second spectrum, the fluorescence intensity is monitored while the emission wavelength is scanned at a constant excitation wavelength. If the constant emission and excitation wavelengths have been chosen correctly, the two spectra will cross at the point of the maximum fluorescence intensity for the compound. A modified system for monitoring fluorescence is already in use in which excitation and emission wavelengths are scanned at the same time with a constant difference between them (Rahn *et al.*, 1980). Because of the high specific fluorescence of many DNA adducts of chemical carcinogens, a complete monitoring of the fluorescence intensities to obtain a 3-D fluorescence spectrum should improve the detection and determination of such adducts. The efficiency of 3-D fluorometry has recently been demonstrated by the determination of concentrations of benzo[α]pyrene in shale oil (Hersberger *et al.*, 1981).

A computer-automated system is required to record and analyse the large amount of data generated by 3-D fluorometry. The present study shows that interfacing a standard laboratory fluorometer with a flexible computer system facilitates the sensitive detection of unknown DNA adducts by 3-D fluorometry.

MATERIALS AND METHODS

Instrumentation

A fluorometer (SPF 500, Aminco) and a desk-top computer (HP 9845B-2/5, Hewlett-Packard) were interfaced to one another *via* four commercially available cards in an interface box (Multiprogramer, 6942A, Hewlett-Packard). One card (HP 69771A, Hewlett-Packard), designed to convert binary-coded data to 16-bit words, was used to send fluorescence intensity data from the fluorometer to the computer. The second and third cards (HP 69751A, Hewlett-Packard) were analogue-to-digital converters and reported the position of the excitation and emission wavelength, respectively, to the computer. The fourth card (HP 69735A, Hewlett-Packard) sent pulses from the computer to the fluorometer. These pulses caused the fluorometer to read fluorescence intensity data and to determine how far, and in which direction, the emission and/or excitation wavelength must be changed.

Program structure

The entire program was written in BASIC. It is composed of four parts: (1) registration of the spectrum, (2) storage, (3) display and (4) interpretation.

In the first part of the program, the fluorescence intensity is monitored for as many combinations of excitation and emission wavelengths as requested. Every adjustment is automatically controlled for the correct position of the wavelength within a range of 0.5 nm. The fluorescence intensities are determined as often as requested at the start of the program. For each combination of excitation and emission, the mean of all fluorescence intensities measured is then stored in the memory as the fluorescence intensity at that point. The fluorescence intensities of all points in a spectrum are combined as elements in a two-dimensional matrix in which the colums contain the fluorescence intensities of the excitation spectra and the rows, the fluorescence intensities of the emission spectra. Due to the capability of the computer system, the total number of elements in the matrix can be up to 32 000. With the convention that all steps in excitation and emission scans represent equal distances, only the initial wavelength and the steps for emission and excitation have to be stored to recalculate the corresponding wavelength settings for each point in the matrix. The fluorescence intensities can be determined either in single-beam or in ratio mode. Corrections for changes in light intensities of the excitation lamp at different wavelengths can be introduced in spectra taken in single-beam mode. The comparison of fluorescence intensities stored on tape of the same standard can be used to correct for changes in the sensitivity of the fluorometer. For solutions which contain additional absorbing material, fluorescence intensities can be corrected by measuring the transmission of the sample in the reference beam using the excitation wavelength. Finally, all fluorescence intensities are corrected for the range settings of the fluorometer.

The resulting matrix is stored on tape together with all the information concerning instrument settings and the sample being analysed. In particular, the following parameters are stored: the sample designation including further parameters such as concentrations, solvent, etc., the starting excitation wavelength, the final excitation wavelength, the step width between two points in the excitation scan, the starting emission wavelength, the final emission wavelength, the step width between two points in the emission scan, the number of determinations of fluorescence intensity at each point, the range settings of the fluorometer, the band-pass for excitation and emission, and the gain of the reference beam. The instrument settings can be stored separately as routine methods to facilitate and to control the start of new spectral analyses under comparable conditions.

The stored matrix, with all necessary additional parameters, is used to display the 3-D fluorescence spectrum with orthogonal coordinates. To obtain a reasonable impression of the spectrum in a two-dimensional plot, the plane constructed by the axes for the excitation and emission wavelengths is filled with zero values for fluorescence intensity at all positions and the elements of the matrix with the measured fluorescence intensities are then added to this plane. Neighbouring points are connected by straight lines at the same excitation wavelength. A cursor can be moved within the display to give an optimal print-out of the excitation and emission wavelengths and the fluorescence intensity at the cursor position.

The matrices of comparable spectra can be used for the interpretation of a complex spectrum. By analysing the spectra of markers, the amount of the markers can be determined in mixtures. By combining various spectra, artificial spectra of mixtures can be created.

RESULTS

The system described above for obtaining 3-D fluorescence spectra proved useful for qualitative purposes and for quantitative determinations of fluorescent DNA adducts.

Three-dimensional fluorescence spectra of DNA samples resulting from treatment of DNA with chemicals *in vitro*, in tissue culture or *in vivo* provided a rapid screening method for determining whether fluorescent DNA adducts were formed or not. We found that, for this qualitative application, the main program can be easily modified by omitting variable parameters and deleting features which are only necessary for quantitative determinations. For example, such overview spectra were not corrected for parameters such as changes in the light intensities or DNA absorption. The time required to obtain these spectra was between one and two hours. In Figure 1, a 3-D fluorescence spectrum of chloroacetaldehyde-treated DNA is shown as an example.

At present, quantitative determinations are being carried out to improve those parts of the program concerned with correcting the spectra and to allow for their interpretation. When standards of known concentrations were checked for the

Fig. 1. Three-dimensional fluorescence spectrum of chloroacetaldehyde-treated DNA. Emission wavelength, 350–675 nm in steps of 5 nm; excitation wavelength, 200–330 nm in steps of 2 nm; band-pass of emission, 10 nm; band-pass of excitation, 8 nm.

The presence of etheno adducts was confirmed by high-performance liquid chromatographic analysis of an enzymatic DNA digest (Oesch *et al.*, these proceedings, p. 373). A control DNA sample showed no fluorescence signals at the same instrument settings; i. e., the 3-D spectrum appeared as a flat plane (data not shown).

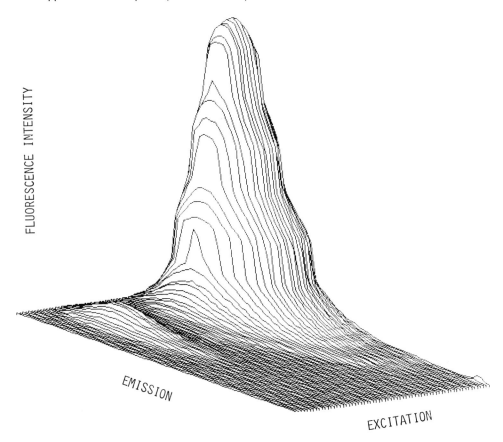

reproducibility of their fluorescence intensities, deviations of up to 10% were observed in their fluorescence maxima even after short times. The reproducibility can be improved when the fluorescence intensities of a solid standard are stored on tape and compared with the fluorescence intensities of the same standard before or after quantitative determination. It has been found that the mean of about five points with high fluorescence intensities and about five points with background fluorescence are sufficient to reduce the deviations to less than 3%. Because this correction for the changing sensitivity of the fluorometer is necessary for quantitative determinations, it was introduced as subroutine in the program and runs automatically after the operator has been asked by the computer to put a certain solid standard in the fluorometer.

In order to correct fluorescence intensities for the absorption of excitation light by additional material (for example, DNA) the following equation was used:

$$F_{corr} = \frac{F}{T_{sol} \times (1 - (T_{sol} - T_{sam}) \times e^{-0.5})}$$

F: measured fluorescence intensity
F_{corr}: corrected fluorescence intensity
T_{sol}: transmission of the solvent in the reference beam
T_{sam}: transmission of the sample in the reference beam

This equation simulates a half-concentrated sample in the reference beam of the fluorometer when fluorescence intensity is monitored in ratio mode. Due to the limitation in sensitivity of the reference photomultiplier at constant gain, a half-concentrated sample in the reference beam does not correct an absorption of the excitation light accurately. With our method, the gain of the reference photomultiplier can be changed and kept in a sensitive range with the result that corrected fluorescence intensities remain almost constant in the presence of DNA concentrations of up to 50 µg/ml.

DISCUSSION

Three-dimensional fluorescence spectra contain more information than single excitation and emission scans. However, commercially available systems are very limited with regard to their capacity to analyse the data from 3-D fluorescence spectra or to change parts of the program. The system described in this study was built from commercially available parts and was programmed in BASIC. Both of those features permit this system to be set up and used in other laboratories. In addition, since the program is written in BASIC, a common computer language, it can be easily adapted for future applications.

Our experience suggests that it is advantageous to have all data of the spectrum present in one matrix. Due to the consecutive determination of fluorescence intensities at equal distances of the emission and excitation wavelength in our program, the matrix requires only one third of the memory space compared to the situation where corresponding wavelengths are stored together with the fluorescence intensities.

The technical and theoretical considerations presented here, as well as the practical example given, indicate the advantages of 3-D fluorometry over high-performance liquid chromatographic separations for the detection of unknown fluorescent DNA adducts. Compared to the time required for a chromatographic separation, the time it takes to generate a 3-D fluorescence spectrum is negligible. However, additional research is still necessary to establish 3-D fluorometry as a standard method.

ACKNOWLEDGEMENT

This study was supported by the Bundesministerium für Forschung und Technologie and the Deutsche Forschungsgemeinschaft.

REFERENCES

Rahn, R.O., Chang, S.S., Holland, J.M., Stephens, I.J. & Smith, L.H. (1980) Binding of benzo(a)pyrene to epidermal DNA and RNA as detected by synchronous luminescence spectrometry at 77 K. *J. biochem. biophys. Methods,* **3,** 285–291

Hersberger, L.W., Callis, J.B. & Christian, G.D. (1981) Liquid chromatography with real-time video fluorometric monitoring of effluants. *Anal. Chem.,* **53,** 971–975

QUANTITATION OF ETHENO ADDUCTS BY FLUORESCENCE DETECTION

M.A. BEDELL[1], M.C. DYROFF[2], G. DOERJER[3] & J.A. SWENBERG[4]

Department of Biochemical Toxicology and Pathobiology, Chemical Industry Institute of Toxicology, Research Triangle Park, NC, USA

SUMMARY

Etheno adducts can be formed by the reaction of vinyl chloride metabolites with DNA and may play a role in the carcinogenicity of this chemical. These adducts are highly fluorescent and may be quantitated by sensitive photometric methods in conjunction with high-performance liquid chromatographic (HPLC) separation. Three HPLC systems were evaluated on the basis of maximal fluorescence intensity and resolution of two etheno adducts, ethenodeoxycytidine and ethenodeoxyadenosine. Analyses were conducted with enzymatically digested DNA that had been incubated with chloroacetaldehyde, a vinyl chloride metabolite which may cause etheno adduct formation *in vivo*. All three known etheno adducts of DNA were tentatively identified in DNA reacted *in vitro*. The sensitivity of the method was highest for the ethenodeoxyadenosine adduct, with the limit of detection (1 pmol per injection in the HPLC system) being similar to that for O^6-methylguanine, another promutagenic DNA adduct for which quantitation by HPLC with fluorescence detection has been reported. The method described here may be useful for the analysis of DNA from animals or humans exposed to vinyl chloride.

INTRODUCTION

The reaction of carcinogens or their metabolites with specific sites in nucleic acids involves a number of chemicals possessing vastly different structures and reactivities. In the case of monofunctional alkylating agents, strong correlations have been found

---

[1] Present address: Department of Molecular Genetics and Cell Biology, University of Chicago, Chicago, IL, USA
[2] Present address: Drug Metabolism Section, Wyeth Laboratories, Philadelphia, PA, USA
[3] Present address: Institute for Toxicology, University of Mainz, Mainz, Federal Republic of Germany
[4] To whom correspondence should be addressed

between target organ/cell susceptibility to cancer development and the formation and/or lack of repair of promutagenic alkylation products in DNA, such as O^6-alkylguanine and O^4-alkylthymidine (Bedell *et al.*, 1982; Pegg, 1983; Swenberg *et al.*, 1984).

In contrast to monofunctional chemicals that have only one reactive site, an important class of compounds has two reactive sites and can therefore react with two nucleophilic centres. An example of a bifunctional reaction within a single nucleoside is the formation of etheno adducts from metabolites of vinyl chloride (VC). Vinyl chloride is a potent carcinogen in both humans and animals (Heath *et al.*, 1975; Maltoni & Lefemine, 1975) and is metabolically activated *via* microsomal mixed-function oxidases primarily to chloroethylene oxide (CEO) and chloroacetaldehyde (CAA) (Bartsch & Montesano, 1975; Guengerich *et al.*, 1979, 1981). The major DNA adduct formed with CEO is a direct alkylation product, 7-oxoethylguanine (Scherer *et al.*, 1981), rather than an etheno adduct. In contrast, CAA reacts with DNA, nucleic acid polymers and deoxyribonucleosides to form etheno adducts of deoxycytidine (dCyd), deoxyadenosine (dAdo) and deoxyguanosine (dGuo) (Green & Hathway, 1978; Laib *et al.*, 1981; Kuśmierek & Singer, 1982; Oesch & Doerjer, 1982). The hypothesis that VC metabolites form DNA adducts that are capable of causing mutations is supported by the demonstration that base-pair substitution mutations, but not frame-shift mutations, occur after treatment of certain bacterial strains with these chemicals (Malaveille *et al.*, 1975). Additionally, two of the three known etheno adducts, namely $3,N^4$-ethenodeoxycytidine (etheno-dCyd) and $1,N^6$-ethenodeoxyadenosine (etheno-dAdo), direct the misincorporation of noncomplementary nucleosides in cell-free translation and transcription systems (Hall *et al.*, 1981; Spengler & Singer, 1981). Similarly, misincorporation by N^2, 3-ethenodeoxyguanosine (etheno-dGuo), although not demonstrated, is theoretically possible (Oesch & Doerjer, 1982). In contrast, 7-oxoethylguanine does not cause misincorporation (Barbin & Bartsch, these proceedings[5]). The N-7 position of guanine is not involved in hydrogen bonding of the DNA helix and the analogous reaction product from monofunctional alkylating agents (7-alkylguanine) is also not considered promutagenic (Pegg, 1983). Evidence for a third kind of adduct which leads to interstrand cross-linking of DNA has recently been reported from studies on the neutral reaction of polymers with CAA (Singer *et al.*, these proceedings[6]). These lesions have potential biological importance; however, no information on their formation or removal in cells or tissues is available.

The major DNA adduct formed in livers of rats acutely exposed to radioactive VC has been reported to be 7-oxoethylguanine (Laib *et al.*, 1981), while both etheno-dCyd and etheno-dAdo, but not 7-oxoethylguanine, were found in liver DNA of rats chronically exposed to VC (Green & Hathway, 1978). However, quantitative comparisons of the results of these studies are difficult because of the low levels of adducts formed and the relative insensitivity of the methods used. More recently, etheno-dGuo was demonstrated in liver DNA from young rats at a concentration

---

[5] See p. 345.
[6] See p. 45.

approximately 100-fold lower than that of 7-oxoethylguanine (Laib, these proceedings[7]). In all of these studies, whole liver DNA was analysed. The results would, therefore, favour the detection of adducts formed in hepatocellular DNA. However, it is the nonparenchymal cell (NPC) population, comprising only a small part of the total liver DNA, that primarily develops tumours following VC exposure (Health et al., 1975; Maltoni & Lefemine, 1975). Although metabolism of VC occurs mainly in the hepatocyte and not in the NPC (Ottenwalder & Bolt, 1980), CAA and to a lesser extent CEO diffuse out of the hepatocyte and react with extra-hepatocellular nucleophiles (Guengerich et al., 1981). Etheno adduct formation is greatest in single-stranded DNA polymers (Kuśmierek & Singer, 1982). This should favour their formation in the NPC population, which has a greater percentage of replicating cells than does the hepatocyte population (Lewis & Swenberg, 1982).

Our investigations have focused on the detection of etheno adducts since it is possible that they are formed in the target cell population and are promutagenic. We have evaluated three HPLC systems utilizing fluorescence detection of the adducts, because this type of analysis does not require administration of radioactive VC. In contrast to unmodified bases and nucleosides which do not fluoresce, etheno-dCyd and etheno-dAdo have fluorescent properties, which were first demonstrated by Barrio et al. (1972). More recently, etheno-dGuo formation in CAA-treated DNA and the repair of etheno bases by cell-free homogenates (Oesch & Doerjer, 1982; Oesch et al., these proceedings[8]) have been detected using a fluorescence/HPLC method. We have optimized our methods using commercially available etheno markers, deoxyribonucleosides and DNA which has been reacted with CAA *in vitro*.

MATERIALS AND METHODS

Chemicals

Etheno-dAdo ($1,N^6$-ethenodeoxyadenosine), etheno-A ($1,N^6$-ethenoadenine, etheno-C ($3,N^4$-ethenocytosine), unmodified deoxyribonucleosides and calf thymus DNA were purchased from Sigma Chemical, St. Louis, MO. Etheno-dCyd ($3,N^4$-ethenodeoxycytidine) was synthesized by the method of Barrio et al. (1972) and purity was greater than 95%, as determined by HPLC. CAA (45% solution in water) was obtained from ICN Pharmaceuticals, Plainview, NY. DNAse I, bacterial alkaline phosphatase and adenosine deaminase were from Sigma and venom phosphodiesterase was from Worthington Biochemical, Frechold, NJ.

Reaction of DNA with CAA

Calf thymus DNA (30 mg) was dissolved in 0.1 M sodium phosphate (pH 7) to give 2 mg/ml. CAA was added to a final concentration of 0.5% and incubations were

---

[7] See p. 101.
[8] See p. 373.

carried out at 37 °C. Five ml aliquots were taken at 0 (no CAA), 2, 8 and 24 hours. NaCl was added and DNA was precipitated with cold ethanol, dissolved and precipitated two more times, washed two times with ethanol and dried under a stream of nitrogen. The DNA was dissolved in 2 ml of water and stored at $-20\,°C$ until hydrolysis.

Enzymatic digestion of DNA

DNA was diluted to 1 mg/ml and digested with DNAse I (13 µl) in 10 mM bis-Tris, 10 mM $MgCl_2$, pH 7.1 for 4 hours at 37 °C. Tris-HCl (pH 8.3) was added to a final concentration of 0.1 M, followed by venom phosphodiesterase (0.1 unit) and bacterial alkaline phosphatase (0.1 unit) and the digestion was continued for an additional 18 hours at 37 °C. Adenosine deaminase (10 units) was then added and the incubation continued for 20 minutes. The samples were stored at $-20\,°C$ until analysis.

HPLC analysis

Three chromatographic systems were used; all utilized Waters solvent delivery systems, a Rheodyne injector, a Water 440 UV-absorbance detector, a Perkin-Elmer 650-10 fluorescence detector, a Shimadzu RF-530 fluorescence detector and a LDC Chromatograph Control Module.

Reverse-phase, pH gradient elution (RP-I): A Spherisorb ODS column (5 µ, 25 × 0.9 cm, Altex) connected to a C18 precolumn was eluted over 5 minutes at room temperature with a linear gradient of 5 to 40% buffer B (50% methanol in 10 mM ammonium phosphate pH 7) in buffer A (10 mM ammonium phosphate, pH 3.5). The flow rate was 1 ml/min and peak detection was at 254 nm for ultraviolet (UV) absorption and 225 nm excitation and 410 nm emission, or 290 nm excitation and 340 nm emission, for fluorescence.

Reverse-phase, neutral elution (RP-II): The same conditions were used as for RP-I except that buffer A was 10 mM ammonium phosphate pH 7 and a linear gradient of 10 to 40% buffer B was run over 6 minutes.

Strong cation-exchange (SCX): A Partisil-10 SCX column (10 µ, 25 × 0.9 cm, Whatman) was connected to a C18 precolumn and eluted with a linear gradient of 0 to 5% methanol in 1 M ammonium formate, pH 6.5. All other conditions were the same as for RP-HPLC.

In both RP- and SCX-HPLC, fluorescent and UV-absorbing peaks were integrated and areas compared to peak areas of known standard solutions.

RESULTS

We compared three chromatographic separation systems using unmodified deoxyribonucleosides and some of the known etheno adducts to determine their suitability for quantitation of adducts by fluorescence. All three systems (two on reverse-phase and one on cation-exchange columns) gave adequate separations of

Fig. 1. Reverse-phase separation of unmodified deoxyribonucleosides and etheno deoxyribonucleosides by pH gradient elution (RP-I). 10 nmol each of deoxyadenosine (dAdo), deoxycytidine (dCyd), deoxyguanosine (dGuo), deoxythymidine (dThd) and 3,N^4-ethenodeoxycytidine (etheno-dCyd) and 1 nmol of 1,N^6-ethenodeoxyadenosine (etheno-dAdo) were applied to a C18 high-performance liquid chromatographic column and eluted with 10 mM ammonium phosphate pH 3.5 with a linear gradient of 5 to 40% buffer (50% methanol in 10 mM ammonium phosphate pH 7) at 1 ml/min. Fluorescence intensities were recorded by two detectors simultaneously, one set at 290 nm excitation, 340 nm emission (A) and the other, at 225 nm excitation, 410 nm emission (B). (The etheno bases, ethenoadenine and ethenocytosine, elute between dCyd and dGuo but are not shown.)

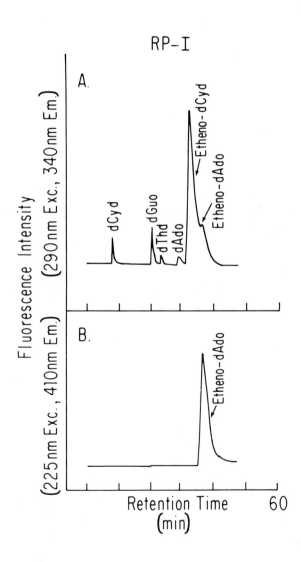

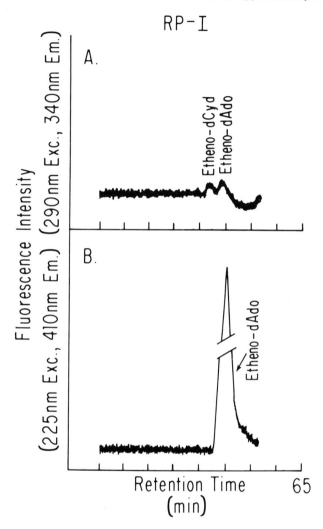

Fig. 2. Reverse-phase chromatography of 10 pmol each of 3,N^4-ethenodeoxycytidine (etheno-dCyd) and 1,N^6-ethenodeoxyadenosine (etheno-dAdo). All conditions were the same as in Figure 1 except that the sensitivity of both detectors was increased approximately 10-fold.

most of the peaks; however, optimum conditions for maximum fluorescence of etheno-dCyd and etheno-dAdo were observed with RP-I, a system using a pH gradient elution.

Typical chromatograms of unmodified deoxyribonucleosides, etheno-dCyd and etheno-dAdo, recorded with two fluorescence detectors in series, are shown in Figures 1 and 2. One of the detectors was set for the fluorescence maxima of etheno-dCyd (290 nm excitation and 340 nm emission) while the other was optimal for etheno-dAdo (225 nm excitation and 410 nm emission). Chromatographie conditions for the two figures were the same; detector sensitivities were approximately 10-fold greater in Figure 2 than in Figure 1.

In RP-I chromatography, a gradient of methanol in 10 mM ammonium phosphate was generated with a concomitant pH gradient from 3.5 to 7. This pH gradient was necessary in order to achieve maximal fluorescence of etheno-dCyd, which has a pH optimum for fluorescence in acidic solutions; in contrast, etheno-dAdo fluorescence is maximal at neutral pH (Barrio et al., 1972). Under these chromatographic conditions, peak areas at 290 nm excitation were approximately equal when 10 pmol each of etheno-dCyd and etheno-dAdo were injected (Fig. 2). Fluorescence of etheno-dCyd at this concentration was not detectable in the other two HPLC systems. We have also found that buffer concentrations of greater than 100 mM decrease the fluorescence intensities of both etheno-dCyd and etheno-dAdo (data not shown). The combination of acidic pH and low buffer concentration was therefore ideal for simultaneous detection of both adducts.

A shown in Figure 3, separation of etheno bases, their corresponding deoxyribonucleosides and unmodified deoxyribonucleosides was also achieved by elution on either RP (Panel A, B) or SCX (Panel C, D) columns at neutral pH. However, fluorescence of etheno-dCyd at 290 nm excitation was minimal in both systems due to the neutral pH used (detection at this wavelength not shown). A dramatic decrease in peak areas was observed for both etheno-A and etheno-dAdo at the high buffer concentration (1 M) used in the SCX system as compared to the low buffer concentration (10 mM) used in the RP-11 system. This effect is evident in Figures 3B and 3D, where fluorescence detector sensitivities are identical.

The neutral-pH RP system (RP-II) did offer some advantages over RP-I. Etheno bases did not chromatograph well in the acidic buffer of RP-I. Even though they eluted earlier, the peaks were very broad. The glycosidic bond of etheno-deoxyribonucleosides is stable so etheno bases should not be formed under the conditions of enzymatic hydrolysis. In addition, etheno-dAdo has maximum fluorescence at neutral pH and a comparison of the peak areas in Figures 1B and 3B indicated that the sensitivity of RP-II for etheno-dAdo detection is approximately two-fold greater than that of RP-I. However, it may still be advantageous to be able to detect both etheno-dCyd and etheno-dAdo, even with a slight loss of sensitivity for the latter, in which case, RP-I would be the system of choice.

Although the separation of etheno-dCyd and dAdo is sufficient at equimolar concentrations, dAdo is expected to be present in DNA at concentrations thousands of times greater than etheno-dCyd. We therefore found it necessary to remove dAdo by incubating with adenosine deaminase. Under the conditions used, dAdo was completely converted to the more polar nucleoside, deoxyinosine, and therefore did not interfere with the detection of etheno-dCyd. Adenosine deaminase had no effect on any of the etheno adducts.

Enzymatic digests of DNA incubated with and without CAA are shown in Figure 4. Fluorescent peaks not associated with unmodified deoxyribonucleosides were observed in CAA-treated DNA with both 225 and 290 nm excitation. One peak was tentatively identified as etheno-dGuo on the basis of RP chromatographic behaviour

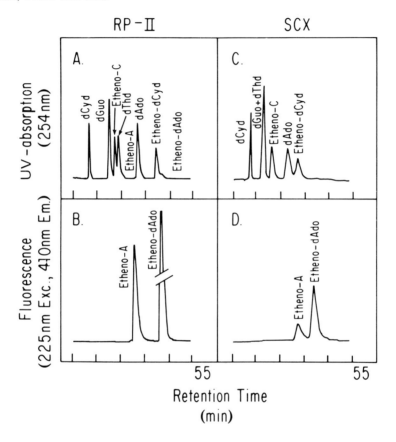

Fig. 3. High-performance liquid chromatography (HPLC), with neutral pH elution, of deoxyribonucleosides, etheno deoxyribonucleosides and etheno bases. Ultraviolet (UV) absorption detection was at 254 nm (A, C) and fluorescence detection, at 225 nm excitation with 410 nm emission (B, D). HPLC system used in A and B: reverse-phase (RP-II), elution with 10 mM ammonium phosphate pH 7, with a linear gradient of 10 to 40% buffer B (50% methanol in 10 mM ammonium phosphate pH 7). HPLC system used in C and D: strong cation-exchange (SCX), elution with 1 M ammonium formate pH 6.5, with a linear gradient of 0 to 5% methanol in 1 M ammonium formate pH 6.5. Abbreviations for compounds are as given in the legend to Figure 1. 10 nmol of each were applied to the columns, except for etheno-A and etheno-dAdo, in which cases, 1 nmol was used.

(see Oesch *et al.*, these proceedings[9]) and a fluorescence spectrum similar to that published for ethenoguanine (Oesch & Doerjer, 1982). The other two peaks had the same retention times as authentic standards of etheno-dCyd and etheno-dAdo. The amounts of these adducts were calculated from standard amounts injected and found to be approximately 20 and 180 pmoles, respectively. Quantitation of etheno-dGuo was not possible because of unavailability of authentic standard.

---

[9] See p. 373.

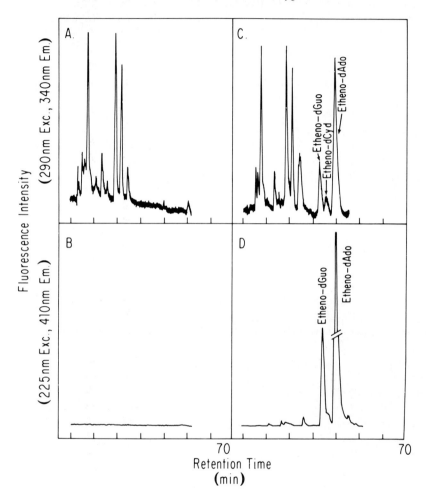

Fig. 4. Reverse-phase chromatography, with pH gradient elution, of enzymatically digested DNA reacted without (A, B) or with (C, D) chloroacetaldehyde. Approximately 50 μg of DNA was digested to deoxyribonucleosides and chromatographed as described for Figure 1. Fluorescence detection was at 290 nm excitation, 340 nm emission (A, C) and at 225 nm excitation, 410 nm emission (B, D). Etheno-dAdo, $1,N^6$-ethenodeoxyadenosine; etheno-dCyd, $3,N^4$-ethenodeoxycytidine; etheno-dGuo, $N^2,3$-ethenodeoxyguanosine

DISCUSSION

Chromatographic separation of etheno adducts and unmodified deoxyribonucleosides was possible on three different column systems. Optimal fluorescence detection of etheno-dCyd and etheno-dAdo was achieved by eluting a reverse-phase column with a linear gradient of methanol and increasing basicity of the eluting buffer. With this system, near optimal pH for fluorescence of each etheno adduct was utilized. Furthermore, simultaneous use of two fluorescence detectors allowed each of these compounds to be detected at their wavelength maxima. The limit of detection for etheno-dCyd was approximately 10 pmol, while that for etheno-dAdo was approximately 1 pmol per injection.

When CAA-treated DNA was enzymatically digested and chromatographed, three fluorescent peaks not corresponding to unmodified nucleosides were observed. On the basis of the known reactivity of CAA with DNA (Kuśmierek & Singer, 1982; Oesch & Doerjer, 1982), and of fluorescent spectra and retention times similar to those of available markers, it was tentatively concluded that etheno-dCyd, etheno-dGuo and etheno-dAdo were formed. When etheno-dAdo and etheno-dCyd were quantitated by comparing peak areas to known standards, the former was found to be present in calf thymus DNA at approximately 10 times the concentration of the latter. A similar fluorescence/HPLC method for quantitating etheno bases has been reported by Oesch et al. (these proceedings[10]); however, etheno-dCyd could not be quantitated. The present method appears suitable for the analysis of all three known etheno adducts in DNA.

The most widely used method for quantitation of DNA adducts is the determination of radioactive adducts after administration of radiolabelled compound. Disadvantages of this methodology include the limit availability of radiolabelled carcinogens in sufficient quantities and with specific activities high enough to afford detection, as well as the prohibitive cost of conducting studies with continous exposure regimens. Immunoassays for DNA adducts using both poly- and monoclonal antibodies have recently been developed and are generally more sensitive than radioassays. In addition, this type of analysis may be conducted on DNA from animals chronically exposed to carcinogen or on human samples. An additional method for quantitating DNA adducts is based on the fluorescent properties of some of the adducts. An example of the usefulness of fluorescent analysis of DNA was the quantitation of O^6-methylguanine in different liver cell populations of animals continuously exposed to methylating agents (Bedell et al., 1982; Lindamood et al., 1982). Using this method, it was demonstrated that the NPC population of both rat and mouse liver accumulated O^6-methylguanine during the exposure regimen, while hepatocytes effectively removed this promutagenic adduct. Similar experiments could be conducted on DNA of different liver cells during VC-exposure using the methodology reported here. The limits of detection for O^6-methylguanine and etheno-dAdo are similar (approximately 1 pmol per injection). Refinement of the detection of the other etheno adducts may result in similar sensitivities. If etheno adducts were formed to a greater extent or repaired to a lesser extent in the NPC population than in hepatocytes, the presence of these promutagenic DNA adducts could constitute one mechanism for cell specificity in VC-induced carcinogenicity.

---

[10] See p. 373.

ACKNOWLEDGEMENT

This research was funded, in part, by grant No. 20023-01 to Duke University from the National Institutes of Health.

REFERENCES

Barrio, J.R., Secrist, J.A. & Leonard, N.J. (1972) Fluorescent adenosine and cytidine derivatives. *Biochem. biophys. Res. Commun., 46,* 597–604

Bartsch, H. & Montesano, R. (1975) Mutagenic and carcinogenic effect of vinyl chloride. *Mutat. Res., 32,* 93–114

Bedell, M.A., Lewis, J.G., Billings, K.C. & Swenberg, J.A. (1982) Cell specificity in hepatocarcinogenesis: preferential accumulation of O^6-methylguanine in target cell DNA during continuous exposure of rats to 1,2-dimethylhydrazine. *Cancer Res., 42,* 3079–3083

Green, T. & Hathway, D.E. (1978) Interactions of vinyl chloride with rat-liver DNA in vivo. *Chem.-biol. Interactions, 22,* 211–224

Guengerich, F.P., Crawford, W.M. & Watanabe, P.G. (1979) Activation of vinyl chloride to covalently bound metabolites. *Biochemistry, 18,* 5177–5182

Guengerich, F.P., Mason, P.S., Stott, W.T., Fox, T.R. & Watanabe, P.G. (1981) Roles of 2-haloethylene oxides and 2-haloacetaldehydes derived from vinyl bromide and vinyl chloride in irreversible binding to protein and DNA. *Cancer Res., 41,* 4391–4398

Hall, J.A., Saffhill, R., Green, T. & Hathway, D.E. (1981) The induction of errors during in-vitro DNA synthesis following chloroacetaldehyde-treatment of poly(dA-dT) and poly(dC-dG) templates. *Carcinogenesis, 2,* 141–146

Heath, C.W., Jr., Falk, H. & Creech, J.L., Jr. (1975) Characteristics of cases of angiosarcoma of the liver among vinyl chloride workers in the United States. *Ann. N.Y. Acad. Sci., 246,* 231–236

Kuśmierek, J.T. & Singer, B. (1982) Chloroacetaldehyde-treated ribo- and deoxyribopolynucleotides. 1. Reaction products. *Biochemistry, 21,* 5717–5722

Laib, R.J., Gwinner, L.M. & Bolt, H.M. (1981) DNA alkylation by vinyl chloride metabolites: etheno derivatives or 7-alkylation of guanine? *Chem.-biol. Interactions, 37,* 219–231

Lewis, J.G. & Swenberg, J.A. (1982) The effect of 1,2-dimethylhydrazine and diethylnitrosamine on cell replication and unscheduled DNA synthesis in target and nontarget cell populations in rat liver following chronic administration. *Cancer Res., 42,* 89–92

Lindamood, C., Bedell, M.A., Billings, K.C. & Swenberg, J.A. (1982) Alkylation and de novo synthesis of liver cell DNA from C3H mice during continuous dimethylnitrosamine exposure. *Cancer Res., 42,* 4153–4157

Malaveille, C., Bartsch, H., Barbin, A., Camus, A.M., Montesano, R., Croisy, A. & Jacquignon, P. (1975) Mutagenicity of vinyl chloride, chloroethylene oxide, chloroacetaldehyde and chloroethanol. *Biochem. biophys. Res. Commun.*, **63**, 363–370

Maltoni, C. & Lefemine, C.T. (1975) Carcinogenicity bioassays of vinyl chloride: current results. *Ann. N.Y. Acad. Sci.*, **246**, 195–218

Oesch, F. & Doerjer, G. (1982) Detection of N^2,3-ethenoguanine in DNA after treatment with chloroacetaldehyde *in vitro*. *Carcinogenesis*, **3**, 663–665

Ottenwalder, H. & Bolt, H.M. (1980) Metabolic activation of vinyl chloride and vinyl bromide by isolated hepatocytes and hepatic sinusoidal cells. *J. environ. Path. Toxicol.*, **4**, 411–417

Pegg, A.E. (1983) *Alkylation and subsequent repair of DNA after exposure to dimethylnitrosamine and related carcinogens.* In: Hodgson, E., Bend, J.R. & Philpot, R.M., eds, *Reviews in Biochemical Toxicology, Vol. 5*, New York, Elsevier Biomedical, pp. 83–133

Scherer, E., Van der Laken, C.J., Gwinner, L.M., Laib, R.J. & Emmelot, P. (1981) Modification of deoxyguanosine by chloroethylene oxide. *Carcinogenesis*, **2**, 671–677

Spengler, S. & Singer, B. (1981) Transcriptional errors and ambiguity from the presence of 1,N^6-ethenoadenosine or 3,N^4-ethenocytidine in polyribonucleotides. *Nucleic Acids Res.*, **9**, 365–373

Swenberg, J.A., Dyroff, M.C., Bedell, M.A., Popp, J.A., Huh, N., Kirstein, U. & Rajewsky, M.F. (1984) O^4-Ethyldeoxythymidine, but not O^6-ethyldeoxyguanosine, accumulates in hepatocyte DNA of rats exposed continuously to diethylnitrosamine. *Proc. natl Acad. Sci. USA*, **81**, 1692–1695

DETECTION OF DNA BASE DAMAGE BY ^{32}P-POSTLABELLING: TLC SEPARATION OF 5'-DEOXYNUCLEOSIDE MONOPHOSPHATES

M. HOLLSTEIN, J. NAIR & H. BARTSCH

International Agency for Research on Cancer, Lyon, France

B. BOCHNER

Department of Fermentation Research and Process Development, Genentech, Inc., South San Francisco, California, USA

B.N. AMES

Department of Biochemistry, University of California, Berkeley, California

SUMMARY

Traces of damaged bases in DNA can be detected without the use of radioactive chemical agent or radiolabelled DNA substrate by incorporation of radiolabel into DNA *after* exposure to the chemical. This postlabelling approach to carcinogen adduct analysis is potentially useful for adduct detection in the DNA of humans exposed to carcinogens. We describe here the postlabelling analysis of DNA by chromatography of deoxynucleoside monophosphates. Separation and characterization of modified residues in nucleotide digests of DNA treated *in vitro* with chloroacetaldehyde is demonstrated.

INTRODUCTION

The use of radioactive test chemicals has provided a powerful approach to the study of covalent interaction between DNA and chemical carcinogens (Singer & Grunberger, 1983). Recently, postlabelling techniques have been developed for detection of base modifications in DNA; these techniques combine the sensitivity achievable with radioisotopes with a methodology requiring neither radioactive test chemical nor radioactive DNA substrate (Randerath *et al.*, 1981; Gupta *et al.*, 1982). The postlabelling analysis of DNA base damage, developed by Randerath and co-workers, is accomplished by digestion of microgram amounts of DNA to

3′-deoxynucleoside monophosphates, followed by ^{32}P-labelling to [3′,5′-^{32}P]-bisphosphates; the chemically altered bisphosphates are separated by thin-layer chromatography (TLC), and detected by autoradiography. A key factor in the sensitivity of this approach is the extent to which trace amounts of modified residues can be resolved chromatographically from the unmodified residues that comprise the bulk of radioactivity in the labelled digest. Total isolation of residues with bulky DNA adducts from normal nucleotides has been made possible by new chromatographic procedures (Reddy et al., 1984), greatly enhancing the usefulness of postlabelling analysis for DNA treated with aromatic carcinogens. Detection of alkylated bases in DNA requires a different procedure, however, because the migration characteristics of nucleotides with alkylated bases can be similar to those of unmodified nucleotides. Therefore we analysed base alkylation by separation of ^{32}P-labelled deoxynucleoside monophosphates rather than bisposphates to enhance migration differences caused by changes in base composition. The deoxynucleoside monophosphates can be separated by a simple, two-dimensional TLC system (Bochner & Ames, 1982); this system allows some predictions to be made regarding migration of modified nucleotides based on the chemistry of the modification. The method is described here and its application in the analysis of chloroacetaldehyde-modified DNA is demonstrated.

MATERIALS AND METHODS

Chemicals

Calf thymus DNA, 5′-deoxynucleoside monophosphate standards, micrococcal endonuclease (grade IV) and spleen phosphodiesterase were obtained from Sigma (St. Louis, MO). T$_4$ polynucleotide kinase was from P-L Biochemicals. [γ-^{32}P]-ATP (adenosine triphosphate) was purchased from New England Nuclear (Boston, MA) and [5′-^{32}P]-deoxynucleoside monophosphates were prepared from [α-^{32}P]-dCTP (deoxycytidine triphosphate) and [α-^{32}P]-dATP (deoxyadenosine triphosphate) (Amersham, Buckinghamshire, England). Chloroacetaldehyde was purchased from Fluka (Buchs, Switzerland), and was also prepared fresh, immediately prior to use, from 4-chloro-1,3-dioxolan-2-one (Fluka) according to published procedures (Gross, 1963) Plastic-backed polyethyleneimine (PEI) cellulose TLC sheets with a UV-254 fluorescent background were from Merck (Darmstadt, FRG). Autoradiography was performed with Dupont (Frankfurt, FRG) cassettes containing X-OMAT intensifying screens, and Dupont Cronex 4 NIF film.

Treatment with chloroacetaldehyde (CAA)

Calf thymus DNA (2 mM, in 0.1 M Tris-HCl buffer pH 8) was denatured by heating, treated with freshly prepared CAA (20 mM) for one hour at 37 °C, and dialysed against 10 mM Tris-HCl pH 7.5. Individual mononucleotide standards were reacted with CAA as follows: 20 µl of [5′-^{32}P]-dCMP (deoxycytidine monophosphate) or [5′-^{32}P]-dAMP (deoxyadenosine monophosphate) (2.5 µM; specific activity,

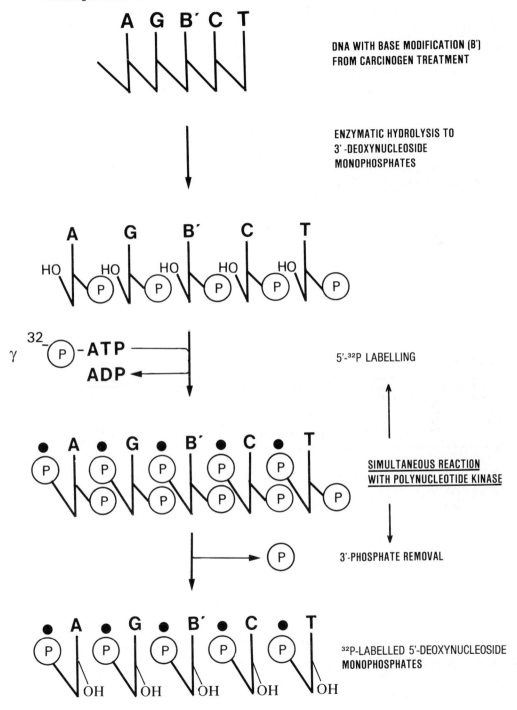

Fig. 1. Postlabelling technique for preparation of [5'-^{32}P]-deoxynucleoside monophosphates from carcinogen-modified DNA

Fig. 2. Two-dimensional thin-layer chromatography (TLC) system for separation of nucleoside monophosphates (modified from Bochner & Ames, 1982). The positions of selected monophosphates are indicated. See Bochner & Ames, 1982, for abbreviations.

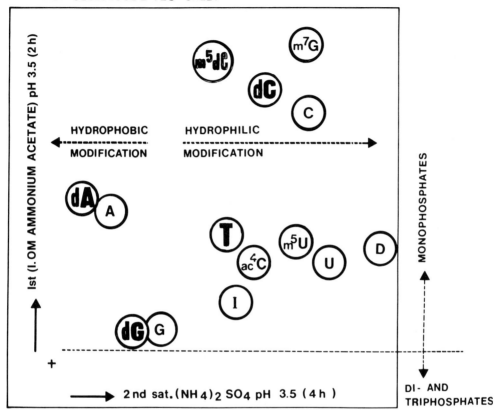

25 Ci/mmol), with or without addition of 1 µl CAA (Fluka) were incubated for 15 minutes to 3 hours at 50 °C and immediately chromatographed as described below. Unlabelled 5′-mononucleotides (5 mg/ml) were incubated for 2 to 4 hours at 50 °C in 1% CAA, lyophilized, resuspended in 1/10 volume water, and chromatographed on TLC sheets. Ultraviolet (UV)-absorbing material was scraped from the sheets, eluted with water, and the pH adjusted to 6.75 before recording the UV absorption spectra.

Enzymatic digestion and postlabelling of DNA

The procedure is outlined in Figure 1, and is essentially that described by Randerath et al., (1981). One microgram of DNA was digested to 3′-deoxynucleoside monophosphates with micrococcal nuclease (2 µg) and spleen phosphodiesterase (4 µg) in 12 µl of 20 mM Tris-HCl pH 6.8, 10 mM $CaCl_2$, for one to three hours at 37 °C.

Digested DNA (30–100 pmol) was then reacted at 37 °C for two hours with 5–20 units of T_4 polynucleotide kinase in a solution of 50 mM Tris-HCl pH 6.8, 10 mM $MgCl_2$, 10 mM DTT, containing 2–4 µCi [γ-^{32}P]-ATP (60–120 pmol)[1]. Samples were either chromatographed immediately or stored at −20 °C after addition of disodium ethylenediaminetetraacetic acid to 25 mM.

Chromatography and autoradiography

The [5′-^{32}P]-deoxynucleoside monophosphate mixture containing trace levels of nucleotides with carcinogen-derived base modifications was separated into its components by two-dimensional thin-layer chromatography (Bochner & Ames, 1982): one-tenth to one-half the reaction mixture was applied near the lower left-hand corner (origin) of a PEI-cellulose sheet, which was then developed to the top with 1M acetic acid (adjusted to pH 3.5 with ammonium hydroxide), soaked for 5 minutes in methanol, dried, and developed in the second dimension with saturated ammonium sulfate, pH 3.5. The separation of some nucleoside monophosphate standards is indicated in Figure 2. The bottom and right-hand margins of each sheet were trimmed 2–3 cm to remove unreacted ATP and inorganic phosphate before exposure to X-ray film at –70 °C for 2–24 hours. Sections of the TLC sheet corresponding to spots on films were cut out and the radioactivity (cpm) determined by liquid scintillation.

RESULTS

Modified nucleotides were detected in CAA-treated DNA by ^{32}P-postlabelling and two-dimensional separation of digested DNA (Fig. 3B). The two major modifications (adducts 1 and 2) are absent in the postlabelled digest of untreated DNA (Fig. 3A) and are well-resolved from the normal nucleotides that constitute mammalian DNA. The percent radioactivity contributed by each monophosphate component in the mixture is shown in Table 1. CAA reaction with DNA causes a substantial loss of dAMP and dCMP, which is accompanied by the formation of comparable amounts of modified derivatives 1 and 2. This indicates that the two major novel nucleotides are derived from cytosine and adenine residues. Also, treatment of ^{32}P-dAMP with CAA results in formation of a modified nucleotide that comigrates with adduct 1, and treatment of ^{32}P-dCMP results in a derivative that comigrates with adduct 2 (Fig. 4).

Reaction of CAA with individual monophosphates was repeated with sufficient unlabelled dAMP and dCMP to obtain UV-absorbing spots at the positions of adducts

---

[1] Under these conditions, the 5′-kinase and 3′-phosphatase activities of T_4 polynucleotide kinase are both high and the 3′-deoxynucleoside monophosphates are converted to radiolabelled 5′-deoxynucleoside monophosphates (Cameron & Uhlenbeck, 1977). 5′-Labelled monophosphates can also be obtained from two-step procedures, for example, by using conditions that favour the 5′-kinase activity, followed by adjustment of the pH to achieve 3′-phosphate removal. In this preliminary work, however, we chose the simplest approach in which both activities were present simultaneously.

Fig. 3. Autoradiograms from two-dimensional thin-layer chromatography of 2 nanograms postlabelled [5′-^{32}P]-deoxynucleoside monophosphates from calf thymus DNA, (A) untreated, (B) chloroacetaldehyde-treated

A

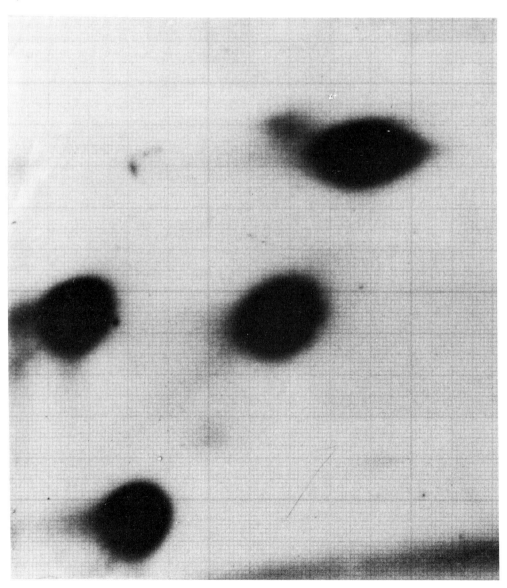

CONTROL

Fig. 3 (contd)

B

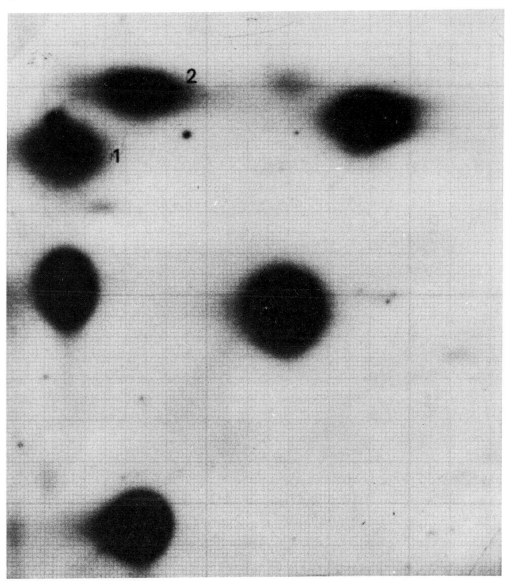

CAA

Table 1. Distribution of ^{32}P-label in the deoxynucleoside monophosphate digests of untreated (control) and chloroacetaldehyde (CAA)-treated DNA[a]

Treatment	Monophosphate	% Radioactivity
Control[b]	dGMP	20 (± 5.2)
	dAMP	29 (± 2.3)
	dCMP & m^{5}dCMP	23 (± 2.2)
	dTMP	28 (± 6.4)
		100
CAA	dGMP	28
	dAMP	21
	dCMP & m^{5}dCMP	14
	dTMP	22
	Adduct 1	8
	Adduct 2	7
		100

[a] Quantitation was achieved by removing sections of the thin-layer chromatographic sheets corresponding to radioactive spots and counting by liquid scintillation.
[b] Average of three experiments; the standard deviation of the mean is in parentheses. Quantitation might be more precise if kinase and phosphatase reactions were performed sequentially. For simplicity, we selected reaction conditions which allowed both activities (see Materials and Methods).

1 and 2 following two-dimensional chromatography. These were subsequently eluted for UV spectral analysis. The UV spectra of adducts 1 and 2 (Fig. 5) correspond to the published spectra of 1,N^6-ethenoadenosine and 3,N^4-ethenocytidine, respectively (Kuśmierek & Singer, 1982). In addition to the two major adducts, we noted three minor radioactive spots on some chromatograms of CAA-treated DNA (data not shown). These may be attributable to the hydrated forms of ethenodeoxyadenosine and ethenodeoxycytidine and/or to other novel adducts (e.g., ethenodeoxyguanosine), but these possibilities were not investigated.

The sensitivity of the postlabelling method depends on the chromatographic resolution and on the specific activity of the ^{32}P-ATP used to postlabel the DNA. An adduct that migrates sufficiently far from normal nucleotides (e.g., adduct 1, from CAA treatment) was detectable at levels as low as 1 adduct/10^3 nucleotides, when labelled with ^{32}P-ATP at 30 Ci/mmol (data not shown). Use of ^{32}P-ATP having a 100-fold higher specific activity would be expected to increase the limit of detection to $1/10^5$.

DISCUSSION

The potential utility of the postlabelling technique for monitoring damage to DNA *in vivo* has been demonstrated by the contributions of the Randeraths and their co-workers (Randerath *et al.*, 1981). They showed that impressive sensitivity ($1/10^7$ to

Fig. 4. Autoradiograms from two-dimensional thin-layer chromatography of chloroacetaldehyde (CAA)-treated [5′-^{32}P]-deoxynucleoside monophosphates. The derivatives produced migrate to the same positions as adducts from CAA-treated DNA. The deoxyadenosine monophosphate (dAMP) derivative comigrates with adduct 1 and the deoxycytidine monophosphate (dCMP) derivative, with adduct 2. (The additional spot to the right of dCMP is the predominant initial CAA reaction product; it is converted to adduct 2 as incubation continues, indicating it may be the hydrated intermediate of adduct 2.)

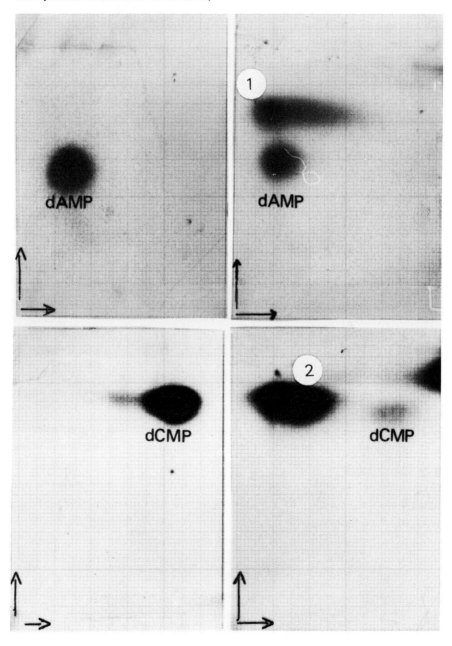

Fig. 5. Absorption spectra of ultraviolet-absorbing derivatives formed by reaction of chloroacetaldehyde with unlabelled deoxynucleoside monophosphates and isolated by two-dimensional thin-layer chromatography (TLC): (a), deoxyadenosine monophosphate derivative, which comigrates in the TLC system with DNA adduct 1; (b), deoxycytidine monophosphate derivative, which comigrates with DNA adduct 2. Spectrum (a) corresponds to the published spectrum for $1,N^6$-ethenoadenosine, and spectrum (b) to that for $3,N^4$-ethenocytidine (Kuśmierek & Singer, 1982).

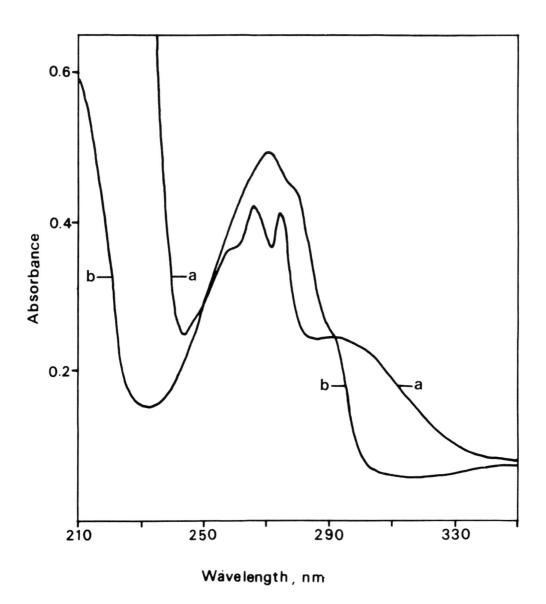

$1/10^8$ adducts/total nucleotides) can be achieved for some types of damage, and as little as nanogram quantities of DNA are sufficient for ^{32}P-postlabelling analysis. High sensitivity, the low sample size required, and introduction of radiolabel after DNA exposure make this technique uniquely suitable for measuring DNA damage in people. Also, the method may reveal formation of novel adducts.

The deoxynucleoside monophosphate analysis described here is an alternative to bisphosphate analysis for certain postlabelling studies and offers the following advantages:

1) The two-dimensional chromatographic system for separation of nucleoside monophosphates (Bochner & Ames, 1982) is simple and well-characterized; separation is complete in six hours and involves a minimum of manipulations. Conveniently, removal of unreacted ATP and inorganic phosphate prior to chromatography is unnecessary because, with this system, ATP remains at the base of the chromatogram and inorganic phosphate runs off into the right margin; these edges can be clipped off prior to film exposure. Migration behaviour follows interpretable patterns: the first dimension separates on the basis of charge and the second, by degree of hydrophobicity. For example, each of the etheno adducts from CAA-treated DNA migrates to the left of the more hydrophilic parent compound. The positions of many nucleoside monophosphates have already been determined (Bochner & Ames, 1982).

2) Labelled and unlabelled monophosphate standards of normal nucleotides can be purchased, and a variety of commercially available nucleoside monophosphates can be used to indicate the approximate position of the corresponding deoxynucleoside monophosphate; the deoxynucleoside monophosphates consistently migrate slightly farther in the first dimension and less far in the second dimension as compared to the corresponding nucleoside monophosphate (Figure 2).

3) The phosphate groups present in DNA during carcinogen treatment are removed by polynucleotide kinase during the postlabelling procedure (Figure 1). Thus, the modified labelled residues detected are only those with modifications of the deoxynucleoside moiety. Postlabelled monophosphates can also be separated by high-performance liquid chromatography (Franklin & Haseltine, 1983; Bodell & Rasmussen, 1984), or obtained from bisphosphates before TLC separation (Reddy *et al.*, 1984).

4) Removal of one of the two phosphates from bisphosphates before chromatography should enhance resolution of the modified residues from the normal parent nucleotide. Minor alterations in charge or hydrophobicity would otherwise be masked more readily by the presence of two phosphates rather than one.

Precise quantitation (number of adducts per 10^n nucleotides) may be difficult with the postlabelling procedure in cases where digestion nuclease or polynucleotide kinase substrate preferences are significant (Cameron & Uhlenbeck, 1977; Garner *et al.*, 1985). Nevertheless the postlabelling method may become an important tool in epidemiology studies for comparing the extent of a particular type of DNA damage in groups of individuals with different exposure histories.

ACKNOWLEDGEMENTS

This work was started at the University of California, Berkeley, and finished at the International Agency for Research on Cancer, Lyon, France. At U.C.B. it was supported by Department of Energy Contract DE-AT03-76EV70156 to B.N.A. and by National Institute of Environmental Health Sciences Center Grant ES01896. J.N.'s work was undertaken during tenure of a research training fellowship awarded by the International Agency of Research on Cancer. Special thanks to A. Barbin for helpful discussions and to W. Bodell for advice on DNA hydrolysis.

REFERENCES

Bochner, B.R. & Ames, B.N. (1982) Complete analysis of cellular nucleotides by two-dimensional chromatography. *J. biol. Chem.*, **257**, 9759–9769

Bodell, W.J. & Rasmussen, J. (1984) A ^{32}P-postlabelling assay for determining the incorporation of bromodeoxyuridine into cellular DNA. *Analyt. Biochem.*, **142**, 525–528

Cameron, V. & Uhlenbeck, O.C. (1977) 3'-Phosphatase activity in T4 polynucleotide kinase. *Biochemistry*, **16**, 5120–5126

Franklin, W. & Haseltine, W. (1983) *Postlabelling methods to detect and characterize infrequent base modifications in DNA*. In: Friedberg, E.C. & Hanawatt, P.C., eds, *DNA Repair: A Laboratory Manual*, Vol. 2, New York, Marcel Dekker, pp. 161–171

Garner, C., Ryder, R. & Montesano, R. (1985) Monitoring of aflatoxins in human body fluids and application to field studies (Meeting Report), *Cancer Res.*, **45**, 922–928

Gross, V.H. (1963) Monochloroacetaldehyde and derivatives of glycolaldehyde and glyoxal, from α-halogen ethers (Ger.). *J. prakt. Chem.*, **21**, 99–102

Gupta, R.C., Reddy, M.V. & Randerath, K. (1982) ^{32}P-Postlabelling analysis of nonradioactive aromatic carcinogen-DNA adducts. *Carcinogenesis*, **3**, 1081–1092

Kuśmierek, J.T. & Singer, B. (1982) Chloroacetaldehyde-treated ribo- and deoxyribopolynucleotides. *Biochemistry*, **21**, 5717–5722

Randerath, K., Reddy, M.V. & Gupta, R.C. (1981) ^{32}P-Postlabeling test for DNA damage. *Proc. natl Acad. Sci. USA*, **78**, 6126–6129

Reddy, M.V., Gupta, R.C., Randerath, E. & Randerath, K. (1984) ^{32}P-Postlabeling test for covalent DNA binding of chemicals *in vivo*: application to a variety of aromatic carcinogens and methylating agents. *Carcinogenesis*, **5**, 231–243

Singer, B. & Grunberger, D. (1983) *Molecular Biology of Mutagens and Carcinogens*, New York, Plenum Press

AUTHOR INDEX

Adamkiewicz, J., *403*
Adler, S., *373*
Ahrens, O., *403*
Allen, E. B., *129*
Ames, B. N., *437*
Averbeck, D., *299*

Bar-Adon, R., *37*
Barbin, A., *57, 345*
Bartsch, H., *3, 57, 345, 437*
Basu, A. K., *175*
Bedell, M. A., *425*
Bhat, U. G., *129*
Bochner, B., *437*
Bodell, W. J., *147*
Boiteux, S., *381*
Bolt, H. M., *261*

Cadet, J., *247*
Chauhan, S. M. S., *129*
Chung, F.-L., *207*
Conner, M. K., *313*
Cruickshank, K. A., *33*

Day, N. E., *327*
Demeunynck, M., *241*
Dipple, A., *235*
Doerjer, G., *373, 419, 425*
Dyroff, M. C., *425*
Eberle, G., *403*
Eder, E., *197*
Emmelot, P., *109*
Essigman, J. M., *393*
Everett, D. W., *165*

Fraenkel-Conrat, H., *45*
Friesen, M., *57*
Fujita, Y., *283*

Gaborian, F., *247*
Gibson, N. W., *155*
Golding, B. T., *227*
Green, C. L., *393*
Guengerich, F. P., *255*
Gulyas, S., *185*

Hall, J. A., *339*
Hayatsu, H., *293*
Hecht, S. S., *207*
Henschler, D., *197*
Hogy, L. L., *255*
Holbrook, S. R., *45*
Hollstein, M., *437*
Hudgins, W. R., *235*

Inskeep, P. B., *255*

Jaskólski, M., *75*
Kaldor, J. M., *327*
Kasai, H., *413*
Koganty, R. R., *129*
Kohn, K. W., *155*
Kosuge, T., *283*
Krzyżosiak, W. J., *75*
Kundu, S. K., *165*
Kuśmierek, J. T., *45*

Laib, R. J., *101*
Larson, K., *387*
Laval, F., *321*

Laval, J., *381*
Leonard, N. J., *33*
Lhomme, J., *241*
Lhomme, M. F., *241*
Liebler, D. C., *255*
Lilley, D. M. J., *83*
Loechler, E. L., *393*
Lown, J. W., *129*
Ludlum, D. B., *137*

Marnett, L. J., *175*
Mertes, J., *419*
Moschel, R. C., *235*

Nagao, M., *283*
Nair, J., *437*
Negishi, K., *293*
Nehls, P., *403*
Nies, E., *419*
Nishimura, S., *413*

Oesch, F., *373, 419*
O'Hara, S. M., *175*
Ohara, Y., *293*
O'Neill, I., *57*
Oohara, K., *293*
Osterman-Golkar, S., *269*

Palladino, G., *207*
Papadopoulo, D., *299*
Politzer, P., *37*

Rabkin, S., *387*
Rajewsky, M. F., *403*
Rettelbach, R., *373*

AUTHOR INDEX

Saffhill, R., *339*
Sapse, A.-M., *129*
Saracci, R., *15*
Sagher, D., *387*
Sahm, J., *387*
Scherer, E., *109*
Shapiro, R., *165*
Singer, B., *45, 359*
Slaich, P. K., *227*
Sodum, R. S., *165*
Spengler, S. J., *359*

Strauss, B., *387*
Sugimura, T., *283*
Svensson, K., *269*
Swenberg, J. A., *425*

Tohme, N., *241*

Urushidani, H., *293*

Vainio, H., *15*
Varghese, A. J., *185*

Vigny, P., *247*
Voituriez, L., *247*

Watson, W. P., *227*
Whitmore, G. F., *185*
Wiewiórowski, M., *75*
Winterwerp, H., *109*

Yamaizumi, Z., *413*

Zilles, B. A., *37*

SUBJECT INDEX

A

Acetaldehyde, 17, 21
N-Acetoxy-N-4-acetamidostilbene, 7
1-Acetoxy-4(acetoxyimino)-1,4-dihydroquinoline, 241–245
N-Acetoxy-N-2-acetylaminofluorene, 404
N-Acetoxyethyl carbamate, 120
Acetylacrolein, 7
Acetylaminofluorene, 322
N-Acetyl-2-bromoacetamide, 36
N^2-Acetylhistidine, 270
α-Acetoxy-N-nitrosopyrrolidine, 207–223
N-Acetyl-S-carboxyethylcysteine, 120
N-Acetyl-N-hydroxyethyl carbamate, 120
Acridine dyes, 181
Acrolein, 5, 19, 22, 25, 165–172, 197–205, 207, 223
    β-substituted, 175–182
Acrylic acid, 171
Acrylonitrile (Vinyl cyanide), 17, 22, 25, 255–259, 266, 267
Acrylonitrile oxide (Oxirane carbonitrile), 258, 259, 266, 267
Actinomycin D, 328
N^2-Acylguanine derivative, 166, 167
Adaptive response, 321–324, 396
1:1 Adducts, 5, 227–231
2:2 Adducts, 227–231
3:1 Adducts, 5, 180, 181
Adenine,
    reaction with,
        N-acetoxy-N-4-acetamidostilbene, 5
        N-acetyl-2-bromoacetamide, 36
        acrolein, 165–172
        acrylonitrile, 8
        acrylonitrile oxide, 257
        1,3-bis(2-chloroethyl)-1-nitrosourea, 7
        1,3-bis(2-fluoroethyl)-1-nitrosourea, 7
        bromoacetaldehyde, 7, 45–54, 86
        1-bromo-2-chloroethane, 139
        α-bromopropionaldehyde, 76
        N-(t-butoxycarbonyl)-2-bromoacetamide, 33, 34
        chloroacetaldehyde, 7, 33, 45–54, 75–80, 101–106
        chloroethylene oxide, 33
        2-chloroketene diethylacetal, 33, 35
        1,2-dibromoethane, 139
        1,2-dihaloethanes, 257
        2-haloacetaldehydes, 256
        2-haloethylene oxides, 256
        haloethylnitrosoureas, 137–145
        malondialdehyde (and substituted derivatives), 5, 178, 180, 181

8-methoxypsoralen, 247–251
β-(*p*-nitrophenoxy)acrolein, 180, 181
4-nitroquinoline-*N*-oxide, 5, 241–245
β-propiolactone, 5
urethane, 8, 33
vinyl bromide, 33
vinyl chloride, 8, 33, 101–106
1,N^6-derivatives, 4, 7, 8
nitrogen mustard derivative, 4
1,N^6-Adenine adducts of acrolein, 169
Adenosine, tosyl derivative, 4
Adenosine deaminase, 431
β-(N^6-Adenosyl)acrolein, 180, 181
S-Adenosylmethionine, 378
Adriamycin, 328
Aflatoxins, 120, 322, 404
Alcohol dehydrogenase, 130, 197–205, 256, 257, 263
Aldehydes, 18, 21, 25, 57–72
  adducts, 53, 106, 121
Alkaline elution, 155–161, 299–309
Allyl carbamate, 314
Allyl alcohols, 198, 202, 204
Allyl bromide, 197, 202
Allyl compounds, 203, 204
Allyl halides, 197, 202
Ames test, 197–205
N^4-Aminocytidine, 293
2-Amino-3,4-dimethylimidazo[4,5-*f*]quinoline, 413
2-Amino-3,8-dimethylimidazo[4,5-*f*]quinoxaline, 413
Aminoethenoribonucleosides, 33, 36
Aminoimidazo[1,5-*a*]-1,3,5-triazinones, 239
2-Amino-3-methylimidazo[4,5-*f*]quinoline, 413
12-Amino-3-propyl(2′,3′:5,4)quinolinyl(1,2-*i*)imidazo purine, 241–245
2-Aminopurine, 389, 390
12-Amino-3-ribofuranosyl(2′,3′:5,4)quinolinyl(1,2-*i*)imidazo purine, 241–245
Aneuploidies, 18
Angelicin, 300
Apurinic/apyrimidinic sites (*see also* Depurination, Depyrimidation), 9, 318, 321, 341, 350–353, 374, 381–383, 389, 390
Arylamines, 235, 239
Autoradiography, 83–96, 387–391, 393–398, 437–448
Azathioprine, 328
Azomycin, 185–195

B

Base-pairing, 83–96, 101–106, 121, 144, 223, 351–353, 359–369, 397
Base-pair substitutions, 175–182, 285, 289, 293–296, 345–355, 359–369, 393–397, 426
Base-site selectivity, 129–135

Base-stacking, 181
Benzene, 331
Benzo[a]pyrene, 7, 404, 419
Benzyl bromide (substituted benzyl bromide), 236
Benzyl chloride (substituted benzyl chloride), 235–239
4-(p-Y-Benzyl)-5-guanidino-1-β-D-ribofuranosylimidazole, 235–239
N^2-(p-Y-Benzyl)guanosine, 238
O^6-(p-Y-Benzyl)guanosine, 238
7-(p-Y-Benzyl)guanosine, 238
Bis(2-chloroethyl)amines, 155
N,N-Bis(2-chloroethyl-2-naphthylamine (Chlornaphazine), 328
1,3-Bis(2-chloroethyl)-1-nitrosourea (BCNU), 7, 19, 20, 22, 25, 53, 129, 130, 134, 137–145, 147, 153, 223, 328, 347, 382
Bis(chloromethyl)ether, 346
1,3-Bis(2-fluoroethyl)-1-nitrosourea (BFNU), 7, 10, 137–145
Bis-haloalkylnitrosoureas, 7
Bleomycin, 328
Borohydride, 109–123, 168, 222, 259, 271, 273
Bromoacetaldehyde, 7, 45–54, 83–96, 102, 256, 257, 261–267, 296
2-Bromoacrolein, 200, 201
1-Bromo-2-chloroethane, 139
2-Bromocinnamaldehyde, 201
Bromodeoxyuridine, 316, 317
S-(2-Bromoethyl)glutathione, 257
α-Bromopropionaldehyde, 76
Bulky cyclic adducts of *(see also individual compounds)*
   benzo[a]pyrene, 7
   N,N-dimethyl-4-aminoazobenzene, 7
   N-acetoxy-N-4-acetamidostilbene, 7
   4-nitroquinoline-N-oxide, 7
   furocoumarins, 7
   8-methoxypsoralen, 7
1,4-Butanediol dimethanesulfonate (Busulfan), 328, 347
But-3-en-1,2-diol, 231
N-(t-Butoxycarbonyl)-2-bromoacetamide, 33, 34
O^6-n-Butyl-2′-deoxyguanosine, 403–410

C

Carbamates (Carmabate esters), 130, 313–318
8-Carbamyl-3-(2-chloroethyl)imidazo[5,1-d]-1,2,3,5-tetrazin-4(3H)one *(see also* Mitozolomide), 157, 159, 161
4-(Carbethoxynitrosamino)butanal, 207–223
2-(Carbethoxynitrosamino)ethoxyacetaldehyde, 207–223
3-Carbethoxypsoralen, 247–251, 300
Carbinolamine intermediate, 53
Carbonyl compounds, α,β-unsaturated, 7
Carboxyesterase, 266, 267
N^6-(2-Carboxyethyl)adenine derivative, 169

N^4-(2-Carboxyethyl)cytosine derivative, 169
1-(2-Carboxyethyl)-2′-deoxyadenosine, 382
S-Carboxymethyl glutathione, 256
Carcinogenicity, 15–25, 120, 283, 285, 289, 290, 327–333
Carcinogenic potency, 346–347
Case-control studies, 329
Catalase, 287
Chemical probes, 83–96, 165, 168
Chemotherapeutic agents (Antineoplastic drugs, Cytotoxic drugs), 19, 129–162, 185–196, 327–333, 382
Chemotherapy, 327–333, 382
Chloral, 256
Chlorambucil, 328, 347
2-Chloroacetaldehyde, 7, 8, 9, 10, 16, 21, 25, 33, 45–54, 57–72, 75–80, 101–106, 112, 120, 223, 257, 261–267, 269, 275, 276, 277, 283–296, 339–342, 345–355, 366, 373, 375, 377, 383, 421, 422, 425–434, 437–448
2-Chloroacrolein, 200, 201, 204
2-Chloroallyl alcohol, 204
2-Chlorocinnamaldehyde, 201
α-Chloroepoxides, 72
Chloroethanol, 25, 131, 263, 264, 275, 276, 277
1-Chloroethyladenosine, 141
2-Chloroethylamine, 25
1-(2-Chloroethyl)-3-cyclohexyl-1-nitrosourea (CCNU), 19, 20, 22, 137–145, 328, 382
S-(2-Chloroethyl)cysteine, 269–278
3-(2-Chloroethyl)deoxycytidine, 134
N^4-(2-Chloroethyl)deoxycytidine, 131
2-Chloroethyldiazohydroxide, 130, 132, 133, 135, 159
2-Chloroethyldiazotates, 129–135
Chloroethylene oxide, 8, 16, 33, 46, 47, 57–72, 101–106, 112, 113, 116, 117, 118, 120, 122, 261–267, 275, 276, 277, 339, 341, 345–355, 383, 426, 427
S-(2-Chloroethyl)glutathione, 269–278
O^6-(2-Chloroethyl)guanine (O^6-(2-Chloroethyl)deoxyguanosine), 132, 133, 137–145
2-Chloroethyl methanesulfonate, 147–153
N^4-(2-Chloroethyl)-1-methylcytosine, 53
2-Chloroethylmethylsulfonyl methanesulfonate, 157–161
Chloroethyl nitrosoureas, 19, 25, 155–161, 328, 382
S^6-(2-Chloroethyl)-6-thiodeoxyguanosine, 152, 153
(2′-Chloroethyl)thioethyldiazotates, 129–135
Chloroethyltriazenoimidazolecarboxamide, 157, 159, 160, 161
gem-Chlorohydrin, 276
2-Chloroketene diethylacetal, 33, 35, 36
Chloromethyl ethoxyimidate intermediate, 36
m-Chloroperbenzoic acid, 231
1-Chloroso-2-chloroethane, 275–276
N^4-(2-Chlorovinyl)cytosine, 345, 354, 355
N^4-(2-Chlorovinyl)-3-methylcytosine, 57–72, 353–355
Chromatography (Liquid chromatography, Thin-layer chromatography), 103, 109–123, 138, 171, 229, 236, 243, 269–278, 294, 339, 340, 407
Chromosomal aberrations, 17, 18, 20, 331

SUBJECT INDEX

Cinnamaldehyde, 199, 201, 206
Circular dichroism, 179, 249
Cisplatin, 328
Citral, 199
Cohort studies, 329
Conjugates,
   metabolic, 255–259, 263, 267, 275, 289
   nucleoside-protein, 403, 404
Cross-links (see also Ethano bridge), 7, 10, 25, 33, 36, 45–54, 105, 106, 109, 112, 121, 122, 123, 129–135, 137–145, 147–153, 155–161, 168, 178, 223, 248, 278, 299–309, 382, 426
Crotonaldehyde, 5, 19, 25, 181, 199, 200, 202, 207–223
18-Crown-6 ether complexes, 129–135
Cruciform, 83–96
Cyclic N-nitrosamines, 5, 207–223
2-Cyclohexen-1-one, 199, 223
Cyclophosphamide, 328
Cysteine, 283–290
7-[S-(2-Cysteine)ethyl]guanine, 269–278
N^{τ}-[S-(2-Cysteine)ethyl]histidine, 269–278
Cytochrome P-450, 256, 258, 261–267, 350, 382
Cytokinin, 4
Cytosine
   reaction with,
      N-acetyl-2-bromoacetamide, 36
      acrolein, 165–172
      benzo[a]pyrene, 7
      1,3-bis(2-chloroethyl)-1-nitrosourea, 7
      1,3-bis(2-fluoroethyl)-1-nitrosourea, 7
      bromoacetaldehyde, 7, 45–54, 86
      1-bromo-2-chloroethane, 139
      α-bromopropionaldehyde, 76
      N-(t-butoxycarbonyl)-2-bromoacetamide, 33, 34
      chloroacetaldehyde, 7, 33, 45–54, 57–72, 75–80, 101–106
      2-chloroethyldiazohydroxide, 134
      chloroethyldiazotate, 131
      chloroethylene oxide, 33, 57–72
      (chloroethyl)nitrosoureas, 155
      2-chloroketene diethylacetal, 33, 35, 36
      1,2-dibromoethane, 139
      1,2-dihaloethanes, 257
      N,N-dimethyl-4-aminoazobenzene, 7
      2-haloethylnitrosoureas, 129–135, 137–145
      osmium tetroxide, 91
      urethane, 8, 33
      vinyl bromide, 33
      vinyl carbamate, 33
      vinyl chloride, 8, 33, 101–106
   $3,N^4$-derivatives, 7
$3,N^4$-Cytosine adducts of acrolein, 169

D

Dacarbazine, 328
Deamination, of nucleic acid bases, 340, 341, 364, 368, 383
Dehalogenation (Reductive dehalogenation, Dehydrohalogenation), 275
Densitometry, 83–96, 406
1-(3-Deoxycytidyl)-2-(1-deoxyguanosyl)ethane (*see also* Ethano bridge), 132, 137, 142
2′-Deoxy-2′-(2″,3″-dihydro-2″,4″-diphenyl-2″-hydroxy-3″-oxo-1″-pyrrolyl)guanosine, 413–418
Deoxyinosine, 340, 431
Deoxynucleoside bisphosphates, 438, 447
Deoxynucleoside monophosphates, 437–448
3-(β-D-2-Deoxyribosyl)-7,8-dihydropyrimido[2,1-*i*]purine-9-one, 382
Deoxyuridine, 340
Depurination (*see also* Apurinic/apyrimidinic sites), 113, 114, 116, 117, 318, 339, 375, 377, 383
Depyrimidation (*see also* Apurinic/Apyrimidinic sites), 60
Desmethylmisonidazole, 185–195
Diacetyl, 289
Diadducts, with
  acrolein, 165, 169–171
  chloroacetaldehyde, 57–72, 355
  chloroethylene oxide, 57–72, 355
  psoralens, 7, 299–309
1,2-Dibromo-3-chloropropane, 257
1,2-Dibromoethane, 139, 257
Dicarbonyl compounds (Diketo compounds, Dialdehydes), 4, 5, 7, 25, 175–182, 283–290
1,3-Dichloroacetone, 204
2,3-Dichloroacrolein, 200, 201
2,2′-Dichlorodiethyl ether, 261–267
1,2-Dichloroethane, 256, 257, 269–278
S-(2,2-Dichloro-1-hydroxy)ethyl glutathione, 256
2,3-Dichloro-1-propene, 204
1,2-(Diguanosin-7-yl)ethane (Di(7-guanyl)ethane, Diguanyl ethane cross-links; *see also* Ethano bridge), 53, 145, 382
vic-Dihaloalkanes, 255–259
4,5-Dihydroxy-4,5-dihydroimidazole, 186
N,N-Dimethyl-4-aminoazobenzene, 7
1,N^2-Dimethylguanosine, 212
Dimethylsulfate, 147–153, 387–391
Dithiothreitol, 283–290
DNA
  conformations (DNA topology, DNA structure), 78, 83–96
  hydrolysis (Enzymatic digestion of DNA), 103, 112, 114, 138, 148, 149, 212, 249, 250, 271, 339–342, 375, 377, 404, 405, 408, 422, 425, 428, 431, 433, 434, 437–441, 447
  polymerases (*see also* DNA synthesis), 103, 104, 339–342, 352, 353, 359–369, 374, 383, 387–391, 395, 396
  repair, 155–161, 185–195, 285, 309, 315–317, 321–324, 360, 373–378, 381–383, 393–398, 426, 427, 434
  sequencing, 387–391, 393–398
  strand breaks, 299–309, 373, 418

supercoiling, 83–96
synthesis, 9, 103, 104, 106, 109, 121, 339–342, 352, 353, 359–369, 373, 383, 387–391, 393–398
Double-stranded DNA (Cellular DNA, Native DNA), reaction with, 9, 47, 51, 52, 57, 58, 71, 72, 101–106, 109–123, 129–135, 137–145, 147–153, 155–161, 185–195, 250, 251, 262–264, 269–278, 299–309, 345–355, 387–391
Drosophila melanogaster, 18

E

Electrocyclic rearrangement (*see also* Secondary lesions), 239
Electronic structure, 37–41, 75–80
Elemental analysis (Combustion analysis), 171, 228, 294
Enals, 175–182
Endonucleases, 352, 381–383
Epichlorohydrin, 331, 347
Epidemiology, 15–25, 327–333
Epoxidation, 203, 204, 255–259
Epoxides, 8, 16, 17, 18, 21, 37, 38, 255–259, 261–267, 275, 346
Epoxide hydrolase, 256
Epoxyethylcarbamate, 109
S-Epoxyethyl homocysteine, 119
3-(β-D-Erythropentofuranosyl)pyrimido[1,2-*a*]purin-10(3H)-one, 179
Escherichia coli, 18, 293, 339, 345–355, 359–369, 382, 383, 387–391, 393–398
Esterase, 207–223
1,N^6-Ethanoadenine, 7, 137–145
Ethano bridge (*see also* 1-(3-Deoxycytidyl)-2-(1-deoxyguanosyl)ethane, 1,2-(Diguanosin-7-yl)ethane), 155
3,N^4-Ethanocytosine, 7, 130, 131, 133–135, 137–145, 153
1,O^6-Ethanoguanine, 7, 10, 137–145
(S^6,7-Ethano)-6-thiodeoxyguanosine, 147–153
(1,S^6-Ethano)-6-thiodeoxyguanosine, 147–153
3,O^4-Ethanothymidine, 144, 145
1,N^6-Ethenoadenine, 8–10, 33, 45–54, 75–80, 86, 101–106, 112, 113, 114, 116, 256–259, 264, 293–296, 318, 339–342, 345–355, 359–369, 373–378, 383, 425–434, 444–446
Ethenoadenosines, substituted, 33–36
Ethenocytidines, substituted, 33–36
3,N^4-Ethenocytosine, 8–10, 33, 45–54, 57–72, 75–80, 86, 101–106, 112, 114, 116, 257, 264, 270, 271, 293–296, 318, 339–342, 345–355, 359–369, 374, 383, 425–434, 444–446
Etheno derivatives, 7–10, 25, 33, 36, 45–54, 57–72, 75–80, 422, 425–434, 444–447
1,N^2-Ethenoguanine, 7, 46, 53, 54, 102, 229, 354
N^2,3-Ethenoguanine, 7, 8, 46, 53, 101–106, 350, 351, 352, 354, 373–378, 383, 425–434, 444
(3,N^4-Etheno)-5-methylcytosine, 57–72
Ethidium bromide, 83–96
DL-Ethionine, 119, 120
Ethoxyethenoribonucleosides, 33, 36
Ethyl carbamate (Urethane), 8, 18, 22, 33, 109–123, 223, 313–318
O^6-Ethyl-2′-deoxyguanosine, 403–410
O^4-Ethyl-2′-deoxythymidine, 403–410
Ethylene (Ethene), 265, 266, 275, 276

S,S'-(1,2-Ethylene)bis-cysteine, 269–278
Ethylene glycol, 131
Ethylene glycol bis[3-(2-ketobutyraldehyde)ether], 168
Ethylene oxide (Ethene oxide), 129, 131, 266, 274–277, 331
Ethylglyoxal, 283–290
Ethyl N-hydroxycarbamate, 314, 315
3-Ethyl-6-methyl-1,N^2-ethenoguanine, 227
N-Ethyl-N-nitrosourea, 347, 404
Ethylquinoxaline, 284
Exonucleases, 387, 389, 394

F

Fibre diffraction, 83
Fidelity, of replication, transcription, 360, 361, 363, 364, 368, 387
Fluorescence, 5, 9, 33, 36, 53, 179, 208, 221, 222, 249, 251, 259, 340, 375, 376, 377, 403, 413–418, 419–423, 425–434
1-Fluoroethyladenosine, 141
1-Fluoroethyl-3-cyclohexyl-1-nitrosourea (FCNU), 137–145
N^4-Fluoroethylcytidine, 141
2-Fluoroethyldiazotates, 129–135
O^6-Fluoroethylguanine, 137, 143
Fluorometry, three-dimensional, 419–423
5-Fluorouracil, 328
Formaldehyde, 227
N^6-Formyladenine, 167
Frameshift mutations, 175–182, 359–369, 426
Free energy, 41, 84, 94, 132
Free radicals (*see also* Oxygen radicals), 275, 276
Furocoumarins (*see* Psoralens)
2-(2-Furyl)-3-(5-nitrofuryl)acrylamide, 416

G

Gas chromatography/mass spectrometry, 57–72, 284, 354
Gel electrophoresis, 83–96, 387–391
Gene conversion, 283–290, 299–309
Glucuronide, 256
Glutathione, 120, 194, 202, 204, 255–259, 263, 266, 267, 275, 288, 289, 290
Glutathione S-transferase, 257
S-(2-Glutathionyl)acetyl glutathione, 256
Glycidaldehyde, 5, 19, 23, 165, 223, 227–231, 347
Glycidonitrile, 17
Glycolaldehyde, 57–72
Glycosidic bond, 78, 431
N-Glycosylases (DNA glycosylases), 321–324, 373–378, 381–383, 389, 390
Glycosyl cleavage (*see also* Apurinic/apyrimidinic sites), 165, 179, 350

Glyoxal, 4, 11, 83–96, 165–172, 185–195, 207–223, 283–290, 413–418
　　substituted, 166, 167, 283–290
Glyoxalase, 289
Guanidinoimidazole, 235–239
Guanine,
　　reaction with
　　　α-acetoxy-N-nitrosopyrrolidine, 207–223
　　　acetylaminofluorene, 368, 369, 389, 391
　　　acrolein, 5, 165–172, 207–223
　　　acrylic acid, 171
　　　acrylonitrile oxide, 259
　　　benzyl halides, 235–239
　　　1,3-bis(2-fluoroethyl)-1-nitrosourea, 7
　　　4-(carbethoxynitrosamino)butanal, 207–223
　　　2-(carbethoxynitrosamino)ethoxyacetaldehyde, 207–223
　　　chloroacetaldehyde, 7, 8, 46, 47, 52–54, 101–106
　　　2-chloroethyldiazohydroxide, 132
　　　chloroethylene oxide, 46, 47, 101–106, 112, 117, 118, 122
　　　1-(2-chloroethyl)-1-nitrosoureas, 155
　　　crotonaldehyde, 5, 207–223
　　　1,2-dihaloethanes, 257
　　　ethyl carbamate, 109–123
　　　glycidaldehyde, 5, 19, 227–231
　　　glyoxal, 4, 90, 185–195, 207–223, 413–418
　　　haloethylnitrosoureas, 137–145
　　　heated glucose, 413–418
　　　2-hydroxyethyldiazohydroxide, 134
　　　malondialdehyde, 5, 178, 181
　　　mesoxaldialdehyde, 413–418
　　　methylglyoxal, 4, 290
　　　2-nitroimidazole, 185–195
　　　β-(p-nitrophenoxy)acrolein, 178, 179
　　　4-nitroquinoline N-oxide, 241
　　　N-nitrosomorpholine, 207–223
　　　N-nitrosopyrrolidine, 5, 207–223
　　　oxygen radical, 413–418
　　　vinyl chloride, 101–106
　　bicyclic 1,N^2,7,8-diadduct of acrolein, 170
　　1,N^2-derivatives, 5, 7, 19, 207–223
　　N^2,3-derivatives, 7
1,N^2-Guanine adducts of acrolein, 169
[^{14}C]-Guanine incorporation, 185–195
Guanosines, 5-substituted, 237
S-[2-(7-Guanyl)ethyl]glutathione, 257

H

2-Haloacetaldehydes, 45–54, 255–259, 261–267
Haloacroleins, 7

Haloaldehydes, 7
2-Haloethanol, 130
Haloethylcytidine, 141
2-Haloethyldiazohydroxides, 129–135
Haloethyldiazotates, 129–135
2-Haloethylene oxides, 255–259
O^6-Haloethylguanine, 137–145
Haloethyl nitrosoureas, 19, 129–135, 137–145
Halogenated alkenes, 38
Halogenated ethylenes, 264, 267
Halomethylbenz[a]anthracenes, 239
Heated glucose, 413–418
Hemoglobin, 269–278
Hepatocytes, 255–259, 427, 434
High-performance liquid chromatography, 48–51, 147–153, 158, 178, 179, 180, 188, 189, 207–223, 227–231, 243, 247–251, 258, 259, 284, 294, 295, 296, 361, 375, 376, 377, 403, 405, 409, 413–418, 422, 425–434, 447
Hybridoma cells, 403, 405
Hydrated aldehydes, 46, 57, 61, 62, 64, 67, 69–71
Hydrogen bonding, 37, 41, 47, 52, 54, 78, 79, 86, 90, 91, 134, 359–369, 426
Hydrogen peroxide, 283–290
β-Hydroxyacrolein, 175, 176
4-Hydroxyalkenals, 175, 176
4-Hydroxyamino quinoline oxide, 241–245
    diacetyl ester, 241–245
    monoacetyl ester, 241–245
N^6-(1-Hydroxy-2-chloroethyl)-1-methyladenine, 45–54
N^4-(1-Hydroxy-2-chloroethyl)-3-methylcytosine, 45–54, 57–72
1,N^6-(N^6-Hydroxyethano)adenine (see also Intermediates of adducts), 101–106, 350
3,N^4-(N^4-Hydroxyethano)cytosine (see also Intermediates of adducts), 101–106, 350, 364
O^6,7-(1'-Hydroxyethano)guanine (see also 7-(2-Oxoethyl)guanine), 101–106, 109, 112, 121, 350
1-Hydroxyethyladenosine, 141
Hydroxylamine derivatives, 185–195
N-Hydroxyethylcarbamate, 120
S-(2-Hydroxyethyl)cysteine, 269–278
3-Hydroxyethylcytidine, 140, 141
O^6-(2-Hydroxyethyl)deoxyguanosine, 132, 134
7-(2-Hydroxyethyl)guanine, 109–123, 134, 269–278
1-Hydroxyethylguanosine, 137, 143
$N^π$-(2-Hydroxyethyl)histidine, 269–278
$N^τ$-(2-Hydroxyethyl)histidine, 269–278
N^2-(2-Hydroxyethyl)valine, 269–278
8-Hydroxyguanine, 413–418
N^2-(3-Hydroxy-1-methylpropyl)guanine, 216
5-Hydroxymethyluracil, 418
4-Hydroxynonenal, 177
N^2-(3-Hydroxypropyl)guanine, 222
2-Hydroxytetrahydrofuran, 222
Hypoxanthine, reaction with acrolein, 168
Hypoxia, 185–195

I

Image, intensification, analysis, 403, 407
Imidazole ring opening, 165, 170, 350–353, 381, 383
Immunoanalytical methods, 10, 403–410
Immunocytological analysis, 403–410
Immuno electron-microscopy, 403–410
Immunofluorescence, 403, 407
Immuno-slot-blot, 403–410
Immunostaining, 403
Intercalation, 181, 248, 300
Intermediates,
   of adducts, 10, 36, 45–54, 57–72, 76, 101–106, 116, 129–135, 141, 143, 144, 147–153, 237, 342, 350–352, 359–369, 444, 445
   of alkylating agents, 129–135
Inverted repeats, 83–96
Isomers, 57–72, 129–135, 171, 179, 212, 214–216, 230, 242, 244, 247–251, 355, 359–369
N^6-Isopentenyl adenine, 4
Isophosphamide, 328
Isopropyl carbamate, 314
O^6-Isopropyl-2'-deoxyguanosine, 403–410
2',3'-O-Isopropylideneguanosine, 413–418

K

Keyhole limpet hemocyanin, 404, 408

L

Lipid peroxidation, 5, 18, 175, 378
Liver, nonparenchymal cells, 427, 434
Lymphocytes, 313–318

M

Macrophages, 313–318
Malonaldehyde, 18, 23, 25
Malondialdehyde, 5, 175–182, 378
   substituted, 5, 223
Mammalian cells (see also Hepatocytes, Liver, nonparenchymal cells, Lymphocytes, Macrophages), 16–18, 22, 283–290, 299–309, 313–318, 321–324, 373–378, 382, 383, 390, 403, 405
Mass spectrometry (see also Gas chromatography/mass spectrometry), 102, 139, 147–153, 179, 207–223, 227–231, 236, 243, 247–251, 416
Melphalan, 328, 330, 331, 332
Mer$^{+/-}$ phenotype (Mex$^{+/-}$ phenotype), 142, 144, 145, 155–161
6-Mercaptopurine, 328
Mercapturic acids, 202, 205, 266
Mesoxaldialdehyde, 413–418

Metabolism, 255–259, 261–267, 269–278
Methotrexate, 328
β-Methoxyacrolein, 176
N^6-Methoxyadenine, 363, 364, 365
N^4-Methoxycytosine, 72, 354, 355, 363, 364, 365
5-Methoxypsoralen, 21, 23, 25, 299–309
8-Methoxypsoralen, 6, 7, 20, 23, 25, 247–251, 299–309
2-Methylacrolein, 199, 200
3-Methyladenine, 147–153, 321–324, 381–383, 389
9-Methyladenine, reaction with acrolein, 165–172
6-Methyladenosine, reaction with chloroacetaldehyde, 76
α-Methylcinnamaldehyde, 202
4-Methylcytidine, reaction with chloroacetaldehyde, 76
1-Methyladenosine, reaction with chloroacetaldehyde, 52, 53
N-Methylbenzothiazolone hydrazone, 229
4-(p-Methylbenzyl)-5-guanidinoimidazole, 235–239
Methyl-bis(2-chloroethyl)amine hydrochloride (HN2), 347
Methyl carbamate, 314
1-Methyl-3-(2-chloroethyl)cytosine, 131, 135
1-Methyl-N^4-(2-chloroethyl)cytosine, 131, 134, 135
Methyl-CCNU, 19, 20
3-Methylcytidine, reaction with
 chloroacetaldehyde, 52, 53, 57–72, 353–355
 chloroethylene oxide, 57–72, 353–355
1-Methylcytosine, reaction with
 acrolein, 165–172
 2-chloroethyldiazohydroxide, 134, 135
3-Methylcytosine, 361, 363, 364
5-Methylcytosine, reaction with
 chloroacetaldehyde, 57–72
 chloroethylene oxide, 57–72
Methyldiazohydroxide, 132
1-Methyl-3,N^4-ethanocytosine, 131, 134, 135
2-Methylfuran, 7
Methylglyoxal, 4, 175–182, 283–290, 416
1-Methylguanine, reaction with glyoxal, 168
3-Methylguanine, reaction with glyoxal, 165–172
7-Methylguanine (7-Alkylguanines), 147–153, 321–324, 352, 381, 383, 389, 426
 reaction with phosphorus pentasulfide, 150
9-Methylguanine, reaction with glyoxal, 167
O^6-Methylguanine (O^6-Alkylguanines), 147–153, 156, 321–324, 352, 363, 364, 365, 373, 374, 375, 382, 383, 393–398, 403–410, 425, 426, 434
 reaction with glyoxal, 168
O^6-Methylguanine-DNA methyltransferase (O^6-Alkylguanine-DNA-alkyltransferase, Alkyl acceptor protein), 145, 156, 321–324, 382, 393, 396–398
Methyl methanesulfonate, 322–324, 347
S^6-Methyl-7-methyl-6-thioguanine, 152
1-Methyl-S^6-methyl-6-thioguanine, 152
S^6-Methyl-X-methyl-6-thioguanine, 151, 152
N-Methyl-N'-nitro-N-nitrosoguanidine, 156, 321–324, 347, 393, 396, 397

N-Methyl-N-nitrosourea, 140, 147–153, 156, 321–324, 347, 397
N-Methyl-N-nitrosourethane, 347
7-Methylpyrido(3,4-c)psoralen, 299–306
Methylquinoxaline, 284, 289
S^6-Methyl-6-thioguanine, 147–153
1-Methyl-6-thioguanine, 150
7-Methyl-6-thioguanine, 147–153
O^4-Methylthymine (O^4-Alkylthymines), 156, 360, 365, 366, 368, 373, 375, 382, 403–410, 426
O^4-Methyluracil, 360, 363, 364
3-Methyluridine, reaction with
   chloroacetaldehyde, 57–72, 355
   chloroethylene oxide, 57–72, 355
Methyl vinyl ketone, 199, 200, 223
Michael addition, 168, 171, 177, 197–205, 207, 223
Micrococcus luteus, 352, 381
Micronuclei, 17, 18
Microsomes, 16, 102, 207–223, 255–259, 261–267, 275, 346, 382, 426
Misonidazole, 185–195, 223
Mispair (Miscoding, Misincorporation, Mismatch), 9, 57, 58, 72, 103, 104, 106, 112, 121, 122, 223, 318, 339–342, 345–355, 359–369, 373, 374, 383, 387–389, 393–398, 426
Mitomycin C, 178, 193, 328, 347
Mitozolomide, 157, 159, 161
Mixed-function oxidases (Monooxygenases), 16, 203, 426
Molecular dosimetry, 10
Monoadducts, non-cyclic, 7, 10, 37–41, 45–54, 57–72, 129–135, 137–145, 147–153, 353–355
Monoclonal antibodies, 403–410
Mutagenic effectiveness, 345, 346, 348
Mutagenicity (Mutagenesis), 16–20, 25, 120, 197–205, 209, 283–290, 293–296, 299–309, 321–324, 345–355, 393–398
Mutagens, detection of, 413–418
Mutational specificity, 10, 175–182, 345–355

N

Nearest-neighbour analysis, 342, 352, 361
4-(p-Nitrobenzyl)pyridine, 205
Nitrogen mustard, 328
2-Nitroimidazoles, 4, 185–195
β-(p-Nitrophenoxy)acrolein, 178–180
4-Nitroquinoline N-oxide, 7, 241–245, 416
N-Nitrosodiethylamine, 119, 120
N-Nitrosodimethylamine, 285, 322
N-Nitrosomorpholine, 207–223
N-Nitrosooxazolidine, 130
N-Nitrosopyrrolidine, 5, 207–223
Non-instructional lesions, 387–391
Nuclear magnetic resonance spectroscopy, 57–72, 129–135, 167, 169, 171, 179, 207–223, 227–231, 236, 243, 247–251
Nucleases (*see also* Exonucleases, Restriction enzymes, Endonucleases), 83–96, 102, 373
Nucleophilic selectivity (*see also* Swain-Scott constants), 346, 347

O

Osmium tetroxide, 83–96
1,2,3-Oxadiazoline, 129, 130, 131, 133
1,O^6-Oxazoliniumdeoxyguanosine, 132–134
Oxygen radicals, 413–418
3-(2-Oxoethyl)cytosine, 57–72
7-(2-Oxoethyl)guanine, 8, 37–41, 46, 47, 71, 101–106, 109–123, 259, 264, 269–278, 318, 345–355, 426, 427
    cyclic hemiacetal, 8, 37–41, 104, 109–123, 350–352

P

Palindromes, 84
Peroxidase staining, 403
Phages, 83–96, 130, 131, 132, 383, 387–391, 393–398
Pharmacokinetics, 261–267
O-Phenylenediamine, 284, 289
Phosphorus pentasulfide, 150
Photoadducts, 247–251, 299–309
Plasmids, 83–96, 176, 285
Point mutations (*see* Base-pair substitutions)
Polycyclic aromatic hydrocarbons, 235, 239
Polynucleotides, reactions with, 45–54, 101–106, 137–145, 339–342, 352–354, 359–369
Polynucleotide kinase, 83–96, 437–448
^{32}P-Postlabelling methods, 10, 83–96, 437–448
Preneoplastic foci, 110, 118, 120, 264
Procarbazine hydrochloride, 328
Promutagenic lesions, 103, 109, 112, 121, 122, 339–342, 345–355, 359–369, 373, 383, 393–398, 425, 426, 427, 434
1,3-Propane sultone, 347
1,N^2-Propanodeoxyguanosine adducts, 207–223
β-Propiolactone, 223, 347, 382
9-Propyladenine, 241–245
Propylglyoxal, 285
Prostaglandin endoperoxide metabolism, 5, 18, 175
Protein adducts, 269–278, 333
Psoralens (Furocoumarins), 7, 20, 25, 247–251, 299–309
Pyrimidine dimers, 382
Pyrimidine ring cleavage, 235–239

Q

Quinoxaline, 284

R

Radiation sensitizers, 185–195, 299–306

Radiolabelled:
  acrylonitrile oxide, 259
  bromoacetaldehyde, 256
  1,2-dibromoethane, 257
  2,2′-dichlorodiethylether, 264
  1,2-dichloroethane, 269–278
  ethyl carbamate, 109–123
  ethylene oxide, 277
  glutathione, 257
  5-methoxypsoralen, 299–309
  8-methoxypsoralen, 299–309
  methylmethane sulfonate, 277
  N-methyl-*N*-nitrosourea, 147–153, 277
  7-methyl-pyrido(3,4-*c*)psoralen, 299–309
  N-nitrosopyrrolidine, 208
  polynucleotides, 339–342
  vinyl chloride, 101–106, 262, 277, 426
Radiotherapy, 328, 330
Recombinant DNA techniques, 394
Recombination (*see also* Gene conversion), 318, 349
Reductive titration (*see also* Borohydride), 109–123
Relaxation, topological, 83–96
Resolvase, 83, 84
Restriction enzymes, 83–96, 387–391, 393–398
Ribose adducts (Deoxyribose adducts), 57–72, 354, 355
Ribose pucker, 78, 83
RNA,
  reactions with, 8, 33, 75–80, 101, 102, 111–113, 116, 121, 168, 264, 353, 354
  hydrolysis, 138, 168
  polymerase (*see also* RNA, transcription), 103, 359–369, 395, 396
  transcription, 9, 318, 341, 359–369, 395, 426

S

Saccharomyces cerevisiae, 283–290, 299–309
Safrole, 120, 239
Salmonella typhimurium, 16–19, 120, 175–182, 197–205, 209, 283–290, 293–296, 413, 416
Schiff's base, 121, 197–205, 348
Secondary lesions (*see* Apurinic/apyrimidinic sites, Imidazole ring opening, Pyrimidine ring cleavage)
Secondary structure, 79, 353
Second primary tumours, 327–333
Sequence selectivity, 83–96, 300, 387
Single-stranded nucleic acids, reactions with (*see also* RNA, reactions with), 8, 47, 48–50, 78, 83–96, 168, 318, 353, 354, 387, 427
Sister chromatid exchanges, 16, 20, 120, 148, 313–318, 331
Site-directed mutagenesis, 10
Site-selectivity, 168, 235, 389
S9 mix, 197–205, 416
SOS repair, 9, 349

Spectro-analytical methods, 10
SR-2508 (*see also* 2-Nitroimidazoles), 187
Stacking interactions, 79, 360
Stereoisomers (*syn, anti*), 359–369, 389, 418
Structure-activity relationships, 10, 197–205
Styrene, 331
Sulfite, 283–290
Sulfoxide derivatives, 132
Sulfur mustards, 153
Swain-Scott constants (*s*-values; *see also* Nucleophilic selectivity), 275, 277, 346

T

Tautomeric equilibrium, 41, 175, 176, 236, 241, 355, 359–369
TD_{50}-values, 347
Tertiary structure, of nucleic acids, 79, 418
5,6,7,8-Tetrahydropyrimido[1,2-*a*]purine, 222
Thin-layer chromatography, two-dimensional, 437–448
6-Thiodeoxyguanosine, reactions with dimethylsulfate, *N*-methyl-*N*-nitrosourea, 2-chloroethylmethanesulfonate, 147–153
Thiodiazotates, 132
Thioether-diazohydroxides, 133
Thio-TEPA (Tris(1-aziridinyl)phosphine sulfide), 328, 347
Thromboxan biosynthesis, 18
Thymine, reaction with
    3-carbethoxypsoralen, 247–251
    haloethylnitrosoureas, 137–145
    8-methoxypsoralen, 6, 7, 247–251
    osmium tetroxide, 91
Thymine glycol, 418
Topoisomers, 83–96
Topoisomerase I, 83–96
Topological changes, 83–96
Torsion angles, 78, 83
Transcriptases (Reverse transcriptase), 360, 368, 369, 389, 390
Transitions (*see* Base-pair substitutions)
Transversions (*see* Base-pair substitutions)
Treosulphan, 328
Triacanthine, 4
Tri-*O*-acetyladenosine, 33, 34, 35
Tri-*O*-acetylcytidine, 33, 34, 35
Trichloroacetic acid, 256
2,3,3-Trichloroacrolein, 200, 201
Trichloroethanol, 256
Trichloroethylene, 256
Trichlorotriethylamine hydrochloride, 328
Triose reductone, 223
2,4,6-Tris(1-aziridinyl)-*s*-triazine, 328
Tris(2,3-dibromopropyl)phosphate, 257

U

Ultraviolet light, 7, 20, 21, 23, 25, 188, 192, 247–251, 299–309, 387
Ultraviolet spectra, 45–54, 139, 142, 150, 165–172, 207–223, 236, 243, 249, 294, 416, 440, 444, 446
α,β-Unsaturated carbonyl compounds, 175–182, 197–205, 207–223
Unscheduled DNA synthesis, 258
Uracil, reaction with acrolein, 168
Uracil mustard, 328
Urethane (*see* Ethyl carbamate)
Urinary metabolites, 120, 202, 264, 269–278, 403

V

Vinblastine sulphate, 328
Vincristine sulphate, 328
Vinyl acetate, 16, 21, 23, 266, 267
Vinyl alcohol, 17
Vinyl bromide, 8, 16, 21, 23, 33, 261–267
Vinyl carbamate, 18, 23, 33, 109–123, 259, 313–318
Vinyl chloride, 8, 9, 10, 15, 21, 23, 25, 33, 37, 38, 101–106, 109–123, 223, 259, 261–267, 269, 275, 276, 277, 327, 331, 339, 342, 345–355, 366, 382, 425, 426, 427, 434
Vinyl compounds, 15, 21
Vinyl fluoride, 265, 266
Vinyl halides, 47, 54, 255–259, 261–267
S-Vinyl homocysteine, 119
Vinylidene chloride, 256

X

X-ray diffraction, 79, 83, 236

PUBLICATIONS OF THE INTERNATIONAL AGENCY FOR RESEARCH ON CANCER

SCIENTIFIC PUBLICATIONS SERIES

(Available from Oxford University Press)

No. 1 LIVER CANCER (1971)
176 pages

No. 2 ONCOGENESIS AND HERPES VIRUSES (1972)
Edited by P.M. Biggs, G. de Thé & L.N. Payne
515 pages

No. 3 N-NITROSO COMPOUNDS - ANALYSIS AND FORMATION (1972)
Edited by P. Bogovski, R. Preussmann & E.A. Walker
140 pages

No. 4 TRANSPLACENTAL CARCINOGENESIS (1973)
Edited by L. Tomatis & U. Mohr,
181 pages

No. 5 PATHOLOGY OF TUMOURS IN LABORATORY ANIMALS. VOLUME 1. TUMOURS OF THE RAT. PART 1 (1973)
Editor-in-Chief V.S. Turusov
214 pages

No. 6 PATHOLOGY OF TUMOURS IN LABORATORY ANIMALS. VOLUME 1. TUMOURS OF THE RAT. PART 2 (1976)
Editor-in-Chief V.S. Turusov
319 pages

No. 7 HOST ENVIRONMENT INTERACTIONS IN THE ETIOLOGY OF CANCER IN MAN (1973)
Edited by R. Doll & I. Vodopija,
464 pages

No. 8 BIOLOGICAL EFFECTS OF ASBESTOS (1973)
Edited by P. Bogovski, J.C. Gilson, V. Timbrell & J.C. Wagner,
346 pages

No. 9 N-NITROSO COMPOUNDS IN THE ENVIRONMENT (1974)
Edited by P. Bogovski & E.A. Walker
243 pages

No. 10 CHEMICAL CARCINOGENESIS ESSAYS (1974)
Edited by R. Montesano & L. Tomatis,
230 pages

No. 11 ONCOGENESIS AND HERPESVIRUSES II (1975)
Edited by G. de-Thé, M.A. Epstein & H. zur Hausen
Part 1, 511 pages
Part 2, 403 pages

No. 12 SCREENING TESTS IN CHEMICAL CARCINOGENESIS (1976)
Edited by R. Montesano, H. Bartsch & L. Tomatis
666 pages

No. 13 ENVIRONMENTAL POLLUTION AND CARCINOGENIC RISKS (1976)
Edited by C. Rosenfeld & W. Davis
454 pages

No. 14 ENVIRONMENTAL N-NITROSO COMPOUNDS — ANALYSIS AND FORMATION (1976)
Edited by E.A. Walker, P. Bogovski & L. Griciute
512 pages

No. 15 CANCER INCIDENCE IN FIVE CONTINENTS. VOL. III (1976)
Edited by J. Waterhouse, C.S. Muir, P. Correa & J. Powell
584 pages

No. 16 AIR POLLUTION AND CANCER IN MAN (1977)
Edited by U. Mohr, D. Schmahl & L. Tomatis
331 pages

No. 17 DIRECTORY OF ON-GOING RESEARCH IN CANCER EPIDEMIOLOGY 1977 (1977)
Edited by C.S. Muir & G. Wagner,
599 pages; out of print

No. 18 ENVIRONMENTAL CARCINOGENS - SELECTED METHODS OF ANALYSIS
Editor-in-Chief H. Egan
Vol. 1 - ANALYSIS OF VOLATILE NITROSAMINES IN FOOD (1978)
Edited by R. Preussmann, M. Castegnaro, E.A. Walker & A.E. Wassermann
212 pages

SCIENTIFIC PUBLICATIONS SERIES

No. 19 ENVIRONMENTAL ASPECTS OF N-NITROSO COMPOUNDS (1978)
Edited by E.A. Walker, M. Castegnaro, L. Griciute & R.E. Lyle
566 pages

No. 20 NASOPHARYNGEAL CARCINOMA: ETIOLOGY AND CONTROL (1978)
Edited by G. de-Thé & Y. Ito,
610 pages

No. 21 CANCER REGISTRATION AND ITS TECHNIQUES (1978)
Edited by R. MacLennan, C.S. Muir, R. Steinitz & A. Winkler
235 pages

No. 22 ENVIRONMENTAL CARCINOGENS - SELECTED METHODS OF ANALYSIS
Editor-in-Chief H. Egan
Vol. 2 - METHODS FOR THE MEASUREMENT OF VINYL CHLORIDE IN POLY(VINYL CHLORIDE), AIR, WATER AND FOODSTUFFS (1978)
Edited by D.C.M. Squirrell & W. Thain,
142 pages

No. 23 PATHOLOGY OF TUMOURS IN LABORATORY ANIMALS. VOLUME II. TUMOURS OF THE MOUSE (1979)
Editor-in-Chief V.S. Turusov
669 pages

No. 24 ONCOGENESIS AND HERPESVIRUSES III (1978)
Edited by G. de-Thé, W. Henle & F. Rapp
Part 1, 580 pages
Part 2, 522 pages

No. 25 CARCINOGENIC RISKS - STRATEGIES FOR INTERVENTION (1979)
Edited by W. Davis & C. Rosenfeld,
283 pages

No. 26 DIRECTORY OF ON-GOING RESEARCH IN CANCER EPIDEMIOLOGY 1978 (1978)
Edited by C.S. Muir & G. Wagner,
550 pages; out of print

No. 27 MOLECULAR AND CELLULAR ASPECTS OF CARCINOGEN SCREENING TESTS (1980)
Edited by R. Montesano, H. Bartsch & L. Tomatis
371 pages

No. 28 DIRECTORY OF ON-GOING RESEARCH IN CANCER EPIDEMIOLOGY 1979 (1979)
Edited by C.S. Muir & G. Wagner,
672 pages; out of print

No. 29 ENVIRONMENTAL CARCINOGENS - SELECTED METHODS OF ANALYSIS
Editor-in-Chief H. Egan
Vol. 3 - ANALYSIS OF POLYCYCLIC AROMATIC HYDROCARBONS IN ENVIRONMENTAL SAMPLES (1979)
Edited by M. Castegnaro, P. Bogovski, H. Kunte & E.A. Walker
240 pages

No. 30 BIOLOGICAL EFFECTS OF MINERAL FIBRES (1980)
Editor-in-Chief J.C. Wagner
Volume 1, 494 pages
Volume 2, 513 pages

No. 31 N-NITROSO COMPOUNDS: ANALYSIS, FORMATION AND OCCURRENCE (1980)
Edited by E.A. Walker, M. Castegnaro, L. Griciute & M. Börzsönyi
841 pages

No. 32 STATISTICAL METHODS IN CANCER RESEARCH
Vol. 1. THE ANALYSIS OF CASE-CONTROL STUDIES (1980)
By N.E. Breslow & N.E. Day
338 pages

No. 33 HANDLING CHEMICAL CARCINOGENS IN THE LABORATORY - PROBLEMS OF SAFETY (1979)
Edited by R. Montesano, H. Bartsch, E. Boyland, G. Della Porta, L. Fishbein, R.A. Griesemer, A.B. Swan & L. Tomatis,
32 pages

No. 34 PATHOLOGY OF TUMOURS IN LABORATORY ANIMALS. VOLUME III. TUMOURS OF THE HAMSTER (1982)
Editor-in-Chief V.S. Turusov,
461 pages

No. 35 DIRECTORY OF ON-GOING RESEARCH IN CANCER EPIDEMIOLOGY 1980 (1980)
Edited by C.S. Muir & G. Wagner,
660 pages; out of print

SCIENTIFIC PUBLICATIONS SERIES

No. 36 CANCER MORTALITY BY OCCUPATION AND SOCIAL CLASS 1851-1971 (1982)
By W.P.D. Logan
253 pages

No. 37 LABORATORY DECONTAMINATION AND DESTRUCTION OF AFLATOXINS B_1, B_2, G_1, G_2 IN LABORATORY WASTES (1980)
Edited by M. Castegnaro, D.C. Hunt, E.B. Sansone, P.L. Schuller, M.G. Siriwardana, G.M. Telling, H.P. Van Egmond & E.A. Walker,
59 pages

No. 38 DIRECTORY OF ON-GOING RESEARCH IN CANCER EPIDEMIOLOGY 1981 (1981)
Edited by C.S. Muir & G. Wagner,
696 pages; out of print

No. 39 HOST FACTORS IN HUMAN CARCINOGENESIS (1982)
Edited by H. Bartsch & B. Armstrong
583 pages

No. 40 ENVIRONMENTAL CARCINOGENS. SELECTED METHODS OF ANALYSIS
Editor-in-Chief H. Egan
Vol. 4. SOME AROMATIC AMINES AND AZO DYES IN THE GENERAL AND INDUSTRIAL ENVIRONMENT (1981)
Edited by L. Fishbein, M. Castegnaro, I.K. O'Neill & H. Bartsch
347 pages

No. 41 N-NITROSO COMPOUNDS: OCCURRENCE AND BIOLOGICAL EFFECTS (1982)
Edited by H. Bartsch, I.K. O'Neill, M. Castegnaro & M. Okada,
755 pages

No. 42 CANCER INCIDENCE IN FIVE CONTINENTS. VOLUME IV (1982)
Edited by J. Waterhouse, C. Muir, K. Shanmugaratnam & J. Powell,
811 pages

No. 43 LABORATORY DECONTAMINATION AND DESTRUCTION OF CARCINOGENS IN LABORATORY WASTES: SOME N-NITROSAMINES (1982) Edited by M. Castegnaro, G. Eisenbrand, G. Ellen, L. Keefer, D. Klein, E.B. Sansone, D. Spincer, G. Telling & K. Webb
73 pages

No. 44 ENVIRONMENTAL CARCINOGENS. SELECTED METHODS OF ANALYSIS
Editor-in-Chief H. Egan
Vol. 5. SOME MYCOTOXINS (1983)
Edited by L. Stoloff, M. Castegnaro, P. Scott, I.K. O'Neill & H. Bartsch,
455 pages

No. 45 ENVIRONMENTAL CARCINOGENS. SELECTED METHODS OF ANALYSIS
Editor-in-Chief H. Egan
Vol. 6: N-NITROSO COMPOUNDS (1983)
Edited by R. Preussmann, I.K. O'Neill, G. Eisenbrand, B. Spiegelhalder & H. Bartsch
508 pages

No. 46 DIRECTORY OF ON-GOING RESEARCH IN CANCER EPIDEMIOLOGY 1982 (1982)
Edited by C.S. Muir & G. Wagner,
722 pages; out of print

No. 47 CANCER INCIDENCE IN SINGAPORE (1982)
Edited by K. Shanmugaratnam, H.P. Lee & N.E. Day
174 pages

No. 48 CANCER INCIDENCE IN THE USSR Second Revised Edition (1983)
Edited by N.P. Napalkov, G.F. Tserkovny, V.M. Merabishvili, D.M. Parkin, M. Smans & C.S. Muir,
75 pages

No. 49 LABORATORY DECONTAMINATION AND DESTRUCTION OF CARCINOGENS IN LABORATORY WASTES: SOME POLYCYCLIC AROMATIC HYDROCARBONS (1983)
Edited by M. Castegnaro, G. Grimmer, O. Hutzinger, W. Karcher, H. Kunte, M. Lafontaine, E.B. Sansone, G. Telling & S.P. Tucker
81 pages

No. 50 DIRECTORY OF ON-GOING RESEARCH IN CANCER EPIDEMIOLOGY 1983 (1983)
Edited by C.S. Muir & G. Wagner,
740 pages; out of print

SCIENTIFIC PUBLICATIONS SERIES

No. 51 MODULATORS OF EXPERIMENTAL CARCINO-GENESIS (1983)
Edited by V. Turusov & R. Montesano
307 pages

No. 52 SECOND CANCER IN RELATION TO RADIATION TREATMENT FOR CERVICAL CANCER: RESULTS OF A CANCER REGISTRY COLLABORATION (1984)
Edited by N.E. Day & J.C. Boice, Jr,
207 pages

No. 53 NICKEL IN THE HUMAN ENVIRONMENT (1984)
Editor-in-Chief, F.W. Sunderman, Jr,
529 pages

No. 54 LABORATORY DECONTAMINATION AND DESTRUCTION OF CARCINO-GENS IN LABORATORY WASTES: SOME HYDRAZINES (1983)
Edited by M. Castegnaro, G. Ellen,
M. Lafontaine, H.C. van der Plas,
E.B. Sansone & S.P. Tucker,
87 pages

No. 55 LABORATORY DECONTAMINATION AND DESTRUCTION OF CARCINOGENS IN LABORATORY WASTES: SOME N-NITROSAMIDES (1984)
Edited by M. Castegnaro,
M. Benard, L.W. van Broekhoven,
D. Fine, R. Massey, E.B. Sansone,
P.L.R. Smith, B. Spiegelhalder,
A. Stacchini, G. Telling & J.J. Vallon,
65 pages

No. 56 MODELS, MECHANISMS AND ETIOLOGY OF TUMOUR PROMOTION (1984)
Edited by M. Börszönyi, N.E. Day,
K. Lapis & H. Yamasaki
532 pages

No. 57 N-NITROSO COMPOUNDS: OCCURRENCE, BIOLOGICAL EFFECTS AND RELEVANCE TO HUMAN CANCER (1984)
Edited by I.K. O'Neill, R.C. von Borstel,
C.T. Miller, J. Long & H. Bartsch,
1013 pages

No. 58 AGE-RELATED FACTORS IN CARCINOGENESIS (1985)
Edited by A. Likhachev, V. Anisimov
& R. Montesano
288 pages

No. 59 MONITORING HUMAN EXPOSURE TO CARCINOGENIC AND MUTAGENIC AGENTS (1984)
Edited by A. Berlin, M. Draper,
K. Hemminki & H. Vainio
457 pages

No. 60 BURKITT'S LYMPHOMA: A HUMAN CANCER MODEL (1985)
Edited by G. Lenoir, G. O'Conor
& C.L.M. Olweny
484 pages

No. 61 LABORATORY DECONTAMINATION AND DESTRUCTION OF CARCINOGENS IN LABORATORY WASTES: SOME HALOETHERS (1984)
Edited by M. Castegnaro,
M. Alvarez, M. Iovu, E.B. Sansone,
G.M. Telling & D.T. Williams
55 pages

No. 62 DIRECTORY OF ON-GOING RESEARCH IN CANCER EPI-DEMIOLOGY 1984 (1984)
Edited by C.S. Muir & G.Wagner
728 pages

No. 63 VIRUS-ASSOCIATED CANCERS IN AFRICA (1984)
Edited by A.O. Williams, G.T. O'Conor,
G.B. de-Thé & C.A. Johnson,
773 pages

No. 64 LABORATORY DECONTAMINATION AND DESTRUCTION OF CARCINOGENS IN LABORATORY WASTES: SOME AROMATIC AMINES AND 4-NITROBIPHENYL (1985)
Edited by M. Castegnaro, J. Barek,
J. Dennis, G. Ellen, M. Klibanov,
M. Lafontaine, R. Mitchum,
P. Van Roosmalen, E.B. Sansone,
L.A. Sternson & M. Vahl
85 pages

No. 65 INTERPRETATION OF NEGATIVE EPIDEMIOLOGICAL EVIDENCE FOR CARCINOGENICITY (1985)
Edited by N.J. Wald & R. Doll
232 pages

No. 66 THE ROLE OF THE REGISTRY IN CANCER CONTROL (1985)
Edited by D.M. Parkin, G. Wagner
& C.S. Muir
155 pages

No. 67 TRANSFORMATION ASSAY OF ESTABLISHED CELL LINES: MECHANISMS AND APPLICATIONS (1985)
Edited by T. Kakunaga & H. Yamasaki
225 pages

SCIENTIFIC PUBLICATIONS SERIES

No. 68 ENVIRONMENTAL CARCINOGENS — SELECTED METHODS OF ANALYSIS. VOL. 7: SOME VOLATILE HALOGENATED ALKANES AND ALKENES (1985)
Edited by L. Fishbein & I.K. O'Neill
479 pages

No. 69 DIRECTORY OF ON-GOING RESEARCH IN CANCER EPIDEMIOLOGY 1985 (1985)
Edited by C.S. Muir & G. Wagner
756 pages

No. 70 THE ROLE OF CYCLIC NUCLEIC ACID ADDUCTS IN CARCINOGENESIS AND MUTAGENESIS (1986)
Edited by B. Singer & H. Bartsch
467 pages

No. 71 ENVIRONMENTAL CARCINOGENS. SELECTED METHODS OF ANALYSIS VOL. 8:. SOME METALS: As, Be, Cd, Cr, Ni, Pb, Se, Zn (1986)
Edited by I.K. O'Neill, P. Schuller & L. Fishbein (in press)

No. 72 ATLAS OF CANCER IN SCOTLAND 1975-1980: INCIDENCE AND EPIDEMIOLOGICAL PERSPECTIVE (1985)
Edited by I. Kemp, P. Boyle, M. Smans & C. Muir
282 pages

No. 73 LABORATORY DECONTAMINATION AND DESTRUCTION OF CARCINOGENS IN LABORATORY WASTES: SOME ANTINEOPLASTIC AGENTS (1985)
Edited by M. Castegnaro, J. Adams, M. Armour, J. Barek, J. Benvenuto, C. Confalonieri, U. Goff, S. Ludeman, D. Reed, E.B. Sansone & G. Telling
163 pages

No. 74 TOBACCO: A MAJOR INTERNATIONAL HEALTH HAZARD (1986)
Edited by D. Zaridze and R. Peto
(in press)

No. 75 CANCER OCCURRENCE IN DEVELOPING COUNTRIES (1986)
Edited by D.M. Parkin
(in press)

No. 76 SCREENING FOR CANCER OF THE UTERINE CERVIX (1986)
Edited by M. Hakama, A.B. Miller & N.E. Day
(in press)

No. 77 HEXACHLOROBENZENE. PROCEEDINGS OF AN INTERNATIONAL SYMPOSIUM (1986)
Edited by C.R. Morris & J.R.P. Cabral
(in press)

No. 78 CARCINOGENICITY OF CYTOSTATIC DRUGS (1986)
Edited by D. Schmähl & J. Kaldor
(in press)

No. 79 STATISTICAL METHODS IN CANCER RESEARCH, VOL. 3, THE DESIGN AND ANALYSIS OF LONG-TERM ANIMAL EXPERIMENTS (1986)
By J.J. Gart, D. Krewski, P.N. Lee, R.E. Tarone & J. Wahrendorf
(in press)

No. 80 DIRECTORY OF ON-GOING RESEARCH IN CANCER EPIDEMIOLOGY 1986 (1986)
Edited by G. Wagner & C. Muir
(in press)

No. 81 ENVIRONMENTAL CARCINOGENS. SELECTED METHODS OF ANALYSIS, VOL. 9, PASSIVE SMOKING (1986)
Edited by I.K. O'Neill, K.D. Brunnemann, B. Dodet & D. Hoffmann
(in press)

NON SERIAL PUBLICATIONS

(Available from IARC)

ALCOOL ET CANCER (1978)
By A.J. Tuyns (in French only)
42 pages

CANCER MORBIDITY AND CAUSES OF DEATH AMONG DANISH BREWERY WORKERS (1980)
By O.M. Jensen
145 pages

IARC MONOGRAPHS ON THE EVALUATION OF THE CARCINOGENIC RISK OF CHEMICALS TO HUMANS
(English editions only)

(Available from WHO Sales Agents)

Volume 1
Some inorganic substances, chlorinated hydrocarbons, aromatic amines, N-nitroso compounds, and natural products (1972)
184 pp.; out of print

Volume 2
Some inorganic and organometallic compounds (1973)
181 pp.; out of print

Volume 3
Certain polycyclic aromatic hydrocarbons and heterocyclic compounds (1973)
271 pp.; out of print

Volume 4
Some aromatic amines, hydrazine and related substances, N-nitroso compounds and miscellaneous alkylating agents (1974)
286 pp.

Volume 5
Some organochlorine pesticides (1974)
241 pp.; out of print

Volume 6
Sex hormones (1974)
243 pp.; US$7.20; Sw.fr. 18.-

Volume 7
Some anti-thyroid and related substances, nitrofurans and industrial chemicals (1974)
326 pp.

Volume 8
Some aromatic azo compounds (1975)
357 pp.

Volume 9
Some aziridines, N-, S- and O-mustards and selenium (1975)
268 pp.

Volume 10
Some naturally occurring substances (1976)
353 pp.

Volume 11
Cadmium, nickel, some epoxides, miscellaneous industrial chemicals and general considerations on volatile anaesthetics (1976)
306 pp.

Volume 12
Some carbamates, thiocarbamates and carbazides (1976)
282 pp.

Volume 13
Some miscellaneous pharmaceutical substances (1977)
255 pp.

Volume 14
Asbestos (1977)
106 pp.

Volume 15
Some fumigants, the herbicides 2,4-D and 2,4,5-T, chlorinated dibenzodioxins and miscellaneous industrial chemicals (1977)
354 pp.

Volume 16
Some aromatic amines and related nitro compounds - hair dyes, colouring agents and miscellaneous industrial chemicals (1978)
400 pp.

Volume 17
Some N-nitroso compounds (1978)
365 pp.

Volume 18
Polychlorinated biphenyls and polybrominated biphenyls (1978)
140 pp.

Volume 19
Some monomers, plastics and synthetic elastomers, and acrolein (1979)
513 pp.

Volume 20
Some halogenated hydrocarbons (1979)
609 pp.

Volume 21
Sex hormones (II) (1979)
583 pp.

Volume 22
Some non-nutritive sweetening agents (1980)
208 pp.

IARC MONOGRAPHS SERIES

Volume 23
Some metals and metallic compounds (1980)
438 pp.

Volume 24
Some pharmaceutical drugs (1980)
337 pp.

Volume 25
Wood, leather and some associated industries (1981)
412 pp.

Volume 26
Some antineoplastic and immunosuppressive agents (1981)
411 pp.

Volume 27
Some aromatic amines, anthraquinones and nitroso compounds, and inorganic fluorides used in drinking-water and dental preparations (1982)
341 pp.

Volume 28
The rubber industry (1982)
486 pp.

Volume 29
Some industrial chemicals and dyestuffs (1982)
416 pp.

Volume 30
Miscellaneous pesticides (1983)
424 pp.

Volume 31
Some food additives, feed additives and naturally occurring substances (1983)
314 pp.

Volume 32
Polynuclear aromatic compounds, Part 1, Environmental and experimental data (1984)
477 pp.

Volume 33
Polynuclear aromatic compounds, Part 2, Carbon blacks, mineral oils and some nitroarene compounds (1984)
245 pp.

Volume 34
Polynuclear aromatic compounds, Part 3, Industrial exposures in aluminium production, coal gasification, coke production, and iron and steel founding (1984)
219 pp.

Volume 35
Polynuclear aromatic compounds, Part 4, Bitumens, coal-tar and derived products, shale-oils and soots (1985)
271 pp.

Volume 36
Allyl Compounds, aldehydes, epoxides and peroxides (1985)
369 pp.

Volume 37
Tobacco habits other than smoking; betel-quid and areca-nut chewing; and some related nitrosamines (1985)
291 pp.

Volume 38
Tobacco smoking (1986)
421 pp. (in press)

Supplement No. 1
Chemicals and industrial processes associated with cancer in humans (IARC Monographs, Volumes 1 to 20) (1979)
71 pp.; out of print

Supplement No. 2
Long-term and short-term screening assays for carcinogens: a critical appraisal (1980)
426 pp.; US$25.00; Sw.fr. 40.-

Supplement No. 3
Cross index of synonyms and trade names in Volumes 1 to 26 (1982)
199 pp.

Supplement No. 4
Chemicals, industrial processes and industries associated with cancer in humans (IARC Monographs, Volumes 1 to 29) (1982)
292 pp.

INFORMATION BULLETINS ON THE SURVEY OF CHEMICALS BEING TESTED FOR CARCINOGENICITY

(Available from IARC)

No. 8 (1979)
Edited by M.-J. Ghess, H. Bartsch
& L. Tomatis
604 pp.

No. 9 (1981)
Edited by M.-J. Ghess, J.D. Wilbourn,
H. Bartsch & L. Tomatis
294 pp.

No. 10 (1982)
Edited by M.-J. Ghess, J.D. Wilbourn
H. Bartsch
326 pp.

No. 11 (1984)
Edited by M.-J. Ghess, J.D. Wilbourn,
H. Vainio & H. Bartsch
336 pp.